Advances and Applications in Model-Driven Engineering

Vicente García Díaz
University of Oviedo, Spain

Juan Manuel Cueva Lovelle
University of Oviedo, Spain

B. Cristina Pelayo García–Bustelo
University of Oviedo, Spain

Oscar Sanjuán Martinez
University of Carlos III, Spain

A volume in the Advances in Systems
Analysis, Software Engineering, and High
Performance Computing (ASASEHPC)
Book Series

Information Science
REFERENCE
An Imprint of IGI Global

Managing Director:	Lindsay Johnston
Editorial Director:	Joel Gamon
Production Manager:	Jennifer Yoder
Publishing Systems Analyst:	Adrienne Freeland
Development Editor:	Christine Smith
Assistant Acquisitions Editor:	Kayla Wolfe
Typesetter:	Travis Gundrum
Cover Design:	Jason Mull

Published in the United States of America by
Information Science Reference (an imprint of IGI Global)
701 E. Chocolate Avenue
Hershey PA 17033
Tel: 717-533-8845
Fax: 717-533-8661
E-mail: cust@igi-global.com
Web site: http://www.igi-global.com

Library of Congress Cataloging-in-Publication Data

Advances and applications in model-driven engineering / Vicente Garcia Diaz,Juan Manuel Cueva Lovelle, Begona Cristina Pelayo, and Oscar Sanjuan Martinez,editors.
      pages cm
  Includes bibliographical references and index.
  Summary: "This book explores this relatively new approach in software development that can increase the level of abstraction of the development of tasks, bridging the gaps between various disciplines within software engineering and computer science"-- Provided by publisher.
  ISBN 978-1-4666-4494-6 (hardcover) -- ISBN 978-1-4666-4496-0 (print & perpetual access) -- ISBN 978-1-4666-4495-3 (ebook) 1. Engineering models. I. Garcia Diaz, Vicente, 1981- editor of compilation. II. Cueva Lovelle, Juan Manuel, editor of compilation. III. Pelayo Garcia-Bustelo, Begona Cristina, 1971- editor of compilation. IV. Sanjuan Martinez, Oscar, editor of compilation.

 TA177.A37 2013
 620.001'1--dc23
                              2013014345

This book is published in the IGI Global book series Advances in Systems Analysis, Software Engineering, and High Performance Computing (ASASHEPC) (ISSN: 2327-3453; eISSN: 2327-3461)

British Cataloguing in Publication Data
A Cataloguing in Publication record for this book is available from the British Library.

Advances in Systems Analysis, Software Engineering, and High Performance Computing (ASASHEPC) Book Series

Vijayan Sugumaran
Oakland University, USA

ISSN: 2327-3453
EISSN: 2327-3461

MISSION

The theory and practice of computing applications and distributed systems has emerged as one of the key areas of research driving innovations in business, engineering, and science. The fields of software engineering, systems analysis, and high performance computing offer a wide range of applications and solutions in solving computational problems for any modern organization.

The **Advances in Systems Analysis, Software Engineering, and High Performance Computing (ASASEHPC) Book Series** brings together research in the areas of distributed computing, systems and software engineering, high performance computing, and service science. This collection of publications is useful for academics, researchers, and practitioners seeking the latest practices and knowledge in this field.

COVERAGE

- Computer Graphics
- Computer Networking
- Computer System Analysis
- Distributed Cloud Computing
- Enterprise Information Systems
- Metadata and Semantic Web
- Parallel Architectures
- Performance Modeling
- Software Engineering
- Virtual Data Systems

IGI Global is currently accepting manuscripts for publication within this series. To submit a proposal for a volume in this series, please contact our Acquisition Editors at Acquisitions@igi-global.com or visit: http://www.igi-global.com/publish/.

Titles in this Series

For a list of additional titles in this series, please visit: www.igi-global.com

Advances and Applications in Model-Driven Engineering
Vicente García Díaz (University of Oviedo, Spain) Juan Manuel Cueva Lovelle (University of Oviedo, Spain) B. Cristina Pelayo García-Bustelo (University of Oviedo, Spain) and Oscar Sanjuán Martinez (University of Carlos III, Spain)
Information Science Reference • copyright 2014 • 424pp • H/C (ISBN: 9781466644946) • US $195.00 (our price)

Service-Driven Approaches to Architecture and Enterprise Integration
Raja Ramanathan (Independent Researcher, USA) and Kirtana Raja (Independent Researcher, USA)
Information Science Reference • copyright 2013 • 411pp • H/C (ISBN: 9781466641938) • US $195.00 (our price)

Progressions and Innovations in Model-Driven Software Engineering
Vicente García Díaz (University of Oviedo, Spain) Juan Manuel Cueva Lovelle (University of Oviedo, Spain) B. Cristina Pelayo García-Bustelo (University of Oviedo, Spain) and Oscar Sanjuán Martínez (University of Oviedo, Spain)
Engineering Science Reference • copyright 2013 • 388pp • H/C (ISBN: 9781466642171) • US $195.00 (our price)

Knowledge-Based Processes in Software Development
Saqib Saeed (Bahria University Islamabad, Pakistan) and Izzat Alsmadi (Yarmouk University, Jordan)
Information Science Reference • copyright 2013 • 318pp • H/C (ISBN: 9781466642294) • US $195.00 (our price)

Distributed Computing Innovations for Business, Engineering, and Science
Alfred Waising Loo (Lingnan University, Hong Kong)
Information Science Reference • copyright 2013 • 369pp • H/C (ISBN: 9781466625334) • US $195.00 (our price)

Data Intensive Distributed Computing Challenges and Solutions for Large-scale Information Management
Tevfik Kosar (University at Buffalo, USA)
Information Science Reference • copyright 2012 • 352pp • H/C (ISBN: 9781615209712) • US $180.00 (our price)

Achieving Real-Time in Distributed Computing From Grids to Clouds
Dimosthenis Kyriazis (National Technical University of Athens, Greece) Theodora Varvarigou (National Technical University of Athens, Greece) and Kleopatra G. Konstanteli (National Technical University of Athens, Greece)
Information Science Reference • copyright 2012 • 330pp • H/C (ISBN: 9781609608279) • US $195.00 (our price)

Principles and Applications of Distributed Event-Based Systems
Annika M. Hinze (University of Waikato, New Zealand) and Alejandro Buchmann (University of Waikato, New Zealand)
Information Science Reference • copyright 2010 • 538pp • H/C (ISBN: 9781605666976) • US $180.00 (our price)

DISSEMINATOR OF KNOWLEDGE

www.igi-global.com

701 E. Chocolate Ave., Hershey, PA 17033
Order online at www.igi-global.com or call 717-533-8845 x100
To place a standing order for titles released in this series, contact: cust@igi-global.com
Mon-Fri 8:00 am - 5:00 pm (est) or fax 24 hours a day 717-533-8661

Table of Contents

Detailed Table of Contents

Chapter 1
Jon Davis, Curtin University, Australia
Elizabeth Chang, Curtin University, Australia

Customizing Enterprise Information Systems (EIS) scale applications can be very expensive, also incurring additional costs during their lifecycle when customizations may need to be re-engineered to suit each EIS upgrade. The ongoing development of a temporal meta-data framework for EIS applications seeks to overcome these issues, with the application logic model supporting the capability for end users to define their own supplemental or replacement application logic meta-data, as what the authors term Variant Logic, to become a variation of the core application logic. Variant Logic can be applied to any defined model object whether visual objects, logical processing objects, or data structures objects. Variant Logic can be defined by any authorized user, through modeling rather than coding, executed by any user as an alternative to the original application logic, and is available for immediate execution by the framework runtime engine. Variant Logic is also preserved during automated meta-data application updates.

Chapter 2
Rachit Mohan Garg, IT Engineer, India

This chapter enables the learners to envisage the prevailing glitches of concern in programming and enables them to identify the concerns that may be incurred in the development of the software. How the designing phase of the product is molded separates these concerns from the core business logic, thereby providing a product with an ability to incorporate any change without actually touching the core business logic unless necessary. Moreover, creating a good design leads to less rework in the latter stages of the SDLC (Software Development Life Cycle).

Chapter 3

Tong-Ying Yu, Independent Consultant, China

How to bridge the gap between business and Information Technology (IT) has always been a critical issue for both the developers and IT managers. The individualized, differentiated demands by different customers and situations, the constantly changing in both business and IT are great challenges to the applications for enterprises. In this chapter, the authors respectively discuss the left side (computer) in software engineering, with Object-Orientation (OO), Model-Driven Engineering (MDE), Domain-Driven Development (DDD), Agile, etc., and the right side (the business) in Enterprise Engineering (EE) with Enterprise Modeling (EM), and Enterprise Architecture (EA) of the gap. It is shown there are some fundamental problems, such as the transforming barrier between analysis and design model, the entanglement of business change and development process, and the limitation to the enterprise engineering approaches such as EA by IT. Our solution is concentrated on the middle, the inevitable model as a mediator between human, computer, and the real world. The authors introduce Model-Driven Application (MDApp), which is based on Model-Driven Mechanism (MDM), operated on the evolutionary model of the target thing at runtime; it is able to largely avoid the transforming barrier and remove the entanglement. Thus, the architecture for Enterprise Model Driven Application (EMDA) is emerged, which is able to strongly support EE and adapts to the business changing at runtime.

Chapter 4

Cristian González García, University of Oviedo, Spain
Jordán Pascual Espada, University of Oviedo, Spain

There has been a rise in the quantity of Smart Things present in our daily life: Smartphones, smart TVs, sensor networks, smart appliances, and many other home and industry automation devices. The disadvantage is that each one is similar but very different from the others because they use different kinds of connections, different protocols, different hardware, and different operative systems; even if they belong to a similar kind, and depending on the manufacturer, two Smartphones or two humidity sensors can be totally different. In spite of this, they all have something in common: All can be connected to Internet. This fact is what gives power to objects, because it allows them to interact with each other and with the environment by intercommunication, a joint and synchronized work. That is why the key resides in creating connections among different objects. This requires technical and qualified personal, although the majority of people who use it do not meet these conditions. These people require objects that are easy to interact with and with as many automations as possible.

Chapter 5

Firas Bacha, University of Valenciennes, France
Káthia Marçal de Oliveira, University of Valenciennes, France
Mourad Abed, University of Valenciennes, France

User Interface (UI) personalization aims at providing the right information, at the right time, and on the right support (tablets, smart-phone, etc.). Personalization can be performed on the interface elements' presentation (e.g. layout, screen size, and resolution) and on the content provided (e.g., data, information, document). While many existing approaches deal with the first type of personalization, this chapter explores content personalization. To that end, the authors define a context-aware Model Driven Architecture (MDA) approach where the UI model is enriched by data from a domain model and its mapping to a context model. They conclude that this approach is better used only for domains where one envisions several developments of software applications and/or user interfaces.

Chapter 6

István Dávid, Budapest University of Technology and Economics, Hungary
László Gönczy, Budapest University of Technology and Economics, Hungary

This chapter introduces a novel approach for design of Domain-Specific Languages (DSL). It is very common in practice that the same problems emerge in different application domains (e.g. the modeling support for complex event processing is desirable in the domain of algorithmic trading, IT security assessment, robust monitoring, etc.). A DSL operates in one single domain, but the above-mentioned cross-domain challenges raise the question: is it possible to automate the design of DSLs which are so closely related? This approach demonstrates how a family of domain-specific languages can be developed for multiple domains from a single generic language metamodel with generative techniques. The basic idea is to refine the targeted domain with separating the problem domain from the context domain. This allows designing a generic language based on the problem and customizing it with the appropriate extensions for arbitrary contexts, thus defining as many DSLs and as many contexts as one extends the generic language for. The authors also present an ontology-based approach for establishing context-specific domain knowledge bases. The results are discussed through a case study, where a language for event processing is designed and extended for multiple context domains.

Chapter 7

Steven Gibson, California State University – Northridge, USA

This chapter highlights one concept representing the human role in requirements engineering and analysis for model synthesis. The production of design documentation to support model development requires elicitation of user requirements. The process of requirement elicitation plays a primary role in all Model-Driven Software Engineering (MDSE). Issues addressed include how requirements are gathered by the use of surveys, interviews, and questionnaires, and the importance of using validated constructs when gathering user information during requirement elicitation. Survey constructs, as used in requirements engineering, are analogues to the models in the final engineering product. A solution to improving the use of survey methods in the gathering of requirements is introduced. A small application is shown that suggests an example use of this proposed solution. This review of current practices explores areas where challenges are faced in the field with a concluding discussion that points to future trends in this research field.

Chapter 8

Iwona Dubielewicz, Wroclaw University of Technology, Poland
Bogumiła Hnatkowska, Wroclaw University of Technology, Poland
Zbigniew Huzar, Wroclaw University of Technology, Poland
Lech Tuzinkiewicz, Wroclaw University of Technology, Poland

Agile methodologies have become very popular. They are defined in terms of best practices, which aim at developing good quality software faster and cheaper. Unfortunately, agile methodologies do not refer explicitly to quality assurance, which is understood as a planned set of activities performed to provide adequate confidence that a product conforms to established requirements, and which is performed to evaluate the process by which products are developed. The chapter considers the relations of agile practices with software life cycle processes, especially those connected to quality assurance, and tries to answer the question of which agile practices ensure software quality. Next, agile practices associated with quality assurance are assessed from different perspectives and some recommendations for their usage are given. It is observed that modeling has a particular impact on quality assurance.

Model-Driven Development (MDD) is an initiative proposed by the Object Management Group (OMG) to model centric software development. It is based on the concepts of models, metamodels, and automatic transformations. Models and metamodels are in constant evolution. Metamodel evolution may cause conforming models to become invalid. Therefore, models must be migrated to maintain conformance to their metamodels. Metamodel evolution and model migration are typically performed manually, which is an error-prone task. In light of this, the authors propose a framework for metamodel evolution and model migration that combine an operator-based approach with refactoring and specification matching techniques. They combine classical metamodeling techniques with formal specifications to reason about transformations, metamodels, and their evolution. The authors describe foundations for MDA-based metamodel evolution that allow extending the functionality of the existing CASE tools in order to improve the MDA-based process quality.

The Rosetta specification language aims to enable system designers to abstractly design complex heterogeneous systems. To this end, Rosetta allows for compositional design to facilitate modularity, separation of concerns, and specification reuse. The behavior of Rosetta components and facets can be viewed as systems, which are well suited for coalgebraic denotation. The previous semantics of Rosetta lacked detail in the denotational work, and had no firm semantic basis for the composition operators. This thesis refreshes previous work on the coalgebraic denotation of Rosetta. It then goes on to define the denotation of the composition operators. Several Rosetta examples using all types of composition serve as a demonstration of the power of composition as well as the clean modular abstraction it affords the designer.

Transformation design is a key step in model-driven engineering, and it is a very challenging task, particularly in context of the model-driven data warehouse. Currently, this process is ensured by human experts. The authors propose a new methodology using machine learning techniques to automatically derive these transformation rules. The main goal is to automatically derive the transformation rules to be applied in the model-driven data warehouse process. The proposed solution allows for a simple design of the decision support systems and the reduction of time and costs of development. The authors use the inductive logic programming framework to learn these transformation rules from examples of previous projects. Then, they find that in model-driven data warehouse application, dependencies exist between transformations. Therefore, the authors investigate a new machine learning methodology, learning dependent-concepts, that is suitable to solve this kind of problem. The experimental evaluation shows that the dependent-concept learning approach gives significantly better results.

The integration between design models of software systems and analytical models of non-functional properties is an ideal framework on which lay the foundation for a deep understanding of the architectures present in software systems and their properties. In order to reach this integration, this chapter proposes a parameterized transformation for a model of performance properties derived from a system model in the MDE context. The idea behind a parameterized term is to leave open the transformation framework to adopt future improvements and make the approach reusable. The authors believe that this kind of integration permits the addition of analysis capabilities to the software development process and permits an early evaluation of design decisions.

Solving the business "software problem" of inflexibility with poor user experience was at the heart of the original R&D that started over 20 years ago. Any solution had to acknowledge how people work and most importantly remove the interpretation gap between users and "IT," thus removing the need for programmers in the build process. It was quickly recognised that in reality business logic does not really change, and it was therefore important to separate from the ever-changing technologies, such as operating systems and browsers, to "deliver." After 20+ years of research and development and working with early adopters, this approach has achieved the objectives set. As is explained, generic task objects and the important links were built and displayed in a Graphical Model where the building of custom applications takes place with no change to the core code and no code generation or compiling. This approach has opened a new perspective on capability in software yet does it by adopting simplicity in the core design. Indeed the driving philosophy for all development was to produce enabling technology that was business friendly, simple to use, and generic in application.

The increasing use of software is giving rise to the development of highly complex software systems. Further, software systems are required to be of high quality as a defect can have catastrophic effect on business as well as human life. Testing is defined as the process of executing a program with the intention of finding errors. Software testing is an expensive process of the software development life cycle consuming nearly 50% of development cost. Software testing aims not only to guarantee consistency in software specification but also to validate its implementation meeting user requirements. On the whole, it is observed that in general, errors in software systems set in at the early stages of the software development cycle (i.e. while gathering user requirements and deciding on specification of intended software). Even though formal specification in B and Z assures a provable system, its use has become less popular due to mathematical rigor. The Unified Modeling Language (UML), a semi-formal language with graphical notations consisting of various diagrams has caught software developers' imaginations and, it has become popular in industry. UML, with its several diagrams, helps to develop a model of intended software, and the model behaviour is simulated and tested to the satisfaction of both developer as well as users. As a UML model includes specifications of different aspects of a software system through several diagrams, it is essential to maintain consistency among diagrams so that quality of the model is maintained, and through inconsistency checking and removal, the model moves toward completeness. The works reported in literature on this topic are reviewed here.

Viewpoint-based modeling is an important recent development in software engineering. It is likely to boost the wider use of modeling techniques because it allows the tailoring of existing tools with respect to the different stakeholders in software design. This chapter reports the results of the project ViBaM, in which viewpoint concepts are investigated. In doing so, the authors give an overview of the most important contributions from literature regarding viewpoint concepts, from which they derive the position that they take in the ViBaM project. After presenting ViBaM's position, the authors derive features that they consider important for tools that support viewpoints. Afterwards, they present use cases, in which the viewpoint concepts are illustrated and discuss which of the viewpoint features are relevant in these use cases.

Chapter 16

Chris Thompson, Vanderbilt University, USA

Jules White, Vanderbilt University, USA

Douglas C. Schmidt, Vanderbilt University, USA

Smartphones are mobile devices that travel with their owners and provide increasingly powerful services. The software implementing these services must conserve battery power since smartphones may operate for days without being recharged. It is hard, however, to design smartphone software that minimizes power consumption. For example, multiple layers of abstractions and middleware sit between an application and the hardware, which make it hard to predict the power consumption of a potential application design accurately. Application developers must therefore wait until after implementation (when changes are more expensive) to determine the power consumption characteristics of a design. This chapter provides three contributions to the study of applying model-driven engineering to analyze power consumption early in the lifecycle of smartphone applications. First, it presents a model-driven methodology for accurately emulating the power consumption of smartphone application architectures. Second, it describes the System Power Optimization Tool (SPOT), which is a model-driven tool that automates power consumption emulation code generation and simplifies analysis. Third, it empirically demonstrates how SPOT can estimate power consumption to within ~3-4% of actual power consumption for representative smartphone applications.

Foreword

MODELS, MODELS EVERYWHERE

Models are pervasive in engineering and serve multiple useful purposes. In general, a model, being an abstract representation of some system that highlights aspects of interest while omitting those that are not, provides a comprehensible view of a system whose complexity might otherwise overwhelm our cognitive abilities. This makes it suitable for reasoning about such systems, helping us identify its key characteristics. Through extrapolations based on such models, it may also be possible to predict the interesting properties of the represented systems even if they are not explicitly rendered in the models. For instance, we may use a mathematical model of a bridge design to compute its load-bearing capacity. Equally important is the ability of models based on some shared representational conventions to facilitate communication between stakeholders.

In addition to these *prescriptive* purposes, models can also serve a *prescriptive* role, that is, they can be used as specifications that guide designers and implementers; i.e., "models as blueprints." In the case of software systems, this is a particularly interesting aspect, because the model and the modeled system share the same medium: the computer. Since computers are eminently flexible automation devices, this creates a number of rich possibilities including the potential to *gradually evolve a model through a succession of refinements into the system, which it was modeling*. This has the potential to mitigate or even eliminate the troublesome transition that is often the curse of traditional engineering: the discontinuities that stand between the idealized world of engineering design and the messy, complex, and unrelenting world of physical implementation.

The possibility of automatically generating implementations from high-level specifications that have evolved from abstract models is certainly one of the primary features of what is sometimes called Model-Based (Software) Engineering (MBSE). However, as discussed above, this is just icing on the cake, since models have many uses. For instance, computer-based specification models have a number of ancillary uses, notably as a basis for automatically generating various artifacts required in design and development: analysis models (e.g., performance, timing, security), test cases, design documents, etc.

The chapters in this volume provide an excellent sampling of the rich possibilities of models in software engineering. They cover the full range of roles that models play in the design of software and systems in general. However, in addition to capturing much of the state-of-the-art in MBE, these contributions provide valuable insight into the future directions of MBE. Researchers, doctoral students, industrial architects, and tool developers will undoubtedly find inspiration in the chapters included here.

The use of *models as specifications*, either as blueprints to guide implementers or as source artifacts for automated code generation, is represented in this volume by a number of contributions. Several of them focus on automating critical functions as a means of increasing productivity and quality. Essaidi et al. ("Model-Driven Data Warehouse Automation: A Dependent-Concept Learning Approach") describe how machine learning can be used to automatically derive model transforms for data warehouse applications. Davis and Chang ("Variant Logic for Model-Driven Applications") describe how metamodeling can be exploited to support the rapid evolution of software systems in a way that bypasses programming, thereby allowing end users to perform the necessary modifications. A similar theme is represented in the chapter by Chassels ("Object Model Development/Engineering: Proven New Approach to Enterprise Software Building"), which argues for separating business logic from implementation technology to allow applications to evolve without the need for code generation.

The use of metamodels is also the subject of the contributions by David and Gonczy ("Ontology-Supported Design of Domain-Specific Languages: A Complex Event Processing Case Study"), which deals with automated generation of custom domain-specific languages from an ontology model, and Pereira et al. ("A Rigorous Approach for Metamodel Evolution"), which addresses the problem of modeling language evolution using formal model-based specifications. Peck and Alexander ("Rosetta Composition Semantics") describe Rosetta, a model-based system specification language that supports compositional system design.

Garg ("Aspect-Oriented Sculpting") explains how model-based specifications of systems can be constructed by exploiting methods used in aspect-oriented programming. Techniques for specifying systems for the emerging "Internet of things" using a generic model-based API are presented by C. Garcia and Espada in their contribution ("Using Model-Driven Architecture Principles to Generate Applications Based on Interconnecting Objects and Sensors").

The use of *models for communications* is a principal theme for several papers in the volume. Yu ("Model-Driven Applications") describes how models can be used effectively as mediators to bridge the gaps between humans, computers, and the environment in which software systems operate. Fischer et al. ("Viewpoint-Based Modeling: A Stakeholder-Centered Approach for Model-Driven Engineering") explain how viewpoints on models can be used to facilitate communication between various stakeholders. Model-based methods for eliciting system requirements from stakeholders are also discussed by Gibson ("The Human Role in Model Synthesis"). This chapter also explains how models can help improve productivity, which is also the theme of the contribution by Dubielewicz et al. ("Quality Assurance in Agile Software Development").

Models as *tools for analysis (reasoning) and prediction* are the topic of several of the contributions. Thompson et al. ("Analyzing Mobile Application Software Power Consumption via Model-Driven Engineering") demonstrate how models and simulation can help in predicting so-called non-functional properties of systems, while G. Garcia et al. focus on performance analysis using models ("Parameterized Transformation Schema for Non-Functional Properties Model in the Context of MDE"). The use of model-based simulation is also the subject of the chapter by Sapna et al. ("Consistency Checking of Specifications in UML"), which describes the use of a semi-formal variant of UML to ensure consistency between a design model and corresponding test cases.

In summary, we can see from these contributions that the future of model-based engineering is a very promising one. It seems that we are at the threshold of a new generation of software engineering methods—methods that are, at last, bringing the design and development of software in line with more established engineering disciplines in terms of dependability and quality. In time, we can expect that the model-based approach will be accepted as standard practice, with models and model-based technologies and methods recognized as an integral and essential part of any systems development. We will know that this has been achieved when the term "model-based" (or "model-driven," if you prefer)—chosen initially to draw attention to models and related technologies—becomes redundant because it is self-evident. The results presented in this book are solid steps that take us firmly in this direction.

Bran Selic
Malina Software Corp., Canada & Simula Research Laboratory, Norway
    & University of Toronto, Canada

Bran Selic *is President and Founder of Malina Software Corp., a Canadian company that provides consulting services to clients worldwide. He is also a part-time Research Scientist at Simula Research Laboratory, in Oslo, Norway, as well as an adjunct at the University of Toronto. Bran has over 35 years of practical industrial experience in designing and implementing large-scale software systems and has pioneered the application of model-based methods in real-time and embedded applications. In the past, he has led several international standards efforts related to modeling technologies, including the widely used UML 2 modeling language. A frequently invited and keynote speaker at various technical events, he is on the editorial board of several mainline scientific journals and has been the general and technical program chair of a number of technical conferences. He has lectured widely on the topic of model-based engineering in various universities and research institutions around the world.*

Preface

A common problem in recent years is the growth of software development complexity due to customer demand for more features and fewer errors. Furthermore, due to recent advancements in technology, it has become necessary to utilize software in multiple domains and professional areas. This leads to problems such as development teams becoming experts in one particular area, necessitating an adjustment period when the team starts new projects in other professional areas.

However, software engineering continually offers new tools that, when properly used, can help in the difficult task of developing software complying with the triple constraint of project management (scope, time, and cost) that is cited in numerous sources. Thus, a relatively new software development approach called Model-Driven Engineering (MDE) has appeared.

MDE is an important and emerging approach in software engineering to increase the level of abstraction of the development tasks. In recent years, Model-Driven Engineering has become a critical area of study, as companies and research institutions have started to emphasize the importance of using models as first-class artifacts in the software development process of complex systems.

The mission of this book is to bring researchers, practitioners, and students to one of the most promising fields in computer science, helping all to understand the current state-of-the-art and to know what the future will bring. The objectives include:

- Bringing together the most relevant research on model-driven engineering.
- Updating the scientific literature on model-driven engineering.
- Identifying and address the complexities encountered in the application of model-driven engineering technologies.
- Identifying the most important issues to be addressed by scientists in the coming years.

The target audience of this book is composed of professionals and researchers working in the field of model-driven software engineering in various disciplines. Moreover, the book will also be a reference for researchers, professionals, and students in computer science and related fields. The book will provide a much needed reference on the state-of-the-art of advanced tools and techniques that are either available or under development to support the maximization of the efficacy and efficiency of model-driven software development. It will also provide foundations to professionals, researchers, and academics on the underlying theory and current applications for use in the future advancement of the existing body of knowledge. This combination of theory, applications, and success stories will provide the reader with an important and detailed view of recent developments in the field and lay the background for future research.

Regarding the distribution of chapters, they are distributed as follows:

- Chapter 1 focuses on the development of a temporal meta-data framework for enterprise information systems applications, avoiding large costs during their lifecycle when customizations may need to be re-engineered.
- Chapter 2 focuses on concerns in programming and tries to identify the ones that may be incurred in the development of software, such as the designing phase, when the product is molded.
- Chapter 3 focuses on how to bridge the gap between business and information technology, using a model as a mediator between humans, computers, and the real world.
- Chapter 4 focuses on the possibility of using models to create applications for smart things such as smartphones or sensor networks, obtaining similar but at the same time very different applications due to the huge number of options.
- Chapter 5 focuses on user interface personalization, providing the right information, at the right time, and on the right support. It proposes a context-aware Model-Driven Architecture approach for content personalization.
- Chapter 6 focuses on a novel approach for designing domain-specific languages to automate their design, which are very closely related, preventing repetitions of work.
- Chapter 7 focuses on the process of requirement elicitation, which plays a primary role in model-driven software engineering, trying to improve the use of survey methods in the gathering of requirements.
- Chapter 8 focuses on agile methodologies that are defined in terms of best practices and their relation with modeling, which has a particular impact on quality assurance.
- Chapter 9 focuses on the evolution of models and metamodels because they are in constant evolution. Metamodel evolution may cause conforming models to become invalid so a new proposed framework can help deal with this issue.
- Chapter 10 focuses on the Rosetta specification language, which aims to enable system designers to abstractly design complex heterogeneous systems. This work improves it by defining the denotation of the composition operators.
- Chapter 11 focuses on transformations design, a key step in model-driven engineering, proposing a new methodology by using machine-learning techniques to automatically derive transformation rules in data warehousing.
- Chapter 12 focuses on the integration between design models of software systems and analytical models of non-functional properties, proposing a parameterized transformation for a model of performance properties.
- Chapter 13 focuses on solving the business "software problem" of inflexibility with poor user experience, proposing a graphical-based solution after 20+ years of research and development.
- Chapter 14 focuses on the consistency among UML diagrams so that quality of the model is maintained, and through inconsistency checking and removal, the model moves towards completeness.
- Chapter 15 focuses on viewpoint-based modeling, an important recent development in software engineering that is likely to boost the wider use of modeling techniques. The results of the project ViBaM are reported.
- Chapter 16 focuses on Smartphones, mobile devices that should conserve battery power to provide maximum possible autonomy. Three contributions to the study of applying model-driven engineering to analyze power consumption are provided.

As a conclusion, we think that the book can be used to learn new the challenges related to software modeling and new lines of research in which we will work in the coming years regarding the Model-Driven Engineering.

Vicente García Díaz
University of Oviedo, Spain

Juan Manuel Cueva Lovelle
University of Oviedo, Spain

Begoña Cristina Pelayo García-Bustelo
University of Oviedo, Spain

Oscar Sanjuán Martínez
University of Carlos III, Spain

Chapter 1
Variant Logic for Model Driven Applications

Jon Davis
Curtin University, Australia

Elizabeth Chang
Curtin University, Australia

ABSTRACT

Customizing Enterprise Information Systems (EIS) scale applications can be very expensive, also incurring additional costs during their lifecycle when customizations may need to be re-engineered to suit each EIS upgrade. The ongoing development of a temporal meta-data framework for EIS applications seeks to overcome these issues, with the application logic model supporting the capability for end users to define their own supplemental or replacement application logic meta-data, as what the authors term Variant Logic, to become a variation of the core application logic. Variant Logic can be applied to any defined model object whether visual objects, logical processing objects, or data structures objects. Variant Logic can be defined by any authorized user, through modeling rather than coding, executed by any user as an alternative to the original application logic, and is available for immediate execution by the framework runtime engine. Variant Logic is also preserved during automated meta-data application updates.

INTRODUCTION

The great majority of software applications in practical use are the result of hard coded program logic that has been compiled and deployed for use as part of the developer's release schedule.

Whether the result of a developer producing a commercial application for widespread release or an internal development team producing software to suit a specific internal purpose or process, the development path will follow similar traditional processes.

DOI: 10.4018/978-1-4666-4494-6.ch001

Externally developed third party software typically provides minimal scope for end users to greatly influence the design and functionality of the application – such influence is usually minor and limited to providing suggestions or advice to the developers, or via bug reporting feedback. While identified bugs in a commercial application may be a priority and receive a higher level of attention in terms of feedback from users and the response of developers it is more typical that user requests for change will suffer long periods before they are introduced into production application releases, if ever.

Expensive alternatives that are often employed by organizations that use large scale third party Enterprise Information System or Enterprise Resource Planning (ERP) style solutions are to engage the vendor or other authorized third parties to develop specific customizations for an organization's requirements to become embedded within a new localized version of the application to support that determination.

Internally developed software may often offer some "time to delivery" opportunities in effecting new desired functionality due to a potentially higher focus on satisfying the organization's specific requirements. The overall cost effectiveness of internal development vs. the use of third party applications with customizations requires a suitable business case for each organization.

In either case, any ongoing customizations or internal development efforts over the lifecycle of the application can be a significant additional cost. An alternative option to choose to utilize commercial-off-the-shelf applications and limit modifications can minimize direct costs but impose internal organizational workflow and inefficiency costs that can also become significant and need to be identified and assessed as part of an overall business case to assist in solution decision making.

The overall lifecycle costs of maintaining an EIS style application are further compounded when accounting for the effort and costs of all major version upgrades, updates, patches and field fixes that may be released by the application vendor. These costs can be significantly magnified when the organization has employed customizations as they need to be reviewed and tested and may require re-engineering using traditional hard coding techniques during each update event to ensure compatibility. Where organizations often choose to defer or skip upgrades to reduce these update costs and any associated application downtime they still incur internal organization inefficiency costs due to delaying the uptake of the otherwise provided updated application benefits to their organization. A suitable review should be conducted to assess these effects.

Our ongoing development of a temporal meta-data framework (Davis 2004) for EIS style applications seeks to overcome these issues as an example of the model driven engineering paradigm. A meta-data EIS (MDEIS) application is fully defined and stored as a model, without the need for application coding, for direct execution by an associated runtime engine.

How do we define MDEIS applications? Firstly, we consider the class of EIS applications that we summarize as visual and interactive applications that prompt for the entry of appropriate transaction data and user events from the application users, use rules based workflow sequences and actions and utilize database transactions in a (relational) database environment to complete the actions. They are typically structurally repetitive and tend to be a technically simpler subset of possible software applications. They generally consist of EIS and Enterprise Resource Planning (ERP) style applications such as; logistics, human resource, payroll, project costing, accounting, customer relationship management and other general database applications. The collective application design requirements are stored and available in a suitable meta-model structure and supported by an execution framework that will allow the EIS application models to be executed automatically and directly from the model, thus the transformation to the MDEIS application. The temporal aspect of the model represents the transaction tracking nature of the application data and meta-data to permit rollback or rollforward through any time period, regardless of the application logic version (the correct state of which is always maintained).

A specific objective of the framework is to also provide the capability for any end user to define and create their own application logic to supplement or replace a vendor's pre-defined MDEIS application logic as what we term Variant Logic - to become an alternate variation of the application logic (Davis, 2011a).

Variant Logic can be applied to any object defined in a meta-data EIS application model whether visual objects, logical processing objects, or as data structures objects. Variant Logic can be defined by any authorized user acting as a Logic Definer via additional object definition in the model and can be assigned for execution to any user as an alternative to the original application logic

The term Variant Logic applies to the overall concepts and supporting framework. A specific instance of Variant Logic is defined as an individual Logic Variant. A defined Logic Variant becomes available for immediate execution by a framework runtime engine and is preserved during any automated meta-data application updates that may be provided by the originating vendor or developer.

Separate detailed analyses (Davis, 2011b) have shown that MDEIS applications can have proportionally significantly lower lifecycle costs compared to traditionally developed EIS applications (circa 18%). This modeling was based on a comparison to a traditional EIS lifecycle where the software is not customized but is regularly updated as per a typical vendor's recommendations. We will demonstrate that an active usage and management of the Variant Logic capability by securely permitting the widespread use of efficient MDEIS customizations to all levels of end users can further provide very substantial additional tangible efficiency savings that will maximize an organization's workflow processes and operations in the most timely and efficient manner.

It is the aim of this chapter to substantiate the application of Variant Logic, by reviewing traditional and emerging software development and customization methods, examining the application of Variant Logic to a meta-data based application model, and provide examples demonstrating the practical application of Variant Logic.

BACKGROUND

Improving the quality, flexibility, and adaptability of software has been a continual process of technological innovation and refinement. Whilst often constrained by hardware development and processing capability the problem has generally been twofold. Firstly there is the available functionality of the software development environment itself; the underlying source code specification, supporting framework, editor functionality and the runtime environment. Secondly is the application flexibility that has been specifically coded by the particular instance of the source code and then made available as the executing application.

We review the following related issues of software flexibility technologies that have guided the overall software development field and subsequently to assist in defining the required Variant Logic capabilities of the temporal meta-data framework for EIS applications.

Object Polymorphism

The polymorphism aspect of object oriented programming has provided a helpful simplification to logic localization and streamlined coding practices. The literal meaning of polymorphism is 'many forms" and in the Object Oriented (OO) world it allows different subclasses of objects to respond differently to the same message, or invoking call, with each subclass invoking its own unique method for resolution.

As a coding aid, polymorphism is of direct use to the OO programmer in developing local object variants. An approximate analogy in the meta-data EIS application framework is the definition of multiple instances of a meta-data model object, each defined with different features as specific instances of defined Logic Variants.

OMG, MDA, and Reusable Objects

The aim of the Object Management Group (OMG) is to "provide an open, vendor-neutral approach to the challenge of business and technology change." The OMG represent one of the largest proponent groups for Model Driven Engineering (MDE) with the goal for their Model Driven Architecture (MDA) initiative to "separate business and application logic from underlying platform technology" (OMG, 2012).

The OMG approach is predicated on the design of platform independent models defined primarily with Unified Modeling Language (UML), which can be rendered into a platform specific model with interface definitions to describe how the base model will be implemented on the target platform.

The OMG also manages the standards, primarily:

- UML 2,
- MetaObject Facility (MOF) where models can be stored, shared and transformed,
- XML Metadata Interchange (XMI) for defining, interchanging, manipulating and integrating XML objects and data, and
- Common Warehouse Metamodel (CWM) as a standard interface to interchange metadata between warehouse tools, platforms, and repositories.

A primary goal of the OMG is interoperability and the tools and technologies are primarily aimed at highly technical analysts and developers. The OMG supports industry developers of supporting toolsets as well as users developing with the technologies. While UML is a widely adopted standard for aiding software development it is a semi-formal language which lacks the precisely defined constructs to fully define application logic (Mostafa, 2007) and has been more commonly used as a coding accelerant rather than a coding replacement.

Complementary strategies aim for more plug'n'play style solutions where components are constructed to an accessible interface standard which allows the component objects to more readily interact with minimal recoding such as by (Talevski, 2003). In Yan (2009) support for multiple simultaneous versions of an application would be provided by a hybrid interim data schema evolution that satisfies both the current and proposed application functionality as a stepping stone to the final version and schema.

The temporal meta-data framework for EIS applications is a MDA class of solution that not only seeks to separate the application logic from the technology platform but to also make the application logic accessible to a less technically skilled base of application modelers that we define as Logic Definers. Specifically we aim to reduce the competency entrance for EIS style application development from technical programmers down to knowledgeable power users, to around the level of competency of medium to advanced spreadsheet creators.

Model Driven Engineering

An important alternative to the common process of hard coded application logic is provided by ongoing Model Driven Engineering (MDE), which is a generic term for software development that involves the creation of an abstract model and how it is transformed to a working implementation (Schmidt, 2006).

A significant proportion of the works to date have involved modeling which contributes more directly to streamlining code generation, processes that are directly aimed for and dependent on highly technical programmers such as (Ortiz, 2009) who specifies alternate aspects and (Cicchetti, 2007) who identifies model insertion points for code insertion. (Fabra, 2008) base their works on the UML 2 specification to seek to reduce coding and transform models of business processes into executable forms.

Zhu (2009) takes a strong model generation approach that seeks to identify customizations to a base model but then implements and maintains each new customized model as separate models executed as individual application instances.

In France (2007), they argue for future MDE research to focus on runtime models, where these executing models can also be used to modify the models in a controlled manner. Such a direction provides not only more manageable change control but also necessarily shifts the target of the change agent closer to the knowledgeable business end user rather than relying solely on the technical programmer.

Such a model is the goal of our temporal meta-data framework for EIS applications. This framework concept is based on our assertion that performance of the analysis and efficient collection of this information can also perform the bulk of the design phase for an EIS application largely as a simultaneous activity. With the collective design requirements stored and available in a meta-model structure, MDEIS applications can be executed automatically from the model with the availability of the framework runtime components.

Every aspect of the EIS application functionality is defined as an object of the meta-data model whether it is identified as core application meta-data produced by the original vendor, or whether it is a modification or extension produced by a user or third party as Variant Logic. Meta-data based version updates will always be clearly identified by a comparison of the meta-data between two time states and then re-producing the sequence of meta-data changes to apply to the meta-data model to be updated.

It is the application of Variant Logic to this meta-data model structure that can extend the scope of the source of the application logic to any authorized user who may act as a logic definer modifying any aspect of the underlying application model.

User Configuration

Most applications will provide some level of user configuration, whereby some aspects of the application can be readily defined by authorized users that will modify some application behavior. Users of business applications would be familiar with the types of application configuration provided in some applications such as: the ability for users to select their own colour scheme for aspects of the user interface; the option to set some environmental options e.g. international locale to adjust some items' display attributes; or save the screen positions of user positioned screens.

Some features may apply to only a specific user or user group whilst others may affect the global user base which may not be a desirable outcome for other affected users. More advanced configuration features of some applications may also allow users to: create simple reports or user defined data extractions and save them as accessible visual objects; or create simple template based user entry screens based on a data table.

There is no common or minimum capability for which user configuration options are provided in applications. Some applications may provide extensive configuration options whilst others may provide minimal flexibility options. While varying in complexity the generally available configurable content for end users tends to be limited to simplistic features (Rajkovic, 2010) with application customization being required for more advanced features. Hagen (1994) long ago identified the need to focus on more configurable software to benefit users and developers as a joint initiative of software development to merge configuration and customization aspects.

Every feature of the meta-data EIS application from the simple to the complex can be optionally configured by authorized users and specified to apply to only specific users or user groups. Simple features will require only basic knowledge to configure whilst more complex features will require a necessarily deeper understanding of function and capability.

User Customization

EIS applications generally consist of three common layers or at least conceptual considerations for development; user interface, business logic and the database repository. Traditional EIS application development almost exclusively requires highly trained developers fluent in the various and often multiple languages, protocols and technologies that constitute the EIS application that may have been progressively developed over many years, containing multiple legacy technologies at any time.

The scope of the typical end user to extend functionality changes beyond that permitted by the commonplace user configuration capabilities is typically quite minimal as traditional software customizations will require:

- Access to the entire or at least partial application source code,
- Documentation and knowledge of the Original Equipment Manufacturer's (OEM) internal software structure, design, directory and Application Programming Interfaces (APIs),
- Appropriate software development licenses and editor environments, correctly configured to produce compatible output software,
- Required technical skills to implement the specific customizations.

These requirements effectively need to duplicate much of the software development environment of the OEM software developer which is always going to be complex and expensive to establish and maintain, and is exclusively the domain of the skilled technical programmer. Whether the OEM software developer is willing to expose any level of its often proprietary intellectual property to third parties or customers is purely at the behest of the individual OEM.

All features of the MDEIS application can be permitted to be customized by authorized end users, acting as Logic Definers, aiming at the knowledgeable business user or power user rather than solely restricted to technical experts. Defining complex or new application segments will require a correspondingly higher understanding of MDEIS functionality which a technical programmer can certainly fulfill as can a knowledgeable power user.

In a traditional organization environment, usually only the highest priority customizations, if any, are likely to be implemented. A key aspect for MDEIS applications is that these customization features are then available to be created by any authorized user, for any particular purpose, drastically increasing the scope for rapid organizational and personal workflow improvement.

Application Customization and Rework

Expensive alternatives that are often employed by organizations that use large scale EIS/ERP solutions are to engage the vendor or authorized third parties to develop specific customizations for the user requirements to become embedded within a new localized version of the application.

Customizations to EIS/ERP systems require developers fluent in the development languages and in the detailed structure of the application logic and structure and depending on the scale can become significant software engineering exercises. (Hui, 2003) provides a model to optimize capturing the requirements for major customizations.

Whilst it may have a suitable overall business case for an organization to take this option it is often expensive and time consuming (Dittrich, 2009). Determining the final scope of a customization is a balancing act between the initial and ongoing costs to maintain the customization against the expected benefits. As the costs are often very high accordingly often only the highest priority subset of features would typically be implemented. Further review and potential re-engineering is also required for each customization whenever the EIS is upgraded or updated by the OEM to ensure ongoing compatibility of the customizations, which adds often considerable further time and expense to each upgrade.

Customization of EIS systems for the local environment has become a fact of life for many end user organizations and reducing the impact of the use of customizations through the maintenance lifecycle is another major objective of our research. As a MDEIS application is based purely on defined meta-data they will always be automatically updated by the framework as a sequence of updated meta-data (Davis, 2011c). Where customizations have been defined, as Variant Logic, many customizations will continue to work directly and the MDEIS framework will report exactly on where any potential conflicts may arise, minimizing any potential re-work. In any case the core OEM MDEIS logic is secure and cannot be compromised.

Software Version Management

Version control is the goal of software configuration management, to ensure the controlled change or development of the software system (De Alwis, 2009), to track the development of the components and manage the baseline of software developments (Ren, 2010) including throughout the various phases of a project (Kaur, 2009).

In traditional software development the atomic level to which version control can be applied varies on the version control systems used but can be as high as individual source code files or database table definitions. The atomic level to apply version control for a MDEIS application model is each individual object's attribute definition within the MDEIS application model. Meta-data version

control needs to be managed at the lowest levels as it is also fundamentally tied to maintaining model integrity and permitting direct dynamic execution.

Version control in an MDEIS application is an automatic function closely related to the management and support of the temporal execution capability of the framework (Davis, 2011d) which maintains the temporal status of data and meta-data to support the rollback and rollforward execution to any point in time. An associated technique for identifying changes between versions of software (Steinholtz, 1987) is a classical key approach when applied to the meta-data model and is instrumental to allowing an automated update approach to be applied to MDEIS applications. The MDEIS application framework will process these automatic updates as a sequence of new meta-data update commands.

VARIANT LOGIC META-DATA MANAGEMENT

What are the Core Impediments to Application Flexibility?

Many of the issues facing software application flexibility and modification have been introduced in the previous section. Key concerns that need to be addressed to significantly improve the current widespread organizational operational inefficiencies, software development duplication efforts, and imposed restrictions on software capability include:

- **Application Logic Openness:** Software developers can and do readily lock access to the underlying application and effectively achieve closed systems via compiled software and proprietary code libraries and technologies. MDEIS applications expose the application logic and workflow definitions within the meta-data model and allow users to ignore the supporting technologies.

Review and analysis of the application logic becomes much simpler and can be supported by the automatic generation of logic structure analysis reports i.e. MDEIS application will self-document - this makes the analysis, understanding, and extension of MDEIS applications much simpler and supports a speedier process to assist in ongoing operation and logic expansion.

- **Data Structure Availability:** The internal data structures of an EIS system can become very complex, especially over multiple generations where legacy data structures have been maintained and where new technologies have been progressively added to the EIS functionality, often resulting in severe internal data segmentation. MDEIS applications define all data objects within the same consistent meta-data model so that all data is always well defined and readily available for access and processing via the same logic definition mechanisms.

- **Configurability vs. Customization:** Software developers do not apply consistent capability or definitions of the level of flexibility that their applications deliver. Some may define that a configuration is based on an attribute definition change executed via a provided user interface feature, whilst others may extend the definition to supporting but requiring user developed application code to be created and integrated. The common aspect is that there is usually only a small subset of application features that can be readily modified in most applications as a configuration. In MDEIS applications every aspect of the defined application logic can be modified from the simplest to the most complex user interface, workflow or data objects.

- **Personal Flexibility:** EIS applications are often the core workflow mechanism for many users and user groups to perform their operational roles; however, the same

core functionality of the EIS, whether customized or not, may not best suit all users or even user groups. Different workflows may be required to achieve maximum efficiency; however, it may not in a traditional software development sense be considered worthwhile to implement any personal or localized customizations. MDEIS applications will provide the same flexible and functional customization capability via Variant Logic to any authorized users as per any organization wide customizations, readily allowing minor or major changes to be defined as is appropriate for any specific individual or user group need.

- **Organizational Flexibility:** EIS applications tend to often provide only the most common functionality and rarely will the "out of the box" workflows suit all customers and all of their users. High levels of customization are commonplace for customers to achieve a suitable level of compatibility with their organisational workflows. MDEIS applications cannot avoid this problem entirely but the MDEIS application will provide great flexibility, economy, and speed of delivery via Variant Logic as an efficient solution.

- **Lifecycle Costs:** the initial implementation cost of an EIS application is typically only a small fraction of the overall system lifecycle cost. The size and scope of customizations vary with associated proportional costs to develop and implement. Every future EIS update from the OEM needs to be reviewed and tested for ongoing compatibility with all customizations and may often involve significant effort and re-work to ensure that the customizations will maintain compatibility with the new core application changes. The MDEIS application can be shown to potentially deliver comparative lifecycle costs below 20% of traditional EIS applications. The efficient and widespread application of Variant Logic could be expected to lower this comparison further when a high number of individually minor but collectively effective Logic Variants are implemented, particularly those that would likely never be of a high enough individual priority to warrant implementation as part of a traditional customization.

- **Speed to Deliver Customizations:** EIS software developers have many considerations when candidate customizations are proposed resource conflicts with their ongoing core development teams, a wide customer base generating many individual customization requests, analysis of customizations for consideration as core application functionality inclusion, plus commercial engagement processes. Customers themselves have many considerations such as prioritization of often many customization requests, budgetary constraints, internal agreement on scope and the ultimate scheduling of associated testing and updates. The net effect can cause significant delays from conception to eventual implementation, with prioritization and business case determinations excluding an often majority of individually smaller but overall significant set of candidate improvements. The Variant Logic capability of MDEIS applications can not only significantly lower the individual cost of each logic definition change to allow a corresponding increase in overall scope but through de-centralization can drastically reduce the delivery time, as knowledgeable users can create their own Logic Variants, and complex Logic Variants can be outsourced to any suitable technical resource such as power users, third parties or the OEM - all as simultaneous asynchronous activities. Additionally the automated update capa-

bility of MDEIS application will greatly speed any update or upgrade processes.

- **Effort and Expense:** The natural market forces of competition do not often strongly apply when dealing with an OEM or restricted set of authorized third party developers which contributes to the typically high level of expense to develop traditional EIS customizations. Whilst it may be prudent to engage the most knowledgeable and technical resources of the OEM for the more complex of customizations, or Logic Variants in a MDEIS application, there is a wide range of modifications that can be effectively performed locally by internal users, subject matter experts and technical resources. The ready use of appropriate internal resources would typically provide much lower cost and turnaround particularly for the often large number of potential Logic Variants that may be individually minor and relatively trivial to implement but overall can be representative of a potentially major collective opportunity benefit.

- **Organization Efficiency:** Due to the high lifecycle costs of customizations usually only the highest priority modifications are ever implemented. Accordingly, there is an indirect organizational cost incurred for every identified but unfulfilled improvement due to continuing to operate with the identified inefficiencies of the current available workflows and processes. MDEIS applications reduce the overall lifecycle costs compared to traditional EIS applications, below 20% in comparison, providing increased profitability and scope for significant Variant Logic customization. Additionally similar comparative savings apply to the creation of each Logic Variant making each modification significantly cheaper to implement again providing further scope to fund further improvements. The potential comparative floor price for Variant Logic definition can be even lower for individual Logic Variants as simpler modifications may often be more directly implemented locally and immediately whereas even trivial customizations can incur a minimum but proportionately higher level of cost and effort from external developers.

These key issues have been in existence for the history of large scale commercial software applications such as EIS style applications. Their resolution requires a paradigm shift away from the traditional software development lifecycle. We believe that the use of Variant Logic as a major aspect of a model driven approach can provide a pathway to providing major improvements.

Using Variant Logic for Customizations in a Model Driven Context

Our ongoing development of a temporal meta-data framework for EIS applications seeks to remove the need for hard coding by technical developers (other than in the creation of the runtime engine and meta-data editors), and transform the responsibility of defining application logic to business analysts, knowledge engineers and even business end users.

Once the MDEIS application logic has been defined as meta-data and stored as a model the model can then be used for direct execution by a runtime engine. The model allows the application logic to be defined as a set of high-level objects (which provide the greatest functionality from the smallest definition) through to allowing for low-level functions such as mathematical expressions. The model is abstracted away from any underlying framework code and relies on specifying the relationships between defined application objects and the required actions.

The model abstraction is also aimed at allowing less technical users such as knowledgeable or power users from the business side of the user organization to define the requirements for and thus the actual model objects for the application logic. The model also supports a specific objective of the framework which is to provide the capability for end users to define and create their own additional application logic to supplement or replace a vendor's pre-defined application logic, as what we term Variant Logic, to become a variation of the application logic for their own specific purposes.

EIS applications typically consist of three common layers or at least conceptual considerations for development; the user interface, business logic and database repository. Traditional EIS application development almost exclusively requires highly trained developers fluent in the various and often multiple languages, protocols and technologies that constitute the EIS application. Variant Logic for MDEIS applications can be defined to specify changes in each of these three areas:

- **User Interface:** The meta-data definitions for user interfaces and logical workflow can be modified and defined with additional application features to operate with, enhance, or optionally replace existing application functionality – without coding, and for immediate execution. The MDEIS application model readily permits the full range of application feature set changes to interaction with existing or new data stores, user interfaces and logical workflows - without limitation other than that imposed by logical integrity and authorization. At the personal user level (single or permitted group access) changes can be made to the application meta-data that have not been flagged as core or mandatory by the MDEIS application's higher level designers. Within this scope logic definers can within their authorization limits;

- Remove from display (not delete) non-mandatory features (e.g. Remove an entry field object that is typically not used by a particular user or role on a particular screen),
- Relocate any features between user interface locations (e.g. Re-arrange a user interface screen or re-arrange objects between multiple screens),
- Modify non-mandatory features (e.g. Change the text for a screen object to be more specific to that user, or re-define a text entry field to a drop-down selection where there may typically be only a limited choice of entry for that user),
- Define any new features to support new or existing workflows or data stores (e.g. Define the user interface entry requirements for newly defined or personal data storage).

- **Logical Processing:** Data manipulation, workflow, and transaction processing features are supported by logical functions that are defined for processing data, similar in nature and format to many of the functions in popular spreadsheet applications. Functions in the MDEIS application are used to provide the following:
  - Can be individual or compound functions,
  - To provide specific processing and verification of data and events,
  - As inline or to be a user-defined function that can be used throughout the MDEIS application,
  - To modify the display, value of or storage of data, or to create data,
  - To perform an evaluation to be used for logical workflow execution,
  - To access and manipulate any model object attribute,
  - To execute methods of any model object,
  - To execute any object method of the MDEIS framework i.e. access any API feature.

- **Database:** The definitions for data storage, management and transaction workflow in the MDEIS application are defined in the model hence the authorized user is also able to define additional new data objects that can be accessed along with existing defined MDEIS application data. Any existing data definitions can be managed under new definer specific rules for creating additional aliases, collections and transactions.

The use and specification of new functions arguably represent the most complex technical knowledge that a competent user needs in order to change existing or define new EIS functionality within the MDEIS application. Given the widespread usage and adoption of tools such as spreadsheets throughout the business world this level of skillset has been considered to be a reasonable acceptable level for the meta-data EIS application meta-data definer which provides potential orders of magnitude of increased accessibility to define application logic over mastering the alternate skill sets required as a traditional EIS code developer.

Variant Logic can be defined by any authorized user and assigned for execution by any user as an alternative to the standard application logic subject to the defined security authorizations of the model. The logic changes will also be available for immediate execution by a framework runtime engine.

All Variant Logic will be preserved during automated meta-data application updates that may be provided by the originating vendor or developer. Any potential Logic Variant conflicts that may arise during the update will be precisely identified to help minimize any re-definition changes that may be required to the user meta-data customizations. By providing such clear direction to potential instances of logic conflict this can avoid wholesale re-engineering efforts and greatly reduce any required rectification efforts. Importantly where no logic conflicts can be identified the Logic Variant can be declared as fully operational with no further review required.

Defined Logic Variants

All of the meta-data that defines the application logic is stored in the meta-data model regardless of whether the meta-data was defined by the original vendor or has been added by any other Logic Definer as a Logic Variant. In our implementation, the source meta-data model is stored entirely as data in a RDBMS, albeit as a complicated schema with many hundreds of tables representing the model – the better to clarify the pure meta-data nature of the model. Any meta-data update commands processed in memory are ultimately managed and committed as standard database transactions similarly to the common database transactions that are composed from the meta-data model structures representing the application logic and data structures at model runtime.

Only the original application model definer will be authorized to modify the application model's core logic – any updates will be applied using the automated update process as a batch sequence of meta-data changes (Davis, 2011c). Any other model changes by users or third parties will be applied as new or updated Logic Variants using a meta-data editor, batch-processing script, or via Web service commands.

A Logic Variant is a set of tagged model changes, composed of a group of meta-data changes with each change tagged by a Logic_Variant_Identifier to indicate that it is not core meta data but part of the designated Logic Variant. Some of the Logic Variant's objects' meta-data will be newly defined meta-data to provide the additional functionality the Logic Variant wants to achieve, whilst some will be copies or replacements of the original meta-data – these copies are particularly important as they indicate a specific alternate execution branch of the Logic Variant, separate in execution from the original core model meta-data object.

The tagging of Logic Variant meta-data is very simplistic using the additional Logic_Variant_Identifier column or field to identify the specific Logic Variant – if it is blank then the meta-data object is core model meta-data. A Logic Variant will also proceed through various versions known as "releases" when they are considered complete by the editor. Only the latest "released" version of

Box 1.

```
FOR each required meta-data object Oi in any object reference request
    IF no Logic Variants exist for meta-data object Oi
    THEN use meta-data object Oi
    ELSE
        SELECT (this Logic Variant) of meta-data object Oi
        WHERE
            SELECT (the newest "released") Logic Variant (for object Oi)
                SELECT (the highest authorized) Logic Variants (for object Oi)
                    SELECT (all "released") Logic Variants (for object Oi)
                        WHERE the current user Uj is assigned access to Logic Variant
                    WHERE equal to the maximum defined Authorisation Level of the Logic Variants
                WHERE is the most recently "released" Logic Variant
```

a Logic Variant will ever be executed by assigned users. The owner or editor of the Logic Variant must manually "release" any new version under edit to make it available for execution.

The following algorithm is used to select the appropriate "released" Logic Variant version of an object whenever any meta-data object is requested as part of object retrieval in Box 1.

Note that this retrieval algorithm assumes asynchronous Logic Definer access to the MDEIS application model, which necessarily allows for any Logic Variant modifications to be made to the MDEIS application instance at any time.

More highly optimized options to potentially further improve runtime performance of a framework runtime engine could restrict this asynchronous capability and pre-determine the available Logic Variant objects and execution pathways based on an analysis of the currently defined Logic Variants and assigned users. Optimizations such as these can be extended to an automated batch or Just-In-Time (JIT) style compilation although in an extreme case it then becomes more of an example of a traditionally developed static functionality application, simply based on the MDEIS

model as source, thus losing many benefits of the real-time MDEIS approach.

Figure 1 illustrates how Logic Variants introduce at least one alternate copy of a core model meta-data object to provide the alternate logic pathways which can then be comprised of any combination of referencing existing meta-data or new meta-data objects defined explicitly for the Logic Variant.

In Figure 1 the core meta-data model object Oi references (and is referenced by) other model objects that were defined as part of the core model by the original application modelers. The LV1 variant of object Oi references some of object Oi's referenced objects and also references other newly defined LV objects as part of the overall Logic Variant definition for variant LV1 – if object Oi represented a user interface form or Canvas then the LV1 variant might be a form with only some of the original objects that object Oi contained plus some new objects defined for variant LV1. The LV2 variant of object Oi references some of object Oi's referenced objects and also references some of variant LV1's newly defined LV objects as well as its own newly defined LV2 objects as part of its overall Logic Variant defini-

Figure 1. Illustration of alternate Logic Variant execution pathways and shared model objects

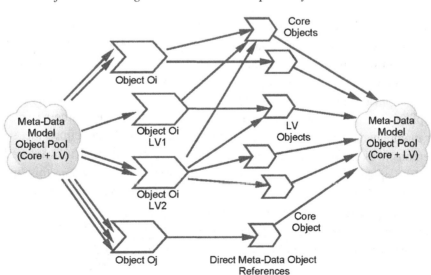

tion – it may be yet another derivative form with some shared and additional objects. Object Oj as shown here does not yet have any defined Logic Variants.

Specifying a Logic Variant is a runtime model management function performed while exercising authorized change control to a permitted subset of the MDEIS application model. Once a Logic Variant editing mode is invoked any subsequent changes to model objects is performed within the operating authorizations of the current Logic Definer and are flagged as part of the identified Logic Variant.

Authorization Framework for Variant Logic

In traditional software development the source code that is compiled to produce the application does not usually have any particular significance in terms of its level of security or role in the application. Rather it is the structure of how the software is designed to work and the appropriate runtime authorizations that may then define such security roles. In terms of modularity an increasing number of software systems will allow the development of third party plug-in modules that can provide user defined functionality to interoperate with the core software system – these modules tend to be themselves examples of additional traditional software development to provide the functionality.

The MDEIS application model supports a range of potential application logic owners to be defined with their relative authorization levels such as; OEM, authorized application partner, corporate IT, regional application maintainer, site owner, group and user maintainers.

Figure 2 demonstrates a potential hierarchy of defined MDEIS application Logic Definer Authorization Levels, from the highest-level original vendor or supplier down through to the authorization for meta-data definition changes that can be made available by end users.

Figure 2. Example hierarchy of Logic Definer Authorization Levels

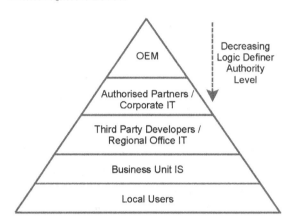

A flexible hierarchy of logic ownership can be defined within the meta-data and managed via the standard application integrity processes of the model framework. These functional authorization processes are governed by the following principles:

- All original meta-data is created by the identified core logic definer (whether internally defined or externally procured) and represents the highest level of authorization for that MDEIS application.
- Additional logic owners can be defined as (usually) lower level authorizations which can include additional external vendors, corporate or local users.
- Meta-data created by one logic definer cannot be modified by a different logic definer, to ensure application semantic integrity.
- Any logic definer can define new meta-data, reference and invoke meta-data owned by other logic definers (where authorized), and modify undefined meta-data attributes of meta-data that was created by other logic definers where this functionality has not been restricted by the creating logic definer.
- Meta-data defined by a higher level logic definer always over-rides any other identical meta-data definition created by a lower-

level logic definer – this may occur during an automated meta-data application update and the update engine precisely identifies changes that may then be required to be made by lower-level logic definers.

Hence, the core meta-data will be able to be extended by any combination of authorized users acting as logic definers to provide enhanced functionality in the MDEIS application, while also maintaining application integrity and while maintaining the ability for the core and other higher-level logic definers to securely provide valid updates to their own meta-data definitions.

This dynamic and distributed editing feature of the MDEIS application framework, Variant Logic, is drastically different from the traditional development lifecycle, further extending the provision of genuine real-time and distributed rapid application development capability and greatly reducing the incidence of locked down or closed

EIS application eco-systems that may be restricted to vendor only modification.

Figure 3 provides an overview of the design for the Logic Definer Access component of the MDEIS application model which allows for assigning hierarchical authorizations, governing the relative priority of aspects of the meta-data definitions that can be modified, and assigning the change permissions based on these aspects.

The Logic Definer Access uses the following classes to model the definition of the change access processes. Note that these main classes are depicted in the central to lower right area of the diagram:

- **Logic Definer Authorization Level:** Is a simple list of authorization levels where a higher level of authorization always has a higher priority and authorization over all lower levels. An example of the highest level of authorization downwards to the lowest is provided in Figure 1.
- **Logic Definer Role:** Are the individual groups or roles that can be assigned to designate an identified group of functional logic definers. A Logic Definer Role is always assigned to a Logic Definer Authorization Level, which defines its relative overall authorization level over other Logic Definer roles. A Logic Definer Role may also be assigned to an Application Security Role to clarify the basic object access for the role.
- **Logic Variant:** Is a designated identifier to group all of the individual logic changes together into a practical set as an instance of Variant Logic. The use of a Logic Variant is to group the associated changes of a set of new functionality for a specific purpose. A Logic Variant is always the assigned responsibility of a designated Logic Definer Role.

Figure 3. Overview class diagram of the Logic Definer Access components

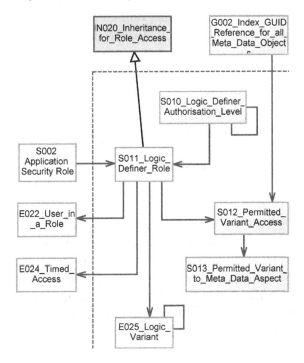

- **Permitted Variant Access:** Identifies the meta-data objects that the Logic Definer Role has access to change as Variant Logic. Permission may be assigned to follow through to all child objects of that object.
- **Permitted Variant to Meta Data Aspect:** Identifies which aspects of the meta-data for that object can be changed as Variant Logic for that Logic Definer Role. Meta-Data Aspects are the internal groups of meta-data attributes for each object and the effect of the change ranges from minor through to major aspects. Examples of meta-data aspects include:
  - **Text:** Allowing a variant to rename non-identifying textual characteristics.
  - **Colours:** Allowing different colours to be assigned to objects and backgrounds.
  - **Help:** Update a variant's version of the help and other documentation information.
  - **Sizes:** Change the default display sizes of objects.
  - **Position:** Move objects to different locations.
  - **Alignment:** Adjust the alignment rules between visual objects.
  - **Access:** Modify whether an object is accessed.
  - **Type:** Modify the type of an object or major attribute.
  - **Function:** Modify the function definitions that an object may use.
  - **Assign:** Change any relationship assignments that are available for an object.
  - **Validation:** Modify any validation rules.

Figure 3 also references multiple other standard model classes that provide additional core functionality. These classes are depicted in the left and upper areas of the diagram outside of the main area bounded by the spaced lines:

- **Inheritance for Role Access:** Is a common inheritance for each of the role based access groups such as; Security, Logic Definer and Distributed Execution Requests. This is shown at the top of the diagram.
- **Index GUID Reference for all Meta Data Objects:** Is a combined global register of the identifier for all instances of the defined meta-data objects (visual and non-visual) as classified by their Entity Object Type. This reference will identify the objects for determining the level of change access available.
- **Application Security Role:** Is the list of the available application roles. A Logic Definer Role may be assigned a default Security Role to provide the appropriate access.
- **User in a Role:** Is the list where Users are assigned to Application Security Roles, Logic Definer Roles and/or Distributed Execution Requests Roles.
- **Timed Access:** Is used to define periods of allowed or denied access for various access types, in this instance, for when Logic Definer Roles can operate.

Some aspects of meta-data objects can be defined to be restricted from permitting any Variant Access, a feature which would be applied to semantically critical application logic components.

The Logic Definer Access process offers a powerful capability to the MDEIS application by empowering authorized groups of users to act as logic definers to configure and customize the MDEIS application to any permitted degree and with the greatly reduced times and expense that the meta-data based application lifecycle provides over traditional software development.

Selection and Execution of Logic Variants

Once the Logic Definers have defined the required application logic changes the Logic Variant will become available in the model for access by specified users however the ongoing default access is to the original meta-data application logic. Access to subsequently execute any defined Logic Variant must be specified and can be assigned in several ways using any combinations of:

- **Application:** The changes are to be applied to become the standard access for all users of the application.
- **Security Roles:** The changes will apply to all users that belong to the specified Security Role.
- **Security User:** The changes will apply to individual users.

As Logic Variants are defined the new meta-data objects will become part of the overall pool of meta-data objects in the application model and thus are subject to the same security access mechanisms to provide access. A runtime engine will then manage the ongoing access to the objects during execution as an additional query to be performed against the meta-data model objects as they are retrieved (as previously described). If there are any Logic Variants in existence for a particular object then any assigned Logic Variant object for that user will be returned for execution.

Figure 4 provides an overview of the design for the Variant Logic Access process. It allows for assigning access by users, by roles or on an application wide basis.

The Variant Logic Access process uses the following classes to model the definition of the alternate logic access. Note that these main classes are depicted below the spaced line in the diagram:

- **User Use Logic Variant:** Assigns individual users to access the designated Logic Variant as their new default for those objects.
- **Security Role Use Logic Variant:** Assigns all users from the Security Role to access the designated Logic Variant as their new default for those objects.

Figure 4 also references multiple other standard model classes that provide additional core functionality. These classes are depicted above the spaced line in the diagram:

- **Security User Account:** Is the list of users that are defined in the application runtime execution environment. These users may be granted individual access.

Figure 4. Overview class diagram of the Variant Logic Access components

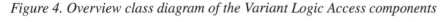
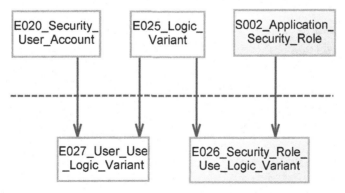

- **Logic Variant:** Is a designated identifier to group all of the logic changes together into a practical set as an instance of Variant Logic.
- **Application Security Role:** Is the list of the available application roles. All users assigned to a Security Role may be granted access.

The Variant Logic Access process is a simple mechanism to assign the ongoing runtime access of users to Logic Variants as an alternative to the existing original logic and any other available logic options. At runtime the execution engine identifies available Logic Variant options for a user's session and on the basis of simple queries will select the appropriate defined Logic Variant and associated objects for execution instead of the originally defined core application objects.

Illustrative Examples of Defining Logic Variants

Starting with a simple Logic Variant that only involves an authorized user re-arranging UI Objects on a Canvas (analogous to a UI screen or form) consider the standard application logic vs. an alternate Logic Variant depicted in in Figure 5.

In this simple example the original form is used to enter customer names and email addresses as displayed on the left of Figure 5. To the right is the newly customized form as an instance of a Logic Variant. The user has made some UI reduction and display modifications to more closely reflect how they use the data entry form. In this example the user has:

- Renamed the form (Canvas),
- Removed the non-mandatory "Initials" fields (UI Objects) as the user never uses this entry,
- Renamed the "Surname" field to "Last Name,"
- Expanded the visual display of all fields, particularly the "Email" field, and
- Resized the Freeform Panel (a display grouping of UI Objects) and Canvas to contain all of the modified objects.

Once the Logic Variant is defined it is assigned by the Logic Definer for execution to the relevant users. The Logic Variant will then be immediately available for use by all assigned users. Any other users will execute the originally defined logic or perhaps another Logic Variant that may have been separately defined.

A more complex example involves defining additional data to be captured with significant modifications to the UI and existing data entry formats to complete a Logic Variant that could then be applied to all users. Figure 6 shows the original Employee Details form (Canvas) that

Figure 5. Simple user interface Logic Variant example

Figure 6. Original application version of Employee Details form

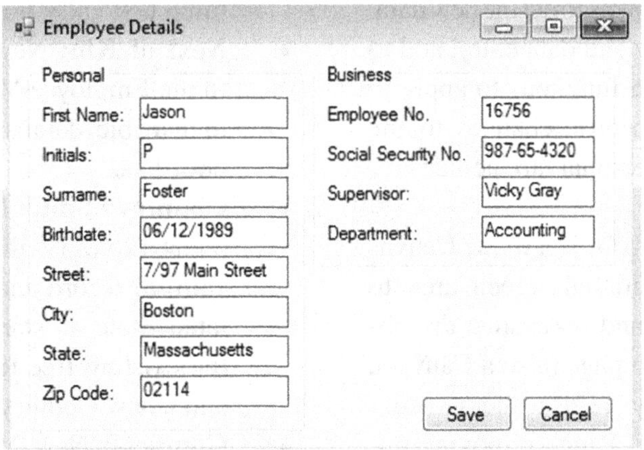

is subsequently expanded upon. Note that the depicted UI Objects layouts have been somewhat condensed from a preferred screen layout to fit readably into the publication format.

To provide a more in-depth data capture of the Employee data the Logic Variant must include the definition of multiple new View Columns

(analogous to data table columns) and associated UI Objects to support their data entry. As the additional required objects will not readily fit on the one Canvas a Tab Object is now defined on the Canvas as a replacement to separate the objects. The revised Canvas with the first tab for Personal data is shown in Figure 7.

Figure 7. Logic Variant for Personal tab of Employee Details form

The changes to be made for this Logic Variant include changes to data objects to define new data, UI Objects for the display and data entry, and to aspects of the verification functions to apply to the entered data. The specific changes to the original Canvas for the Personal tab include:

- **New Tab:** Add a Tab Object to the Canvas to provide the additional screen area to add the new fields and re-arrange the objects. On the first tab page (also a Canvas), "Personal", initially re-use the existing "Personal" Freeform Panel, removing unwanted UI Objects, and additionally adding new Freeform Panels for "Next of Kin", "Address" and "Contacts" – the original "Personal" Freeform Panel could also be re-used as a starting point for the new "Address" Freeform Panel if desired,
- **Birth Date:** Re-define the entry type (UI Object) for the existing data to become a

Date selection instead of the mask structured text entry field,
- **Next of Kin:** New View Columns added to the Employee View Table (analogous to updateable database views) to record the new data,
- **Country / State:** New View Columns and View Tables (with master / detail relationship) to record and select Country and related State as standard entries to replace the existing free text entry State field and add a new Country field,
- **Contacts:** New View Columns added to the Employee View Table to record the new data.

Figure 8 illustrates the remaining changes that have been reflected in the new Business tab.

The additional Logic Variant changes to the Business tab include:

Figure 8. Logic Variant for Business tab of Employee Details form

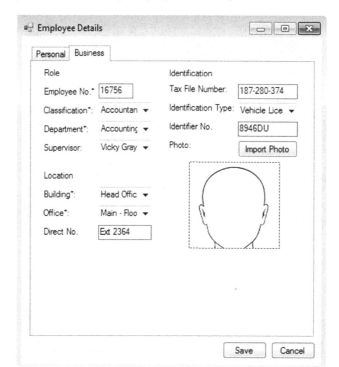

- **Business Tab:** Add a second tab page, "Business," initially re-use the existing "Business" Freeform Panel removing unwanted UI Objects and additionally adding the new Freeform Panels for "Location" and "Identification,"

- **Classification:** A new View Column and View Table to record and select role classifications as new data,

- **Department:** A new View Column and View Table to record internal departments to replace the existing free text entry field,

- **Supervisor:** A new View Column based on existing data, concatenating the names from the Employee View Table to replace the existing free text entry field,

- **Location:** New View Columns and View Tables (with master / detail relationship) to record and select Office and Building locations plus new View Column added to the Employee View Table to record the new Direct No. data,

- **Tax File Number:** Modify the definition of the Social Security No. object to reflect local conditions or create a new View Column.

- **Identification:** Define new View Columns in the Employee View Table for Identification Type and No. defining a Selection list for the Identification Type entry.

- **Photo:** Define new View Columns in the Employee View Table for a Photographic object plus a button to invoke loading a new photo file from disk.

Additional modifications and new definitions are required for various verification functions used by the UI Objects to ensure appropriate formatting and content for each data entry and to ensure overall record integrity.

While there is not enough space to list out all of the individual meta-data object model changes the following major classes of model objects are modified during the definition of this Logic Variant:

- **Canvases:** Are analogous to a screen or form comprised of any Freeform Panels and Navigation Panels. The base "Employee Details" Canvas would be replicated as part of the Logic Variant with new Canvases created for each tab page.

- **Freeform Panels:** Are a grouping of any UI Objects and treated as a single unit on a Canvas. Some existing panels would be replicated as a starting point for the new Logic Variant while other new panels would be created as an integral component of the Logic Variant e.g. Location, Identification.

- **UI Objects:** Consist of any common user interface style controls used for display, data entry, or manipulation. Some UI objects would remain unchanged in definition with most in this example being modified plus new UI objects added to address the new data.

- **View Columns:** Are defined within the data layer of the model and relate directly to either other View Columns (as redefined aliases) or the underlying columns' physical database columns. New View Columns would be created for each new data attribute. Unused View Columns e.g. Social Security No., can also be aliased and redefined to a new View Column where the underlying data types are compatible, e.g. for use by Tax File No.

- **View Tables:** Are collections of any View Columns. The ultimate database transactions are determined on the relative relationships between the View Columns. New View Tables would usually be created to also match new database tables e.g. Classification.

- **Functions:** The most common usage in this example is to perform any additional verification upon data entry. As functions can access and invoke every data and object attribute and method they can be used to perform any simple or complex task.

To illustrate more complex Logic Variants involving complex functions, application workflow and data processing requires a significantly more detailed tutorial approach to demonstrate. A brief entrée into additional examples of Logic Variants is provided in the next section.

Practical Examples of Variant Logic in MDEIS Applications

To further illustrate some of the operational benefits that Variant Logic can bring by allowing any defined component of the MDEIS application to be changed, consider the following scenario examples of how authorized users could directly tailor a MDEIS application to more closely suit their local operational processes. In increasing complexity:

- **Application example:** A payroll clerk may modify the initial starting Canvas of a financial application so that instead of the same global application start-up screen that may apply to all users, their more regularly used Payroll Processing Canvas is always displayed first allowing an immediate start to work.
- **Canvas example:** When performing reviews of staff timesheet entries on the Timesheets Canvas the Payroll Supervisor may choose to reorder the positioning and change the sizing of various Freeform Panels on the screen to suit the review. E.g. as he has a very large monitor, he greatly increases the size of the standard Freeform Panel which displays the multi-record timesheet data grid to re-order, expand and show more columns together as well as moving the corresponding individual record Freeform Panel from below the data grid Freeform Panel to the side. The Logic Variant of the Timesheets Canvas object now uses the modified Freeform Panels'

new positioning and location rules to dynamically manage the layouts instead of the original Canvas object layouts, for that user.

- **Freeform Panel example:** A personnel officer continually uses the New Employee screen (Canvas) to enter new employees centrally for a large organization. The default entry for each employee's Home Base has been defined as a free text data entry object however there are usually only a few options ever entered so the personnel officer has defined a Logic Variant that changes the UI Object to a UI Selection (analogous to any selector such as Drop Down Box etc.) and defined a short list of items with the most common as the initial default entry.
- **Freeform Panel example:** A data entry clerk has to transcribe the contents of hundreds of timesheets that are faxed from remote offices. The format of the timesheets does not match the order of entry on the Freeform Panel and some new additional information is now required from remote offices. A new Logic Variant is created that re-arranges the UI Objects on the Freeform Panel to better suit the manual procedure including resizing the objects to their most common entry size. New View Columns have been created to record the latest information requirements and updated validation functions have been defined to help minimize the occurrence of data entry miss-keying on both existing and new UI Objects. This Logic Variant has been assigned to all other clerks who are entering the remote timesheets.
- **Freeform Panel example:** The financial system was procured from a vendor in the United States and includes by default references to an employee's 401K retirement savings. The Finance Manager has

altered the names of several 401K related objects to more relevant local superannuation terms as well as changing the supporting text and help files. There are also some data columns that are not relevant at all to local conditions so these have been removed completely from all Freeform Panels and from the View Tables that are typically used. This Logic Variant has been assigned to all users of the finance application to ensure conformity.

- **Navigation Panel example:** The Finance Manager has created several new reports that are regularly reviewed as well as some new display screens (Canvases) that she has defined herself to help with workflow approvals for overtime payments. The common Navigation Panel (analogous to menus or toolbars) that is used is updated with reference to the new objects to include these new Reports and Canvases. The General Manager also finds these very useful additions so has also now been granted access by the Finance Manager to the new changed Navigation Panel Logic Variant.

- **Function example:** A new category of employee payment was defined to allow for payments to be made under a federal parental leave scheme. The Payroll Supervisor defined a Logic Variant with the new data columns (View Columns) to track these hours and the Functions to calculate the payable amounts plus changed the existing overall collation calculation Function to include the new amounts in each person's payroll totals. This Logic Variant was assigned to be Application wide.

- **Workflow example:** The original Workflow to authorize overtime payments needs to be changed. There is a new role in the organization called Divisional Manager that needs to approve any record where the overtime hours are more than 50% of an employee's normal hours. A Logic Variant to the Workflow was created where a new workflow stage was defined and inserted into the current Workflow to achieve this.

- **Module example:** The original financial application did not have any timesheet entry or project costing module. The local business analysts worked with the project and accounting team to model and define a new timesheet entry module including; new data objects, all UI screens, reports and workflows for entry and approval, integrating the costing outputs as entries into the standard general ledger module. This large and complex set of objects was defined as a single Logic Variant and assigned to be Application wide.

- **Complex Functions:** As part of a new project costing module various new data and UI objects have been defined to capture the data. Some data processing can be performed by defining smaller functions as required however there is often the need to perform various event driven batch processing of data. More complex Functions will need to be defined to perform this processing. In the MDEIS framework base functions have been defined to allow conditional processing and looping, utilizing many similar function definitions as would be used in popular spreadsheet applications. Functions can be performed inline as required or saved as a new openly accessible function call.

By extending the above simplistic examples with more complexity, including adding entirely new functionality, the accessibility, power and immediacy of Variant Logic becomes a key capability of the MDEIS application framework particularly with the empowerment of a much wider base of new Logic Definers including knowledgeable end users.

Implementation and Execution of Variant Logic

The features as described for Variant Logic have been defined and modeled as part of our original MDEIS application framework (Davis, 2004) which has undergone major enhancements to add the temporal and distributed capability to the core model to become what we now refer to as the distributed temporal MDEIS application framework.

The Variant Logic functionality is a fundamental concept embedded into the modeling and execution framework and will provide the user logic modification features as a real-time capability to define and execute any new third party application logic. Application modelers and administrators would follow these steps to establish and maintain the Variant Logic execution environment:

- **Establish MDEIS Instance:** Variant Logic as defined in this chapter requires a MDEIS application instance to be established with an appropriate application model as the basis of the application logic. User accounts also need to be setup and assigned to defined roles to provide users access to the MDEIS application instance.

- **Assign Logic Definer Role Authorizations:** Logic Definer roles need to be created and assigned a relative authorization level in the Logic Definer authorization hierarchy. Users need to be assigned to Logic Definer roles to provide the appropriate authorization for them to be able to define or create new Logic Variants as well as defining the range of model objects that they are able to optionally modify. These authorizations are additional to any other Security Access or Distributed Execution Request roles that users may have.

- **Define Logic Variants:** Authorized Logic Definers define a new Logic Variant or select an existing one to be used for sub-

sequent model logic editing. The MDEIS model object based objective version control will manage every model change with full audit tracking and temporal management. Logic Definers can optionally exercise subjective version control management at will throughout the lifecycle of the Logic Variant to "release" discrete new versions of the Logic Variant. Changes cannot be made to a "released" Logic Variant – either an "unreleased" Logic Variant (editing in progress), a new version of an existing "released" Logic Variant or a new Logic Variant itself need to be defined to enable changes as a Logic Variant.

- **Modify and Extend Application Logic as a Logic Variant:** Once an "unreleased" version of a Logic Variant has been selected for editing, all subsequent modifications to the meta-data model will be associated with this version of the selected Logic Variant. The complete set of associated meta-data modifications forms the full definition of the Logic Variant. Execution of an "unreleased" version of a Logic Variant during testing on a live MDEIS application instance will execute on live data in the same way that testing application changes on a live system will – hence, it would be recommended to trial (at least) major Logic Variant changes in a test MDEIS instance initially until the semantic functionality is verified.

- **Release a Logic Variant:** Only the latest "released" version of a Logic Variant will ever be available for execution. The exception to this rule is when executing the MDEIS application during a temporal rollback – in this case the latest "released" versions of Logic Variants at that particular point in time will be executed in conjunction with the temporal state of the rest of the MDEIS application model and data at that point in time.

- **Assign Logic Variants to Users:** Logic Variants need to be assigned to users in order to be executed by them. Logic Variants can be assigned for global access by all users, or assigned as any combination of roles and individual users. Assigned users will only ever execute the latest "released" version of a Logic Variant.

- **Logic Variant Execution:** Once a Logic Variant is "released", either as a first release of a new Logic Variant or as a new version of an existing Logic Variant, the meta-data objects of the Logic Variant will become available for execution by all assigned users. Whenever a model object is requested the retrieval process will scan for any Logic Variants that have been defined for the requested object and if the requesting user is assigned any of the available Logic Variants then an alternate copy of the requested model object will be returned – this is the retrieved alternate Logic Variant copy of the object. As the Logic Variant copy of the object will have some changed properties and relationships to other model objects, either core or Logic Variants, then subsequent application execution will follow the links from the retrieved Logic Variant object instead of the originally requested object, thus executing the alternate logic – for only the assigned users.

Following the above listed sequences to apply Variant Logic, users of MDEIS applications can gain the full benefits of using Logic Variants – the ability for any authorized users and third parties to modify the application logic easily and readily, and to permit other users access to execute the application customizations, virtually instantly and without having to rely solely on technical solution providers

Organizational Business Case for Variant Logic

The paradigm asserted by our meta-data modeling and lifecycle methodology is that performance of the requirements analysis and efficient collection of this information can also perform the bulk of the design phase for a MD EIS application largely as a simultaneous activity. With the collective design requirements recorded as a model in the described meta-data structures the MDEIS applications will be executed automatically from the meta-data model by a runtime engine of the temporal meta-data framework.

Expected major savings in time, resources and effort will be further recognized by the virtual removal of the development, test and deployment phases which commonly account for over 50% of the current system development effort (Pressman, 2004). It would generally not be prudent to fully ignore testing however significant effort savings will be realized by the MDEIS application due to; automatic generation of test plans, reduction in testing efforts focused on only the required application logic, rather than all aspects of the runtime engine for the temporal meta-data framework which can be considered as common and pre-tested. Deployment and upgrade management will also be greatly minimized due to the automated update capability.

Accordingly, we have demonstrated that the MDEIS application lifecycle can be expected to be as low as 18% of the corresponding traditional development lifecycle. For larger customization efforts where complex Variant Logic modules are required a similar figure of comparative savings can be expected as the development would then be following the more efficient MDEIS application lifecycle in the development of the Logic Variant maintaining a similar overall level of savings.

To achieve an even greater reduction in overall lifecycle savings is possible when taking into account the effects of a widespread implementation of Variant Logic throughout an organization using MDEIS applications. In such a scenario the organization is able to implement many useful smaller scale Logic Variants that would likely never be considered to be of suitable priority as customizations in a traditional development environment. This collective effect of many smaller Logic Variants has the potential to drastically lower this effective comparison savings figure due to the following factors:

- **Wide Scope:** The ability to readily implement minor Logic Variants locally for a single user or group permits process optimization throughout every aspect of the organization that is serviced by the MDEIS application. The cumulative efficiencies provided could readily provide significant return on investment.

- **Large Scale:** A large organization with hundreds or thousands of MDEIS application users could generate correspondingly large numbers of smaller Logic Variants for personal usage, departmental groups, and globally across the organization.

- **Lower Cost:** Minor Logic Variants implemented by local resources will usually be much cheaper due to an average lower cost rate of internal staff. E.g. it is common that internal staff cost rates may be only 25-50% of an outsourced technical development resource. Additionally minor changes with only a limited user or functional scope could be quickly implemented and tested, often immediately and interactively, whereas in an outsourced development would often attract a minimum external resource charge to allow for the full set of traditional software development lifecycle processes. E.g. an internal resource may spend 1 hour defining and testing a minor Logic Variant

for their own use whereas the corresponding minimum commercial quotation for a similar traditional customization could be expected to incur costs based on days of overall lifecycle effort. Note that this is not advocating any less stringent development processes, rather that the MDEIS application framework will simplify and localize the scope of Logic Variants.

- **Inefficiency Reduction:** Every customization that is not performed maintains an ongoing indirect cost to the organization for the life of the application in use. Every Logic Variant that is subsequently implemented in an MDEIS application will reduce or remove these inefficiencies across the remaining life of the application. If a minor Logic Variant saved a few percent of effort for a group of staff the cumulative savings can become significant from dozens, hundreds or thousands of Logic Variants.

The basic savings comparisons between the traditional software development lifecycle and the MDEIS application lifecycle were performed on the basis that no customizations were made i.e. maintaining an "out of the box" implementation throughout the lifecycle and performing regular OEM updates to achieve any new functionality. When outsourcing major customizations, as Variant Logic, we expect that a similar savings factor would be experienced due to the procedural similarity with general MDEIS application development.

When an organization actively permits and undertakes a large scale usage of minor Logic Variants we expect that the comparative costs will actually reduce further. However it is more difficult to accurately determine the additional levels of savings that can be made as there are many factors that influence the potential savings for any particular MDEIS application user:

- **Cost Base:** The original implementation and ongoing costs of the application forms the basis of any costs and savings comparison.
- **User Base:** The number of users and groups that can be affected by Logic Variants.
- **Customization:** The number, scale and effect of Logic Variants implemented.
- **Performance:** Accurate baseline information about organizational efficiency and performance to determine the measurement of improvements as well as the cost of inefficient procedures.
- **Policy:** The level of capability training and permissibility to create Logic Variants across the organization.

While the savings cannot be readily determined other than for very specific application, (in)efficiency, organization and scenario examples, consider the following basic example: A large organization with thousands of users (say 1000 users) could readily generate thousands of minor Logic Variants over the lifecycle (say 1000 created). Each Logic Variant might affect one user, a group of users or even every user earning efficiency savings from each affected user (say an average Logic Variant provides 10 users with a 1% improvement). 1000 x 10 x 1% = 10000% overall, equivalent to 100 staff at 100%, or 10% improvement for all staff. Without knowing the cost base the return on investment is variable but could be very significant to one enterprise let alone many. If many minor Logic Variant might be created locally in a few hours rather than a few days or weeks of external efforts (say 50% of the rate and 10% of the effort / time). This could mean further comparative costs of 50% x 10% = 5%, incrementally well below the overall estimated 18% (of traditional costs) MDEIS comparison.

Even at this simplistic analysis the potential return on investment by implementing widespread use of minor Logic Variants, in addition of course to major Logic Variants, will contribute to a higher rate of savings and thus an even lower comparative cost rate compared to traditional software developments.

FUTURE RESEARCH DIRECTIONS

The many industry examples of successful Model Driven Engineering technologies are primarily employed by technical staff to accelerate the coding and maintenance of traditional software applications. Full application modeling and execution initiatives similar to the MDEIS application framework have not yet made a strong impact in the commercial software environment and largely remain as academic test beds with limited usage by stalwart partner organizations.

A unique aspect of the MDEIS application model is the Variant Logic capability permitting logic changes to be made in parallel to existing defined logic, without coding, and in providing immediate runtime access to the modified application model definition. This key aspect has the ability to provide the greatest overall efficiency capability to organizations and users that widely exploit the capability.

Major impediments to the adoption of any new software development technology such as the MDEIS application framework no matter how promising must always include:

- **Development Environment:** Providing a full suite of supporting application development and runtime tools so that application developers are fully able to readily and efficiently define, test and deploy MDEIS applications. The models are currently largely manually defined with scripted assistance on a single platform. A design sequence is underway to iteratively develop a MDEIS application editor as an MDEIS application itself. This involves implementing a sequence of separate stages commencing initially with a manual definition of basic MDEIS editor functions, from which progressively higher function editors are themselves defined as data for the model's meta-data objects, to become meta-data in the model itself.

- **Platforms:** Providing a globally accessible application defining capability best requires globally accessible development and execution platforms even if limiting full global access to only the deployed user interface layers. The MDEIS application framework development currently consists of a relational database for the model and a thick client execution platform. Global access in the era of the Internet, smartphones and tablets requires several appropriate platform user interfaces accessing cloud-based model application and data servers.

- **Connectivity:** Full data and logic interaction is to be provided by secure Web services that allow access to all data, attributes and methods of all model and framework objects.

- **Performance:** Ensuring potential developers and customers that the runtime performance of the deployed MDEIS application will be adequate and comparable to existing available platforms is a mandatory requirement. As expected, on-demand model execution is much slower compared to traditional pre-compiled code although modern multi-processor hardware capabilities provide enhanced performance opportunities. Options to migrate to a managed pre-compiled or efficient JIT environment need to be prototyped with particular consideration to maintaining Variant Logic capability,

- **Legacy Migration:** Providing application logic migration and reverse engineering tools from existing business modeling and application design tools source code for-

mats, libraries and repositories common APIs database structures and merging with other MDEIS application models will facilitate migration to MDEIS application development. Reverse engineering from a relational data structure via scripted wizards is an initial step to populating the application structures.

- **Support:** No technology is fool-proof or guaranteed to work flawlessly nor complete and perfect in its early iterations. To adequately guide prospective developers and users requires an appropriate level of education and assistance resources. Ensuring adequate future strategies and standards are adopted and appropriate features added and improved requires ongoing partnership and liaison with influential and supportive partners.

To fully satisfy these commercial realities is beyond the scope of any academic research program, requiring either strongly managed open source projects or well-funded commercialization ventures to ensure that an appropriate future strategy is realized.

If ultimately successful in a commercial sense, technologies like the MDEIS application framework will result in major wins for end users of EIS style applications through lower development costs and greatly expanded Logic Variant customizations. Many developers may face a more cautionary evolution though with the potential for traditional EIS application developers facing an unviable business model in the face of widespread and rapid competition from an open application model environment.

CONCLUSION

The Variant Logic capability in combination with the MDEIS application framework could allow any level of application customization to be securely and flexibly applied to an existing meta-data based application model by a much wider range of Logic Definers, extended to include non-technical staff and users rather than only technical programmers.

While our separate analyses (Davis, 2011b) have shown that MDEIS applications can have proportionally significantly lower lifecycle costs compared to traditionally developed EIS applications (circa 18%), the widespread application of the Variant Logic capability will provide additional significant lifecycle savings that could generate even higher order workflow efficiencies and returns on investment.

The key drivers to these efficiencies and savings include:

- Providing an open standard of application logic definition (for MDEIS application models) that minimize the level of vendor dependence and greatly increases the availability of competitive additional application logic definers.

- The internal mapping between meta-data objects, as the application logic definition, in a MDEIS application fully defines all aspects of the application, will self-document and fully expose all model structures to all Logic Definers.

- Allowing organization business analysts and knowledgeable and power end users to directly and securely define or modify the application logic to suit local business functions has virtually unlimited potential to offer increased efficiency due to rapid concept turnaround into concrete solutions, while reducing the high cost of using dependent vendor resources.

- When combined with the ability to merge multiple MDEIS applications with their common functions progressively mapped to a single logical application structure

facilitating rapid data and application integration, without coding (Davis, 2005), logic modules can be swapped, shared, merged and integrated from any open source or third party to readily extend local functionality.

- The ability to make language, culture or localized logic changes in addition to the inherent multi-lingual translations of textual aspects of the MDEIS application framework will provide immediately globally deployable applications.

- All Variant Logic will also be preserved during an automated meta-data application update that may be provided by the originating vendor or developer. Any logical conflicts that may arise during the update will be precisely identified to minimize any re-definition changes that may be required to the user meta-data customizations to avoid wholesale re-engineering efforts, reducing the effort in updating the customizations. Logic Definers can be given advance notice of any potential issues based on the results of update simulations to ensure the timely availability of their updated Logic Variants to complete the final MDEIS application update.

Progressing to model driven applications development can provide a paradigm shift to drastically reduce ongoing software development costs. The MDEIS application framework with features such as Variant Logic is an example of how this paradigm shift does not just address the initial development economics but also extends the role of technical application customization to widespread Logic Definer changes that can be deployed by orders of magnitudes more business staff in order to directly address organization inefficiency.

REFERENCES

Cicchetti, A., Di Ruscio, D., & Di Salle, A. (2007). Software customization in model driven development of web applications. In *Proceedings of the 2007 ACM Symposium on Applied Computing* (pp. 1025-1030). ACM Press.

Davis, J., & Chang, E. (2011a). variant logic meta-data management for model driven applications applications - Allows unlimited end user configuration and customisation of all meta-data EIS application features. In *Proceedings of the 13th International Conference on Enterprise Information Systems (ICEIS 2011)* (pp. 395-400). SciTePress.

Davis, J., & Chang, E. (2011b). Lifecycle and generational application of automated updates to MDA based enterprise information systems. In *Proceedings of the Second Symposium on Information and Communication Technology (SoICT '11)* (pp. 207-216). ACM Press.

Davis, J., & Chang, E. (2011c). Automatic application update with user customization integration and collision detection for model driven applications. In *Proceedings of the World Congress on Engineering and Computer Science 2011* (pp. 1081-1086). Newswood Limited.

Davis, J., & Chang, E. (2011d). Temporal meta-data management for model driven applications - Provides full temporal execution capabilities throughout the meta-data EIS application lifecycle. In *Proceedings of the 13th International Conference on Enterprise Information Systems (ICEIS 2011)* (pp. 376-379). SciTePress.

Davis, J., Tierney, A., & Chang, E. (2004). Meta data framework for enterprise information systems specification - Aiming to reduce or remove the development phase for EIS systems. In *Proceedings of the 6th International Conference on Enterprise Information Systems (ICEIS 2004)* (pp. 451-456). SciTePress.

Davis, J., Tierney, A., & Chang, E. (2005). Merging application models in a MDA based runtime environment for enterprise information systems. In *Proceedings of INDIN 2005: 3rd International Conference on Industrial Informatics, Frontier Technologies for the Future of Industry and Business* (pp. 605-610). IEEE Computer Society.

De Alwis, B., & Sillito, J. (2009). Why are software projects moving from centralized to decentralized version control systems? In *Proceedings of the 2009 ICSE Workshop on Cooperative and Human Aspects on Software Engineering* (pp. 36-39). IEEE Computer Society.

Dittrich, Y., Vaucouleur, S., & Giff, S. (2009). ERP customization as software engineering: knowledge sharing and cooperation. *IEEE Software, 26*(1), 41–47. doi:10.1109/MS.2009.173.

Fabra, J., Pena, J., Ruiz-Cortez, A., & Ezpeleta, J. (2008). Enabling the evolution of service-oriented solutions using an UML2 profile and a reference Petri nets execution platform. In *Proceedings of the 3rd International Conference on Internet and Web Applications and Services* (pp. 198-204).). IEEE Computer Society.

France, R., & Rumpe, B. (2007). Model-driven development of complex software: A research roadmap. In *Proceedings of Future of Software Engineering (FOSE 2007)* (pp. 37-54). IEEE Computer Society.

Hagen, C., & Brouwers, G. (1994). Reducing software life-cycle costs by developing configurable software. In *Proceedings of the Aerospace and Electronics Conference,* (Vol. 2, pp. 1182-1187). IEEE Press.

Hui, B., Liaskos, S., & Mylopoulos, J. (2003). Requirements analysis for customisable software: A goals-skills-preferences framework. In *Proceedings of the 11th IEEE International Requirements Engineering Conference,* (pp. 117-126). IEEE Press.

Kaur, P., & Singh, H. (2009). Version management and composition of software components in different phases of the software development life cycle. *ACM Sigsoft Software Engineering Notes*, *34*(4), 1–9. doi:10.1145/1543405.1543416.

Mostafa, A., Ismall, M., El-Bolok, H., & Saad, E. (2007). Toward a formalisation of UML2.0 metamodel using Z specifications. In *Proceedings of the 8th International Conference on Software Engineering, Artificial Intelligence, Networking, and Parallel* []. IEEE Computer Society.]. *Distributed Computing*, *1*, 694–701.

OMG. (2012). *OMG model driven architecture*. Retrieved May 13, 2012, from http://www.omg.org/mda/

Ortiz, G., & Bordbar, B. (2009). Aspect-oriented quality of service for web services: A model-driven approach. In *Proceedings of the IEEE International Conference on Web Services* (pp. 559-566). IEEE Computer Society.

Pressman, R. (2004). *Software engineering: A practitioner's approach* (6th ed.). Columbus, OH: McGraw-Hill Science/Engineering/Math.

Rajkovic, P., Jankovic, D., Stankovic, T., & Tosic, V. (2010). Software tools for rapid development and customization of medical information systems. In *Proceedings of 12th IEEE International Conference on e-Health Networking Applications and Services* (pp. 119-126). IEEE Computer Society.

Ren, Y., Xing, T., Quan, Q., & Zhao, Y. (2010). Software configuration management of version control study based on baseline. In *Proceedings of 3rd International Conference on Information Management, Innovation Management and Industrial Engineering*, (Vol. 4, pp. 118-121). IEEE Press.

Schmidt, D. (2006). Introduction: Model-driven engineering. *IEEE Computer Science*, *39*(2), 25–31. doi:10.1109/MC.2006.58.

Steinholtz, B., & Walden, K. (1987). Automatic identification of software system differences. *IEEE Transactions on Software Engineering*, *13*(4), 493–497. doi:10.1109/TSE.1987.233186.

Talevski, A., Chang, E., & Dillon, T. S. (2003). Meta model driven framework for the integration and extension of application components. In *Proceedings of the 9th IEEE International Workshop on Object-Oriented Real-Time Dependable Systems* (pp. 255-261). IEEE Computer Society.

Yan, J., & Zhang, B. (2009). Support multi-version applications in SaaS via progressive schema evolution. In *Proceedings of the IEEE 25th International Conference on Data Engineering* (pp. 1717-1724). IEEE Computer Society.

Zhu, X., & Wang, S. (2009). Software customization based on model-driven architecture over SaaS platforms. In *Proceedings of International Conference on Management and Service Science* (pp. 1-4). CORD.

ADDITIONAL READING

Abeywickrama, D. B., & Ramakrishnan, S. (2012). Context-aware services engineering: Models, transformations, and verification. *ACM Transactions on Internet Technology*, *11*(3). doi:10.1145/2078316.2078318.

Aquino, N. (2009). Adding flexibility in the model-driven engineering of user interfaces. In *Proceedings of the 1st ACM SIGCHI Symposium on Engineering Interactive Computing Systems (EICS '09)* (pp. 329-332). ACM Press.

Avila-Garcia, O., Estevez Garcia, A., & Sanchez Rebull, E. V. (2007). Using software product lines to manage model families in model-driven engineering. In *Proceedings of the 2007 ACM Symposium on Applied Computing (SAC '07)* (pp. 1006-1011). ACM Press.

Bellavista, P., Corradi, A., Fontana, D., & Monti, S. (2011). Off-the-shelf ready to go middleware for self-reconfiguring and self-optimizing ubiquitous computing applications. In *Proceedings of the 5th International Conference on Ubiquitous Information Management and Communication (ICUIMC '11)*. ACM Press.

Bollati, V. A., Atzeni, P., Marcos, E., & Vara, J. M. (2012). Model management systems vs. model driven engineering: A case study. In *Proceedings of the 27th Annual ACM Symposium on Applied Computing (SAC '12)* (pp. 865-872). ACM Press.

Chanda, J., Sengupta, S., Kanjilal, A., & Bhattacharya, S. (2010). Formalization of the design phase of software lifecycle: A grammar based approach. In *Proceedings of the International Workshop on Formalization of Modeling Languages (FML '10)*. ACM Press.

Cicchetti, A., Di Ruscio, D., & Pierantonio, A. (2009). Model patches in model-driven engineering. In *Proceedings of the 2009 International Conference on Models in Software Engineering (MODELS'09)* (pp. 190-204). Berlin: Springer-Verlag.

Cirilo, C. E., do Prado, A. F., de Souza, W. L., & Zaina, L. A. M. (2010). Model driven RichUbi: A model driven process for building rich interfaces of context-sensitive ubiquitous applications. In *Proceedings of the 28th ACM International Conference on Design of Communication (SIGDOC '10)* (pp. 207-214). ACM Press.

Coutaz, J. (2010). User interface plasticity: model driven engineering to the limit! In *Proceedings of the 2nd ACM SIGCHI Symposium on Engineering Interactive Computing Systems (EICS '10)* (pp. 1-8). ACM Press.

Floch, A., Yuki, T., Guy, C., Derrien, S., Combemale, B., Rajopadhye, S., & France, R. B. (2011). Model-driven engineering and optimizing compilers: A bridge too far? In *Proceedings of the 14th International Conference on Model Driven Engineering Languages and Systems (MODELS'11)* (pp. 608-622). Berlin: Springer-Verlag.

Fraternali, P., Comai, S., Bozzon, A., & Carughi, G.T. (2010). Engineering rich internet applications with a model-driven approach. *ACM Transactions Web, 4*(2).

Gray, J., White, J., & Gokhale, A. (2010). Model-driven engineering: raising the abstraction level through domain-specific modeling. In *Proceedings of the 48th Annual Southeast Regional Conference (ACM SE '10)*. ACM Press.

Hutchinson, J., Rouncefield, M., & Whittle, J. (2011). Model-driven engineering practices in industry. In *Proceedings of the 33rd International Conference on Software Engineering (ICSE '11)* (pp. 633-642). ACM Press.

Jackson, E. K., & Sztipanovits, J. (2006). Towards a formal foundation for domain specific modeling languages. In *Proceedings of the 6th ACM & IEEE International Conference on Embedded Software (EMSOFT '06)* (pp. 53-62). ACM Press.

Jakob, M., Schwarz, H., Kaiser, F., & Mitschang, B. (2006). Modeling and generating application logic for data-intensive web applications. In *Proceedings of the 6th International Conference on Web Engineering (ICWE '06)* (pp. 77-84). ACM Press.

Kulkarni, V., Reddy, S., & Rajbhoj, A. (2010). Scaling up model driven engineering-experience and lessons learnt. In *Proceedings of the 13th International Conference on Model Driven Engineering Languages and Systems: Part II (MODELS'10)* (pp. 331-345). Berlin: Springer-Verlag.

Loniewski, G., Insfran, E., & Abrahao, S. (2010). A systematic review of the use of requirements engineering techniques in model-driven development. In *Proceedings of the 13th International Conference on Model Driven Engineering Languages and Systems: Part II (MODELS'10)* (pp. 213-227). Berlin: Springer-Verlag.

Manolescu, I., Brambilla, M., Ceri, S., Comai, S., & Fraternali, P. (2005). Model-driven design and deployment of service-enabled web applications. *ACM Transactions on Internet Technology, 5*(3), 439–479. doi:10.1145/1084772.1084773.

Marx, F., Mayer, J.H., & Winter, R. (2012). Six principles for redesigning executive information systems—Findings of a survey and evaluation of a prototype. *ACM Transactions Management Information Systems, 2*(4).

Mili, H., Tremblay, G., Bou Jaoude, G., Lefebvre, E., Elabed, L., & El Boussaidi, G. (2010). Business process modeling languages: Sorting through the alphabet soup. *ACM Computing Surveys, 43*(1). doi:10.1145/1824795.1824799.

Papazoglou, M. P., & Heuvel, W. J. (2007). Service oriented architectures: approaches, technologies and research issues. *The VLDB Journal, 16*(3), 389–415. doi:10.1007/s00778-007-0044-3.

Rashid, A., & Moreira, A. (2006). Domain models are NOT aspect free. In *Proceedings of the 9th International Conference on Model Driven Engineering Languages and Systems (MoDELS'06)* (pp. 155-169). Berlin: Springer-Verlag.

Schauerhuber, A., Wimmer, M., & Kapsammer, E. (2006). Bridging existing web modeling languages to model-driven engineering: A metamodel for WebML. In *Proceedings of the Sixth International Conference on Web Engineering (ICWE '06).* ACM Press.

Sousa, K., Mendonça, H., Vanderdonckt, J., Rogier, E., & Vandermeulen, J. (2008). User interface derivation from business processes: A model-driven approach for organizational engineering. In *Proceedings of the 2008 ACM Symposium on Applied Computing (SAC '08)* (pp. 553-560). ACM Press.

Steimann, F. (2005). Domain models are aspect free. In *Proceedings of the 8th International Conference on Model Driven Engineering Languages and Systems (MoDELS'05)* (pp. 171-185). Berlin: Springer-Verlag.

Valderas, P., & Pelechano, V. (2011). A survey of requirements specification in model-driven development of web applications. *ACM Transactions Web, 5*(2).

KEY TERMS AND DEFINITIONS

Distributed Temporal MDEIS Application Framework: Refers to major features of the framework. Distributed refers to the distributed operations as described in this chapter. Temporal relates to audit processes that manage the ongoing change of all meta-data and data to enable temporal execution of MDEIS applications, i.e. the framework then permits full application and data rollback and rollforward execution throughout the entire history to maintain the exact application and database state at any point.

Enterprise Information System (EIS): Large-scale computing system that offers high quality of service, managing large volumes of data and capable of supporting large organizations.

Logic Variant: A specific instance of Variant Logic as an identified selection of associated logic changes.

Model Driven Architecture (MDA): is an initiative of the Object Management Group to "separate business and application logic from underlying platform technology."

MDEIS Application Framework: Is the Meta-Data based Enterprise Information Systems application framework. It is an example of a MDA based application modeling and execution environment for EIS applications.

Meta-Data: Computing information that is held as a description of stored data.

Meta-Model: Defines the language and processes from which to form a model.

Variant Logic: The overall concepts supporting allowing supplemental application logic changes to be managed within the MDEIS application framework.

Chapter 2
Aspect Adapted Sculpting

Rachit Mohan Garg
IT Engineer, India

ABSTRACT

This chapter enables the learners to envisage the prevailing glitches of concern in programming and enables them to identify the concerns that may be incurred in the development of the software. How the designing phase of the product is molded separates these concerns from the core business logic, thereby providing a product with an ability to incorporate any change without actually touching the core business logic unless necessary. Moreover, creating a good design leads to less rework in the latter stages of the SDLC (Software Development Life Cycle).

INTRODUCTION

Success at developing simpler systems leads to aspirations for greater complexity. Managing complex software system is one of the most important problems for software engineering to handle. Increase in the complexity of the software requirements and the initiation of innovative technology makes it indispensable for software packages to integrate and accommodate the countless assortments of special concerns which primarily include concurrency, distribution, real-time constraints, location control, persistence, and failure recovery. Hiding among these special purpose concerns is the core concern responsible for the fundamental

business logic. These non-functional concerns help in either fulfilling special requests of the application (real-time, persistence, distribution), or to manage and to optimize the core business logic (location control, concurrency).

Even though the concerns may be separated conceptually and incorporated correctly, interspersing them in the code brings about a number of problems:

- Interpreting and maintaining a tangled code is rigid and multifaceted as all the concerns have to be maintained at the same time and at the same level due to very strong coupling among them.

DOI: 10.4018/978-1-4666-4494-6.ch002

- Tangled code is hard to understand because of the lack of abstraction as depicted in the above point.

Also in systems designed using OOPs concepts, tangled code gives rise to inheritance incongruities as it becomes impossible to re-factorize a method or a concern without re-factorizing both, which again is the byproduct of strong coupling of different concerns.

Modeling is considered to be one of the most successful elements in the progression of computer systems expansion. It is the process of representing real-world concepts in the computer domains as a blueprint for the purpose of software development. Recent trends in software development have also revealed the value of developing systems at higher levels of abstraction.

The significance of modeling and abstraction becomes further strengthened in the process of Model Driven Software Engineering (MDSE) (Bran Selic, 2003), where they are considered the heart and soul of software development.

The core objective of this chapter is to enable the learners to visualize the above mentioned glitches in the vastly followed OOP practice. A better understanding on the concept of concerns will allow the learners to separate out the concerns from the core logic while developing a software product in the initial phase itself so that the glitches are not transferred to the latter phases which in turn will lead to less rework in the latter stages of the SDLC.

BACKGROUND

Aspects Case

Jacobson (2003) describes how the design aspects can be developed using use cases. It uses the concept of composition techniques that maps the work directly into the program level aspects.

The proposed work does not provide details about models transformations, composition rules, structural relations, etc.

Using Meta-Models and UML Profiles for Separating Out Concerns (Garg, 2011)

UML is the most popular standard used for describing systems in terms of object concepts and mostly used in software design. UML allows for extensions to augment the meta-model in two ways:

- Direct extension of the UML meta-model through MOF elements and
- Extension of the meta-model through UML Profiles.

The first (also called 'heavy-weight') approach is more flexible, since it allows generous modification of the UML meta-model. Nevertheless, one is generally discouraged to use it, since the resulting UML meta-model does not conform to the OMG standard breaking, among other things, interoperability between tools.

The latter ('light-weight') approach uses Profiles, a built-in extension mechanism, to adapt UML to the desired domain. According to Kozak Ayse (2002), UML is not defined as a modeling language but rather as a language family, where Profiles are the individual family members, that is, and the actual languages. Nevertheless, neither approach allows for specific parts of the meta-model to be hidden, if they are not required for the specific purpose. If one wants just a glass of milk, one has to take the whole cow.

The UML Profile mechanism in particular is used heavily in Aspect Oriented Modeling (AOM) to introduce platform specific annotations as well as to define platform-independent language constructs. In AOM based MDSE, UML profile plays a very important role in expressing the PIM models, the PSM models, and the transformation

Figure 1. UML stereotypes

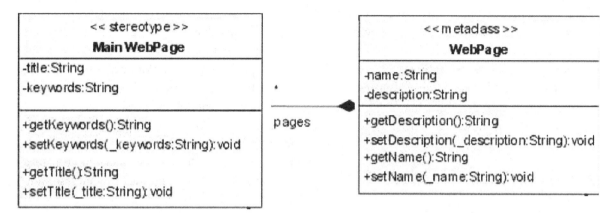

rules. This profile can also be used as a semantic profile which enables a model to express specific information. It can also be used for tagging purpose so as to supply more information during model transformation and code generation.

A UML profile is a combination of stereotype, tagged value and constraints. It uses stereotypes to assign special meaning to designated model elements i.e. how an existing meta-class may be extended. Whenever a stereotype is applied to a model element, this is shown by either an icon or the stereotype name between guillemets, i.e. <<>>. Figure 1 shows an example of this.

When a stereotype is applied to a model element, the values of the properties may be referred to as tagged values. The profile constraints are used to specify the domain restrictions.

There are four types of constraints defined in OCL (Alhir S, 2003), which are shown in Table 1.

Here a brief description some of the proposed works in the field of MDSE is provided.

Reina, Torres, and Toro (2004) proposed an application of meta-models and UML profiles to separate out the concerns. The main problem addressed in the proposed methodology was that different meta-models were required for each of the concerns.

Kulkarni and Reddy (2003) proposed the integration of aspect oriented modeling with model driven architecture. They tried to design a

framework to bring models and aspect together. Kulkarni and Reddy (2003) also use the concept of MDSE to separate out the concerns from the business logic by using abstract templates.

Simmonds (2008) and Simmonds, Reddy, Song, and Grant (2009) describe two approaches for integration of the aspects with the model driven development. The first approach Weave-then-Generate creates an integrated design model and from that model the code artifacts are generated. The second approach Generate-then-Weave creates a complete software application by the aspect code and the business logic code generated separately from the aspect model and primary model respectively. The former approach lacks behind the latter one due to the following reasons.

Table 1. OCL constraints types

OCL Constraints Type	Description
Invariant	Additional rules that must be obeyed by all classes, interfaces or other objects that are defined in the class diagram.
Precondition	A condition that must be true before an operation executes.
Post-condition	A condition that must be true after an operation executes.
Guard	A condition that must be true before a transition in a state machine happens.

- **Presence of Large Semantic Gap:** Due to presence of potentially large semantic gap between individual design models and code.
- **Scattering and Tangling:** These are not eliminated completely as aspects are weaved in models which may lead to the occurrence of these problems.

EVOLUTION OF SOFTWARE

Issues, Controversies, Problems

Evolution of the Software Process

Software design processes and encoding languages occur in a conjointly associate affiliation. Design processes break a system down into smaller and smaller units. Programming languages provide mechanisms that allow the programmer to define abstractions of system sub-units, and then compose those abstractions in different ways to produce the overall system. A design process and a programming language work well together when the programming language provides abstraction and composition mechanisms that cleanly support the kinds of units the design process breaks the system into.

In the early days of computer science, developers wrote programs by means of direct machine-level coding. Unfortunately, programmers spent more time thinking about a particular machine's instruction set than the problem at hand. Slowly, we migrated to higher-level languages that allowed some abstraction of the underlying machine. Then came structured languages; we could now decompose our problems in terms of the procedures necessary to perform our tasks. However, as complexity grew, we needed better techniques. Object-Oriented Programming (OOP) let us view a system as a set of collaborating objects. Classes allow us to hide implementation details beneath interfaces. Polymorphism provided a common

behavior and interfaces for related concepts, and allowed more specialized components to change a particular behavior without needing access to the implementation of base concepts.

Programming methodologies and languages define the way we communicate with machines. Each new methodology presents new ways to decompose problems: machine code, machine-independent code, procedures, classes, and so on. Each new methodology allowed a more natural mapping of system requirements to programming constructs. Evolution of these programming methodologies let us create systems with ever increasing complexity. The converse of this fact may be equally true: we allowed the existence of ever more complex systems because these techniques permitted us to deal with that complexity.

Currently, OOP serves as the methodology of choice for most new software development projects. Indeed, OOP has shown its strength when it comes to modeling common behavior. However, as we will see shortly, and as you may have already experienced, OOP does not adequately address behaviors that span over many—often unrelated—modules. In contrast, AOM methodology fills this void. AOM quite possibly represents the next big step in the evolution of programming methodologies.

What Actually Is a Concern?

A concern in the AOSD has several definitions, but IEEE defines it as: "those interests which pertain to the system's development, its operation or any other aspects that are critical or otherwise important to one or more stakeholders."

It is important to realize that the application itself is a stakeholder, having concerns over issues such as:

- Security
- Performance
- Optimization
- Accuracy

- Data Representation
- Data Flow
- Portability
- Traceability
- Profiling

While this last list of concerns are "held" by different people, the application itself has a stake in these issues as it directly affects implementation and rework issues. Regardless of the front-end design work, many of the problems in this last category are not revealed until most of the components have been glued together and the product is nearly complete. Hence the expression, 90% of the work is done in the last 10% of the development. If Aspect Oriented Software Development (AOSD) has anything of value to contribute to software development, it is in the fact that it brings these concerns to consciousness and lets the whole entity involved in the software production consider these concerns from different points of view.

Concerns are basically divided into two broad categories:

1. **Core Concerns:** These comprise of the program elements that constitute towards the business logic. For instance, in the *address book* example the core concern is adding, removing and updating the entries.

2. **Secondary Concerns:** These comprise of the program element related to requirements other than that of the business logic. It primarily constitutes the following concerns:
    a. **Functional Concerns:** It includes those concerns which are related to specific functionality to be included in a system. For example, logging, security etc.
    b. **Quality of Service Concerns:** It includes those concerns which are related to the nonfunctional behavior of a system. For example, complexity, response time etc.
    c. **Policy Concerns:** It includes those concerns which are related to the overall policies that govern the use of the system.
    d. **System Concerns:** It includes those concerns which are related to attributes of the system as a whole such as its maintainability or its configurability.
    e. **Organizational Concerns:** It includes those concerns which are related to organizational goals and priorities such as producing a system within budget, making use of existing software assets or maintaining the reputation of an organization.

Figure 2. Various types of concerns

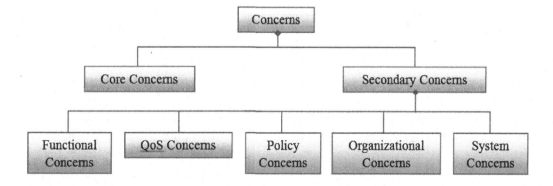

Figure 2 gives a pictorial representation of the concerns discussed above.

Cross-Cutting Concern Problems

Cross-cutting is a term used to describe implementations in which the application interfaces with multiple artifacts. Since artifacts are physical implementations of different concerns, cross-cutting therefore typically involves an operation (or workflow) that is trying to address several concerns. Since the artifacts representing these concerns are typically implemented as autonomous objects so as to promote re-use, we end up with messy code. Often objects are passed into other objects *(void ObjectD::DoSomething(ObjectA * a, ObjectB * b, ObjectC * c))* or are acquired through global mechanisms *(AfxGetMainWnd())*.

Although cross-cutting concerns span over many modules, current implementation techniques tend to implement these requirements using one-dimensional methodologies, forcing implementation mapping for the requirements along a single dimension. That single dimension tends to be the core module-level implementation. The remaining requirements are tagged along this dominant dimension. In other words, the requirement space is an *n*-dimensional space, whereas the implementation space is one-dimensional. Such a mismatch results in an awkward requirements-to-implementation map.

Some of the secondary concerns especially the functional one's depends on many parts of the system and not only the single one i.e. their implementation cuts across many program elements of the software product. These are known as *cross-cutting* concerns (Laddad, 2009; Walls & Breidenbach, 2004) due to their relation with many elements. Since these concerns relate to many elements they cause the problem of *code scattering and tangling* (Walls & Breidenbach, 2004; Laddad, 2009).

1. **Tangling:** It refers to the interdependencies between different elements within a program i.e. when two or more concerns are implemented in the same module, it becomes more difficult to understand it. Changes to implementation of any concern may cause unintended changes to other tangled concerns.
2. **Scattering:** It refers to distribution of similar code throughout many program modules. It creates problem if functionality of that code is to be changed. Changes to the implementation may require finding and editing all affected code.

Figure 3 and 4 represent these issues for an *Address Book* application. In Figure 3 logging and authentication are intermixed with the sole function of the application i.e. to add and delete

Figure 3. Example showing code tangling

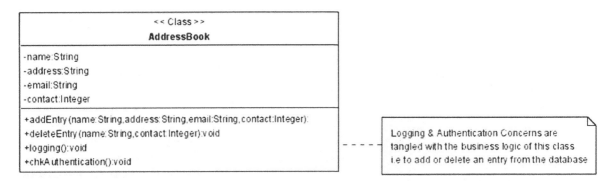

an entry from the database. In Figure 4 two concerns logging and authentication are scattered in different program elements. Both these depict the problems described above.

Efficacy of Concerns

Combined, code tangling and code scattering affect software design and developments in many ways:

- **Meager Traceability:** Simultaneously implementing several concerns obscures the correspondence between a concern and its implementation, resulting in a poor mapping between the two.
- **Inferior Productivity:** Simultaneous implementation of multiple concerns shifts the developer's focus from the main concern to the peripheral concerns, leading to lower productivity.
- **Less Code Reuse:** Since, under these circumstances, a module implements multiple concerns, other systems requiring similar functionality may not be able to readily use the module, further lowering productivity.
- **Meager Code Quality:** Code tangling produces code with hidden problems. Moreover, by targeting too many concerns

at once, one or more of those concerns will not receive enough attention.

- **Additional Difficulties in Evolution:** A limited view and constrained resources often produce a design that addresses only current concerns. Addressing future requirements often requires reworking the implementation. Since the implementation is not modularized, that means touching many modules. Modifying each subsystem for such changes can lead to inconsistencies. It also requires considerable testing effort to ensure that such implementation changes have not caused.

Viewpoint

A viewpoint is a way of looking at the entire system from different positions and determining the concerns based on those positions. Software suppliers and customers have different viewpoints regarding the process of software construction. Within these two groups (and this is not to imply that there are only two groups), the suppliers have different viewpoints (management, engineer, technical support, etc.) as does the customer (management, accountant, data entry person, for example).

Figure 4. Example showing code scattering

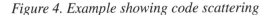
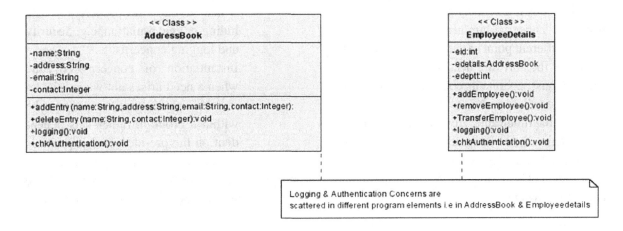

Separation of Concerns

This is an analysis of the concerns that exist in the development of the application. Each stakeholder has a different set of concerns regarding his/her/its involvement in the application. The separation of concerns is an important analysis in that it identifies the areas in which the physical implementation ends up involving multiple stakeholders, be they departmental/organizational interests or objects. Considering concerns from different viewpoints aids in the process of separation of concerns. Requirements often do not explicitly state when a concern involves multiple stakeholders. Instead, a requirement document often states concerns in terms of the specific department's needs. The same is true for object design, in which an object design specifies its concern from the point of view of expected inputs, resulting outputs, and manipulating methods, without regard to the system as a whole. Separating the different concerns is therefore a vital step in later on determining the actual flow of information in the application.

The purpose of separation of concerns is to identify areas that can be designed and implemented in a decentralized way. This allows separate teams to work in parallel and to have independent and autonomous decision making processes. Even simple applications take advantage of this concept. Consider for example a simple network messaging application. The application relies on the fact that network protocols have been developed completely independently and reliably, thus the programmer need not "concern" him/herself with that issue. From a different point of view, large applications (consider the software needed to run the space shuttle) must separate its concerns into manageable "chunks" if work is to be accomplished in a reasonable timeframe. However, this leads to a big issue i.e. the re-integration of these separate concerns since at the end all the different pieces have to be put back together in a meaningful and consistent manner.

Challenges in Designing Concerns

To design and model these concerns, a vast number of approaches have been suggested by numerous researchers (Baniassad & Clarke, 2004; Stein, Hanenberg, & Unland, 2004; Wagelaar & Bergmans, 2002; Clarke, Harrison, Ossher, & Tarr, 1999; Nebulon, 2005; Noble, Schmidmier, Pearce; & Black, 2007). These methodologies can be characterized into following classifications.

- Approaches suggesting design representations and modeling for concerns and their related attributes.
- Approaches suggesting capturing cross-cutting concerns in new modular constructs such as themes (Baniassad & Clarke, 2004), subjects (Clarke, Harrison, Ossher, & Tarr, 1999), and features (Nebulon, 2005).

Some of the problems that aroused from these approaches basically are described as follows.

- **Designing Concerns as Classes:** Concerns should not be considered as classes for numerous reasons.
  - Classes follow object oriented properties viz. Encapsulation, Data Hiding, Inheritance and Object Instantiation. But concerns modeled cannot follow these properties as concerns need to access other classes' private data which violates the property of data hiding or encapsulation, e.g. Security and logging concerns.
  - Instantiation of concerns is done when a need arises not when it is demanded. The instantiation cannot be within a program rather it is dependent on the pre-identified join points during program execution cycle.
- **Designing Pointcuts as Operations of the Concern Classes:** Pointcuts should not be

modeled as operations of concern class due to underlying reasons.

○ Pointcuts like operations cannot have any return value.

○ Pointcuts like operations cannot have local variables as they don't have to perform any local processing. These are just representation of the join-points in the program.

- **Modularity Issue of the System as a Whole and of Concerns in Particular:** The modularization of system and the concerns should be based on the following principles.

  ○ Modules (System modules and the identified Concerns) should be impervious to the remaining system and initialized through a well-known interface.

  ○ Modules should be isolated from each other and from the invoking application.

  ○ Dependencies among different modules should be externally delivered.

  ○ Modules should support plugin or plug-out facility.

The Designer's Predicament

The best designers contemplates current and imminent necessities to avoid an irregular implementation. However therein lays a problem. Predicting the future is a difficult task. If future cross-cutting requirements are overlooked, than the software design need amendments, or possibly contrivance many parts of the system. On the other hand, focusing too much on low-probability requirements can lead to an overdesigned, perplexing, overstuffed system. Thus a predicament for system designers: How much design is too much? Whether to lean towards under design or overdesign?

In summary, the designer seldom knows every possible concern the system may need to address. Even for requirements known beforehand, the specifics needed to create an implementation may not be fully available. These designer thus faces the under/overdesign predicaments.

Solutions and Recommendations

Aspect Oriented Software Development

Aspect Oriented Modelling better separates concerns than previous methodologies, thereby providing modularization of cross-cutting concerns. It improves adaptability, reusability and flexibility. The modularization unit in AOM is called an aspect. AOM involves three distinct development steps: aspectual disintegration, concern enactment and aspectual and concern combination (known as weaving). AOM perform an important role in management of software evolution and software complexity because software complexity and management of its evolution depends on the separation of concerns. The advantages of using this approach are as follows.

1. **Modularized Implementation of Cross-Cutting Concerns:** AOM addresses each concern separately with minimal coupling, resulting in modularized implementations even in the presence of cross-cutting concerns. Such an implementation produces a system with less duplicated code. Since each concern's implementation is separate, it also helps reduce code clutter. Further, modularized implementation also results in a system that is easier to understand and maintain.

2. **Easier-to-Evolve System:** Since the aspect modules can be unaware of cross-cutting concerns, it's easy to add newer functionality by creating new aspects. Further, when you

add new modules to a system, the existing aspects crosscut them, helping create a coherent evolution.

3. **More Code Reuse:** Because AOM implements each aspect as a separate module; each individual module is more loosely coupled. For example, you can use a module interacting with a database in a separate logger aspect with a different logging requirement.

In OOP, the core concerns can be loosely coupled through interfaces, but there is no easy way to do the same for cross-cutting concerns. This is because a concern is implemented in two parts: the server-side piece and the client-side piece. OOP modularizes the server part quite well in classes and interfaces. But when the concern is of a cross-cutting nature, the client part (consisting of the requests to the server) is spread over all the clients.

The fundamental change that AOP brings is the preservation of the mutual independence of the individual concerns. Implementations can be easily mapped back to the corresponding concerns, resulting in a system that is simpler to understand, easier to implement, and more adaptable to changes.

Basic Concepts

Aspects and Other General Terminologies

Aspects can be thought of as blankets (Walls & Breidenbach, 2004) that cover many elements of the software product. At the center lie the modules that perform the core functionalities of the system. This core application can be extended with additional functionalities like logging, security and transaction without even a trace of such collaboration. This is represented diagrammatically in Figure 5.

The basic terminologies (Wampler, 2003) associated with AOP are presented in Table 2.

Thus the issues which arose earlier in the Figure 3 and 4 will be rectified as shown in Figure 6 using the AOM concepts. The highlighted functions in the resultant class are the aspects weaved at the appropriate join points. Before entry authentication is checked and after the execution of any of the function, log of the important information is maintained.

The vision of AOM is to maintain the overall modularity of the system by separating out the cross-cutting concerns, modularizing them as

Figure 5. Collaboration of aspects with the core functionality

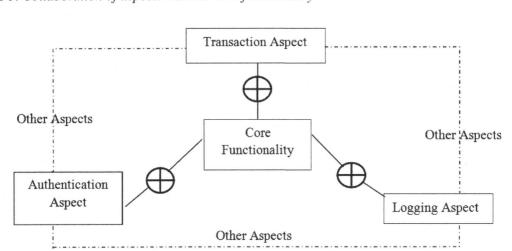

Table 2. AOP basic terminology

Concept	Definition
Advice	Implementation regarding a particular concern. Advice is the code that is executed when the join point specified by the pointcut is reached. Within the advice, you can specify code that is to run before, after, and/or instead of ("around") the code that is intercepted (or cut). Advices, or enhancements, are "woven" into the existing implementation through this mechanism. An advice looks like as follows. *after(Point changed): fieldChanged(changed)* *{* *changed.makeXYZ();* *}*
Aspect	Defines an abstraction cross-cutting concern. It encapsulates definition of the pointcut and the advice associated with that concern.
Join Point	Locations in the program currently in execution where the advice related to an aspect can be implemented
Pointcut	A statement in an aspect, that defines the join points where the corresponding advice should be executed. Given an object Point, a pointcut that specifies the interception of changes to the object Point would look like. *pointcut fieldChanged(Point changed):* *call(* Point.set*(..)) && target(changed);*
Weaving	Incorporating the advice at the specified join points in the program.

aspects, and then weaving them back into the application at well-defined and appropriate locations, called join points.

An Approach to Aspect Modelling (Garg & Dahiya, 2011)

A brief overview of the proposed methodology is as follows.

Initially the whole product is examined rigorously so that all the core components and the relevant aspects are revealed. Then the designing of core components and the separated concepts are done separately using UML Modeling. The point in modeling aspects in isolation with core logic is to avoid code tangling and code scattering problem. The aspects models are stored into a repository so as to make them reusable. The aspects and the code are weaved together so as to generate a complete software product. Figure 7 represents the discussed approach.

The approach is divided into following phases:

1. **Examination Phase:** In this phase the requirements are analyzed to bring out each minute detail. It is itself divided into

Figure 6. Decomposition of problem in to modules and aspect weaving

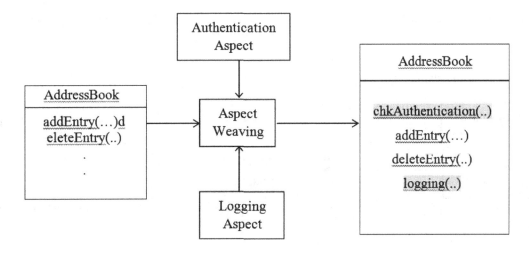

Figure 7. Methodology overview

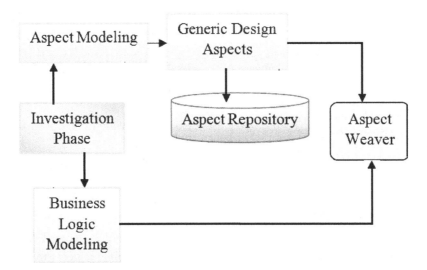

requirement elicitation phase, requirement analysis and requirement recording phase. Many approaches are used to perform this phase as this is the soul of any software development. Any ambiguity and error in this phase will lead to catastrophic results in the end. Some of these approaches are One-to-One interview, Group Interviews, Questionnaires, and Brainstorming etc. The results are combined together which reveals approximately all the features of the intended system. The outputs of this phase are the use case diagrams and the process specification documents. The overall process is depicted in Figure 8.

2. **Designing Phase**: After the completion of the examination phase the designing phase of the core functions and the aspects comes into picture. The designing of core functionality is well versed so will not be discussed in this chapter. Here designing of concerns or aspects will be the main area of concern. To model aspects UML modeling can be used as it is the most widely used standard for designing any OOP system but don't confuse AOSD to be an extension of OOP as unlike OOP AOSD supports both procedural and functional decomposition. To map UML with AOSD what required is a relationship model such a Conceptual

Figure 8. Examination phase

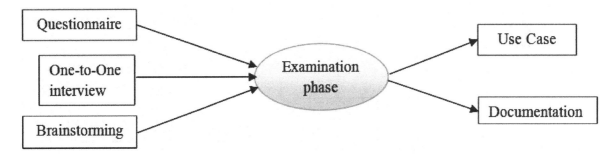

Aspects Model (CAM) (Barra, Genova, & Llorens, 2004). The basic concepts of the CAM are as described below.

a. **Rudimentary Component Concept:** It represents a context which helps in decomposing the problem in terms of some type of unit.

b. **Concept of Join Points:** It represents a context describing join points and the associated constraints. It largely depends on the rudimentary component concept chosen.

c. **Concept of Cross-cutting Boundaries:** It consists of one or more stipulations which include a complete set of join points for an aspect.

d. **Weaving Concept:** It represents the context which defines the different weaving approaches. Weaving can be stagnant or vibrant. In Stagnant weaving aspect and the rudimentary

component concept are weaved before or during accumulation. In Vibrant weaving aspect and the rudimentary component concept are weaved during the execution.

The basic rule followed in AOSD is a "clear separation of aspects and the rudimentary component concept" but while using UML for aspect modeling another rule has to be kept in mind always, "Aspect is an encapsulated module and is isolated from the other aspects and the whole system."

The modeling is done to envisage, postulate and manuscript the conclusions about the realm jargon and about how these rudiments cooperate both architecturally and externally. To model aspects UML 2.0 Components diagram is used. Figure 9 represents component diagram depicting interaction of aspects with the rudimentary functionality and with the weaver.

Figure 9. Aspect modeling

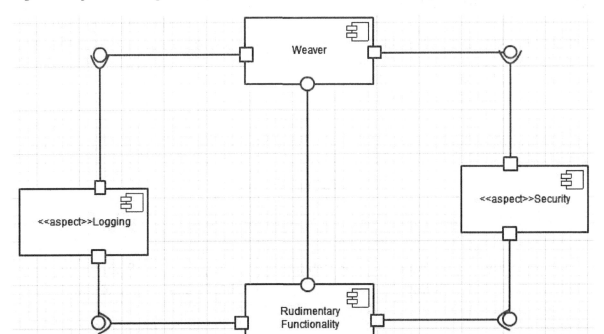

The rudimentary functionality is depicted by component named *Rudimentary Functionality*. Each of the *aspects* is depicted by diverse components. The weaver is shown by the component *Weaver*. The weaver manages the join points which are in isolation of aspects and the main component. The join points are used to indicate the cross cutting relationships between aspects and rudimentary module.

The aspectual interaction illustrations viz. *collaboration* and *sequence* are the important aspects design approach. These describe the specific relationships between aspects and the rudimentary module.

Weaving Phase: After the completion of the modeling phase the weaving phase comes into picture. The weaver is used to weave the aspects to the core functionality based on the join points specified. The output of this phase is the final aspect weaved model for the system which can be used directly to convert to code skeleton using any transformation rules in case of MDSE approach.

FUTURE RESEARCH DIRECTIONS

A direction for future research is studying the applicability of this approach on modeling language in general. Several improvements need to be addressed in future work. These improvements are:

1. The core components and aspects identification is done manually by the model designers. An automation of this feature can be done.
2. Not much tool support for modeling of aspects during design phase. A tool that supports aspect designing during modeling could lead to a revolutionary change in the field of designing.
3. The architecture can be combined with the concept of knowledge based learning for identifying the relevant independent components of the system.

CONCLUSION

We have come across the main anomalies in the widely used Object Oriented Approach and have a brief overview on how Aspect Oriented Approach intends to remove the evils that the OOP was not able to remove. It separates out core business functionality and the aspects thereby removing the problems of code scattering and code tangling.

Then an overview of concerns and how they can be divided into different categories is discussed in the chapter. Then what are the misconceptions related to the concerns modeling is discussed pointing out the problems that arises due to each design technique.

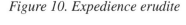

Figure 10. Expedience erudite

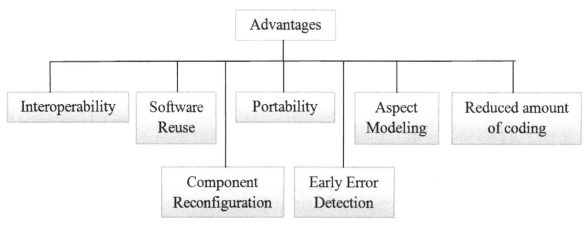

Then moving on to the solution for the already discussed anomalies the aspect oriented approach is discussed with each keyword related to it in a detailed matter.

Moving forward an approach to model aspect separately from core functionality is discussed with appropriate explanation along with the use of UML 2.0 in the approach.

The division of the product into the components enables the concept of software reuse. Aspects are modeled separately from the business logic thereby eliminating the problem of scattering and tangling. Thus the major concepts in software development viz. reusability and aspects get incorporated into the designing approach to make it more robust in nature (see Figure 10).

REFERENCES

Alhir, S. (2003). Understanding the model driven architecture. *Methods & Tools: An international Software Engineering Digital Newsletter, 11*(3), 17-24.

Ayse, K. (2002). *Component-oriented modeling of land registry and cadastre information system using COSEML*. (Master Thesis). Orta Doğu Teknik Üniversitesi, Ankara, Turkey.

Baniassad, E., & Clarke, S. (2004). An approach for aspect- oriented analysis and design. In *Proceedings of the 26th International Conference on Software Engineering*. IEEE.

Barra, E., Genova, G., & Llorens, J. (2004). An approach to aspect modelling with UML 2.0. In *Proceedings of the 5th Wsh. Aspect-Oriented Modeling (AOM'04)*. Lisboa, Portugal: AOM.

Clarke, S., Harrison, W., Ossher, H., & Tarr, P. (1999). Subject-oriented design: Towards improved alignment of requirements, design and code. In *Proceedings of the Object-Oriented Programming, Systems, Languages and Applications*. Denver, CO: IEEE.

Garg, R. M. (2011). *Aspect oriented component based archetype driven ontogenesis: Combining aspect-oriented and model-driven development for better software development*. LAP Lambert Academic Publishing.

Garg, R. M., & Dahiya, D. (2011). *Integrating aspects and reusable components: An archetype driven methodology*. Paper presented in the 4th International Conference on Contemporary Computing (IC3-2011). Noida, India.

Garg, R. M., Dahiya, D., Tyagi, A., Hundoo, P., & Behl, R. (2011). *Aspect oriented and component based model driven architecture*. Paper presented in the International Conference on Digital Information and Communication Technology and it's Applications (DICTAP-2011). Paris, France.

Jacobson, I. (2003). Case for aspects - Part I. *Software Development Magazine*, 32-37.

Jacobson, I. (2003). Case for aspects - Part II. *Software Development Magazine*, 42-48.

Kulkarni, V., & Reddy, S. (2003). Separation of concerns in model-driven development. *IEEE Software, 20*, 64–69. doi:10.1109/MS.2003.1231154.

Laddad, R. (2004). *AspectJ in action* (2nd ed.). Manning Publication.

Modularity. (2009). *Composite application guidance for WPF and silverlight*. Retrieved from http://msdn.microsoft.com/en-us/library/dd490825.aspx

Nebulon. (2005). Retrieved from http://www.featuredrivendevelopment.com/

Noble, J., Schmidmier, A., Pearce, D. J., & Black, A. P. (2007). Patterns of aspect-oriented design. In *Proceedings of European Conference on Pattern Languages of Programs*. Irsee, Germany: IEEE.

Reina, A. M., Toress, J., & Toro, M. (2004). Towards developing generic solutions with aspects. In *Proceedings of the Workshop in Aspect Oriented Modeling*. IEEE.

Selic, B. (2003). The pragmatics of model-driven development. *IEEE Software*, *20*, 19–25. doi:10.1109/MS.2003.1231146.

Simmonds, D. (2008). Aspect-oriented approaches to model driven engineering. In *Proceedings of the International Conference on Software Research and Practice*. Las Vegas, NV: IEEE.

Simmonds, D., Reddy, R., Song, E., & Grant, E. (2009). A comparison of aspect-oriented approaches to model driven engineering. In *Proceedings of Conference on Software Engineering Research and Practice*, (pp. 327-333). IEEE.

Stein, D., Hanenberg, S., & Unland, R. (2002). A UML-based aspect- oriented design notation for aspect. In *Proceedings of Aspect-Oriented Software Development (AOSD 2002)*. AOSD.

Wagelaar, D., & Bergmans, L. (2002). Using a concept-based approach to aspect-oriented software design. In *Proceedings of Aspect-Oriented Design workshop*. AOSD.

Walls, C., & Breidenbach, R. (2004). *Spring in action* (2nd ed.). Dreamtech Press.

Wampler, D. (2003). *The role of aspect-oriented programming in OMG's model-driven architecture*. Aspect Programming Inc..

ADDITIONAL READING

Abmann. (2003). *Invasive software composition*. Secaucus, NJ: Springer-Verlag.

Batory, Sarvela, & Rauschmayer. (2003). Scaling step-wise refinement. In *Proceedings of the 25th International Conference on Software Engineering*. Portland, OR: IEEE.

Bergmans & Aksit. (2001). Composing cross-cutting concerns using composition filters. *Communications of the ACM*, *44*(10), 51–57. doi:10.1145/383845.383857.

Clarke & Walker. (2002). Towards a standard design language for AOSD. In *Proceedings of the 1st International Conference on Aspect-Oriented Software Development*. Enschede, The Netherlands: AOSD.

Czarnecki & Eisenecker. (2000). *Generative programming: Methods, tools, and applications*. New York: ACM Press/Addison-Wesley Publishing Co..

D'Hondt & D'Hondt. (2002). *The tyranny of the dominant model decomposition*. Paper presented at the OOPSLA Workshop on Generative Techniques in the Context of Model-Driven Architecture. Seattle, WA.

Elrad, Aldawud, & Bader. (2002). Aspect-oriented modeling: Bridging the gap between implementation and design. In *Proceedings fo the ACM SIGPLAN/SIGSOFT Conference on Generative Programming and Component Engineering*, (pp. 189-201). ACM Press.

Filman & Friedman. (2000). *Aspect-oriented programming is quantification and obliviousness*. Paper presented at the OOPSLA Workshop on Advanced Separation of Concerns. Minneapolis, MN.

Frankel. (2002). *Model driven architecture: Applying MDA to enterprise computing*. New York: John Wiley & Sons Inc.

Gokhale, Schmidt, Natarajan, Gray, & Wang. (2003). Model-driven middleware. In Q. Mahmoud (Ed.), *Middleware for Communications*. New York: Wiley & Sons.

Gray, Bapty, Neema, & Tuck. (2001). Handling cross-cutting constraints in domain-specific modeling. *Communications of the ACM*, *44*(10), 87–93. doi:10.1145/383845.383864.

Gray, S., Schmidt, B., Neema, , & Gokhale, . (2003). Two-level aspect weaving to support evolution of model-based software. InFilman, R., Elrad, T., Aksit, M., & Clarke, S. (Eds.), *Aspect-Oriented Software Development*. Reading, MA: Addison-Wesley.

Gray & Karsai. (2003). An examination of DSLs for concisely representing model traversals and transformations. In *Proceedings of the 36th Annual Hawaii International Conference on System Sciences (HICSS'03)*. IEEE.

Ho, J. Pennaneac'h, & Plouzeau. (2002). A toolkit for weaving aspect oriented UML designs. In *Proceedings of the 1st International Conference on Aspect-Oriented Software Development*. Enschede, The Netherlands: AOSD.

Katara & Katz. (2003). Architectural views of aspects. In *Proceedings of the 2nd International Conference on Aspect-Oriented Software Development*, (pp. 1-10). Boston, MA: AOSD.

Kiczales, , Hilsdale, , Hugunin, , Kersten, , Palm, , & Griswold, . (2001). Getting started with ASPECTJ. *Communications of the ACM, 44*(10), 59–65. doi:10.1145/383845.383858.

Lédeczi, , Bakay, , Maróti, , Völgyesi, , Nordstrom, , Sprinkle, , & Karsai, . (2001). Composing domain-specific design environments. *Computer, 34*(11), 44–51. doi:10.1109/2.963443.

Lieberherr, , Orleans, , & Ovlinger, . (2001). Aspect-oriented programming with adaptive methods. *Communications of the ACM, 44*(10), 39–41. doi:10.1145/383845.383855.

Nuseibeh, , Kramer, , & Finkelstein, . (1994). A framework for expressing the relationships between multiple views in requirements specification. *IEEE Transactions on Software Engineering, 20*(10), 760–773. doi:10.1109/32.328995.

Parnas. (1972). On the criteria to be used in decomposing systems into modules. *Communications of the ACM, 15*(12), 1053-1058.

Pohjonen & Kelly. (2002). *Domain-specific modeling*. Dr. Dobb's Journal.

Rashid, Moreira, & Araújo. (2003). Modularisation and composition of aspectual requirements. In *Proceedings of the 2nd International Conference on Aspect-Oriented Software Development*, (pp. 11-20). Boston, MA: IEEE.

Stein, , Hanenberg, , & Unland, . (2006). Join point designation diagrams: A graphical representation of join point selections. *International Journal of Software Engineering and Knowledge Engineering, 16*(3). doi:10.1142/S0218194006002811.

KEY TERMS AND DEFINITIONS

Advice: Implementation regarding a particular concern. Advice is the code that is executed when the join point specified by the pointcut is reached.

Aspect: Defines an abstraction cross-cutting concern. It encapsulates definition of the pointcut and the advice associated with that concern.

Concern: Interests which pertain to the system's development, its operation or any other aspects those are critical or otherwise important to one or more stakeholders.

Guard: A condition that must be true before a transition in a state machine happens.

Invariant: Additional rules that must be obeyed by all classes, interfaces or other objects that are defined in the class diagram.

Join Point: Locations in the program currently in execution where the advice related to an aspect can be implemented.

Pointcut: A statement in an aspect, that defines the join points where the corresponding advice should be executed.

Post-Condition: A condition that must be true after an operation executes.

Precondition: A condition that must be true before an operation executes.

Weaving: Incorporating the advice at the specified join points in the program.

Chapter 3
Model–Driven Applications:
Using Model–Driven Mechanism to Bridge the Gap between Business and IT

Tong-Ying Yu
Independent Consultant, China

ABSTRACT

How to bridge the gap between business and Information Technology (IT) has always been a critical issue for both the developers and IT managers. The individualized, differentiated demands by different customers and situations, the constantly changing in both business and IT are great challenges to the applications for enterprises. In this chapter, the authors respectively discuss the left side (computer) in software engineering, with Object-Orientation (OO), Model-Driven Engineering (MDE), Domain-Driven Development (DDD), Agile, etc., and the right side (the business) in Enterprise Engineering (EE) with Enterprise Modeling (EM), and Enterprise Architecture (EA) of the gap. It is shown there are some fundamental problems, such as the transforming barrier between analysis and design model, the entanglement of business change and development process, and the limitation to the enterprise engineering approaches such as EA by IT. Our solution is concentrated on the middle, the inevitable model as a mediator between human, computer, and the real world. The authors introduce Model-Driven Application (MDApp), which is based on Model-Driven Mechanism (MDM), operated on the evolutionary model of the target thing at runtime; it is able to largely avoid the transforming barrier and remove the entanglement. Thus, the architecture for Enterprise Model Driven Application (EMDA) is emerged, which is able to strongly support EE and adapts to the business changing at runtime.

1. INTRODUCTION

The context of this chapter is the applications of Information Technology (IT), mainly, the software which supports the comprehensive operations (business and management) of an enterprise[1], i.e.,

the "applications for enterprises" or "enterprise (business) applications"; another conventional name is Information System (IS). For many software developers, "enterprise applications" means a series of technical characteristics, but our focus on the demand, the applying targets and objec-

DOI: 10.4018/978-1-4666-4494-6.ch003

tives, the environment, as well as the openness and collaboration between computers and humans, in particular the essential functional requirements including the adaptation to business changing.

IT and the applying enterprise / business are mutually independent fields, which have its own ecosystem, though IT should always services business in our context. It is not surprising that, there is an innate gap between them, which can be appeared in anywhere the IT needs to be aligned with the business; this is the basic issue we must face. Empirically, the biggest challenges to enterprise applications come from the differences and changes in the demands, also the running environment and the IT itself. First, there are some conflicts between the different needs from different customers, different situations, and developer's pursuit to generic a solution; the latter is also a factor what makes the product to be bloated and complicated. Second, the uncertainty, constant and rapid changing of business requires the application system has higher flexibility and capability to evolve in runtime; this is a huge challenge to the existing software techniques. Third, IT itself is one of the most complex and rapid changing fields, which leads to an increasing requirements to the maintenance, governance, and the adaptation to the operating environment. Furthermore, in such the applying context, the problems that require solving by the application system is *open*; this will make a significant impact on the development and the whole life cycle of the application. Moreover, between the business and IT camps, and also the different roles in various phases, there are always the problems to the communication and understanding with the different professionals and standpoints; technically, this requires some common languages or media to represent the issues of business and IT they are faced. In addition, it seems we are already in the age of total computer information processing, that is, almost all data and their derivatives (e.g. forms or documents) should be stored, processed by a computer-based way; papers are only the complementary media

in some special circumstances. However, have a review on those problems, it can be seen, such the information islands and the functional silos, which proposed about half a century ago, in our experience, are still not a substantive change, even more serious. Above challenges and issues, with such the cultural and regional differences, and so on, makes that bridging the business-IT gap, establishing strategic alignment or fusion of them, to be still the hard and most important subject for the industry. We believe, there are still more essential approaches that waiting for discovery and validation.

In this chapter, we focus on the role of models for the problem-solving by human with computer, and targeted at the relationship between a certain computer system and the outside world—the people and physical or abstract entities, their states, activities and events—especially in dynamic way at runtime. We start to demonstrate at the inevitable way to connect human, computer, and the real world: models and modeling. And then, around them, have some review on the two side of the gap (see Figure 1), respectively in software engineering and enterprise engineering. Based on the discussions, the Model-Driven Application (MDApp) is introduced, which is based on the principle we called Model-Driven Mechanism (MDM); further, the architecture of Enterprise Model Driven Application (EMDA) is presented.

2. TWO SIDES OF THE GAP AND INEVITABLE MODEL

2.1. Model as Mediator between Computer, Real World, and Human

No matter what challenges the business-IT gap bring us, there is only one of core issue, that is, how to utilize computer to solve the real world problem we faced. Regardless of the means, to solve any problem, we must first know what fact there—by model. This principle is not only suit-

Figure 1. Inevitable model as mediator between human, computer, and the embedding world

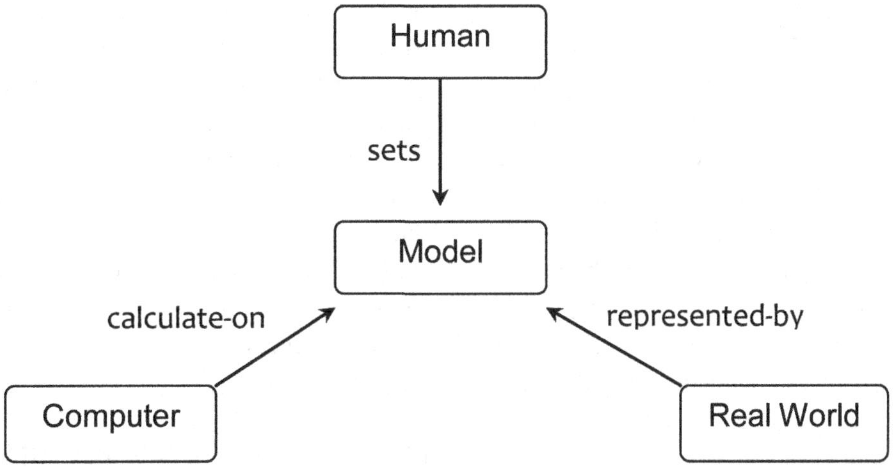

able for human, and is also suitable for computer. Smith (1985) depicted the relationship between computer and the real world and showed us a basic view for the relationship between computers (generally may be a program, computer system, even a description, thought) and real world; he put the computers in the left box and the real world in the right box, where the "inevitable model" is placed as mediator between them[2]. His picture also implied the same relationship between human and real world; for our subject, however, all the three relationships are important, so we depict the basic landscape as Figure 1, wherein the lower half layout is logically equivalent to the Figure 1 in Smith (1985, p. 21), and we labeled the relationships.

Smith implied that, the gap between computers and the real world is able and only able to be connected by model of the real world thing, as well as between human and the world. This view is a background of the exploration to business-IT gap: the IT domain on the left side; the business domain on the right side, and some model as mediator to be connected them.

Another important point shown on Figure 1 is that, under our subject, the question is not only between computer and the world but with human: a computer handles something based on the mediating model setting by human, thus the model is served as a mediator for *both* the human and computer.

In addition, the "real world" is a relative concept; what thing is in the real world depends on your standpoint: e.g., for computer, or human? When a human takes a computer to handle some affair, it can be regarded that there is something in real word for both[3]. In this case, why we cannot use the same model to represent the real word thing, if it is possible, for both the computer and human? Fortunately, it is not just possible but the best way.

Relevant to our subject, based on Figure 1, the efforts to bridge the business-IT gap can be classified into three groups: on the left side such as (mainly) software engineering, on the right side as enterprise engineering (see Sec. 4), and on the middle for bridge over the gap—the solution based on the mediating model.

2.2. Domains and Models for Enterprise Applications

The term "domain" and "domain model" is overloaded in software field; there are many different definitions (Larman, 2004, Sec. 9.2). Here, for effective discussion, we try to give a set of relatively consistent descriptions for some concepts of domain and model involved in the subject of this chapter, as consistent as possible to the common usage. The word "domain" is used against "problem" or "subject," i.e., it is referred to the scope, in which the problem or subject are occurred or involved, and more specifically, it is referred to the set of objects or entities with properties and basic relationships[4]. A point is that the different problems or affairs may be occurred and repeatedly in the same domain; on other hand, the boundary of a domain is usually relative. Accordingly, the models are appeared as two classes: the first class is modeling a target or subject, e.g., a system, a problem, a business process, etc.; the second class is modeling a domain, i.e., "domain model" in general, e.g., the domain of an problem, the domain of an application or system. They reflect two different perspectives, though there may be no need to establish a domain model for such a single, certain problem. To some open problems or subject, however, the relative independent, predetermined domain model is necessary and important. Such the domain model is quite close to the concept of ontology.

The total information processing needs total information modeling; as mentioned above, under our subject, what problems faced by an enterprise application are essentially in changing and uncertain—it is open. The inevitable mediating model thereby should satisfy three conditions: human-understandable / operational, computer-treatable, and evolutionary with the changing of the target as well. Thus, a predetermined domain model, namely *enterprise model*, is necessary for enterprise applications. Figure 2 shows some typical domains and their relationships for an enterprise application, in which the labels are omitted "the domain of," e.g. "the domain of a problem." The dotted line indicates the border is relatively uncertain.

3. SOFTWARE ENGINEERING ON THE LEFT

3.1. Difficulties and Transforming Barrier from Analysis to Design

Kaindl (1999) demonstrated the difficulties in the transition from object-oriented requirement analysis (OOA) to the system design (OOD). He focused on two problems that are particularly difficult in OOD: first, *architectures*: "Finding an architecture that fits well to solve a given problem is both difficult and important." Second, *model of the domain inside the program*: "the OOA model cannot be simply part of the OOD model" (pp.

Figure 2. Some domains and their relationships for an enterprise application

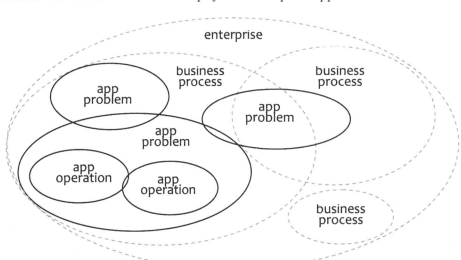

99-100). This is involved in the design decisions about which information about real world objects the deployed system will store and how it will be represented, i.e., how to model the thing in domain inside the program. Such the opinion is appeared as a logical succession from the opinion by Bosh and Molin (1997): "the most complex activity during application development is the transformation of a requirement specification into an application architecture."

In Kaindl (1999), the OOA model roughly is a problem model in which there are both the elements of computer and other things involved in the problem. From a modeling perspective, the problem is the transformation from the analysis models to a design model. Model transformation is the major strategy to improve the system development in MDE, but it does not effectively address the problem even often avoided the issue (see the discussions about MDE and CIM later). Karow et al. (2006) presented that "the transformation results are potentially insufficient and inappropriate for the design process." And "an automation of the design starting from the CIM is not possible by utilising currently existent transformation approaches."

Flowing above accounts, we can describe the problem more generally, that is, in general, there is no simple (liner or automatic) complete transformation from analysis model to design model or code. It can be called *transforming barrier* between analysis model and design model. The analysis model here can be involved in problem model, domain model, business (process) model (as is; to be), and may also the functional model (as a black-box model of the system, e.g. the use case model) of the application system. Against, the design model is somewhat white-box model of the system which are can be implemented in some executable code.

3.2. Object-Orientation and Use-Case Driven

Some one might think that the scenario described in sec. 2.1 is very consistent with the Object-Oriented (OO) approach: Classes / Objects are the models of the real world thing; this relationship is shown in most the examples in OO books or articles. There was a cautious statement in Craig (2004, Sec. 9.3.): "This is a key idea in OO: Use software class names in the domain layer inspired from names in the domain model, with objects having domain-familiar information and responsibilities." However, this issue is indeed relating to the transforming barrier, "The question remains how OOD objects can be both abstractions of something in a problem domain and objects in a solution space" (Kaindl, 1999, p. 95). The direct correspondence between the Classes / Objects in software and the objects in real word is helpful but not enough to solve the transition difficulties. We prefer regard an Object as a container to the model of the real object (if there is—an Object in software can also be not corresponding to any real world thing), i.e., the data of a model is encapsulated in an Object as its private data; in other words, the model is hard-coded, embedded in the code.

Moreover, much OO development methods with UML employ the use-case model as a basis, for example, the unified software development process (UP) directly captures the required functions by the use-case driven, iterative and incremental approach, where domain model and business model are unnecessary but only used as the context to assist in understanding the functionality that captured by use-case model (Jacobson, Booch & Rumbaugh, 1999). The use-case driven, iterative and incremental process is typically a nonlinear process, which is an evidence of the transforming barrier.

3.3. Model-Driven Engineering and Abstract Hierarchy

Model-Driven Engineering (MDE) was regarded as an "important paradigm shift is happening in the field of software engineering"(Bézivin, 2004, p. 21), which might represented "the first true generational shift in basic programming technology since the introduction of compilers" (Selic, 2003, p. 20). MDE extends models and modeling to almost all concerns relevant to the development, which consist of a multi-dimensions modeling space, and lies engineering on their mapping and transformation (Kent, 2002). "Considering models as first class entities and any software artifact as a model or as a model element is one of the basic principles of MDE" (Bézivin, 2006). Although there are a lot of different ways to the research and practice, MDE is mainly developed along some basic principles that outlined by the Model-Driven Architecture (MDA™) of Object Management Group (OMG), including: raising the level of abstraction, transforming the model to lower level, code automated or partially automated, and so forth. As discussed in Frankel (2003), it is mainly for filling the gap for / in software architecture, e.g., the gap between programming language and machine code, or development environment and execution platform—but not yet the gap between computer and the real world.

In last over ten years, represented by MDA, the field had become a hotspot in software engineering, and attracted much research and development. However, so far, compared with the expectations, its development does not appear to be satisfactory. In the COOMP workshop 2011, Jean Bézivin presented the views that "MDE has reached a standstill" (Dubray, 2011), and even used the word "failure." It seems to be disappointed, at least, for MDA (of course there is no more influential instance than MDA)—this is appeared somewhat the common sense in MDE community[5]. This situation is somewhat familiar to what we had been seen, e.g., the Computer-Aided Software Engineering (CASE) in 1980s:

it "attracted considerable attention in the research and trade literature, it wasn't widely adopted in practice" (Schmidt, 2006, p. 25), as well as the case to the fourth-generation programming language (4GL). Related to CASE, however, France and Rumpe (2007) pointed out "MDE can and should be viewed as an evolution of early CASE work" (Sec. 3.3). There may be a lot of reasons leading MDE/MDA to be disappointing, one of them may be related to the issue mentioned above, i.e., existing MDE is not directly touch on the difficulties and the transforming barrier.

Furthermore, we can discuss the problem on the concept of abstraction. CASE, 4GL and MDA are all the efforts that somehow raise the level of abstraction for developer but it appears they are a bit less successful than 3GL or the earlier efforts. One of the reasons might be related to the ultimate of raising abstract level. First, there are *two* abstract hierarchies on the two sides of the gap, one is on the computer operations/constructs, and another is on the real word things. Second, the distinction of the two abstract hierarchies determines the target and the ultimate of the division and degree for abstraction in system development. We argue that there is a most important principle, i.e., it should not only unilaterally raise an abstraction level but *matching* them: to establish appropriate link mechanism between the two abstract hierarchies. We can see, however, either the existence and distinction or the link mechanism of the two abstract hierarchies are often to be confused and ignored.

France and Rumpe (2007) had proposed a conclusion that MDE is more likely to contribute to software development rather than the inherent software complexity; it seems that is in line with the fact we have seen (they also encouraged research on the use of runtime models, see discussion in sec. 5.5). As shown in many literatures, e.g., Miller & Mukerji, 2003; France & Rumpe, 2007), both MDA / MDD or broader MDE / MDSD are all around to the improvements for software development process, that is, on the left of the gap, and in software productivity and quality.

3.4. Some Issues with the Computation Independent Model

We pay more attention to the CIM in MDA, which is put at the top of the abstract levels. First, there are some ambiguous to what is a CIM. In *MDA Guide 1.0.1* (Miller & Mukerji, 2003) defined that "A CIM is a model of a system that shows the system in the environment in which it will operate" (this distinctly refers to the black-box model of the application system, i.e., a system model, the first class model we identified in previous, at sec. 2.2); but in the same document, it also stated that "CIM describing the situation in which the system will be used. Such a model is sometimes called a domain model or a business model." (i.e., as the second class model) This issue is similar to the ambiguous with so-called analysis model, may be seen as it is a general designation of some different types of models; the key question is, however, which one is the object being modeled by a CIM? Is it on the left or right of the gap? If we regard CIM as analysis model, then the transformation from CIM to PIM will be faced the transformation barrier we discussed above; otherwise, it in fact will be overlapped with the concept of PIM.

Second, In fact, it appears that the subject about CIM, including the transformation from CIM to PIM, did not get the enough attention; MDA transformations are only focused on the PIM and PSM (Fouad et al., 2011). The similar views can be seen from more documents (e.g., Kardoš & Drozdová, 2010; Sharifi, Mohsenzadeh, & Hashem, 2010). Nowadays, the situation is still as stated in Karow et al. (2006): "CIM has not achieved much attention so far. [...] there is no model type in MDA covering the requirements for the software system in development, which is the actual source model on the transition from analysis to design." And, CIM "is disregarded by most MDA methodologies and tools, which start their transformation process on the level of

platform independent software models (PIM)." In our views, the most efforts of MDE are limited on the left side but not virtually across over the gap; it in fact a little avoid the essential issue, the analysis and design model transformation barrier; the ambiguous and disregard to CIM may be related to some key factors which cause the MDE/MDA to "miss the boat."

3.5. Domain-Driven Design

Nowadays, many architectural approaches or designs take the domain / business model to a more important position, for example, the Domain-Driven Design (DDD). This is a good example that tried to across the transforming barrier, as stated by Evans (2003): DDD "discards the dichotomy of analysis model and design to search out a single model that serves both purposes" (ch. 3); notice that the model is not use case model or system model but domain model. On the other hand, DDD emphasizes to *bind* model and implementation and suggests a domain (model) layer where "concentrate all the code related to the domain model in" and "isolate it from the user interface, application, and infrastructure code." That is, the domain model is explicitly embedded into system and will be working at runtime. It can be seen that takes domain model into a special layer has become a quite common architectural choice in recent years. Taking some of the analysis model embedding in system and keep it at an independent position inside the system, this is a partial answer to the transforming barrier. However, the model as data is still private, and we still have to face that how to code the models.

Moreover, if there is not a specific independent modeling tool and representations for the models that suitable for business / application analyst but not the system architect / analyst, then the merger of the analysis model(s) to design model will lead the merger of requirements analysis and architecting, design (both the jobs and the roles), this will lead to more serious entanglement of business changing and development process (see next section). From software engineering view, that is not a progress but somehow a regress.

DDD is not current MDE, in comparison, it is *using programming language as modeling language* rather than the opposite[6], and the driven model is domain model but not the system model (this is close to MDApp, see later). On our perspective, by embedding domain model in an isolate layer, DDD has taken a small step to cross the gap – as a step forward toward our approach, the MDApp.

3.6. Agile and Entanglement of the Business Change and Development Process

The primary intention of Agile software development is to adapt to the "changing requirements, even late in development"(Beck et al., 2001). In addition to the part of the attitude, however, its methodological basis is still the "iterative and incremental" that the practices can be found from 1950s (Larman & Basili, 2003). However, the emergence and continued popularity of Agile methods in fact show another problem up: which can be described as *entanglement* of the business change and software development process. In fact, faced the constant and rapid changing business, a faithful implementation to the Agile principle will inevitably lead to a permanent close cooperation between the developers and business people; on the other hand, the different enterprises' different demands inevitably require them make cooperation across different enterprises (i.e. the system developer and the client). To permanently keep such a close collaborative relationship, such development process is only suitable for customized development with a long-term contract, or within the same enterprise. Such passive bundling becomes always heavy shackles on both the client and the developer.

4. ENTERPRISE ENGINEERING ON THE RIGHT

4.1. What is Enterprise Engineering

Enterprise Engineering (EE) was defined as "a body of knowledge, principles, and practices having to do with the analysis, design, implementation, and operation of the enterprise" by SEE[7] in 1995 (Liles, Johnson, & Meade, 1996). For about half a century, in the context of IT applications, there was emergence of many of ideas or approaches for the enterprises towards information age, such as CIM (Computer Integrated Manufacturing) or enterprise integration, Enterprise Modeling (EM), business-IT strategic alignment, Business Process Reengineering (BPR), Enterprise Architecture (EA), Business Process Management (BPM), and so on. Although they respectively have different origin, all have some background of IT application, and to regard an enterprise as complex system and take engineering methods into the whole life-cycle of enterprise, such as the planning, design, construction, governance, and transition, and so on. Therefore, they can be naturally gathered under the general concept of EE (Liles et al., 1996; Yu, 2002b; Cuenca, Boza, & Ortiz, 2010). Naturally, a common feature of such the approaches is the combination of business and IT/IS application. That is what the keynote of the early monograph by James Martin: using EE to align people, IT, and strategy (Martin, 1995).

4.2. Enterprise Modeling

As the necessary foundation for a mature engineering discipline, Enterprise Modeling (EM) is naturally the "major element in Enterprise Engineering" (Liles & Presley, 1996). The conventional EM can be traced back to 1960s-1970s, it is originated from the fields such as the industrial automation, Computer Integrated Manufacturing (CIM), Information Systems (IS), and Enterprise Integration (EI), etc. To 1980s, it has produced rich achievements, including much different EM methods or tools, as well as enterprise modeling languages. Although they are essentially developed as a branch of the IT application, they do not seem to get too much direct use in the general application development. We believe that this is related to the issue of the architecture style of application system so far, as well as all the issues in software engineering discussed previously.

In any case, for our goal—building enterprise application system on evolutionary enterprise models—EM will become one of the most important bases, in the meantime, the new applications will also become the most important use of EM but not just for integration or communication and understanding. Fox and Gruninger (1998) defined that "An *enterprise model* is a computational representation of the structure, activities, processes, information, resources, people, behavior, goals, and constraints of a business, government, or other enterprise." It clearly implied the duality of the model for computer and human; but it seems those refined work has some distance to the general business applications. Our objective, however, is first to achieve an comprehensive information system which is able to be constantly evolving, to support the day-to-day business and management activities in an open, fast changing environment, not the senior activities like as deducing on model.

4.3. Enterprise Architecture

In contrast with EM, the rise of Enterprise Architecture (EA) is a bit later, but the development trends obviously over the former, has become one of the most active fields in EE. It appears that the benefits of EA increasingly being discovered and summarized and it is gained more and more recognition and adoption from the practitioners such as the CIOs. (Tamm, Seddon, & Shanks, et al., 2011).

As a practice-oriented young approach, there are much different understandings about EA; TOGAF (The Open Group Architecture Framework) is relatively representative one. According to TOGAF 9 (2011), EA is divided into four architecture domains: *business*, *data*, *application*, and *technology*, in which the latter two are in fact belonged to IT field. That is, although the name has changed, from early "IT architecture" or "IT planning" etc. becomes current "enterprise architecture," it is actually still a strategic tools to IT application. We can see that (e.g., Baudoin, Covnot & Kumar, et al., 2010), how to build a strong and balance alignment between business and IT, from the top strategic level to the bottom IT infrastructures, are always the major concerns of EA researchers and practitioners.

In addition, EM is the necessary basic tool which be contained in EA, and EA also be used by EE (Cuenca et al., 2010), the problems encountered in EM will still exist within EA.

The EA approach, however, is not able to change the inflexible software, cannot break the gridlock from the entanglement of the business change and development process. Thus, each application *silo* or *island* becomes some indecomposable "granularity" in EA implementation. The application architecture thereby becomes a very macro planning on the big granularity of the silos or islands. The data architecture is also only used for integration, as well as the enterprise modeling. Such the problems largely limit the role of EA, and lead to that EA seems only useful to huge enough enterprise with a highly complicated IT environment. For example, the findings in Tamm, et al. (2011)

5. MODEL DRIVEN APPLICATION AS BRIDGE

5.1. Using Enterprise Model to Drive the Application

In order to adapt to different customers and situations, much generic or semi-generic application software provides a lot of optional parameters to reach a certain extent of customization. Such the parameters are, in fact, described or corresponding to some properties of the enterprise—it is actually some fragments of a complete enterprise model, though it seems usually too simple, from a modeling point of view. This fact prompted us the way which can be used to handle a more complete enterprise model, thereby, to make the user to be able to control or change the function or behavior of an application through modify the models, to adapt to the changing of business and other elements of an enterprise. Based on such thoughts and some attempts in actual application system development, we suggested the idea for a new generation of enterprise application (information system) based on enterprise-model-driven (Yu, 1999). This idea has attracted the attention of some local people in the industry; some successful commercial products appeared in a few years; and some industry researchers classified it as *business system infrastructure platforms* (ChinaLibs, 2002).

5.2. Model-Driven Mechanism

Based on the initial idea, we further found that, the approach that build a system on some evolutionary model is according to certain general principles,

we called as *Model-Driven Mechanism* (MDM): it makes all or part of functions and behaviors (or the structure and form) of a system to be in control of or mastered by model (Yu, 2005, 2007). MDM is shown as three relative parts in a system (see Figure 3):

The *model* (or *applied* model for emphasize) models the target thing related to the functionality or behavior of the system, which is conformed to the modeling knowledge.

The *modeling knowledge* including the specifications, rules, languages, notations or / and formats, etc. to the applied model; it is roughly the *metamodel*[8] of the applied model (see the discussion later).

The *operational device* accesses the applied model and does some actions (functions or behaviors) on / for the target thing being modeled by the applied model, which is according to the modeling knowledge.

As shown on Figure 3, the three parts of MDM consists of a closed triangle relationships, this is related but different to the metamodeling architecture well recognized in MDE which illustrated in e.g. Atkinson and Kühne (2003), Bézivin (2004, especially the Figure 5: The 3+1 MDA Organisation). The modeling knowledge is the key to determine the role or effect of MDM.

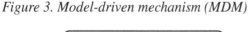

Figure 3. Model-driven mechanism (MDM)

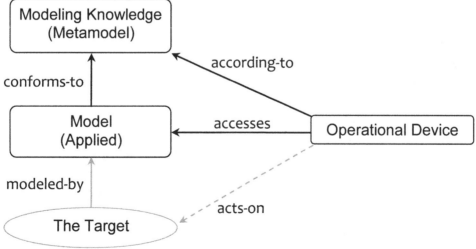

It may include any type of metamodels of / for the applied model. Here, we emphasis on the relation to existing work of MDE, so, in the rest of the chapter we will principally use the term "metamodel" in a relative wide sense, for example, it may be referred to logical data model (scheme), conceptual model such as E-R model, or as the "ontological commitment" (Kurtev, 2007), and so on. According to the needs of a system, the different metamodels (modeling knowledge) may be used either individually or multiply.

Moreover, there are some basic characteristics around the applied model of MDM (Yu, 2002a, 2005, 2007):

- **The Independence of Applied Model:** It is independent of media and the system, and can be modified, reproduced, transmitted or transferred in some general, system-independent way.
- **The Timeliness of Applied Model:** It works at runtime within the system.
- **The Evolvability of Applied Model:** It can be continuous changing when they are working in a system at runtime.
- **The Program-Model Decoupling:** From a programming perspective, the operational device accesses the data of applied model at just runtime; it *never* references them in the programs / code when programming but only the types or the variable name specified in the metamodel.
- **The Metamodel in System:** In contrast to the applied model, the metamodel will be embedded in the system, e.g., to be hard-coded in the program in a single MDM structure, or accessed from a repository in runtime as in such a multiple MDM structure (see latter).

MDM has very universality in many kinds of advanced systems, such as computer / software-intensive systems, intelligent systems, etc. It is a fundamental answer to the basic question about models: how does a model work? More discussion about MDM will be presented in a specific paper.

5.3. Model-Driven Applications

From the standpoint of software, the concept of MDM is very close to the topic of applications. First, We defined *Model-Driven System* (MDS) as a class of systems that the main functions and behaviors are achieved on MDM, namely if all the functions and behaviors of a system can be defined, controlled and changed through MDM at runtime, then the system is a full model-driven system (Yu, 2005, 2007). Further, it is presented that a *Model-Driven Application* (MDApp for short) is a model-driven system, in which the relevant things (the target objects or affairs, such as the business) to the application system are modeled by the applied models that the functions or behaviors of the system will be operated / based on, or in control of, these models. In other words, MDApp is the system which is run on some *evolutionary* models; one can achieve some operations for some target thing via add the model of the thing that conforms to the metamodel. For an MDApp, can say that *what can be done is what can be modeled*.

MDM is the basic structural feature of MDApp. As a software-intensive system, the device may appropriate to be divided into some modules; a basic structure of MDApp is shown as Figure 4.

In Figure 4, the operational device in MDM is separated as three typical modules: the Modeler provide the function to modeling the applied model; the Model Driver (somebody may be used to call it engine) access the data of the applied model and provide the basic operations (e.g., via APIs, services or compiler/interpreter to some language) for it; and the Operator to achieve the functions of the system. For instance, an autopilot system as Figure 5; in this case the operator acts on the real target thing directly.

Figure 4. A basic structure of model-driven application (MDApp)

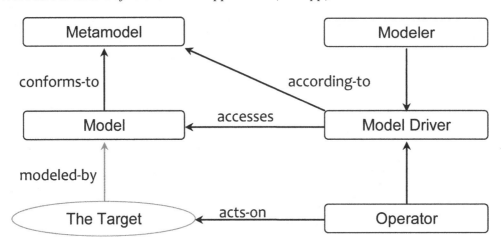

In addition, as we mentioned above, MDM also the fundamental of MDE / MDA[9]; the author believe that this is a useful complement to the current theory of MDE / MDSD. Figure 6 shows that a typical code-generation system is an application of MDM (as an example, it is to be simplified, which shows only one existence of MDM in the basic structure).

Database and the applications are important existing instances of MDApp, as well as some BPMS which support run on the business process model directly. Figure 7 illustrates the database application which the functions are on the data, that is, a simplified architecture of database en-

vironment with MDM. The concepts in the figure are referenced to Elmasri & Navathe (2003, in the Figure 2.3).

Furthermore, Figure 7 also shows a very important aspect of MDApp, i.e., the multiple structure of MDM overlapped: as in the case of database architecture, the stored data are the models of real world which are conforms to some schemes, and they are evolvable that supported by the DB Processor; the schemes (catalog) are models of the data and it is *also* evolvable that supported by the DB Manager—both are can be updated by some users on the Database Management System (DBMS). Hence, the term "applied

Figure 5. Autopilot system based on MDM

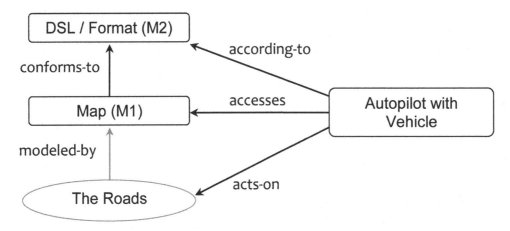

Figure 6. MDM in MDA style code generation system

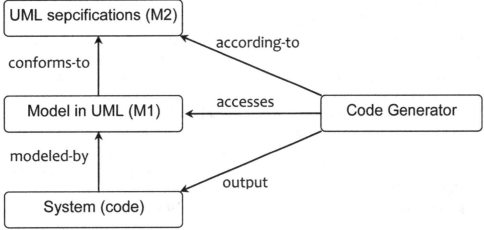

model" for MDApp will not only be limited to the model in a certain MDM structure but can also be metamodel in another MDM structure. The applied model for MDApp is always user-changeable and has its metamodel (perhaps a meta-metamodel in a multiple MDM structure) and the corresponding operational device (the model driver). In addition, the division of metamodel and model may be relative: it indeed depends on the device—how to program for the metamodel, which is the most important job for the software of an MDApp.

Especially, MDM can also be used for the user interface, report format, system deployment and configuration, driving the peripherals or equip-ments which are connected to computer directly, and so on; some of the targets are probably a part of the system *itself*, which are typically not been regarded as so-called real world. To apply MDM on both the model of outside target and model of own target is indeed a very important point for MDApp.

Based on MDM, an MDApp makes the computer operations (functions) to be relied on the models which are resided in the computer and represented the target things (e.g., business) on the right side of the gap; such the models can be created and updated—evolutionary—by human (e.g., business analyst) in the runtime of the application system. What business can be handled

Figure 7. A simplified architecture of database environment with MDM

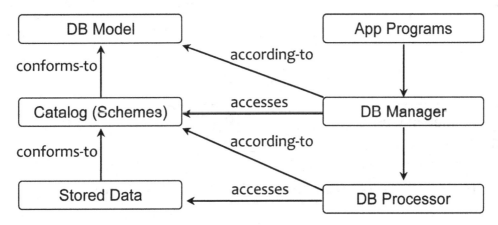

on the MDApp are what can be modeled by the business analyst—according to the metamodels; a well-designed platform may be capable to add new model drivers to support certain new business (models). In addition, the applied models can be classified into different groups strictly depend on the metamodel and be processed by the special model driver as well as the modeling interface / tool. This is the basis to support the separation of the concerns by different stakeholders.

5.4. Enterprise Model Driven Application

On the illustration of MDApp, it is clear that how to build an application on the enterprise model, which can be called *Enterprise Model Driven Applications* (EMDAs). A primary structure for EMDA is illustrated on Figure 8, in which the applied model will mainly include enterprise model. In an actual system, the primary (MDM) structure may be repeatedly used for different model-metamodel pairs; the relationships between them are dependent on the relationships between the models. Around these core construct, the system can still be constructed with different architectural styles, e.g., SOA.

As shown in Figure 8, ignore some conventional part for the similar system, the special parts of an EMDA system are including (Yu, 2005):

- **EMDA Platform:** A software platform, the fundamental functionality is to provide the access mechanism to all the models (data), the enabling, governance and configuration mechanism to the model drives, functional engines and tools, and to the operational interfaces.

Figure 8. The primary structure of enterprise model driven applications

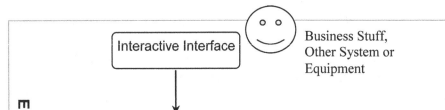

- **Applied Models**: The first class of applied models is the enterprise models, which may comprise of such as business (process) model, organizational model, resource model, and so on. The organization and modeling of enterprise models will be one of most important, special subjects of the development of EMDA—in principle, it is only depended on the metamodels (modeling knowledge) which are supported by the model drivers, without the knowledge about programming or the principles to the work of software; we can inherit the rich heritage of the existing enterprise modeling research. Other applied models may include UI model, report model, configuration model, etc. A practicable solution may have a multi structure of MDM; hence, an applied model may be also a metamodel of other applied model.

- **Metamodels**: Each model has its own metamodel(s). The storage way to a metamodel is depended on the MDM structure and the driver. Whatever, the metamodels are the shared specifications for the modeling of its applied models and the design of its model drivers, functional engines and modeling tools.

- **Model Drivers:** A model driver achieves the basic operations for certain model according to its metamodel(s). It is the key artifact that determines what the system can do—with the applied model.

- **Functional Engines:** A functional engine is a loadable module which is running over the different model drivers to provide special function(s), e.g., to does the operations on / for the target things.

- **Interactive Interface Layer:** Including / supporting the interface to human, equipment, or other computer systems. As mentioned above, we can also apply MDM for this via modeling the interface, e.g., the

human-computer interface. SOA is an obvious way to the implement of this layer.

- **Enterprise Engineering Tools:** The fundamental function of EE tools is modeling. On the basis, it can be extended to add variety of advanced features, such as more advanced supporting to enterprise/business design for the needs from real enterprise engineering or business engineering. Other functions may include model management, import/export, and conversion, etc. In addition, an EMDA can also be provided more special engineering/modeling tools, if MDM is used for, e.g., UI modeling, report modeling, and so on.

5.5. Discussions

Our contribution is to discover the way that how to dynamically connect the behaviors (functions or operations) of a computer system (on the left side) to the target things in real world (on the right side) through the inevitable models of the things, where the models are working as a part of the application system and (can be) setting by human in real time. The models in application system are evolutionary; thereby the behaviors of the system can be evolved to follow the changes of the target things via the models, and can be reflected the understanding and intention of the users in time. As illustrated in MDM, the independence of applied model is one of the most important feature of MDApp, compared to traditional structures (both the structural and object-oriented, the domain model is indeed mixed with code) and MDApp, some newer architecture styles such as DDD displays a transitional features.

For an enterprise, a comprehensive EMDA as an MDApp implements a real separation of the changing of business and the software (the platform); thus, the desired Enterprise Engineering (EE) can be achieved which is based on dynamic enterprise (business) modeling, done by pure

67

enterprise engineers / business analysts, and supported, combining with the enterprise information system. The enterprise models hence become the main role in EE but not only for the integration between application systems or as the reference for understanding the enterprise / business. The Enterprise Architecture (EA) can more deeply be placed on the planning of concrete business process / function (or say, in finer granularity) since they are evolutionary on the EMDA platform by the modification of the enterprise models, that is, it is able to adaptation to the changes of the enterprise (business) without changing the software.

One important issue is the separation of the concerns about business and computing techniques. The business analysts are required for only the knowledge and skills for modeling the business and use the modeler tool; the programmers of the MDApp platform are also no longer to be needed to understand the business but merely the metamodels. These with the runtime feature and the evolvability of the models fundamentally remove the root of the entanglement of business change and development process.

To the development of enterprise applications, compared to the traditional solutions which programmed to specific business functions directly, the transforming barriers will be effectively avoided; this is not by the automatic transformation from some analysis models or CIMs into design models or code but by bringing the operations of an application system directly based on the evolutionary enterprise models. In our views, the transforming barrier is essential; for MDApp, however, the system analyst / architect faced only the models of a general platform and the model drivers but not the huge cases of concrete business from different customers and situations. We can still use any traditional or exiting approaches to developer the software for MDApp, the difference is, the developers will work on a higher abstract level: the meta-level, thus the use-cases or the

functional points they must be programming for will be greatly reduced.

For the topic of raising abstract level, say, a model is also an abstraction[10], and for MDApp with MDM, the abstract level for programmers is not simply, blindly raised but to be *matched* with the abstractions (the applied models) of the real word things, such as the business, via its metamodel, that is, the two abstract hierarchies on the two sides of the gap are be matched in an MDApp; for the multi-structure of MDM (e.g., in the Figure 7), it is more obvious and meaningful.

As mentioned above, the applied models can also be of the (part of) system itself, thus one can evolve the (part of) system by modify the corresponding model. This aspect appears to be relevant to, in principle, some research since about 2006 on *runtime models*, which brings the use of models to the entire life-cycle of software especially during runtime (France & Rumpe, 2007), this research approach more emphasized on software evolution and self-adaptation, the core mechanism is reflection (de Lemos et al., 2011; Vogel, Seibel, & Giese, 2011; Blair, Bencomo, & France, 2009). In contrast, MDApp has *functional* adaptability to the changing of the target, e.g., the business; in the meantime, the system itself can also be evolved by use MDM on the model of system. Furthermore, all the adaptations on MDApp are achieved by the modeler; this may be somewhat different to the self-adaptation. Whatever, it is interesting topic that the relationships between MDM / MDApp and reflection and such the research for models at runtime, as well as model execution.

6. CONCLUSION

What behind business-IT gap is computer-world gap; adhering to the picture depicted by Smith (1985), the approaches for filling the gap are divided into two groups, which can be called *left side approaches* and *right side approaches*. We extended the picture to present human; the inevitable mediating model between computer and the real world therefore to be for the three, which representing the target thing in the real world for both human and computer. For filling the gap, many existing efforts on the two sides but not really cross over the gap. On the left side, in software engineering, there are still some fundamental problems do not be removed, such as the transforming barrier between analysis and design model and the entanglement of the business change and development process, although we have developed a lot of approaches, such as OO, MDE, DDD, Agile. On the right side, the existing EE approaches such as EA and EM are not able to change the problem from the solution systems themselves: the inflexibility, the application silos or islands, etc.; this greatly limits the role of EA or EM, makes them difficult to be grounding—can only be used for such the IT planning, governance, or application integration in giant or special enterprises.

Our target is on the right, the real world; the problems for enterprise application are essentially opened. Thus, we must do modeling for the world continuously. The human-understandable/operational and computer-treatable models therefore become the inevitable core artifacts to establish dynamic connection between the two sides of the gap; the conventional, left or right approaches do not really fully utilize of the duality. Model-Driven Mechanism (MDM) is the fundamental principle to make the model working as a part of the application system at runtime; the Model-Driven Application (MDApp) based on MDM is the bridge cross over the gap, which is able to avoid such the transforming barrier, disengage the entanglement, and thereby release more the productivities on the two sides. An MDApp based on evolutionary enterprise models, that is, an Enterprise Model Driven Application (EMDA), can enable real enterprise/business engineering, and bring such the EE/EA approach to be grounding.

This chapter is just a preliminary introduction to MDM and MDApp; it needs more, specific illustration on almost every aspect. We argue that this is a valuable, significant direction for either study or practice.

ACKNOWLEDGMENT

I would like to thank Vicente García Díaz and the editorial team for inviting me to participate in this book. They might have to pay more efforts to my writings because it was a new, hard work for me; all errors are mine. I would also like to thank Markus Kerz, Vincent Hanniet, Andreas Leue, Rui Curado, Andriy Levytskyy, and more friends from the MDE community and blogs on Internet, who gave wise and inspiring discussions and comments, which were very instructive and useful to the work.

REFERENCES

Atkinson, C., & Kuhne, T. (2003). Model-driven development: A metamodeling foundation. *IEEE Software*, *20*(5), 36–41. doi:10.1109/MS.2003.1231149.

Baudoin, C., Covnot, B., Kumar, A., LaCrosse, K., & Shields, R. (2010). Business architecture: The missing link between business strategy and enterprise architecture. In B. Michelson (Ed.), *EA2010 Working Group, SOA Consortium*. Retrieved October 15, 2012, from http://www.soa-consortium.org/EA2010.htm

Beck, K., Beedle, M., & van Bennekum, et al. (2001). *Principles behind the agile manifesto*. Retrieved October 15, 2012, from http://agile-manifesto.org/principles.html

Bézivin, J. (2004). In search of a basic principle for model driven engineering. *UPGRADE*, *5*(2), 21–24.

Bézivin, J. (2005). On the unification power of models. *Software & Systems Modeling*, *4*(2), 171–188. doi:10.1007/s10270-005-0079-0.

Bézivin, J. (2006). Model driven engineering: An emerging technical space. In Lämmel, Saraiva, & Visser (Eds.), Generative and Transformational Techniques in Software Engineering (LNCS), (vol. 4143, pp. 36-64). Berlin: Springer. doi:doi:10.1007/11877028_2 doi:10.1007/11877028_2.

Blair, G., Bencomo, N., & France, R. B. (2009). Models@ run.time. *Computer*, *42*(10), 22–27. doi:10.1109/MC.2009.326.

Bosch, J., & Molin, P. (1997). *Software architecture design: Evaluation and transformation*. Karlskrona.

ChinaLabs. (2002). *Business system infrastructure platform (Technical report)*. Beijing, China: ChinaLabs.

Cuenca, L., Boza, A., & Ortiz, A. (2010). Enterprise engineering approach for business and IS/IT strategic alignment. In *Proceedings of the MOSIM 2010 Conference*. Lavoisier.

de Lemos, R., Giese, H., Müller, H., Shaw, M., Andersson, J., Baresi, L., & Wuttke, J. (2011). Software engineering for self-adaptive systems: A second research roadmap. Inde Lemos, , Giese, , Müller, , & Shaw, (Eds.), *Software Engineering for Self-Adaptive Systems*. Dagstuhl, Germany: Schloss Dagstuhl - Leibniz-Zentrum fuer Informatik.

Dubray, J.-J. (2011). Why did MDE miss the boat? *InfoQ*. Retrieved October 15, 2012, from http://www.infoq.com/news/2011/10/mde-missed-the-boat

Elmasri, R., & Navathe, S. B. (2003). *Fundamentals of database systems* (4th ed.). Reading, MA: Addison Wesley.

Evans, E. (2003). *Domain-driven design: Tackling complexity in the heart of software*. Reading, MA: Addison-Wesley.

Favre, J.-M. (2004). *Foundations of model (driven) (reverse) engineering - Episode I: Story of the fidus papyrus and the solarus*. Language Engineering for Model- Driven Software Development.

Fouad, A., Phalp, K., Kanyaru, J., & Jeary, S. (2011). Embedding requirements within model-driven architecture. *Software Quality Journal*, *19*(2), 411–430. doi:10.1007/s11219-010-9122-7.

Fox, M. S., & Gruninger, M. (1998). Enterprise modeling. *AI Magazine*, *19*(3), 109–121.

France, R., & Rumpe, B. (2007). Model-driven development of complex software: A research roadmap. In *Proceedings of the 2007 Future of Software Engineering*, (pp. 37-54). Washington, DC: IEEE Computer Society.

Frankel, D. S. (2003). *Model driven architecture: Applying MDA to enterprise computing*. New York: Wiley Publishing, Inc..

Jacobson, I., Booch, G., & Rumbaugh, J. (1999). *The unified software development process*. Reading, MA: Addison-Wesley Professional.

Kaindl, H. (1999). Difficulties in the transition from OO analysis to design. *IEEE Software*, *16*(5), 94–102. doi:10.1109/52.795107.

Kardoš, M., & Drozdová, M. (2010). Analytical method of CIM to PIM transformation in model driven architecture (MDA). *Journal of Information and Organizational Sciences*, *34*(1), 89–99.

Karow, M., Gehlert, A., Becker, J., & Esswein, W. (2006). On the transition from computation independent to platform independent. In *Proceedings of AMCIS 2006*. AMCIS.

Kent, S. (2002). Model driven engineering. In Butler, Petre, & Sere (Eds.), Integrated Formal Methods (LNCS), (vol. 2335, pp. 286-298). Berlin: Springer. doi:doi:10.1007/3-540-47884-1_16 doi:10.1007/3-540-47884-1_16.

Kosanke, K., & Martin, R. (Eds.). (2008). SC5 glossary. ISO/TC 184/SC5 N994 Version 2 (2008-11-02).

Kühne, T. (2006). Matters of (meta-) modeling. *Software & Systems Modeling, 5*(4), 369–385. doi:10.1007/s10270-006-0017-9.

Kurtev, I. (2007). Metamodels: Definitions of structures or ontological commitments? In *Proceedings of the Workshop on Towers of Models*, (pp. 53–63). York, UK: University of York.

Larman, C. (2004). *Applying UML and patterns: An introduction to object-oriented analysis and design and iterative development* (3rd ed.). Reading, MA: Addison Wesley Professional.

Larman, C., & Basili, V. R. (2003). Iterative and incremental developments: A brief history. *Computer, 36*(6), 47–56. doi:10.1109/MC.2003.1204375.

Liles, D. H., Johnson, M. E., & Meade, L. (1996). The enterprise engineering discipline. In *Proceedings of the 5th Industrial Engineering Research Conference*, (pp. 479–484). IEEE.

Liles, D. H., & Presley, A. R. (1996). Enterprise modeling within an enterprise engineering framework. In *Proceedings of the 96 Winter Simulation Conference*. IEEE.

Martin, J. (1995). *The great transition: Using the seven disciplines of enterprise engineering to align people, technology, and strategy*. New York: AMACOM.

Miller, J., & Mukerji, J. (2003). *MDA guide V1.0.1*. OMG.

Morrison, M., & Morgan, M. S. (1999). *Models as mediating instruments*. Cambridge, UK: Cambridge University Press.

Schmidt, D. C. (2006). Model-driven engineering - Guest editor's introduction. *Computer, 39*(2), 25–31. doi:10.1109/MC.2006.58.

Selic, B. (2003). The pragmatics of model-driven development. *IEEE Software, 20*(5), 19–25. doi:10.1109/MS.2003.1231146.

Sharifi, H. R., Mohsenzadeh, M., & Hashemi, S. M. (2012). CIM to PIM transformation: An analytical survey. *International Journal of Computer Technology and Applications, 3*(2), 791–796.

Smith, B. C. (1985). The limits of correctness. *SIGCAS Computer Society, 14*(15), 18–26. doi:10.1145/379486.379512.

Tamm, T., Seddon, P. B., Shanks, G., & Reynolds, P. (2011). How does enterprise architecture add value to organisations? *Communications of the Association for Information Systems, 28*(1).

The Open Group. (2011). *TOGAF version 9*. Retrieved October 15, 2012, from http://pubs.opengroup.org/architecture/togaf9-doc/arch/

Vogel, T., Seibel, A., & Giese, H. (2011). The role of models and megamodels at runtime. In Dingel & Solberg (Eds.), Models in Software Engineering (LNCS), (vol. 6627, pp. 224-238). Berlin: Springer. doi:doi:10.1007/978-3-642-21210-9_22 doi:10.1007/978-3-642-21210-9_22.

Yu, T.-Y. (1999). Toward 21st century new generation of enterprise information systems. *Enterprise Engineering Forum*. Retrieved October 15, 2012, from http://www.ee-forum.org/eis21c.html

Yu, T.-Y. (2002a). Model-driven software architecture and hierarchical principle to complex systems. *Enterprise Engineering Forum*. Retrieved October 15, 2012, from http://www.ee-forum.org/pub/1998-2009/hm.html

Yu, T.-Y. (2002b). Emerging enterprise engineering. In *Proceedings of the Enterprise Engineering Conference 2001*. Retrieved October 15, 2012, from http://www.ee-forum.org/eee/

Yu, T.-Y. (2005). New generation of enterprise information system: From essential requirements analysis and research to model-driven system. *Enterprise Engineering Forum*. Retrieved October 15, 2012, from http://www.qiji.cn/eprint/abs/3641.html

Yu, T.-Y. (2007). Lecture on enterprise engineering, model, and information system. *Enterprise Engineering Forum*. Retrieved October 15, 2012, from http://www.ee-forum.org/pub/1998-2009/downloads/ty_jnu070917a.pps

ENDNOTES

1. In this chapter, "enterprise(s)" refers to "one or more organisations sharing a definite mission, goals, and objectives to offer an output such as a product or service" (defined in ISO15704, cited from Kosanke & Martin, 2008). This meaning is used for some important concepts, such as enterprise architecture, enterprise modeling, and enterprise engineering, etc. Accordingly, we prefer use business (uncountable) in the sense that is closer to the transactions or affairs in/of an enterprise.

2. Smith's statement is: "Mediating between the two is the inevitable model" (Smith, 1985, p. 21); it is quite close to Morrison and Morgan (1999): "models themselves function as mediators"(p. 36). Hence, we adopt the term "mediator" to refer this role of model.

3. This indeed involves some deep topics worth more exploration. There are some preliminary discussions, e.g., *Three Spaces for Entities and Models of Applications*, http://thinkinmodels.wordpress.com/2011/10/28/three-spaces-for-entities-and-models-of-applications/

4. Following this opinion, the model of a domain is quite closed to an ontology.

5. For example, see the discussion on Model Driven Software Network at http://www.modeldrivensoftware.net/forum/topics/have-models-filed

6. For MDE, e.g., said by Frankel (2003): "MDA is about using modeling languages as programming languages rather than merely as design languages" (Preface, xv).

7. SEE, the Society for Enterprise Engineering, was active in the mid-1990s.

8. This is an interesting point waiting for more discussion.

9. The initial illustration for this topic is published here: http://thinkinmodels.wordpress.com/2012/01/19/using-model-driven-mechanism-to-explain-model-driven-software-development/

10. This viewpoint directly appeared in most of the literature (e.g., France & Rumpe, 2007; Kühne, 2006; Favre, 2004).

Chapter 4
Using Model–Driven Architecture Principles to Generate Applications based on Interconnecting Smart Objects and Sensors

Cristian González García
University of Oviedo, Spain

Jordán Pascual Espada
University of Oviedo, Spain

ABSTRACT

There has been a rise in the quantity of Smart Things present in our daily life: Smartphones, smart TVs, sensor networks, smart appliances, and many other home and industry automation devices. The disadvantage is that each one is similar but very different from the others because they use different kinds of connections, different protocols, different hardware, and different operative systems; even if they belong to a similar kind, and depending on the manufacturer, two Smartphones or two humidity sensors can be totally different. In spite of this, they all have something in common: All can be connected to Internet. This fact is what gives power to objects, because it allows them to interact with each other and with the environment by intercommunication, a joint and synchronized work. That is why the key resides in creating connections among different objects. This requires technical and qualified personal, although the majority of people who use it do not meet these conditions. These people require objects that are easy to interact with and with as many automations as possible.

DOI: 10.4018/978-1-4666-4494-6.ch004

INTRODUCTION

Most people have Smartphones or other intelligent mechanisms, like Smart TVs, Smart Labels (NFC, RFID). This opens new doors to know the Internet of things.

Thank to these mechanisms, users can receive totally personalized notifications almost instantly, through an application, a message, a call, a code or by consulting a Web service. Also, this gives the user the possibility to interact with other mechanisms in a much more easy way. For example, he can send a photo from his tablet to a Smart TV with a simple movement of the photo, 'throwing' it towards the Smart TV.

All this give both the user and the developer endless possibilities. In our case, with a view to process automatization, notifications, references, and configurations, all this is achieved by the interconnection and identification of objects using their different sensors and protocols. This way, objects can interact among themselves, with or without the user's interaction, but always performing tasks in an easier way or even automatically, according to the user's preferences.

For example, if this is applied to food, information about it can be given to other mechanisms in a simple way. A fridge would be able to send mails to the cellular phone, informing about the amount of food stored inside or the products that are about to expire, thanks to the reading of the information (Rothensee, 2007). All this is possible if the food has a Smart Label (NFC or RFID) and the fridge has a reader, a computation unit that can perform the suitable actions and an Internet connection (Gu & Wang, 2009).

In the same way that it is applied to a fridge, this technology can be used with other elements, as the house itself and its conditions when we are outside, cities, supermarkets and shops to improve the way to bring their inventory or give information to their clients … all that is needed is to joint those smart object by using the suitable sensors.

Internet of Things does not just present small-scale utilities at houses and shops but there are also some systems and IOT initiatives that include buildings and even integral cities.

Madrid, Santander, Málaga, Barcelona, Luxembourg, Aarhus, Turku, Beijing... (Vienna, 2013) All of these are smart cities (Hao, Lei, & Yan, 2012) and they use sensors and others smart things in order to perform different tasks.

Humidity, temperature, ozone, movement, pressure capacity, gas, noise, light, pollution… There are different kinds of sensors that can obtain real-time information about the city's condition.

Traffic, parking, timetables about Smart city transport (Falvo, Lamedica, & Ruvio, 2012), environmental danger, lack of trash recycling, the quality of the water, light control, traffic control, reduction of CO_2, energy saving, access to hospitals…All this is possible thanks to the combination of Smart Object and sensors in the cities.

For example, when a certain temperature is reached, the air-conditioning can be activated or the windows can be opened if it is not windy or rainy outside. Also, this information can be sent to Internet and the users within the affected zone can be notified by sending information or using smart labels that send the information to users with nearby compatible mechanisms.

By using other sensors (like those that react to movement) and the consignment of information to a process station and the cloud, both the traffic and the parkings can be controlled, and they also can offer or send information to the citizen, who will decide how they want to move.

In order to create a smart city, not everything has to be based on ordinary things, private things, can be also included. Also, each citizen can provide several sensors and install them in different places of their houses. For example, temperature and humidity sensors on a balcony. Other citizen could as well get access to this data if it is public and, following the same system, a page where the city map is selected, it is possible to observe the exact temperature and humidity in each zone. All this could extend and, with the use of the fridge mentioned before, know when certain element is needed (like water in a drought) and so the country could establish measures for any kind of situation.

This brings one problem: not all people have knowledge about sensors or computers to be able to create the program that is needed for interconnection. Because of this, it is necessary to offer a system that can make this difficult task accessible for persons without specific knowledge about sensors and computing. It is necessary to offer it to persons with knowledge about language command, who know what they want to do and, thanks to a simple assembly and a little, easy configuration, can offer the information that they wish.

However, applications created for this case or other similar to this one, as is normal, share some identical program points, they treat similar data and they behave in similar ways. But in spite of that, they are applications created by different people so in the end the inside is different too.

This will result in the existence of several applications that are practically the same and aim to resolve the same problem, but they are used in a different way. Also, they could have been developed in a different way and each one has required a great amount of time for that development.

However, the most important thing for users is to have a great speed and automatisms that make easier to perform a certain task. In this case, connecting the sensor and registering it.

This way, it arose the Model-driven engineering. This fact offered the necessary abstraction layer to create some languages of specific control and improve the usability and accessibility for the creation of applications for those people unconnected to the computer world or just making the use of certain technologies easier. They only need to know the domain.

The productivity and speed were increased in those areas in which the Model-driven engineering was applicated.

In cases like this, the creation of a sensor net with its configuration and the connections with other mechanisms or smart objects is an arduous task. The diversity of objects (telephones, computers, smart labels, sensors…) and the different kinds of connection (Wi-Fi, Wireless, Bluetooth, SMS, …) make administration very difficult. All this, along with the different kinds of sensors (humidity, temperature, pressure, speed, movement…) makes this application, very hard to use for most of the people.

Because of this, the objective is the application of the Model-driven engineering to the field of sensors, and, by doing so, being able to form and create sensor nets easily and allowing the user to choose what objects he wishes to interact with and the actions that he wanted to perform in a clear and easy way.

In this way, we get to reduce the complexity thanks to the layer abstraction that is inserted to represent a specific domain. As well, the portability, interoperability, and reusability created by it are improved.

The objective of this research is the presentation of the basics and the idea that will serve as a base to create the next prototype. This one will consist of a platform Web in which smart object and sensors can be registered through the use of Domain Specific Language (DSL) done with Graphical Modeling Framework (Kolovos, Rose, Paige, & Polack, 2009) and Textual Modeling Framework.

With the creation of this platform, Muspel, we want to solve several existing problems. First, there is the dificulty to interconnect and communicate different platforms, smart objects and existing sensors. Nowadays, because of the many different software and hardware, it's necessary to have some technical knowledge about each element, and a specific application for each mechanism. The problem we want to solve is how to make this easy for any kind of user, because all this requires a very technical knowledge and so it's limited to just a small percentage of people. That's why, by using a framework, we try to make the interconnection of heterogeneous smart objects totally accesible for any user who knows the smart objects.

An example is the creation of a sensor net and its configuration in a graphic form. It is as simple as dragging boxes and writing the graphics we need and we ask for. By doing this, any person, regardless of their computer skills, could create an application based in a minimum knowledge of control.

This makes the same application a good one for modeling, too, without having to worry about different kinds of mechanism connections, because the user would not have to take into account if the mechanism is Bluetooth, Wireless, infrared, Wi-Fi, or HTTP.

All things created with this model are abstracted out of all short and medium configuration processes. In this way, the user—who only knows about the model control—can create sensors and different applications which administer different actions according to the user's condition.

CONTRIBUTION

With this research, we intend to make the following contributions:

- **Simplicity of Software**: It will be offered in a easy and intuitive form, exact or graphic, like the user wants, to create the applications. In this way, the user will only have to know each element (sensor, action, smart object...) and the different configurations he can apply to them, that is, if it is allowed. This way, any user, even without having knowledge in programming or electronics, could create and register an application based in sensors.
- **Reutilization of Used Technologies**: A lot of programs that use the same software can be done.

- **Creation of an Abstraction Layer in a Bigger Level:** To configure and work with different connection kinds in a low level is something to avoid, for example Bluetooth, Wi-Fi, Wireless, SMS, HTTP .... All this will be done clearly and the user will only have to choose the connection that he wants to perform with the object, if this one can be accessed through several connections.
- **Domain Specific Language for Creating of Applications:** A domain specific language will be created to serve as a base for creating applications based in sensors.
- **Domain Specific Language for Exchange of Data:** A domain specific language will be created to serve as a base for exchanging data between smart objects.
- **Metamodel:** A specific model that serve will be created as a base for creating some models for applications based in sensors and which want to solve similar problems.
- **Data Centralization:** A system that will be able to centralize the developed applications will be created. This way, any user will be able to get access to public content published by other people.
- **Connection Abstraction:** Thanks to the application, we will abstract and encapsulate all connections and communications of objects to Internet. This is because all objects that wants to send and receive data might connect or perform a necessary request to the application. This way it will not be necessary to use other protocols that can be slower, spend less battery or aren't available.

The Figure 1 shows that our contribution is an abstraction over API offered by different systems, connection kinds, programming languages and the Hardware itself.

Figure 1. Muspel's application layers

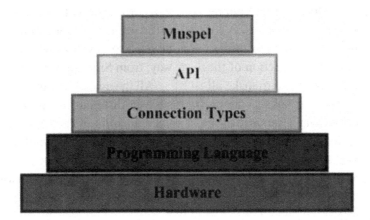

STATE-OF-THE-ART

Smart cities are the pinnacle of sensor nets. In them, the situation of certain zones is watched and also certain mechanisms act accordingly.

The most classical example are the lights. These ones turn on when the sensor does not receive a minimum amount of light previously fixed.

By implementing this system we could make possible for the smart city to send a text message to the subscriber's cellular and inform about a certain situation. For example, the traffic on the road. This example can be extended to parkings where we are notified about free places even before entering.

These configurations are created by the government or specific people thanks to a certain software. All this is not accessible for users who cannot add a sensor to the government's net.

However, there are certain Web sites that offer users the possibility of sharing the measurings done by their own sponsors. This is the case of COSM (previously known Pachube) (LogMeIn, 2013). Other examples are Paraimpu (Piras, Carboni, Pintus, & Features, 2012) o ThingSpeak (IoBridge, 2013), etc.

COSM offers its API and several advices to work over the mother base Arduino. Despite this, the user will have to learn how to use the API and work in a lower lever with Arduino, because it is programmed in a programming language C. It offers data in different forms: XML, JSON, CSV. We can access to them by a free and bidirectionnal API RESTful (upload and download of data) available in several programming languages (C, Ruby, PHP, JavaScript, …),as in modifiable graphics in the Web. It allows users to upload their applications in a public or private way and a total administration of them and their data. It also allows users to chat in the same page where the sensors are.

ThingSpeak offers the same than COSM, but there is a diference. It offers an only API for connecting to the service and it does not give API to work with any system, like Arduino.

There are pages and a lot of different mechanisms to interact according to conditions. Among them, there are some which are intelligent. The most used nowadays is Smartphone. But Smart TVs, tablets, NFC, RFID and also the laptops are getting more presence at homes, in the streets or at establishments. It can interact with each one in several ways: Wi-Fi, Wireless, infrared, Bluetooth, SMS, call, MMS, HTTP, SMTP, IMAP,… The only problem is connecting with them. Sometimes it is simple, for example when we send an electronic message. Other times it can be more complex, as sending a SMS, MMS or connecting Bluetooth; this is due to the absence of information, examples, or the system complexity.

Because of these problems, It is needed to know the sphere and the different programming languages, technologies and mechanisms. A lot of knowledge is required just to do a few things.

Because of that, with the introduction of the Model-driven engineering we intend to make all this process easier for the user thanks to the introduction of an abstraction layer. Users will have to know the sphere and have a little knowledge about each tool. However, this last process will be drastically simplified to offer a very easy system for creating sensor nets next to conditions that are necessary in a certain act.

Our goal is to create a Web site that centralizes all the information sent by users as well as the register of its applications. We will provide a RESTful IAP, too, so a download data can be sent from different notations. We will also provide a framework to make possible to create applications in an easy way for different systems, like Smartphones with sensors and sensor base plates. By doing this, the user, just by using this framework, would be able to create a native application that uses the Smartphone's sensors or base plate and send the desired information to the Web. To send all data a standard DSL will be created, one that will be compatible with any application or any sensor. With all this we intend to make the extension through new systems and sensors quicker and easier.

DESCRIPTION

Muspel

Muspel will be established from the Web site and from different APIs and frameworks. The Web site will centralize all user information and his application's register. It will be responsible for managing this information and, via RESTful service, providing the necessary data required by the applications.

To make the interaction easier for the user, we will put a set of APIs with which he could connect to the service and some frameworks, which could be used to develop the application in an easier way, from Smartphone to base plates.

All the available information would be accessible in different forms, like XML, JSON, and CSV. In spite of that, for direct communication whit applications, it would be necessary to create a standard DSL that could serve to send data. This way, the future extension to more systems and sensors would be easier.

System and Equipment

The main piece for connecting sensors to the environment will be a mother base Arduino (Hribernik, Ghrairi, Hans, & Thoben, 2011). The required sensors of data and several mechanisms are used to perform actions like loudspeakers, LED's, motors, servomotors.... These will be connected to the mother base Arduino (Yamanoue, Oda, & Shimozono, 2012).

The Arduino will be connected to the PC via USB on the first tests: then, through a Wireless unit (Georgitzikis, Akribopoulos, & Chatzigiannakis, 2012). For the performing of all the tests there will be an exclusive computer that will process and save all the data, and it will make the connection with different systems and smart objects.

Several Smartphones will be used, each one with different operating systems like smart objects, sensors, and text messages, multimedia messages, calls or mails according to the configuration.

Other kind of smart objects which will be used are Smart TVs, sensor nets, tablets and computers.

Software

Like IDE we will be using Eclipse. For desing a metamodel and the text and graphic environment we will use Eclipse Modeling Framework (Hegedus, Horvath, Rath, Ujhelyi, & Varro, 2011),

Graphical Modeling Framework (Kolovos et al., 2009) and Textual Modeling Framework. This way we get a development almost for free.

We will use the Java program language; that will create an application which will be easily ported to different known systems. The only requirement is to install the Java virtual machine so it can work with the application.

To create the model it an Ecore metamodel will be used.

Herewith it wants to create a framework to implement applications in a simple way in cellular phones or well to use frameworks of the graphic or textual interface. If the first one is used, a user with programming knowledge should be able to create a new design and implement it. This way, he could use it in a lower level and make a deeper configuration of the application. With the last two, any user without programming knowledge could implement an application because the domain is the only thing he needs to know. With the framework of graphic interface, the user only might drag and drop boxes and fill his data to generate the application. By using textual framework, the user would have to write, according to some rules which would be given in real time, the application in each text.

Architecture

Now we will describe the internal architecture and the operations of the Web service, as well as the way in which sensors and devices interact with it (see Figure 2).

First of all, to explain that the way to interact with the Web service that is sending the text in XML format, we will describe the internal operations of the Web service. This has a series of steps:

Figure 2. Internal architecture of application web service in altogether with sensors and devices

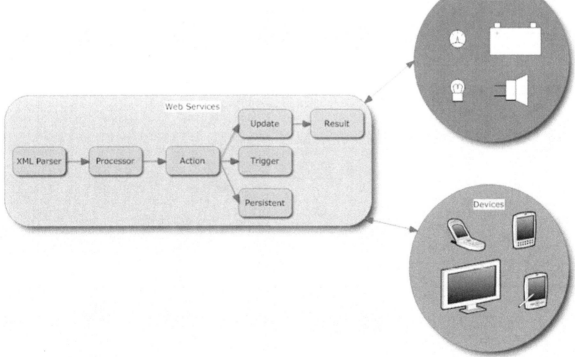

When a XML file is received, it processes that one en the XML Parser platform. Once it processes the file, it sends the data to the processor. This one, from this data, knows if it is a sending request of data or an order, and then it runs a series of actions: Update, Trigger and Persistent. All three will be run at the same time.

If the persistence configuration is activated, one of these actions will be kept (or not) within the data base.

If the processed action must break out a process, the Trigger will be the one which makes it. For example, when a temperature sensor shows 100°C, it will send a call to the firefighters and an e-mail to the cellular of the sensor's owner.

At last, the Update executes the necessary update in the Web service and in the respective URL; it will always keep the RESTful service. It

creates, erases or updates new files when necessary; also, new URL, images or folders…

Once the last process I finished, the final result will be returned. This can be a XML with successful confirmation of the processing of the process trigger. If data or one or more text files are sent with the order, according to the asked format, as well as the respective XML to send the required orders.

Applications

First we will create the Ecore with the Eclipse Modeling Framework and after that, the graphic editor and the text editor will be created thanks to the Graphical Modeling Framework and Textual Modeling Framework (see Figure 3).

Figure 3. Ecore metamodel used to describe elements, connections, and limitations of modeling language that will be used

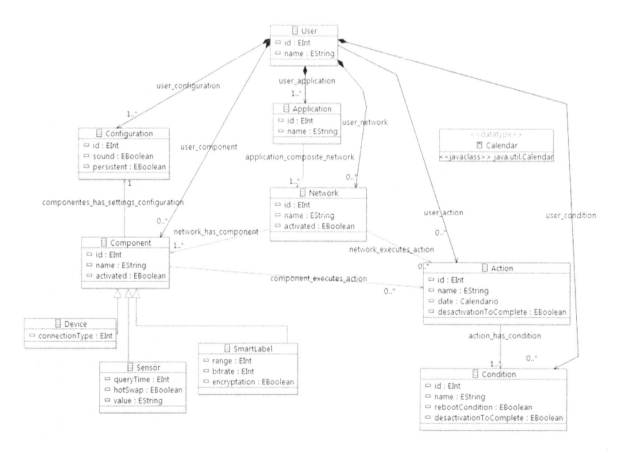

After that, we need to create all the elements required to abstract whatever we want through the model. That is, different kinds of connection, options, configurations, etc.

In other way, its temperature, humidity, movement and light sensors will be connected to the mother base Arduino; as well as mechanisms like loudspeakers, LED's, a motor, a servomotor and a relay.

Next step is programming sensors so they can receive the wanted values at the time we want.

After that, the Java model has to be modified to be connected to Arduino and, this way, the configuration that was created as a model inter-

acts directly with the hardware, so the data can be processed in both directions: from model to Arduino and vice versa.

In Figure 4 We can see an illustrated example of the prototype. Through frameworks of Eclipse, the model is created. To create the Ecore metamodel we use the Eclipse Modeling Framework (1). To create the graphic framework we use the Graphical Modeling Framework (2) and to create the textual framework, the Textual Modeling Framework (3). All that is in a PC that we use to read different systems (4). In this case, the PC (4) has connected 6 mechanisms.

Figure 4. Example of system architecture where it is shown the interconnection of different elements: from sensor interaction, going through the cloud, to arriving at smart objects

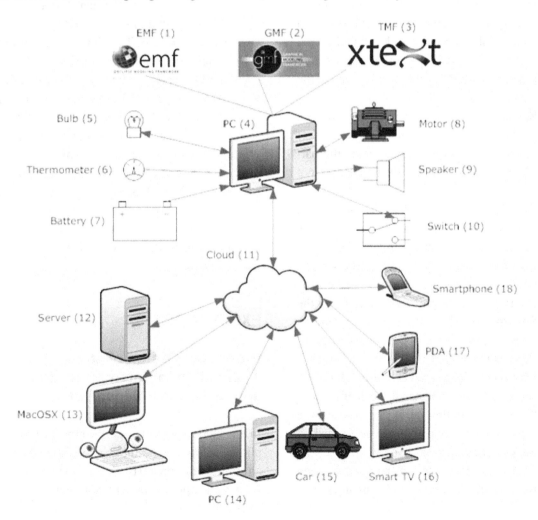

- **Bulb (5):** It has bidirectional connection. We can know if it is switched on or switched off and also turn it on or turn it off. Because of that we can, under certain conditions, turn it on or turn it off or know if we need it to be in the state in which it is at that moment.

- **Thermometer (6):** We can only check its state, that is, the temperature that it gives us. This one can be located at home, outside or in a mobile mechanism.

- **Battery (7):** We can only check its state, that is, the remaining battery. For example, in a cellular phone or a laptop.

- **Motor (8):** It has a bidirectional connection. We can know if it is switched on or switched off and turn it on or turn it off under certain conditions or by direct orders from an IOT mechanism. It can be the load motor of a compressor and, if it is detected that is full, it can stop or inform us and, this way, alter the power that is given to the battery load.

- **Speaker (9):** We only can give orders. In this case, it can be used to alert us of certain states. For example, when the battery is over 5% or when the motor does not work, it sends an alarm warning.

- **Switch (10):** Through its bidirectional communication we can know the state of this element and control it. For example, if this belongs to the heating, we can turn it on or turn it off depending on the exterior temperature and, even from the Smartphone (18) or the PDA (17), send an order to turn it off until the home thermometer has reached 21°C.

- **Cloud (11):** The cloud is Internet. It is a resource which connects all mechanisms, all the data is sent through it, and it is also the calls to other mechanism are performed.

- **Server (12):** Server that can receive and send data. This one can be used to keep mechanism data and send e-mails to dif-

ferent mechanisms when they meet a certain condition, like the compressor and the battery being full and the temperature surpassing 45°C.

- **Mac OS X (13):** A computer with an OS X operative system that can interact in a bidirectional way with the cloud, that is, with different mechanism connected to it.

- **PC (14):** A computer with a Microsoft Windows or a GNU/Linux operative system that can interact in the same way as the Mac OS X.

- **Car (15):** Car with bidirectional connection, that car transmits its GPS position in a public or private form, even other data and emergency signals. Also, it can receive notifications of other IOT mechanisms. For example, from another car that has had an accident and it is within 1 KM.

- **Smart TV (16):** This device has a bidirectional connection. Thanks to it, video calls or VOIP calls can be sent and received.

- **PDA (17):** By PDA the data can be sent and received to operate with different mechanisms. For example, if we notice that our house temperature sensor shows 10°C, we can send an order to turn on the heat.

- **Smartphone (18):** As the same way than PDA (17), we can interact with other smart devices, as well as receiving the GPS position of the car (15) in the case of an accident.

EXPERIMENTS

Prototype: Design of an Application for Interconnection and Its Use in a Sensor Base

In this section, we will describe how the proposed platform is used to define an application that adds an application created by a user. For this example, we have an application with tow sensors (humidity and temperature), a Smartphone, and a NFC label. Everyone will share the same configuration and will perform the same actions. There are two conditions: a temperature beyond 40° and a humidity level below 60%. That is when it will be necessary to call the firemen.

That application is composed of a Network that has got a Smartphone from which we use the temperature and humidity sensors and a Smart Label as NFC. The Smartphone has a configuration where sounds and persistence are allowed. The action created for Smartphone, sensors, the NFC Label calls the firefighters if it happens at 13/12/2012 and the temperature is over 40° C and humidity is less than 60%. The first condi-

tion, once it is done, does not begin again. Even if the temperature decreases, this condition will not be taken into account. The opposite case is the humidity, because, for the action jump, it will have to be less than 60%. If the humidity condition is met first, or it happens that the temperature is over 40° and the humidity is over 60%, the action will not be executed until the humidity levels are below 60% again, the action will not jump until the humidity goes back to normal.

Table 1 shows an example of the application by the text editor. First, it is necessary to write the user's name; then the information that will follow this order:

Table 1. Code snippet of an application made with the textual editor

```
User id: UO206639                                    SmartLabel {
Name: "Cristian"                                         Id: sl4
Application {                                            Name: "NFC Active"
    Id: app1                                             Activated: TRUE
    Name: "Muspel"                                       Range: 50
    Networks: network1                                   Bitrate: 30
}                                                        Encryptation: FALSE
Network {                                                Configuration: conf1
    Id: network1                                         Actions: actionCallFirefighters
    Name: "Muspel One"                                  }
    Activated: TRUE                                  Configuration {
    Components: device1 sensorTemp sensorHum sl4         Id: conf1
    Actions: actionCallFirefighters                     sound: TRUE
}                                                       persistent: TRUE
Device {                                                }
    Id: device1                                      Action {
    Name: "Smartphone 1"                                 Id: actionCallFirefighters
    Activated: TRUE                                      Name: "Call firefighters"
    Connection Type: "SMS"                               Date (dd/mm/yy): "13/12/12"
    Configuration: conf1                                Desactivation to complete: TRUE
    Actions: actionCallFirefighters                     Conditions: conditionTemp40 conditionHum60
}                                                       }
Sensor {                                             Condition {
    Id: sensorTemp                                       Id: conditionTemp40
    Name: "Temperature"                                  Name: "Temperature > 40"
    Activated: TRUE                                      Reboot Condition: FALSE
    Query Time: 10000                                    Desactivation to complete: TRUE
    Hot Swap: TRUE                                       }
    Configuration: conf1                             Condition {
    Actions: actionCallFirefighters                     Id: conditionHum60
}                                                       Name: "Humidity < 60"
Sensor {                                                Reboot Condition: TRUE
    Id: sensorHum                                       Desactivation to complete: FALSE
    Name: "Humidity"                                     }
    Activated: TRUE
    Query Time: 20000
    Hot Swap: FALSE
    Configuration: conf1
    Actions: actionCallFirefighters
}
```

1. One or more applications.
2. One or more networks with ID attributed in the application.
3. One or more components with the ID attributed in networks. All these can be devices (Smartphones, speakers…), sensors, Smart Labels (RFID, NFC).
4. Configuration of components.
5. Their actions.
6. Conditions for the actions that are carried out

In next figure it can be seen the same example, but presented with the graphic framework. In it we can see that the systems are connected by lines and each one contains its information.

In the Figure 5 we can see the same application than the ones in the example of Figure 4. The identifier 1 matches with the application. The identifier 2 is the Network that has got 4 devices. The Smartphone 1, the 3 is the temperature sen-

sor, the 4 is the humidity sensor and the 6 is the Smart Label as NFC. These four devices have the same configuration, one that matches with identifier 7. They have the same actions, identifier 8. This has two conditions: humidity under 60% and temperature over 40°C.

FUTURE WORK

Future research lines derived from this work are divided into the following areas:

- The mobile mechanisms that are connected to the platform need a framework which makes the synchronization and communication with the platform. To make possible for several electronic mechanisms with different features and different operative systems to connect to the platform it's required that this connecting application is imple-

Figure 5. The same application made with the graphic editor

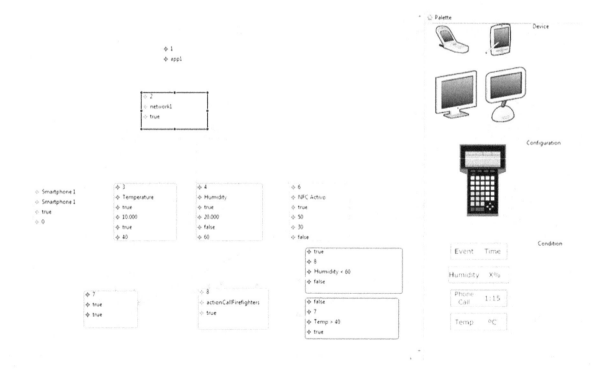

mented in several programming languages; this way the application that adapts better to the specifications could be installed in each mechanism. Because of that, part of the future work could be to create native frameworks in other platforms and in other languages, like: C++, C#, Ruby, PHP, JavaScript, Objective-C.

- To complete the specific language capacities of graphic domain to allow the inexperienced users to specify business processes which involve the combined work of several smart objects and mechanisms. To develop this new specific language of graphic domain we want to evaluate some languages and tools designed to model business processes.

- Thanks to the heterogeneous nature of mechanisms with its own operative systems and its different functioning, it would be interesting to expand to a bigger number of systems: Smartphones (Ubuntu, Windows phone, Firefox OS), video consoles (PS3, Wii, Wii U, Xbox 360, Nintendo DS, Nintendo 3DS), computers (Microsoft Windows, Mac OS X, GNU/Linux), graphic card with sensors (NVidia, ATI), Smarts Cities…

- Every mechanism has several forms of interconnection with the exterior. For example, a Smartphone or cellular phone with infrared, Bluetooth, *Short* Message Service (SMS), Multimedia Messaging System (MMS) and Wireless. A computer can have some of these and even connection ports, like Universal Serial Bus (USB), LPT1 ports o Ethernet ports. Other devices have High Speed Downlink Packet Access (HSDPA), (High-Speed Uplink Packet Access (HSUPA), Global System for Mobile communications (GSM), Worldwide Interoperability for Microwave Access (WiMax), Long Term Evolution (LTE), Universal Mobile Telecommunications System (UMTS), Code Division Multiple Access (CDMA). Because of this many possible forms of data sending, it can be increased to give support to a lot of protocols and give more possibilities in the interaction with other mechanisms or serve as an extra support if there is a failure in one of the used objects.

CONCLUSION

As it has been demonstrated, the way to abstract the configuration, organization, and creation based in sensors has been taken to a very high level thanks to the use of model-driven engineering.

Users can choose between a textual environment that drive it or a graphic environment where they drop the boxes, fill them with data and joint them.

Among data, user only has to choose those one he prefers, choose the kind of connection, conditions, actions, persistence, and he will not have to take care of how it operates in the inside.

All that makes possible for unconnected people (as long as they know the sphere) to create applications of sensor nets for personal use in an easy and quicker way, which can be used and modified again. They can create a net at home or at work, or one net in each room and they can modify it in a little time. In the primary sector, they can create sensor nets to control fields, greenhouses, and automate tasks in a simply way, according to a mechanism that receives the action, for example, a motor or a robot.

REFERENCES

Falvo, M. C., Lamedica, R., & Ruvio, A. (2012). An environmental sustainable transport system: A trolley-buses line for Cosenza city. In *Proceedings of the International Symposium on Power Electronics Power Electronics, Electrical Drives, Automation and Motion*, (pp. 1479–1485). IEEE.

Georgitzikis, V., Akribopoulos, O., & Chatzigiannakis, I. (2012). Controlling physical objects via the internet using the arduino platform over 802.15.4. networks. *Revista IEEE America Lantina*, *10*(3), 1686–1689. doi:10.1109/TLA.2012.6222571.

Gu, H., & Wang, D. (2009). A content-aware fridge based on RFID in smart home for home-healthcare. In *Proceedings of the Advanced Communication Technology Conference*, (vol. 2, pp. 987–990). IEEE.

Hao, C., Lei, X., & Yan, Z. (2012). The application and Implementation research of smart city in China. In *Proceedings of System Science and Engineering* (pp. 288–292). IEEE. doi:10.1109/ICSSE.2012.6257192.

Hegedus, A., Horvath, A., Rath, I., Ujhelyi, Z., & Varro, D. (2011). Implementing efficient model validation in EMF tools. In *Proceedings of the 26th IEEE/ACM International Conference on Automated Software Engineering* (pp. 580-583). IEEE.

Hribernik, K. A., Ghrairi, Z., Hans, C., & Thoben, K. (2011). Co-creating the internet of things - First experiences in the participatory design of intelligent products with arduino. In *Proceedings of Concurrent Enterprising*. IEEE.

IoBridge. (2013). *Thingspeak*. Retrieved from http://www.thingspeak.com

Kolovos, D. S., Rose, L. M., Paige, R. F., & Polack, F. A. C. (2009). Raising the level of abstraction in the development of GMF-based graphical model editors. In *Proceedings of the 2009 ICSE Workshop on Modeling in Software Engineering*, (pp. 13–19). IEEE.

LogMeIn. (2013). *COSM*. Retrieved January 15, 2013, from https://cosm.com/

Piras, A., Carboni, D., Pintus, A., & Features, D. M. T. (2012). A platform to collect, manage and share heterogeneous sensor data. In *Proceedings of the 9th International Conference on Networked Sensing Systems*. IEEE.

Rothensee, M. (2007). A high-fidelity simulation of the smart fridge enabling product-based services. In *Proceedings of the 3rd IET International Conference on Intelligent Environments (IE 07)*, (pp. 529–532). IEE.

University of Vienna. (2013). *European smart cities*. Retrieved November 26, 2012, from http://www.smart-cities.eu

Yamanoue, T., Oda, K., & Shimozono, K. (2012). A M2M system using Arduino, Android, and Wiki software. In *Proceedings of the 2012 IIAI International Conference on Advanced Applied Informatics*, (pp. 123–128). IEEE. Retrieved January 13, 2013, from http://ieeexplore.ieee.org/lpdocs/epic03/wrapper.htm?arnumber=6337170

KEY TERMS AND DEFINITIONS

Domain Specific Language: Language given to solve or involve a specific problem and provide a technique for solving private situations.

Internet of Things: Network of daily objects connected to each other to allow an interaction among them.

Model-Driven Engineering: Methodology of software development that is focused on model creation near the concept of private domain instead of software.

Sensor Platforms: Set of sensors in a same platform to allow the process of interconnection and make it easier.

Sensor: Mechanism that is capable of detecting physic or chemical magnitudes. For example: temperature, light, humidity, CO_2, radiation.

Sensor-Based Applications: Applications created to work and interact with sensors.

Smart City: City that creates an environment where different mechanisms are interconnects among them, all by using sensors and automations.

Smart Objects: Objects that can interact with others. For example, two Smartphones or a Smartphone with the TV.

Web Services: Technology that uses a series of protocols and standards to exchange data between applications.

Chapter 5
A Context–Aware MDA Approach for Content Personalization in User Interface Development

Firas Bacha
University of Valenciennes, France

Káthia Marçal de Oliveira
University of Valenciennes, France

Mourad Abed
University of Valenciennes, France

ABSTRACT

User Interface (UI) personalization aims at providing the right information, at the right time, and on the right support (tablets, smart-phone, etc.). Personalization can be performed on the interface elements' presentation (e.g. layout, screen size, and resolution) and on the content provided (e.g., data, information, document). While many existing approaches deal with the first type of personalization, this chapter explores content personalization. To that end, the authors define a context-aware Model Driven Architecture (MDA) approach where the UI model is enriched by data from a domain model and its mapping to a context model. They conclude that this approach is better used only for domains where one envisions several developments of software applications and/or user interfaces.

INTRODUCTION

Recent years, the proliferation and the continuing growth of different computer devices with several interaction modes, allow users to access information anywhere and at anytime. This flexibility makes the user even more exigent and brings new challenges to those systems. Since the relevance of the delivered information and its adaptation to the users' preferences are key factors for success or rejection of systems, the solution is to conquer

DOI: 10.4018/978-1-4666-4494-6.ch005

users by developing personalized systems adapted to them.

Personalization is the capability to customize communication based on knowledge preferences and behaviors (Dyche, 2002) being able to provide content and services tailored to individuals (Hagen, Manning & Souza, 1999). Moreover, Garía-Barrios, Mödritscher, and Gütl (2005) say that to perform personalization, an internal and individual model about the user is needed. Personalization can take into account several aspects (e.g., navigation, structure, functionalities) and it can be performed on the interface elements' presentation (i.e., layout, colors, sizes, and other design elements), and on the content (data, information, document) provided in the UI (Anli, 2006; Brossard, Abed, & Kolski, 2011). Unlike content personalization, many works explore the adaptation of UI element presentation (called containers adaptation).

This chapter presents an MDA-compliant approach that takes into account the content personalization when designing UIs. Our goal is to generate context-aware UIs that provide users with personalized content, according to their context.

In the following sections we first present briefly definitions of context-awareness and its application for UI development (section 2). Then, in section 3, we describe in detail our MDA context-aware approach for content personalization. At the meantime, we showed an example of its application in the development of software system for medicine recommendation. Sections 4 and 5 present, respectively, some related works and our conclusions.

BACKGROUND

Context-aware software application was first defined by (Schilit, Adams, & Want, 1994) in 1994 as the software that "adapts according to its location of use, the collection of nearby people

and objects, as well as changes to those objects over time" (p. 1). They defend that a context-ware software application can examine the computing environment and react to environment changes. In 2001, Dey (2001) defined that "a system is context-aware if it uses context to provide relevant information and/or services to the user" (p. 1), where relevancy depends on the user's task. Schilit, Adams, and Want (1994) say that three important aspects of context are: where you are? Who you are with? and what resources are nearby? Dey (2001) generalizes those aspects by defining context as "any information that can be used to characterize the situation of an entity" (p. 3) where entity is a person, place, or object relevant to the interaction between a user and an application.

In the Human-Computer Interaction (HCI) domain, our particular interest in this chapter, Calvary et al. (2003) propose the CAMELEON framework. In this framework, the UI generation is performed threw a set of models transformations, considering the context of use. According to the authors, the context of use is composed of three classes of entities: the user of the system, who is intended to use the system; the platform (hardware and software) that is used for interacting with the system, and, the physical environment where the interaction takes place. In CAMELEON, four models are proposed: task and concepts models, that describes the user's tasks to be carried out and the domain-oriented concepts required by these tasks; abstract UI, that describes a UI independently of any modality of interaction; concrete UI, that concretizes an abstract UI for a given context of use; and, the final UI, that is the operational UI running in a specific platform.

Several context-aware approaches and tools were defined based on CAMELEON, such as: UsiXML environment (Limbourg et al., 2005), that supports UI generation using Model Driven Architecture (MDA) (OMG, 2003); TERESA tool (Berti et al., 2004), that supports the semi-automatic generation of multimodal UI; Sottet's proposition of dynamic UI elements' presentation adaptation (Sottet et al., 2007) and the parameterized approach for generating plastic interfaces (Bouchelliga, Mahfoudi & Abed, 2012).

The essential difference of our proposal from all those others is that our one focuses on content personalization that means which pertinent information should be provided in the input/output entries of the UI, while existing approaches focus only on UI containers adaptation.

Despite this difference, as a context-aware approach we identify similarities with the others:

- As proposed by UsiXML (Limbourg et al., 2005; UsiXML, 2007) and (Bouchelliga, Mahfoudi & Abed, 2012), we explicitly define a context model considering the triplet (user, platform environment) proposed by the CAMELEON framework (see sub-section Context Model) ;
- Similar to Berti et al. (2004), Bouchelliga, Mahfoudi & Abed (2012) and Gajos, Weld & Wobbrock (2010), we consider the UI containers adaptation only at design time since our focus was on the content personalization;
- Similar to Hachani, Chessa, and Front (2009) and Clerckx, Luyten, and Coninx (2005), we propose to annotate the UI task model with information related to the context. However, they use CTT for task modeling and we use BPMN (see sub-section Business Process Model [BPM - CIM level]).

Finally, Brossard, Abed, and Kolski (2011) proposed an MDA methodology for the design of personalized information system for the transportation domain that, like us, aims to perform content personalization at runtime. Two main differences from our proposal are identified.

- First, their proposal is specific for transportation software development, while our one is domain-independent.
- Second, the authors presented simply the proposed approach models and they did not develop the models transformation.

AN APPROACH FOR CONTENT PERSONALIZATION IN UI DESIGN

Issues, Controversies, Problems

We recall that our main goal is to include content personalization since the beginning of the UI modeling, that is, to define, when possible, which personalized information should be provided for each UI entry (input/output). To address this goal and considering the literature review, we set the following requirements:

1. To consider information about the domain application to which the system is developed, since the information to be provided should be pertinent to that domain; and, information about the user context since we want a personalized information, that means related to a person (the user);
2. To use a notation/method for the UI modeling that we can deal with the data being changed (the content) and that could be easily integrated with system modeling since providing the personalized content in the UI is directly related to the functional requirements of a system; and,
3. To allow the reuse of the UI models for different platforms, in a way that if we need to generate the same interface for other devices it is not necessary to re-implement all models. This fact is particular interesting nowadays where the same system can be executed in different kind of platforms (ex.: PC and mobile phones) requiring personalized UI.

Solutions and Recommendations

Based on requirements previously presented, we chose the main technical features for our approach. To answer requirement (a), we identified the need of:

- A domain model that captures the concepts of the domain as defended by CAMELEON framework and other approaches for UI design. To define the domain model, one can use class diagram or domain ontology. The importance is that this domain model could models the application data source; and,
- A context model to provide the relevant information, as stated by Dey (2001) and Schilit, Adams, and Want (1994), capturing who is the user, where s-he is and the resources nearby. In the HCI domain, we consider the context as a triplet composed of user; platform and environment (see previous section).

Moreover, in order to assure the personalization, the domain model and context model should be related in some way since the delivered content is depending on context. We need, therefore, to describe "relations of dependency" between context model elements and domain model elements. We call these relationships "mappings."

Several notations exist for UI modeling (e.g., Diane, MAD, Petri networks, etc.). Two of the most recent notations are Concur Task Tree (Paternò, 1999) and the Business Process Model Notation (BPMN) (OMG, 2006) used for this purpose (Bouchelligua et al., 2010; Brossard, Abed, & Kolski, 2011). Based on the requirement (b), we chose BPMN since 1) It is the only task modelling notation which represents a standard; 2) it is closer to the system engineering and, therefore, allows the functional tasks modelling and application business logic description; and 3) It allows following the information flow between tasks, which is especially important for integrating the

content personalization. In fact, tasks in different processes communicate by means of message connections and tasks in the same process coordinate using flow connections for controlling sequencing. Although CTT provides the ability to pass information between tasks, this passage is restricted between neighboring tasks belonging to the same level. Nevertheless, we should highlight here that BPMN is only used to express user interactions as proposed by Brossard, Abed, and Kolski (2011) and it will not be integrated in a workflow engine.

Finally, to answer the last requirement (c), it was clear for us that we should use a MDA approach, as proposed for example by UsiXML (Limbourg et al., 2005), in a way that we could use the same UI model to generate UI for different platforms. This UI generation will not be defined dynamically, at runtime, as defined by Sottet et al. (2007), since our main interest is the content personalization and not the UI containers personalization. In this way, the generation of the final UI is done at design time reusing the defined UI model (requirement [c]). The content personalization, in turn, is provided at runtime.

Being compliant to MDA, we defined therefore the models for the three MDA levels (OMG, 2003): the Computer Independent Model (CIM), to establish de user interaction in a high-level of abstraction; the Platform Independent Model (PIM), where we introduce the UI structure using a UI description language; and Platform Specific Model (PSM), to orientate the UI for a specific platform.

The CIM is named Business Process Model and is expressed using BPMN with some extensions to contemplate the content personalization. PIM and PSM are named respectively Platform Independent Interaction Model (PIIM) and Platform Specific Interaction Model (PSIM). Both models are specified in UIML (User Interface Markup Language) (Helms, 2009). UIML was chosen for many reasons: 1) It is a general language for describing the different UI parts. UIML ensures to separate these aspects in order to facilitate both the reuse of interface definitions and coherence between different platforms; 2) UIML provides facilities to work with dynamic information (the use of UIML variables inside the Behavior part) and to access to any data type based on the invocation of external methods, threw the logic part; and 3) Actually there are several tools that allows either interpreting the UIML code and generating directly the final interface, or transforming the UIML code to another development language code (for example, the toolkit LiquidApps allows the conversion of UIML to Java, HTML, WML, VoiceXML). In our approach, the generation from PIIM to PSIM, is just the inclusion or modification of platform specific characteristics in the generated PIIM code. We preferred, therefore, calling this step a transition instead of transformation.

Figure 1 summarizes our approach (Bacha, Oliveira & Abed, 2011). The content personalization is integrated since the beginning of the UI modeling, threw the mapping between the context and domain model elements that are used in the BPM. Those models are used in the transformation

Figure 1. A context-aware MDA approach for content personalization of UI

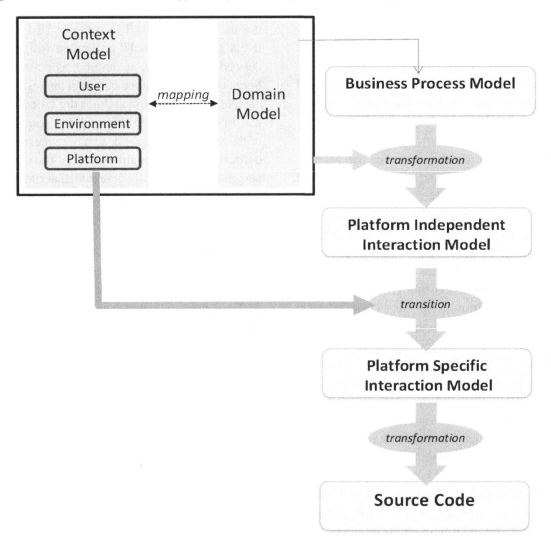

process from BPM to PIIM. Platform information is used for the integration of UIML parts at PSIM. Finally, the source code is generated. Next section presents each one of those models and their transformations.

UI Models Definition

For better explanation of the models, we will show the UI modeling of a software system as we introduce the models. This software system aims to provide medicine recommendation that does not depend of doctor's prescription. We will use in the example, two main user interfaces. The first UI ask the input information from the user: his/her disease area (e.g. head, throat), his/her sex, and some pre-defined symptoms (fever, tiredness, eye pain, cough, dizziness, trembling). The second interface presents the recommended medicines based on the input data and user age. We consider that once the user connects to the system, the system identifies his/her context to fill in some required data (for example, age in the first interface) and to provide personalized results (recommending medicine based on his/her age).

Context Model: As defined in previous section the context model should contain data about the user, platform and environment. To define this model, we did a large literature review (Bacha, Oliveira & Abed, 2011a). Although eighteen context model propositions were found none of them was considered complete. Some propositions considered only one of the context dimensions (user, platform or environment) (Becker & Dürr, 2005; FIPA, 2001; Hobbs & Pan, 2006; Kostadinov, 2008; Rousseau, 2004; UMO, 2003; W3C, 2009). Others were particular for a particular domain (e.g. smart phones, e-commerce) (Chen et al., 2004; Kim & Choi, 2006; Korpipää et al., 2003; Lin et al., 2005; Schmidt, Beigl & Gellersen, 1999; Taconet & Kazi Aoul, 2008; Weißenberg, 2004) and/or not enough detailed (Arabshianand & Schulzrinne, 2006; Preuveneers et al., 2004;

Wang et al., 2004; UsiXML, 2007). We decided, therefore, to integrate the main information from all propositions.

Figure 2 shows a part of the context model we defined with the information about the user, platform and environment. The central element of the context model is the *user profile* that is composed of five major parts allowing to specify the user when interacting with the final interface: *Contact information* that contains personnel data; *Demographic information* that contains basic and unchanged user data; *Preference* to describe user interests and preferences; *User State* to describe the physiological user state and the activity s/he is practicing; and finally *Ability and Proficiency* that specify the user skills and abilities. The information about the platform used by the user is organized in two main classes: *Hardware*, describing the physical aspects; and, *Software*, with description of computer systems. The information about environment is organized in three main classes: *Location*, refers to the place where the user is located at the time of interaction (the geometric, i.e. exact, location and the symbolic that is relative to another location); *Time* that describes the interaction moment (exact time or by a symbolic one - summer, school holidays...) and the *Environmental Condition* identified at the moment of user interaction.

Mapping between Domain and Context Models: The domain model represents the vocabulary used to define the input/output in the UI. As previously described, the domain model can be defined as class diagram or domain ontology. Since the context model is independent of domain and we look for personalization for a specific application domain, we have to find a correspondence between the context model and the domain specific ontology. For that reason, we created the mapping model that defines the relationship between the context elements and the ontology ones. In fact, the domain elements (classes or attributes) should be analyzed against the context model

Figure 2. The context model

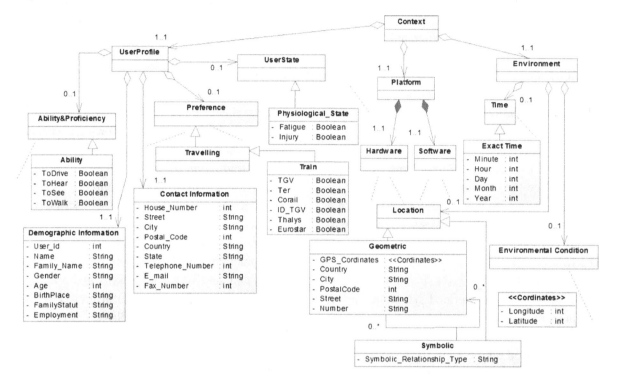

elements looking for each context element that could influence the domain concept. Once we find an element, we should set a mapping with the domain concept. In this way, the information provided in the UI will be personalized considering the context. Analyzing domain models, we identified three main cases of mappings.

The first one refers is used when ontology concept represents exactly the same information present in the context model although, sometimes, with different name. That means, they have the same meaning. This mapping can be done for any one of the attributes in the context model that is not Boolean (for example, age and name of the user). We say, therefore, that we have a *direct mapping*, i.e. the concepts from the domain model are directly associated to the context model.

A second case defined when any information of the context indicates the presence or absence of some information. It is similar to the direct mapping, except that in this case, the context at-

tribute element must have the *Boolean* type. We say, therefore, that we have an *indicative mapping*.

Finally, there are some concepts from context model that could have an indirect influence in the domain model concepts, which means it could probably impact the domain concept. This type of mapping is called *Indirect mapping*. To define an indirect mapping, the designer should verify if there is any information in the domain model that could change depending on some personnel data modeled in the context.

Actually, all mappings are done manually. We present the algorithm, Box 1, that a designer could follow when defining mappings (we refer to an ontology element (classes or attributes) by "o" and to a context model element by "c". "O" represents the set of all ontology elements and "C" all context attributes).

To define the mappings for the example of medicine recommendation system we defined a domain model based on the translation medicine

Box 1.

```
Begin
  For "o" in "O"    do
    For "c" in "C"     do
       If ("o" and "c" have the same meaning) and
       (the type of "c" is different from Boolean)
       then
         create a direct mapping between "o" and "c".
       EndIf
       If ("o" and "c" have the same meaning) and
       (the type of "c" is Boolean)
       then
         create an indicative apping between "o" and "c".
       EndIf
       If (the value of "o" can change according to the value of "c")
       then
         create an indirect mapping between "o" and "c".
       End
    End
  End
End
```

Figure 3. Domain model and mappings with context model elements

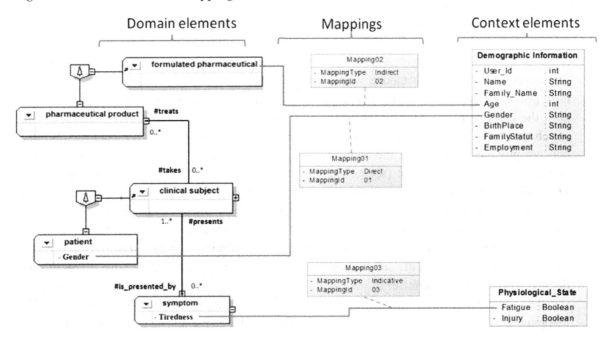

domain ontology (Batchel et al., 2012). Figure 3 shows an extract from the concepts we are interested in. Analyzing the domain model of Figure 3 and the context model, we mapped the *gender* attribute (from *Patient* domain concept) and the *gender* attribute (from *Demographic Information* context class) using the direct mapping, since they have the same meaning. Then, *Tiredness* (from *Symptom*) was mapped with *Fatigue* attribute (from *Physiological_State* context class) using the indicative mapping. For indirect mappings, we set that the *FormulatedPharmaceutical* domain concept is indirectly associated with *Age* attribute from *Demographic Information* class. That means that the chemical substance of a medicine (*PharmaceuticalProduct*) depends on the user age (for example, some drugs are used only for adults).

Business Process Model (BPM - CIM level): As defined in other approaches, the BPMN is used to define the tasks to support the business goal (user tasks to represent the interaction between the user and the system; system tasks to represent functions executed by the system and manual tasks to represent task executed only by the user) and the data flow between tasks. Figure 4 shows the BPM for the medicine recommendation system.

To provide the content personalization for all input/output information, we should annotate each user or system tasks with:

- The concept of the domain model, whenever possible, and its pertinent mapping with the context model. Direct mappings are chosen when we want that the domain concept provides, at runtime, the content of the related context element (see task 6 - Gender in Figure 5). Indicative mappings are chosen when we want to set at runtime the selection/not selection of a domain element from a list of options (see task 8 -.*Tiredness* in Figure 5). This selection is defined by the value (true/false) of the associated context element. Finally, indirect mappings are chosen when we want to show

results that depends on many other domain elements to which we associated indirect mappings with context elements (in Figure 5 the result is *PharmaceuticalProduct* that is searched based on the *Formulated Pharmaceutical* that is indirectly associated with a context model element - age);

- The identification of kinds of input/output manipulated in the task, that we named interaction elements. The interaction elements are an abstract view of types of interaction between user and system. The interaction elements that we are particular interested for the content personalization are the different types of input of information (e.g., informed by the user – named UIFieldManual, selected from a defined set of options – named UIFieldOneChoice, etc.) and output information (named UIFieldOutput). However, others interaction elements were defined to be associated for any task modeled with BPMN (e.g., to represent a group of information - named UIUnit, to represent a UI with several groups of information – named UIGroup, etc.).

We adapted the BPMN meta-model proposed by (OMG, 2006) to be used in our approach, supporting those annotations and the definition of the kinds of tasks.

Platform Independent Interaction Model (PIIM): As defined in previous section, this model uses UIML. A UIML model is composed of two main components: interface and peers. The interface component represents the description of the interface threw four parts: structure, that represents the organization and hierarchies of all UI parts; content that describes the set of the application information that will be displayed (e.g. in different languages), behavior that represents the behavior of the application at the user interaction time, and style that defines all properties specific for each UI element. The peers component links

Figure 4. Business process model for the medicine recommendation system

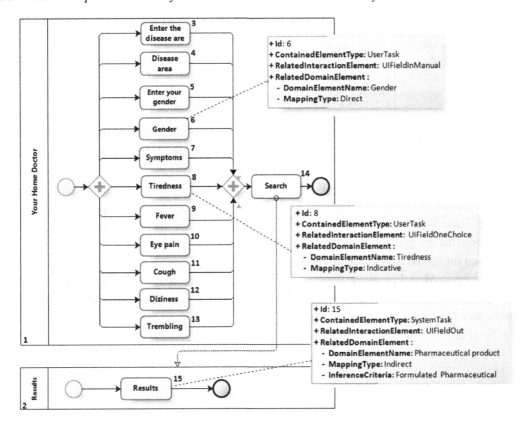

the generic UI elements and their properties, to a specific platform using the presentation part. Indeed, it describes the calling conventions for methods that are invoked by the UIML code in the logic part. The logic part links methods that are used in UIML with other ones used in a platform-specific source code.

The PIIM is composed of the structure, behavior, content, and style parts. To manipulate the content, UIML offers two choices: either to integrate it within the style part, or to separate it under the content part. The second alternative is useful only if designers have several contents for an interaction element and only if the contents are already known. For that reason, we decide to adopt the first choice by integrating the content part within the style one. In the PIIM, the style part contains only properties related to content.

Platform Specific Interaction Model (PSIM): This model is composed of the style, presentation

and logic parts. The style part here contains the layout information using the appropriate style properties based on the chosen target platform. The presentation part serves to map generic UIML classes with platform-specific ones and the logic part contains mappings between the methods used in the behavior part and those that will be used on the platform-specific source code.

UI Generation

We use ATL (ATLAS Transformation Language) to write the transformation from BPM to PIIM and the translation from PIIM to PSM.

From BPM to PIIM: Based on the information from BPM, context model, domain model and the mappings, the three UIML parts (structure, behavior and the content part of style) are generated.

To generate the structure part, the transformation depends on the BPMN element used and its

Table 1. Examples of UIML classes generated in PIIM

BPMN Element	BPM Interaction Element	PIIM - generated UIML Class
Pool	UIGroup	G:TopContainer
SubProcess	UIUnit	G:Area
SubProcess	UISubUnit	G:SplitArea
UserTask	UFieldAction	G:Button
UserTask	UIFieldInManual	G:Text

associated interaction. In general, for each interaction element, a UIML <class> is created under the <part> tag. Table 1 shows some examples of the UIML classes that are generated in the transformations based on the BPMN element and its associated interaction Model element.

The source code in Box 2 presents a part of the ATL code that generates the structure part of the PIIM related to the *Pool* BPMN element.

For each *Pool* element in the Business Process Model, a <part> element is created with the <class> attribute G:TopContainer. Then, we analyze all elements that compose this *Pool*. For each element, a <part> is created with the correspondent class attribute, and with the same *id* from the Business Process Model element.

The behavior part is created for each annotated task of the BPM. To set the behavior, we used the UIML rule statements which are composed of a set of conditions and associated actions. The condition is used to keep the dynamics of the application modeled in the BPM when transforming to UIML. This is done by the use of activation variables that controls when the task (or other elements) will be performed and after that which next elements should be activated to be, then, executed. An action (i.e., when-true statement) is defined for each kind of mapping.

Box 2.

```
rule UIMLStructureFromPool {
  from
  pool:bpmn!Pool
    to
    partpool: uiml!Part (),
    structure: uiml!Structure()
    do{
if
(pool.RelatedStaticElement. oclIsTypeOf(Static_UI!UIGroup){
structure.part <- partpool;
partpool.class <- 'G:TopContainer';
partpool.id <- pool.Id;}
for (x in pool.ContainedElements)
{  if (x.ContainedElementType = #SubProcess
  and x.RelatedStaticElement.oclIsTypeOf (Static_UI!UIUnit))
  {
  thisModule.SubProcessUIUnit(x,partpool); }
  if(x.ContainedElementType = #Task)
  {thisModule.Task (x,partpool);}}
  }
}
```

For direct mappings, the when-true part sets that the g:text property will be filled-in automatically by the value taken from the context by calling the "GetValueFromContext" method. For indicative mappings, the generated when-true statement, set the g:selected property of the created UIML with the value of the context element (true/false). For indirect mappings, a <call> UIML statement is generated in PIIM under the when-true statement. The <call> statement represents a call to an external method or service (that uses a language other than UIML). It defines which information should be returned based on parameters (elements indirectly associated to the domain concept). In the behavior part, the generated UIML code manipulate style properties related to content (such as g:text, g:selected).

Figure 5 shows a part of the generated PIIM for the medicine recommendation system. It presents the structure (Figure 5a) and behavior (Figure 5b) generated from BPM to PIIM. As a direct mapping, the property that contains the content of the element is filled in with the value of the *Birthday* attribute deduced from the context model through the 6GetValueFromContext method. Figure 5(c) presents the PIIM generated for the indirect mapping.

Since this element will serve to display information, in the generated when-true part, the method 15Get-Element is called threw the call statement, and it has as parameters, firstly the searched element (*Pharmaceutical Product*) followed by the parameters that the system should consider during searching process (*Formulated Pharmaceutical*).

Figure 5. Example of PIIM generated by transformation from BPM to PIIM (UIML structure part [a], UIML behavior part for direct mapping [b], and for indirect mapping [c])

```
<UIML:Structure>
..
  <part class="G:TopContainer" id="1">
  ...
    <part class="G:TextField" id="3"/>
    <part class="G:CheckBoxButton" id="8"/>       (a)
    ...
  </part>
...
</UIML:Structure>
...
<UIML:Behavior id="Main Behavior">
  <rule id="Rule6">
    <condition>
      <op name="Equal">
        <variable name="6isactivated"/>
        <constant value="true"/>
      </op>
    </condition>
    <action>
      <whenTrue>
        <property name="g:text" partName="6">
          <call componentId="6Context"
              methodId="6GetValueFromContext">           (b)
            <param name="Birthday"/>
          </call>
        </property>
        <property name="g:visible"
          partName="6">
          <constant value="true"/>
        </property>
        <variable name="16isactivated"
               value="true"/>
      </whenTrue>
    </action>
```

```
<rule id="Rule15">
  <condition>
      <op name="Equal">
          <variable name="15isactivated"/>
          <constant value="true"/>
      </op>
  </condition>
  <action>
    <whenTrue>
      <property name="g:text"
partName="15">
        <call componentId="15Context"
            methodId="15Get-Element">
<param name="Pharmaceutical Product"/>
<param name="Formulated Pharmaceutical/>
        </call>
      </property>
      ...
    /whenTrue>
    </action>
  </rule>
</UIML:Behavior>


              (c)
```

From PIIM to PSIM: The transition from PIIM to PSIM considers characteristics of specific platforms by integrating specific remaining style part layout properties for each UI element. Moreover, for each call generated for the indirect mappings at the PIIM, a <logic> statement will be added with the information about the implemented code for this method. This code is implemented by the software designer to search the required information based on the defined parameters. Figure 6 shows some parts of the PSIM supposing that the target platform will have Java as a programming language. The generated PIIM structure class named G:TextField will be mapped to the JText-Field Swing library class. The generated method named 6GetValueFromContext is mapped to the platform-specific method named Lamih.Context. GetValueFromContext that allows getting information from context. The 15GetElement method sets the method to search an element (param1_15) considering one criterion (param2_15).

This approach was also applied for UI modeling of a travel planning software (Bacha, Oliveira & Abed, 2011).

From PSIM to Source Code: After having a UIML code that contains all parts and in order to generate a final UI, we have to choose between two alternatives: either by using a tool that interprets this code or by transforming it into another development language code. Since the logic UIML part is not really connected with a data source, we must choose the second option. To do so, we used the LiquidApps[1] tool. As starting file, it will take the UIML file and it permits generating Java Swing interfaces; Web interfaces in Eclipse RWT; and mobile applications for Android devices.

Figure 7 illustrates an example of an obtained UI that represents a content and a UI containers' personalization of the medicine recommendation application that is developed using the Java language. In fact, the size adaptation of UI elements such as fonts and widgets represents, for instance, personalization of the containers' presentation based on the specific target (in this case, a PC). These properties were integrated within the UIML code when moving from PIIM to PSIM.

The progress of our work, allows us actually to generate the static aspect of interfaces and a part of the dynamic one. In order to obtain a

Figure 6. Example of UIML code integrated to PSIM

```
<UIML:Peers id="MainPeers">
 <UIML:Presentation id="MainPresentationPart">
  <d-class id="G:TextField" used-in-tag="part" maps-type="class" maps-
     to="javax.swing.JTextField">
     <d-property id="text" maps-type="setMethod" maps-to="setText">
       <d-param type="java.lang.String"/>
     </d-property>
   <d-class
 </UIML:Presentation>
 <UIML:logic id="MainLogicPart">
    <dComponent id="6Context" mapsTo="Lamih.Context">
        <dMethod id= "6GetValueFromContext" mapsTo="Lamih.Context.GetValueFromContext">
          <dParam id= "param6GetValueFromContext" type="String"/>
        </dMethod>
    </dComponent>
     <dComponent id="15Context" mapsTo="Lamih.Context">
        <dMethod id= "15Get-Element" mapsTo="Lamih.Context.Get-Element">
          <dParam id= "param1_15" type="String"/>
          <dParam id= "param2_15" type="String"/>
        </dMethod>
     </dComponent>
 </UIML:logic>
</UIML:Peers>
```

Figure 7. Example of generated personalized UI (a medicine searching system)

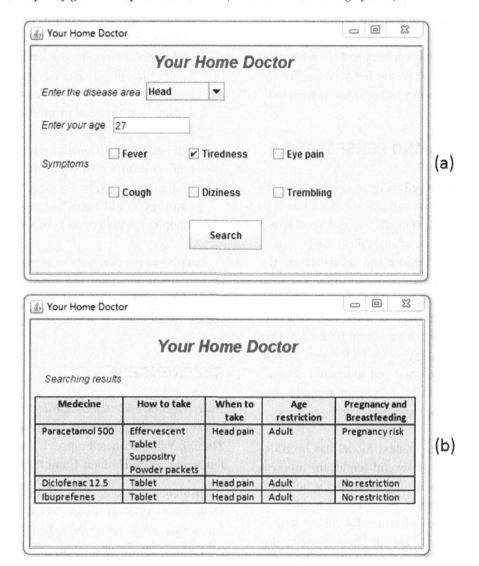

functional application (Figure 7), we have to complete manually the generated Java code. This code enrichment concerns mainly the liaison with the data source and the improve of the application behavior description.

In the example proposed in Figure 7(a), the information about the age and the user state are examples of content personalization. In fact, the user age is automatically filled-in based on the knowledge of his/her age. Same for the automatic ticking of the *Tiredness* option when the system knows the physiological state of the user. The

importance of personalization services that provide automatic forms filling based on user profile has been highlighted by several authors such as (Rukzio et al., 2008).

Another example of content personalization is presented in Figure 7(b). In this example, to provide results to the final user, the system will take into account all the explicit elements of the user query such as the age of the user, the disease area, and the symptoms. But, in order to provide a more personalized result, the application could automatically include some other implicit information deduced from his/her profile. In this case, since the user is a Female, the initial query will be extended (with User gender information) and the system adds the Pregnancy and Breastfeeding column in order to prevent her when using the medicine in pregnancy or breastfeeding period.

CONCLUSION AND PERSPECTIVES

This chapter proposed an approach to take into account content personalization in the UI design. To that end, a context model was proposed based on the consolidation of several existing models, mechanisms to consider the influence of the context in the content provided in the UI were specified; and, models and transformations were defined to support the generation of UI for different platforms from the same user UI model. This approach was used to develop UI for two different domains: medicine and transportation (Bacha, Oliveira & Abed, 2011b). Two important lessons learned from these experiences are:

- Although the context model has been defined based on several (eighteen) propositions, we know that this model is not complete and should be extended to able the content personalization for different software systems. This limitation exists primarily for the user psychological, preferences, abilities and proficiency classes. We investigate the definition of a meta-model to describe those characteristics and what should be adapted in the current approach to use this meta-modeling; and,
- The definition of mappings requires a deep knowledge of the application domain in order to define which context element can influence the domain concepts. It is not profitable, therefore, to use this approach if we have only one software application for one specific platform to develop. In other words, this approach is better used only for

domains where one envisions the development of several software applications or several UI (for different platforms) for the same application. To that end, the domain model should be defined independent of a specific software application. Mappings defined for application independent models can be reused in several UI designs. We believe that domain ontologies can be really useful for this purpose. We are investigating if axioms defined in a domain ontology could help in the definition of mappings with a context model.

Finally, we are currently working in the development of an integrated environment to support all features of this approach, from the definition of the UI model until the generation of the final UI.

REFERENCES

Anli, A. (2006). *Méthodologie de développement des systèmes d'information personnalisés. (PhD Thésis)*. Valenciennes, France: University of Valenciennes et du Hainaut-Cambrésis.

Arabshianand, K., & Schulzrinne, H. (2006). Distributed context-aware agent architecture for global service discovery. In *Proceedings of SWUMA 2006*. Trentino, Italy: SWUMA.

Bacha, F., Oliveira, K., & Abed, M. (2011a). Using context modeling and domain ontology in the design of personalized user interface. *International Journal on Computer Science and Information Systems, 6*, 69–94.

Bacha, F., Oliveira, K., & Abed, M. (2011b). Providing personalized information in transport systems: A model driven architecture approach. In *Proceedings of the First IEEE International Conference on Mobility, Security and Logistics in Transport* (pp. 452-459). Hammamet, Tunisia: IEEE.

Baldauf, M., Dustdar, S., & Rosenberg, F. (2007). A survey on context-aware systems. *International Journal of Ad Hoc and Ubiquitous Computing, 2*(4), 263–277. doi:10.1504/IJA-HUC.2007.014070.

Batchel, O., et al. (2012). *Translational medicine ontology*. Retrieved from http://translational-medicineontology.googlecode.com/svn/trunk/ontology/tmo.owl

Becker, C., & Dürr, F. (2005). On location models for ubiquitous computing. *Personal and Ubiquitous Computing, 9*(1), 20–23. doi:10.1007/s00779-004-0270-2.

Berti, S., Mori, G., Paternò, F., & Santoro, C. (2004). TERESA: A transformation-based environment for designing multi-device interactive applications. In *Proceedings of CHI 2004* (pp. 793-794). Wien, Austria: ACM Press.

Bouchelliga, W., Mahfoudi, A., & Abed, M. (2012). A model driven engineering approach toward user interfaces adaptation. *International Journal of Adaptive, Resilient and Autonomic Systems, 3*, 65–86. doi:10.4018/jaras.2012010104.

Bouchelligua, W., Mahfoudhi, A., Mezhoudi, N., Daassi, O., & Abed, M. (2010). User interfaces modelling of workflow information systems. In Barjis, J. (Ed.), *Enterprise & Organizational Modeling and Simulation (LNBIP)* (*Vol. 63*). Berlin: Springer. doi:10.1007/978-3-642-15723-3_10.

Brossard, A., Abed, M., & Kolski, C. (2011). Taking context into account in conceptual models using a model driven engineering approach. *Information and Software Technology, 53*, 1349–1369. doi:10.1016/j.infsof.2011.06.011.

Calvary, G., Coutaz, J., Thevenin, D., Limbourg, Q., Bouillon, L., & Vanderdonckt, J. (2003). A unifying reference framework for multi-target user interfaces. *Interacting with Computers, 15*(3), 289–308. doi:10.1016/S0953-5438(03)00010-9.

Chen, H., Perich, F., Finin, T., & Jochi, A. (2004). *SOUPA: Standard ontology for ubiquitous and pervasive applications*. IEEE Computer Society, 258-267. doi:10.1109/MOBIQ.2004.1331732.

Clerckx, T., Luyten, K., & Coninx, K. (2005). Dynamo-aid: A design process and a runtime architecture for dynamic model-based user interface development. In *LNCS* (pp. 77–95). Berlin: Springer-Verlag. doi:10.1007/11431879_5.

Dey, A. (2001). Understanding and using context. *Journal of Personal and Ubiquitous Computing, 5*, 4–7. doi:10.1007/s007790170019.

Dyche, J. (2002). *The CRM handbook: A business guide to customer relationship management*. Reading, MA: Addison-Wesley Educational Publishers.

FIPA. (2001). *Device ontology specification*. Retrieved from http://www.fipa.org/specs/fipa00091/PC00091A.html

Gajos, K., Weld, D., & Wobbrock, J. (2010). Automatically generating personalized user interfaces with Supple. *Artificial Intelligence, 174*(12), 910–950. doi:10.1016/j.artint.2010.05.005.

Garía-Barrios, V., Mödritscher, F., & Gütl, C. (2005). Personalisation versus adaptation? A user-centred model approach and its application. In K. Tochtermann & H. Maurer (Eds.), *Proceedings of the International Conference on Knowledge Management* (pp. 120-127). Graz, Austria: IEEE.

Hachani, S., Chessa, S., & Front, A. (2009). Une approche générique pour l'adaptation dynamique des IHM au contexte. In *Proceedings of the 21st International Conference on Association Francophone d'Interaction Homme-Machine* (pp. 89-96). Grenoble, France: IEEE.

Hagen, P., Manning, H., & Souza, R. (1999). *Smart personalization*. Washington, DC: Forrester Research.

Helms, J., Schaefer, R., Luyten, K., Vermeulen, J., Abrams, M., Coyette, A., & Vanderdonckt, J. (2009). Human-centered engineering with the user interface markup language. InSeffah, , Vanderdonckt, , & Desmarais, (Eds.), *Human-Centered Software Engineering*, (pp. 141-173). London: Springer. doi:10.1007/978-1-84800-907-3_7.

Hobbs, J., & Pan, F. (2006). *Time ontology in OWL*. Retrieved from http://www.w3.org/TR/owl-time/

Kim, E., & Choi, J. (2006). An ontology-based context model in a smart home. *Computational Science and Its Applications*, 11-20.

Korpipää, P., Mäntyjärvi, J., Kela, J., Keränen, H., & Malm, E. (2003). Managing context information in mobile devices. *IEEE Pervasive Computing / IEEE Computer Society [and] IEEE Communications Society*, 2(3), 42–51. doi:10.1109/MPRV.2003.1228526.

Kostadinov, D. (2008). *Personnalisation de l'information: une approche de gestion de profils et de reformulation de requêtes*. (PhD Thesis). University of Versailles Saint-Quentin –en-Yvelines, Versailles, France.

Limbourg, Q., Vanderdonckt, J., Michotte, B., Bouillon, L., & Lopez, V. (2005). UsiXML: A language supporting multi-path development of user interfaces. In *Proceedings of 9th IFIP Working Conference on Engineering for Human-Computer Interaction jointly with 11th International Workshop on Design, Specification, and Verification of Interactive Systems EHCI-DSVIS'2004* (LNCS), (vol. 3425, pp. 200-220). Berlin: Springer-Verlag.

Lin, X., Li, S., Xu, J., Shi, W., & Gao, Q. (2005). An efficient context modeling and reasoning system in pervasive environment: using absolute and relative context filtering technology. [LNCS]. *Proceedings of Advances in Web-Age Information Management*, 3739, 357–367. doi:10.1007/11563952_32.

OMG. (2003). *MDA guide*. Retrieved from http://www.omg.org/cgi-bin/doc?omg/03-06-01

OMG. (2006). *Business process modeling notation specification*. Retrieved from http://www.omg.org

Paternò, F. (1999). *Model-based design and evaluation of interactive applications*. Berlin: Springer.

Preuveneers, D., Bergh, J., Wagelaar, D., Georges, A., Rigole, P., Clerckx, T., & Berbers, Y. (2004). Towards an extensible context ontology for ambient intelligence. *Ambient Intelligence*, 148-159.

Rousseau, B., Browne, P., Malone, P., & Ofughlu, M. (2004). User Profiling for content personalisation in information retrieval. In *Proceedings of the ACM Symposium on Applied Computing*. Nicosia, Chypre: ACM Press.

Rukzio, E., Noda, C., De Luca, A., Hamard, J., & Coskun, F. (2008). Automatic form filling on mobile devices. *Pervasive and Mobile Computing*, 4(2), 161–181. doi:10.1016/j.pmcj.2007.09.001.

Schilit, B., Adams, N., & Want, R. (1994). Context-aware computing applications. In *Proceedings of the International Workshop on Mobile Computing Systems and Applications*, (pp. 85-90). IEEE Computer Society.

Schmidt, A., Beigl, M., & Gellersen, H. (1999). There is more to context than location. *Computers & Graphics Journal*, 23(6), 893–902. doi:10.1016/S0097-8493(99)00120-X.

Sottet, J. S., Ganneau, V., Calvary, G., Coutaz, J., Demeure, A., Favre, J. M., & Demumieux, R. (2007). Model-driven adaptation for plastic user interfaces. *Human-Computer Interaction*, 4662, 39–410.

Taconet, C., & Kazi Aoul, Z. (2008). Context-awareness and model driven engineering: Illustration by an e-commerce application scenario. In *Proceedings of ICDIM*. ICDIM.

Tesoriero, R., & Vanderdonckt, J. (2010). Extending UsiXML to support user-aware interfaces. In *Proceedings of 3rd IFIP Conf. on Human-Centred Software Engineering HCSE 2010* (LNCS), (vol. 6409, pp. 95-110). Berlin: Springer-Verlag.

UMO (User Model Ontology). (2003). Retrieved from http://www.u2m.org/2003/02/UserModelOntology.daml

UsiXML. (2007). *User interface extensible markup language) (version 1.8)*. Louvain, Belgium: Université Catholique de Louvain.

W3C. (2009). *Delivery context ontology (DCO)*. Retrieved from http://www.w3.org/TR/2009/WD-dcontology-20090616/

Wang, X., Gu, T., Zhang, D., & Pung, H. (2004). Ontology based context modeling and reasoning using OWL. In *Proceedings of the Second IEEE Annual Conference on Pervasive Computing and Communications Workshops*, (pp. 18-22). IEEE.

Weißenberg, N. (2004). Using ontologies in personalized mobile applications. In *Proceedings of the 12th Annual ACM International Workshop on Geographic Information Systems* (pp. 2-11). ACM Press.

ENDNOTES

[1]   http://liquidapps.harmonia.com/features/

Chapter 6
Ontology–Supported Design of Domain–Specific Languages:
A Complex Event Processing Case Study

István Dávid
Budapest University of Technology and Economics, Hungary

László Gönczy
Budapest University of Technology and Economics, Hungary

ABSTRACT

This chapter introduces a novel approach for design of Domain-Specific Languages (DSL). It is very common in practice that the same problems emerge in different application domains (e.g. the modeling support for complex event processing is desirable in the domain of algorithmic trading, IT security assessment, robust monitoring, etc.). A DSL operates in one single domain, but the above-mentioned cross-domain challenges raise the question: is it possible to automate the design of DSLs which are so closely related? This approach demonstrates how a family of domain-specific languages can be developed for multiple domains from a single generic language metamodel with generative techniques. The basic idea is to refine the targeted domain with separating the problem domain from the context domain. This allows designing a generic language based on the problem and customizing it with the appropriate extensions for arbitrary contexts, thus defining as many DSLs and as many contexts as one extends the generic language for. The authors also present an ontology-based approach for establishing context-specific domain knowledge bases. The results are discussed through a case study, where a language for event processing is designed and extended for multiple context domains.

DOI: 10.4018/978-1-4666-4494-6.ch006

PROBLEM FOUNDATION

Domain-Specific Languages (DSL) are key elements of model driven engineering. They serve for describing the concepts of a given field (domain) in order to enable accurate and compact modeling (Fowler, 2010).

The domain itself is the most major factor in designing DSLs, since it fundamentally determines the concepts of the modeling language and interrelationships among them. Our research points out that the influence of the domain is a complex phenomenon and if investigated from a sufficiently high abstraction level, a more efficient DSL design methodology can be employed.

The targeted domain can be separated into two aspects: the problem domain and the context domain. The problem domain is the set of requirements being actually modeled with the language and is associated with at least one but usually with many context domains. Both aspects are required to characterize a DSL. This can be formally defined in the following way.

Let P denote the set of problem domains and C denote the set of context domains. Given this notation, the set of every possible domain which can be targeted using a domain-specific language would be:

$$D = P \times C.$$

(Of course, not every problem-context combination results in a meaningful domain.) A domain-specific language targets exactly one d domain from the set D. A family of languages for a given problem domain p is a set of languages with a fixed problem but with an arbitrary set of context domains:

$$L_p = p \times C_p,$$

where C_p denotes the subset of context domains which would create a meaningful domain with p.

This level of refinement of the targeted problem is important because it is common that the same problem arises in several contexts (such as online processing of high volume information in the domains of sensor networks, IT infrastructure management, and stock market informatics). Implementing a "general" DSL, which is sufficient for proper modeling in all the contexts, would lead to losing the desired description power. On the other hand, implementing a language family (i.e. a DSL for every context) would produce significant overlaps among the languages' concepts – because of the same targeted problem.

A solution, which allows the separation of reusable language parts and context-dependent elements, would fit well with the overall Model-Driven Architecture (MDA) approach, where platform-independent elements of high abstraction level are getting merged with platform-definition models to yield a more concrete model (Aßmann & Zschaler, 2006).

It is important to understand that from now on, we strictly distinguish one DSL from another by investigating both its problem domain and its context domain. This means extending a common "language-stub" of a given problem for arbitrary contexts would result in as many different DSLs as many contexts we take. This is, of course in accordance with the practice; we just introduced here a different definition for the *domain* itself.

This chapter presents a technique which facilitates semi-automated engineering of families of domain-specific languages closely related to each other because of the same targeted problem but in different context domains. The approach hence aims at the creation of a framework where reusable and custom context-specific language elements help an efficient development. Reusable parts depict the core of the problem; context-specific customizations aim at extending these parts with the knowledge from the targeted context domain. Therefore, reusable elements (e.g. common domain entities, relationships, data types) happen to be generic over different contexts and remain intact despite the context-specific customizations. Furthermore, we do not require the reusable parts being a DSL of full value, instead: generic parts with context-specific extensions shall piece a DSL together.

The approach can also handle the problem of the changing requirements and the consequent changes in the DLS's language structure (Spinellis, 2001).

Context-specific extensions are derived from some kind of domain knowledge related to the targeted context domain. An important question is how domain knowledge can be represented in order to be properly formalized for enabling automated generation of context-specific language artifacts.

We propose an ontology-driven solution for this purpose, which also allows the re-use of existing domain ontologies. This knowledge base will be referred as the *Semantic Knowledge Base* (SKB), since it provides semantics for the model-driven procedures, like mappings and (meta)model transformations, thus facilitating automation. Knowledge from the SKB is getting mapped onto the language level. For that, a fixed meta-schema is required otherwise the mapping functionality would change from case to case with the schema of the ontology. Therefore, we propose a layered architecture for ontologies which consists of (1) the context domain ontology, describing the actual concepts of the context domain and the relations among them; (2) the generic ontology, which defines basic concepts of complex of the problem domain. The latter one serves also as the mentioned meta-schema for the knowledge representation (see Figure 1).

The approach is discussed through this chapter by investigating a textual modeling language, called *Complex Event Description Language* (CEDL), which aims at defining complex event patterns (Dávid, 2011).

Patterns are built up from (1) generic parts of the event processing algebra, (2) structural data of the context domain (e.g. event sources), and

Figure 1. High-level overview of the approach presented in this chapter. Different DSLs are derived from the same generic language by adding context-specific knowledge. The generic language defines elements for the problem itself, which are reused by the context-specific languages.

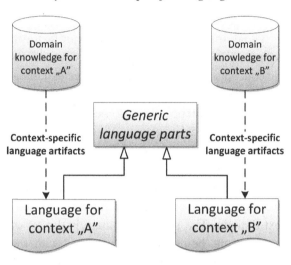

(3) high-level information of the context domain (e.g. we want to describe events of a critical CPU load; here, the interval of the load being critical is a context domain information, as well as the CPU as event source and its load as measurable quantity). By facilitating this modeling layer, a bridge between high-level requirements and the executable code is created, thus the implementation tasks can be partially automated. Moreover, this approach also provides semantics for the workflow, since the models are built up using the domain knowledge; this is a great support for model transformations or code generation as well.

Traceability is a key feature in dependable system design: high-level goals must be properly implemented and there shall be no superfluous implementation exist, since with the growing codebase the number of probable errors increases as well. Traceability in its simplest form can be supported by automated mapping of logical elements of high-level criteria onto the modeling level in form of atomic event patterns, where those can be assembled into complex event patterns implementing the high level criteria. A much more extensive way to support traceability is the automation of assembling complex event patterns from atomic event patterns. With an ontology using the appropriate metascheme, this step is also feasible, though it will not be discussed here, since it is not strictly related to the topic of this chapter.

Ontologies offer the capability to use inductive reasoning; that is, a set of concrete entities can be evaluated based on a set of premises or propositions. Since after creating the SKB no manual coding just automated mappings are employed, this solution also offers the possibility to validate a knowledge base instance against special high-level goals: standards and law regulations, if these are formalized with the given meta-schema as well.

The Approach

Through this chapter, the previously discussed concepts and methods are demonstrated by the practical example of a model-driven approach for processing complex events. First, a modeling language is presented in the necessary details to emphasize the key points of the language being extendable; after that, the concrete technique for extending the language is discussed. Finally, the conclusions and the possible future directions are drawn.

A GENERIC LANGUAGE FOR MODELING COMPLEX EVENTS

In this section, we introduce a framework for modeling complex events. The core part of the framework is a domain-specific modeling language, called *Complex Event Description Language* (CEDL). We will investigate its structure and features from the usual three aspects: (1) the abstract syntax, (2) the concrete syntax, and (3) the semantics.

We consider CEDL to be a *generic* DSL, since different context domains can be targeted without changing the metamodel of the language (IT infra-

structure monitoring, stock trading, etc.); instead, the language must be extended, customized for the given context, using automated procedures discussed later. Customization means that whilst changing the context domain, the syntax changes as well, because certain language elements are defined in the way to be reusable, while others to be replaceable.

Introduction to Complex Event Processing

Complex Event Processing (CEP) aims at processing large amount of distributed data triggered by event-based systems, usually with a slight delay and processing time. Processing events actually means pattern recognition on the event stream, published by the event sources.

Event pattern recognition requires the patterns to be defined prior to execution process; usually a declarative notation is used for this purpose with strict algebraic semantics in the background, also called *process algebra*. It defines an axiom system for describing the different cases on the event stream, e.g. the type of events, the precedence and concurrency operators for event-event pairs, the definition of two events being concurrent, etc.

Each CEP engine ships with a language base on a custom process algebra. These, of course are very similar, however they implicitly contain information about the processing- and event representation model of the given platform. The event representation model is the actual metamodel which the event instances must be conform to. The most common way in practice to represent event instances is using event beans (in case of Java: *event POJO*s) and XML-based notation – we will stick to the former one in this chapter. CEP platforms define different meta-schemas, since there is no common standard for this purpose.

The certain benefit of employing CEP is the ability to define complex patterns, which would not be trivial without the above mentioned algebra. Consider the following example.

Example

Let us assume a large-scale IT infrastructure which needs to be monitored. The specification requires that *if the CPU load for the Webserver A reaches the critical limit and the backup is not ready to take over the job, meaning that either its CPU load reached the critical limit as well, or there's not enough space on the drives, or it's just simply not available, a warning shall be sent to the administrator.*

Although CEP engines offer a framework for defining event patterns and stream processing logic, do not facilitate the high-level design of them. This results in low-level implementation tasks, meaning that the requirements and domain concepts of the targeted problem are not available for the developer, since these are high-level information and the process algebra does not support including these kinds of elements into the pattern language.

Consider the previous example and notice the elements derived from the domain knowledge, like the critical limit of CPU load, or the insuf-ficient free space on the hard drive. It's easy to see how a difficult task would it be to describe these concepts with a low-level algebra of limited elements. Furthermore, conventional approaches would yield ad-hoc models, because of the verbose textual format of requirements instead of a formalized one.

A modeling layer between the requirements and the implementation offers a viable solution to overcome this problem, since it can be supported with domain-specific information and after building the models, executable source code and configuration can be generated for different platforms.

The Modeling Framework

Figure 2 illustrates our approach for modeling complex events from a high-level aspect (Dávid, 2011).

The modeling is accomplished by a textual modeling language, called *Complex Event Description Language (CEDL)*, which conforms to a *General metamodel*. This metamodel defines a

Figure 2. A high-level view of the approach discussed in the chapter

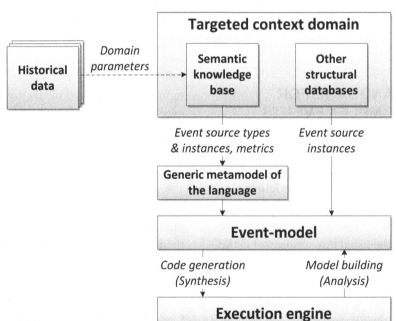

set of terms used in modeling process, such as *Event, ComplexEvent, Source*, etc. These concepts are context-agnostic, i.e. have the same semantics over different context domains and remain intact as the context changes.

Prior to actually defining an event pattern, the general metamodel must be extended by the domain knowledge of the *Targeted context domain* (event source types, metrics, etc.) as well as by the *Structural data* of the domain (the concrete event sources), which will be referable in the patterns of the *Event-model*.

Structural data serves as a *configuration* for the language, after it is extended by the context-specific domain knowledge. The domain knowledge is stored in the *Semantic Knowledge Base* (SKB) which can contain the structural data as well. There is also a possibility to derive structural data from alternative sources, like Configuration Management Databases (CMDB) for example.

Domain parameters refer to a set of values, which can affect the business logic. Consider "the critical CPU load on a server" as an example, which can vary in a broad range and clearly depends on the domain, since no default value can be assigned here without being familiar with the exact problem. Moreover, this definition may change over time. This kind of data can be either determined by a domain expert, or collected by monitoring the functioning system. In the latter case, of course, intelligent analysis needs to be performed on the *historical data*, prior to the derived information being fed back.

At this point, high-level requirements can be modeled, i.e. a modeling layer has been established which executable code and configuration can be generated from for the appropriate *Execution engine*. In our research, we tested our solution on two execution engines: on the open-source CEP platform Esper and on relational databases. In the first case, platform-specific code and configuration elements were generated; in the latter

one, schema definitions and insert statements were produced to store data of recognized event patterns in a special schema.

Domain standards can be interpreted as special domain parameters, since they determine how systems shall behave. If standards can be formalized properly and included into the Semantic Knowledge Base, special event patterns can be generated, which can serve for maintaining the proper system functionality according to the standards.

Consider the following example. A standard prescribes that every service entity (hardware, middleware or software) requires at least one hot backup. If the semantics of the knowledge base and the modeling language allow, this requirement can be formalized and the related extensions for the modeling language can be produced. Since the properly formalized requirements can be used as premises by reasoning tools (usually shipped with ontology editors), this approach also supports assessing the traceability between the requirements and the model. The reasoning shall be applied for a concrete instance of the configuration model, i.e. a conformance check is being executed against the requirements.

Assumptions

We implicitly assumed that the underlying system actually exists and provides measurement points. In our opinion, this assumption is not strong at all, considering that event-driven architectures are common in practice nowadays and even if this assumption would fail, there are several viable options for instrumentation, moreover: there are published solutions explicitly for complex event processing (Gönczy, Csertán, Urbanics, Khelil, Ghani & Suri, 2011).

Another assumption is that building the Semantic Knowledge Base is feasible. The implementation and technology for the knowledge base, however, is not tied: though we use ontologies for

Box 1.

```
Event CPULoadCritical {
    source Server1
    PercentageMeasurement CPULoad Minimum 0.9
}
```

this purpose, an EMF model or a database could work as well, just to mention the most common industrial solutions.

Example

To understand how the modeling language is assembled from a generic and a context-specific part, consider the example in Box 1, taken from the context domain of IT infrastructure management.

This event pattern describes a situation where the CPU load of the resource Server1 is at least 90%.

Here, the emphasized keywords (*Event, source, PercentageMeasurement, Minimum*) are parts of the *general metamodel* of the language. Meanwhile the event source *Server1*, the measurement *CPULoad* and the value of the relevant case of the measurement *0.9* are defined by the domain (e.g. the underlying system, the feasible measurements, high-level goals, such as SLAs, for example) and come from the Semantic Knowledge Base. In some cases we will refer to these elements as *domain information concepts*.

In order to model a complex event, we need at least another simple event pattern. The example in Box 2 is very similar to the one in Box 1, but instead of an at least 90% load of the CPU, it

triggers when there are less than two available backups for the *Server1* server instance.

Here, the *AvailableBackups* type yields the number of backups available for the server. As a measurement, it might look a bit different as e.g. the load of the CPU, since in this case a logical relationship between two server elements must exist, which semantically describes the backup nature of one for another; but with the suitable instrumentation, this can be handled just as trivial as the former case of CPU load (Gönczy et al, 2011).

The support for connecting context-specific domain knowledge to generic event patterns will be discussed in a later section.

In Boxes 1 and 2 we defined atomic events patterns: each of them originates from a single source and depicts a simple phenomenon associated with the corresponding source. In contrast, in Box 3 there is a *complex* event pattern, which depicts the case of two atomic events being concurrently observable for at least 30 seconds.

Looking at the pattern, it is really concise, since our modeling language was intentionally designed to be as concise as possible, e.g. by simplifying the semantics of concurrency compared with Allen's interval algebra (Allen, 1983).

Box 2.

```
Event BackupProblem {
    source Server1
    ScalarMeasurement AvailableBackups LessThan 2
}
```

Box 3.

```
ComplexEvent CriticalServer {
        CONCURRENT_T(CPULoadCriticalBackupProblem; T: Minimum 30)
        action {
                    Action of Type sendWarning
        }
}
```

When a pattern is matched, some kind of a reactive action needs to be executed. Modeling right-hand sides was a marginal aspect of our research, since executing a reactive action might be as easy as e.g. calling an appropriate service in a service-oriented architecture. The keyword *Action of type* handles triggered action in a modest way: here, the previously defined *sendWarning* action is executed, which dumps a message into a log file.

Abstract Syntax: The Complex Event Metamodel

Abstract syntax describes the taxonomy and relationships of model elements, including well-formedness rules as well. It is important to facilitate extendibility of the language in this early stage already, since the appropriate interfaces for language extensions are engineered on the metamodel level.

As Figure 3 shows, an event model which describes event patterns must conform to the general metamodel of the language. This means an *instance-of* relationship between the two levels. In the previous section it was clarified, that a simple instance of the general metamodel would not be enough to actually define patterns, since the model does not contain e.g. event types. These context-specific features come from external sources (e.g. the Semantic Knowledge Base). To facilitate the extending of the language in this way, the proper interfaces are needed to be defined. Extending the language means that the same abstract syntax (or metamodel) element might differ on concrete

syntax (or model) level. The interfaces for context-specific elements are denoted by white boxes. The grey ones are the context-agnostic elements. E.g. the type *Event* can be simply instantiated from the metamodel; but when instantiating a *Source*, it is required to previously extend the language by the measurements of the domain.

A vital part of the metamodel is the *SuperEvent* element, which is abstract, thus it cannot be instantiated; instead, one of its subclasses can be used to describe different kinds of events. The *Event* depicts a simple phenomenon; while on the other hand, a *ComplexEvent* represents a set of *SuperEvents*, which are combined using an event processing algebra. This algebra is used for expressing relations among *SuperEvents*, e.g. orderings, parallelism, etc. The algebraic elements are defined in the *Operator* association class. Also, *Aggregations* are slightly different kinds of operations; they support summing, averaging, etc. A *SuperEvent* may also refer to Aggregations.

There are use-cases when the designer needs to prescribe patterns for all the instances of a given type of *Source*. It is also a frequent use-case, when *Events* are very similar – they just differ in a measurement type. In the latter case, using a classic Object Oriented Programming (OOP) approach, one would resolve the issue with interfaces and/or abstract superclasses. This is expressed by the *EventStub* in CEDL, as discussed in a few sections later (Design patterns). *EventStubs* can be implemented by *Events*.

EventStub also serves for defining measurements without assigning values to them – in this case the implementing Event is obligated

Figure 3. A complex event processing metamodel

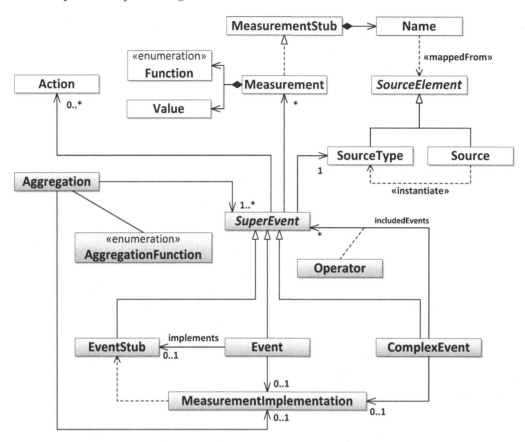

to implement the measurement as well. This is accomplished by defining a *MeasurementImplementation* block and enumerating all the necessary assignments.

As the context-interfaces (the interfaces for context-specific knowledge; denoted by white boxes) are concerned, the *Measurement* is the central element here. It describes a feature that can be evaluated by measurements over the underlying system. As by definition, this type plays an important role in building event patterns, since pattern matching is executed considering the measurable data (and the event source) in the processing phase. Measurement is an abstract entity, which can be instantiated either as numeric measurement (*PercentageMeasurement* or *ScalarMeasurement*) or as a literal measurement (*LiteralMeasurement*).

We think, these types cover the practical cases of using measurements and do not complicate over the language.

The rest of the context-specific metamodel parts will be introduced through a previously used example measurement: PercentageMeasurement CPULoad Minimum 0.9. This pattern means that the related event will be recognized by the CEP platform if: (1) it contains information labeled "CPULoad" and (2) has a value of between 0.9 and 1.0 (Since we are talking about a percentage value, the upper limit of 1.0 seems practically valid).

Every *Measurement* has a *Function* and a *Value*. The *Function* defines arithmetical comparator functions, like "more," "less," "equals," when used with *PercentageMeasurement* or *ScalarMeasurement*; and defines string operators, like "contains," "equals," "starts with," when using *with LiteralMeasurement*. Functions enable comparison of the measurement with the *Value* type. In our example, the *Function* is keyword *Minimum*, while the *Value* is 0.9. Using a proper *Function* and *Value*, one can easily define which interval of the given scale (here: the load of the CPU) is important.

Since the Measurements are derived from the context-specific domain knowledge and their names appear directly in the language, a proper naming is desirable. *MeasurementStub* is an interface for this purpose, which must be implemented by every *Measurement*. It contains a *Name* type reference, which is a single string, mapped from the observed source element (*SourceElement*). This mapping is usually done in the phase of instrumentation, or manually when building the semantic knowledgebase.

The *SourceElement* is an abstract type; the *SourceType* and the *Source* derives from it. There is an instantiation relation between *SourceType* and Source. Usually this kind of relation is handled on different meta-levels of the model stack: the instantiated element is located in a higher abstraction layer. The reason why these elements yet appear on the same meta-level is that both of them are utilized in the language's concrete syntax: one can define event patterns for a concrete *Source* (e.g. Server1), but in some cases it is also possible to bind a pattern to a *SourceType* (e.g. WebServer).

The language also contains Action as a context specific element.

Concrete Syntax: Language Essentials

This section presents the most essential elements of the Complex Event Description Language (CEDL). CEDL has a textual concrete syntax, which was chosen instead of graphical notation because it was a better fit to practical use-cases (UML extensions were also investigated, but preciseness and intuitive modeling could not be achieved using standard profiling mechanism). The language was designed using the Xtext framework which employs the Extended Backus-Naur Form (EBNF) to describe Context-Free Grammars (CFG). Nevertheless, presenting the whole metasyntax would not fit the chapter, thus we aim at unfolding only the most important elements of the language.

In a later section, a suitable Eclipse-based IDE (the *CEPWorkbench*) will be presented for building CEDL-based models with features like code completion, validation and on-the-fly code generation.

The language bears the marks of the object oriented paradigm and the Java programming language. The top element of a model is the package: model parts are organized into packages, which can be imported into each other. This facilitates organizing the model into many separate physical files, which helps the logical grouping of model parts.

There is a recommended initial structure for every CEDL-based modeling project, which consists of three model parts:

Box 4.

```
SourceType WebServer extendsPhysicalElement

Source WebServer WebServer1 {
        name: CPULoad, physicalComponent: "CPU", type: %;
        name: FreeMemory, physicalComponent: "RAM", type: scalar;
}
```

- **Topology:** This model part defines sources and basic measurements tied to them.
- **Defaults:** Contains predefined structures, like relationships among sources or action types.
- **Events:** This is where the event patterns are defined using the previous two model parts.

Model part reusability is achieved by reusing package imports as presented by Java. In order to be importable into another package, a model part shall be encapsulated within a *package* block.

Defining sources is the first step to define patterns. This consists of two phases: (i) defining the source type by extending the appropriate CIM (Common Information Model) metascheme element; (ii) defining the event source by instantiating the source type.

The definitions in Box 4 are located by default in the Topology model part. Basic patterns can be defined by the domain expert (Box 5).

Complex event patterns are built from basic patterns using timeliness operators. In the basic case, this implicates the usage of the operator *EXISTS()*:

```
EXISTS(WebServerProblem)
```

If this definition by the operator gets omitted or the atomic event pattern is not used in at least one complex event pattern, no event POJO from the pattern will be generated. There are additional complex event operators, but these will be discussed later in the section of Semantics.

Implementing event stubs is accomplished via the *implements* keyword (Box 6).

The same event stub can be of course implemented by multiple event patterns. This practical, when e.g. the concrete events are similar and unnecessary redundancy shall be avoided; or when tracking of some kind of change is desirable on a given source. The measurements defined in an event stub can be declarations without assigned value or full-featured measurement definitions. The former ones must be *implemented* in an implementing Event (using the annotation @*ImplementMeasurement*), while the latter ones can be optionally *overridden* (using the annotation @*OverrideMeasurement*).

Since we assume the generated configuration to be loaded and used on distributed systems, it is also desirable to assign some deployment metadata on the model level. For example, when event POJOs are generated, a deployment-dependent attribute is the package name and the file name of the POJO.

This information is handled by *namespaces* in the model.

It is also possible to define negative, restrictive patterns using the keyword *not* with a similar semantics to the *NOT* keyword of SQL.

CEDL defines two fundamental design patterns for event based modeling, these are the *Event stub pattern* and the *Change tracking pattern*.

Box 5.

```
Event WebServerProblem {
        source WebServer1
        PercentageMeasurement CPULoad Minimum 0.9
        ScalarMeasurement FreeMemory Maximum 256
}
```

Box 6.

```
EventStub CPULoadOnWebServer {
        sourceType WebServer
        PercentageMeasurement CPULoad
        ScalarMeasurement freeMemory 512
}

Event NormalLoad implements CPULoadOnWebServer {
        source WebServer1
        @ImplementMeasurement {
                PercentageMeasurement CPULoad Maximum 0.5
        }
        @OverrideMeasurement {
                ScalarMeasurement freeMemory 1024
        }
}
```

The Event Stub Design Pattern

Problem

1. Multiple events logically extend the same event by adding new parameters to measurements or overriding the original ones. One would like to define measurements in the ancestor without assigning values and limits to them, since the inherited events would refer to custom values.

2. The structure of the events is known just partially, or the structure of the inheriting events is identical just partially. An example can be an event where an intensity measurement (i.e. number of requests within a time interval) reaching a threshold is an obligatory part whereas additional conditions can also be added.

3. One would like to define an event pattern for the type of an event source, instead of the concrete event source, in order to allow the pattern either for *all* of the instances, or *any* of the instances.

Proposed Solution

Reusing the object-oriented concepts, the problem is solved in one half by using the *interface* and in another half by using the *superclass*. Case (2) is the clear case of the superclass; case (3) is the clear case of the interface. Case (1), however, utilizes both of the concepts: one would like to define some kind of an ancestor (superclass), but it is needed to leave some attributes without assigning values to them (interface).

The solution is delivered by the event stubs. (Also denoted as *EventStub* in some cases, according to the related element in the metamodel of the language.) By the formal definition, an

EventStub can be derived from an *Event*, using truncation, where the truncation is one of the following procedures.

- **Structure-Truncation:** Some structural elements are removed from the Event, e.g. attributes, actions or measurements.
- **Value-Truncation:** The value of a structural element of the Event is removed, thus remains only a definition referring to the structural element. The implementing event patterns is obligated to assign a value to the element.

The association between an *EventStub* and an *Event* is denoted as implementation, using the keyword implements, since we find this relation closer to interfaces as to superclasses.

By definition, the event stub can refer only to the metalevel of *SourceType*; using concrete sources is not allowed, since they are context dependent. In the implementing class, however either a concrete Source conform to the *SourceType* shall be used, or the keywords *FORALL* or *OFTYPE* which specify the event pattern for every instance of the source type and any instance of source type, respectively. In the former case, the event pattern must be present on every instance; in the latter one, there must be at least one instance of the source type.

Consequences

Using the pattern eliminates the undesirable redundancy on the model level.

The Change Tracking Design Pattern

Problem

One would like to track events which depict changes in a measurement parameter of a given event source.

Proposed Solution

An event stub is defined for the event source which defines the changing parameter. The stub is being implemented by two event patterns: the pre-condition event pattern and the post-condition event pattern. Finally, the two event patterns are tied together in a complex event pattern via a *FOLLOWS* operator, optionally associated with a timeframe.

Consequences

Using the design pattern, changes in a source can be tracked – of course only the ones, which are defined in the event stub as measurements.

Semantics: An Event Processing Algebra

Semantics do not strictly belong to the language, but it facilitates the interpretation of models, since they ensure that every model element has the same meaning for everyone who deals with the model, thus establishes real model-driven development. As the vital part of CEDL is the set of complex event operators, in this section we will provide the semantics for unequivocal interpretation of them.

From the aspect of duration, we distinguish two kinds of events: point and interval events.

Points have no real duration in the dimension of time. Their important time-related parameter is the timestamp of appearance. Intervals, however, possess a duration parameter, which is the amount of time elapsed between the appearance and disappearance of the event.

CEDL defines three basic complex event operators; each one can be annotated with an additional time window parameter – this means six operators in all. Table 1 summarizes the operators by presenting the signature, the applicability for point/interval events and defining the meaning of the operator.

Table 1. Complex event operators of the CEDL

Operator	Event	Meaning
EXISTS(*params*)	point, interval	Parameters should exist.
EXISTS(*params*).timewin(τ)	point, interval	Parameters should exist within a time window with length of τ.
FOLLOWS(*params*)	point, interval	Parameters follow each other.
FOLLOWS_T(*params*; τ)	point, interval	Parameters follow each other regarding a time window with length of τ.
CONCURRENT(*params*)	point, interval	Point events appear in the same time. Interval events are happening concurrently.
CONCURRENT_T(*params*; τ)	interval only	Interval events are happening concurrently regarding an interval with length of τ.

There are, of course, different approaches for defining event processing algebras; the two most significant are *Allen's interval algebra* (Allen, 1983) and the *Event Detection Algebra* by Carlson (2004). However, none of these algebras defines executable semantics, that is, though they can be used for describing intervals and event patterns, the event pattern detection cannot be implemented directly. In contrary, event algebras of event processing platforms, like Esper (Bernhardt & Vasseur, 2007) or Drools Fusion are executable (see Figure 4).

Figure 4 summarizes the basic relations among these three algebras mentioned.

In general, EDA is capable to describe the whole Allen's algebra, as well as the "event-core" of the CEDL's algebra. However, it falls short when dealing with context-specific extensions.

Allen's algebra introduces thirteen relations between two intervals, which is a significant refinement of the CEDL's concurrency operators. On the other hand, CEDL was designed to capture practically relevant use-cases in order to be easy-to-use, that's why concurrency is handled in a simplified way and timing operators were added.

It is clarified now, how context-specific extensions for a generic language work; the next step is the automation of this process, which aims at a higher abstraction level in the MDA process.

Figure 4. The relation of three event algebras: the event detection algebra (EDA), Allen's interval algebra, and the complex event description language (CEDL)

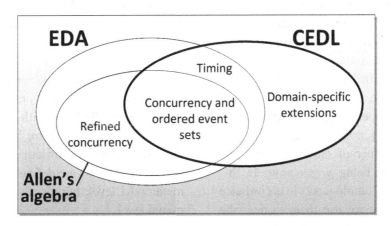

CONTEXT SPECIFIC LANGUAGE-EXTENSIONS USING ONTOLOGIES

When the metamodel of a language supports its extensibility, as we saw in the previous section in the case of CEDL, the question arises: how the context-specific knowledge should be stored and retrieved for further usage? A broad range of technologies are viable here; we chose ontologies since they are essentially intended for storing *knowledge*. The *Semantic Knowledge Base* (SKB), as addressed in our framework, is implemented as a multi-layer architecture of ontologies.

Technical Background

The term "ontology" was coined by medieval philosophers, meaning the study of existence, existing things. Information science reused this expression to describe sets of concepts, a knowledge base of a given domain, created by domain experts and knowledge engineers. Gruber (1993) defined ontologies as "formal explicit specifications of a shared conceptualization", which consist of a vocabulary and a taxonomy, or the so called *A-boxes* (assertion components) and *T-boxes* (terminology components).

According to Aßmann and Zschaler (2006), *an ontology is a model shared by a group of people in a certain domain* (p. 255).

The fundamental difference between models and ontologies is how one handles unspecified things. Traditional models (and databases) work with the *closed-world assumption*, meaning what is not explicitly specified, it does not exists. In contrast, ontologies assume and *open world*, that is, anything unspecified actually *can exist* and hence, its inexistence cannot be stated.

Because of the open-world assumption, Aßmann and Zschaler (2006) also emphasize that the *descriptive* manner of ontologies shall be preferred instead of being *prescriptive*. In this chapter, the usage of ontologies is in accordance with this suggestion: we use them to describe high-level requirements of a domain, as well as structural information about observed underlying systems to facilitate efficient support for generative language design.

Happel and Seedorf (2006) described several use-cases where ontologies can help facilitate efficient software engineering. Using ontologies as *domain object model*s is one of them. The authors underline, in order to reduce costs and increase quality, the domain model shall exists in one instance and this shall be shared by the participants. They also state that automated mapping between the domain model and code are desirable, since this enables dynamic usage by other components.

The Semantic Knowledge Base

The Semantic Knowledge Base (SKB) is intended for storing domain knowledge which is mapped onto the language level. This is in accordance with (Happel & Seedorf, 2006) as the SKB serving as a domain object model and this model being utilized by automated mappings onto a more concrete, executable level.

Binding semantics to domain knowledge establishes higher level of automation, since this way hard-coding the semantics into the mapping logic can be avoided. Without that, one has to define e.g. which elements of the domain data model and the language model belong logically together in order to execute the mapping. (For instance when mapping IT infrastructure resources into sources and source types.) With the Semantic Knowledge Base containing not just the raw context domain model, but also the semantics, the mapping gets straightforward.

Domain data without semantics can be retrieved from e.g. Configuration Management Databases (CMDB), or with a simple discovery over the underlying system. This data, of course, can be found in the knowledge base as well, since several constraints are defined not just on higher metamodel level, but also on concrete instance model level.

It is important to mention: semantics can be depicted with conventional modeling solutions, like databases or EMF models as well. Ontologies are, however, explicitly designed for describing knowledge, thus they fit better into the approach.

With the SKB, we expect a write-once-use-everywhere mapping, i.e. it shall not depend on the structure of the SKB. It is obvious that different context domains and problems would yield different ontology structures and metamodels if no standardization approach was used.

The idea here, as well as in the case of the language, is to separate changing parts from the fixed ones. This is achieved by a layered architecture. The top level consists of one or more *Context domain ontologies*, while on the bottom layer the *Generic ontology* can be found. The former one describes the context domain, latter one describes the concepts of complex event processing with a fixed set of concepts and interrelations among them; hence it is basically driven by the metamodel of the modeling language (CEDL). There is one single Generic ontology reusable for different context domains, which all Domain ontologies shall be mapped onto. Since there is only one instance of the Generic ontology, which remains intact even if context domains change, fixed mapping logic between the SKB and the language can be realized (A similar approach was also proposed in Happel & Seedorf, 2006 and Aßmann & Zschaler, 2006.).

The concepts defined in the Generic ontology are presented by Table 2.

This is a hierarchical representation of knowledge. *SourceType* defines event source types, just as in the metamodel of the CEDL. *Quantifiers* depict measurable values, e.g. throughput, load. *Metrics* are compound elements, since they link a *Qualifier* with a *SourceType*, i.e. define measurable variables for *SourceTypes*. *Range* defines plausible boundaries of the *Metrics* – these two types are linked together in the *Qualifiers* (see Figure 5).

In order to allow depicting complex relationships among event sources elements, the *ComplexGroup* and the *ComplexRelationship* elements are introduced. The *ComplexGroup* is an unordered set of event sources; e.g. a cluster consisting of a certain instances of Web servers and databases. The *ComplexRelationship*, however, is a directed association among event sources. Every *ComplexRelationship* has a target (or many targets) and a source element. To explain the concept with a concrete example, consider a *backup-for* relationship. It is important to distinguish whether the event source WebServer1 serves as a backup for the event source WebServer2, or inversely. If event source WebServer1 is the backup, the

Table 2. The elements of the generic ontology

SourceType	**Defines an event source type.**
SourceMetaType	*Metatype for the SourceType. (Does not belong strictly to the model.)*
Quantifier	A measurable variable.
QuantifierMetaType	*Metatype for the Quantifier. (Does not belong strictly to the model.)*
Metrics	A Quantifier applied for a SourceType.
Range	A range of the possible values of a Quantifier.
Qualifier	A Range applied for Metrics.
ComplexRelationship	A directed relation among SourceTypes.
ComplexGroup	An undirected relation among SourceTypes.
Constraint	A constraint applied for a ComplexRelationship or ComplexGroup.

Figure 5. The hierarchy of generic ontology elements

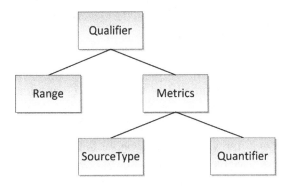

backup-for relationship's source will be the event source WebServer1 and the targeted event source will be the element WebServer2.

SourceMetaType and *QuantifierMetaType* are applicable for *SourceTypes* and *Quantifiers* respectively; and take their value from sets {*PhysicalElement, LogicalElement, Service*} and {*Percentage, Scalar, Literal*} respectively.

As a demonstration of the viability of the layered approach, Table 3 presents two domain ontology examples of different contexts.

As presented in the Stock market example ontology, there are optional parts of the model. These are the *ComplexRelationship*, the *ComplexGroup* and the *Constraint*.

The *Range* element serves for describing concrete value requirements. By definition, if the lower or the upper level is unset, a scalar measurement is generated with a *Maximum* or *Minimum* comparison operator respectively. (E.g. the critical CPU load on the Server1 is assigned with a range of (0.9,-) – this means, no upper bound is specified, therefore the CEDL measurement *CriticalCPU-Load* with a Minimum of 0.9 is generated. The concrete type of the measurement (i.e. whether the measurement is a PercentageMeasurement, a ScalarMeasuerment or a LiteralMeasurement) is determined by the QuantifierMetaType, which is an element in the ontology and can be assigned to Quantifiers to describe its concrete nature.

For assembling the Semantic Knowledge Base, two scenarios are distinguished: (1) using an already created ontology; and (2) building a new SKB from scratch. (In practice, however, neither one of these use cases occurs in this form; practical cases usually employ both scenarios.) Of course, both approaches have their own advantages and drawbacks.

Existing ontologies typically use a metamodel that differs from the Generic ontology, thus certain integration steps are required in order the context domain ontology to be mapped on the generic one. Refactoring the context domain ontology or using intelligent mappings (e.g. via an integration ontology) could be viable solutions for the problem. On the other hand, however, this case implicitly assumes that all the domain knowledge is already collected into an ontology and a certain expertise is available at the user-side.

Table 3. Two examples for context domain ontologies using the generic ontology

Generic ontology element	Context domain: IT infrastructure monitoring	Context domain: Stock market trading
SourceType	Webserver, Database_server...	Stock, Bond...
Quantifier	load, reliability...	variance...
Metrics	m1(Webserver, load)	m2(Stock, variance)
Range	r1(0.9, 1)	r2(0, 0.1)
Qualifier	critical(m1, r1)	reliable(m2, r2)
ComplexRelationship	backup(Webserver, Webserver)	-
ComplexGroup	cluster(Webserver, Database_server)	portfolio(Stock, Stock)
Constraint	c1(cluster, at_least_one_Webserver)	c2(portfolio, suboptimal)

On the contrary, building the SKB from scratch implies the task of collecting domain knowledge and organizing it into an ontology which might be a significant effort in real life scenarios.). The Generic ontology facilitates this task by defining a metamodel for the context domain ontology. Considering the moderate complexity of the generic ontology, with an appropriate ontology design tool (e.g. Protégé) this task should not be elaborate at all.

The design from scratch starts with importing the Generic ontology into the new ontology – this will provide the appropriate metamodel elements for the elements contained by the context domain ontology.

Mapping the Knowledge on the Language Level

After creating the SKB, context-specific knowledge is ready to be mapped onto the language level. This is a two-step process described in the following.

First, *SourceTypes* are generated with the name and the extended source metatype denoted. From every *SourceType* element in the ontology, a source type definition is generated into the *topology.cedl* file.

Following this step, the concrete source types of the context domain are available for use in the language: the parser will recognize them as special language elements and the IDE will include them into the code completion proposals when the context requires it.

This is already a great help for the developers dealing with event patterns, since the code generated from this model contains context-specific metainformation which, therefore, can be omitted from the model-text transformation logic.

The second step of mapping the context knowledge onto the language level is creating event stubs which depict the constraints defined in the knowledgebase with Quantifiers, Qualifiers, Metrics, etc. Here, the previously presented *EventStub* design pattern is reused. The generated stub can be implemented later by *Events* (see Figure 6).

Looking at Figure 6, the structure of the generated event stub is pretty straightforward. The name consists of the quantifier, the type of the source and the qualifier; e.g. *ThroughputWebserverLow* or *TransactionfailurecountDatabaseHigh*. This name depicts the intended goal of the event stub in a concise way. The *sourceType* parameter is derived from the *SourceType* element of the ontology. The *characteristic* parameter depicts the functional or extra-functional manner

Figure 6. Mapping of the knowledge to the language level in form of event stubs

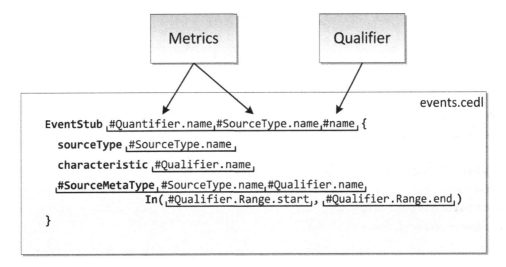

of the requirement as additional information. The type of the measurement i.e. whether it is a percentage, scalar or literal, is determined by the related *QuantifierMetaType*. Its name consists of the source's type and the qualifier; e.g. *Webserver-Critical*. Finally, a range, scalar, percentage or string with the appropriate operator is generated.

From the example in Table 3 the code in Boxes 7 and 8 is generated for the "IT infrastructure monitoring" context, thus creating the *CEDL4IT* language.

Of course, both steps can be accomplished also by manual coding. Omitting the SKB and do these two steps of modeling manually usually fits well in the case of small models, e.g. when there is only one dedicated event source, or the requirements are lack of available metrics.

Technically, the mappings are implemented using one of the many appropriate APIs for ontology querying, called OWLAPI. The API is invoked from a Java source code; the output is generated by model-to-text transformations, using Xtend. (This framework is also used for generating source code from CEDL models.)

Since this is a key step in the development process, our tooling, the *CEPWorkbench* also contains appropriate functionality for simply importing context domain ontologies and to generate language artifacts.

Summary

This section presented how context-specific domain knowledge shall be recorded in order to be reusable for extending a generic DSL and customize it for an arbitrary context domain. The key was the synchronized metamodel of the generic language, i.e. the problem domain and the

Box 7.

```
SourceType Webserver extends PhysicalElement

EventStub LoadWebserverCritical {
        sourceType Webserver
        characteristic critical
        PercentageMeasurement WebserverCritical In (0.9, 1)
}
```

Box 8.

```
In addition, relationships and groups are generated as well:

ComplexGroup {
        type cluster
        topologyElements w1 d1
}
ComplexRelationship {
        type backup
        sourceTopologyElement w1
        targetTopologyElement(s) w2
}
```

general ontology. This created a semantic bridge on metamodel level and facilitated the mapping of knowledge from various contexts onto the language level.

The task of the developer is now reduced to defining implementations for the event stubs, i.e. event patterns and to organize them into complex event patterns if needed. From the models executable platform-specific code is generated; therefore, the implementation tasks are completely eliminated and so are the faults, failures and errors arising from the wrong implementation.

IMPLEMENTATION AND ALTERNATIVES

In this section we briefly present our Eclipse based tooling for modeling complex events and provide an overview about the possible alternatives regarding to modeling and establishing the knowledge base. A recommended architecture for integrating the generated code from the models is also presented.

The CEPWorkbench

After an appropriate modeling language has been designed, it is also important to support the work with feasible tooling. Our Eclipse-based tool, the CEPWorkbench offers a convenient way for modeling.

We chose Xtext (Efftinge & Völter, 2006) for implementing CEDL, a state-of-the-art solution for defining textual modeling languages. After defining the grammar the Xtext framework generates all the necessary components for language interpretation (e.g. parser, linker) as well as a basic IDE that uses validation, problem markers, intelligent code completion, etc.

For code generation we used Xtend, which ships with the Xtext. The IDE executes code generation, triggered by default on every save action. This means, that executable code is being created right in design-time, parallel to building the models.

To ensure model well-formedness, we implemented custom validation rules; these are being checked while modeling. Xtext provides basic validation against the rules inferable from the grammar. E.g. every model element must have a unique name. This is checked by the framework itself. More complex validation rules shall be implemented manually (e.g. to ensure that every measurement stub in an Event stub is properly implemented in the implementing Events.).

In order to facilitate efficient work, we also boosted up the content assist feature and added different wizards to the environment that support for example creating new projects or importing OWL artifacts from the Semantic Knowledge Base.

A graphical notation is planned to be added to the framework, which will support engineering some specific model parts, e.g. dynamics of complex event operators.

Alternative Solutions for Storing Structural Data

The semantic knowledge base is just one of the possible alternatives for storing domain knowledge, including information about the topology of the underlying system, i.e. structural data (Szatmári, Izsó, Polgár & Majzik, 2010). Though in our opinion this might be the best suiting option for storing *knowledge*, practical solutions usually lean on different techniques. One of them is the employing of Configuration Management Databases (CMDB). Of course, mapping the elements of models defined in CMDBs or other sources can be supported as well, but there are a few requirements to meet prior to mapping data to the language level and one has to count on certain limitations.

As an important prerequisite, the source types must be defined before traversing a CMDB model, since CMDBs does not necessarily define types in the meaning CEDL uses them. After that, the mapping can be executed. Mapping from an appropriately organized ontology is quite straightforward, since a set of metainformations is available, e.g. it is obvious which language elements shall be the ontology elements mapped into. This is not the case when dealing with solutions like CMDBs. One has to exactly know how to match elements of the two different metamodels and this knowledge will be hardcoded into the implementation of the mapping logic. In a model-driven approach this is an undesirable side-effect, which should be avoided.

In practical solutions, it is also rare to have an integrated way for handling potential event sources and define metrics with values for them. This drawback might raise two kinds of issues: (1) limitations in the generated language artifacts, e.g. no initial events stubs can be produced because of the lack of defined metrics; or (2) in order to avoid the previous problem and propose a correct domain model, a huge effort is used to integrate different tools, e.g. CMDBs with requirement analysis tools.

As this section emphasized, there are feasible techniques for utilizing ontologies as knowledge sources for language design. Is shall be considered, however, that not all of them are widely used in industry; e.g. close-world assuming modeling approaches are prevalent instead of ontologies.

Regardless the implementation, it is important to remember that the knowledge base is intended to support automation of development; thus establishing the appropriate metascheme is a must.

As model driven techniques evolve and tend to reach higher abstraction levels, suitably built-up knowledge bases ought to be the keys of successful modeling and development.

Modeling Alternatives

UML class diagrams are one of the most frequently used modeling notations. It seems to be a valid idea to investigate, how this notation performs as an alternative to CEDL. We assembled the metamodel of CEDL in UML, using stereotypes for class diagrams. Complex event patterns are defined in special, OCL-like additional model element. In order to properly define event patterns, one needs to define a pattern metamodel with class diagrams and then instantiate an object model from it. The class diagram defines e.g. the source types and measurements used in patterns.

Our experiments showed that the theoretical advantage of graphical models being easier to read was not beneficial at all dealing with models of average complexity (5-10 events, 3-4 complex events). Of course, this was a naïve implementation, meaning that with appropriate optimization the class diagram formalism could be more efficient.

Figure 7. Integration of the generated codebase with existing architectures

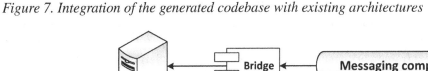

Integration with Existing Architectures

We also present a possible structure for integrating our solution into existing architectures. We stick to the context domain of monitoring enterprise IT systems, but the structure is also feasible in other contexts (embedded systems, e-trading, etc.).

After creating the models, the generated codebase is used as presented on Figure 7.

The monitoring agents collect data from the observed topology and transmit it to a message queue. We tested our solution using the open-source messaging middleware JBoss HornetQ. The queue is being polled for new events on the other end by a component called *Bridge*. It is a custom implementation, which has three fundamental functions:

- Collecting newly appeared event data from the queue;
- Assembling event objects using the *Generated codebase* (from the previously created models) in order to be executable on the employed CEP platform;
- Transmitting event objects to the CEP platform.

The Generated codebase consists of platform-specific artifacts, however they can be categorized into the following types, regardless the platform:

- **Event Descriptors:** The platform-level representation of the events. In a Java-based platform this is an even bean with the appropriate class properties, getters, setters and constructors; given a relational database, this might be a serialized identifier which will be inserted into the appropriate table with a unique id and a timestamp.

- **Event Filters:** Defines the filtering logic which is executed upon the event stream. This is where the actual pattern matching happens.
- **Event Listeners:** A listener for a given type of events. It is responsible for the executable action for an event or complex event.
- **Configuration:** Some platform-specific configuration might be needed when embedding the source code, e.g. SQL-based implementations might require schema or namespace definitions; the CEP-platforms' event processing parameters can be configured, etc.

RELATED WORK

van Amstel, van den Brand and Engelen (2010) examined several use cases of DSL design and identified four major influences on the evolution of modeling languages: the *problem domain*, the *target execution platform*, the *quality of transformations* between the model and the execution environment and the *quality of models*. Our research verified all these aspects while designing and implementing the CEDL. The results of the authors underline the main message of this chapter: investigating domain-specific languages needs to be taken to the next abstraction level in order to provide more efficient techniques for DSL design.

Spinellis (2001) presents best practices for DSL design. One of them is the "Language Extension" creational pattern, which describes the use case when a base language is extended by a DSL with additional features, like new data types, language block interaction mechanisms, semantic elements or syntactic sugar. Though the author recognizes the benefits of the pattern, does not provide any technique for implementation. In this chapter, we presented a feasible way for implementing the "Language Extension" pattern.

Ráth, Ökrös and Varró (2010) propose a mapping model between abstract and concrete syntax elements of domain-specific languages. The approach provides a feasible way to separate conceptual language elements from they representation and automated synchronization and traceability record between the models of abstract and concrete syntax. This kind of abstraction level is not adopted in practical DSL design yet, but points towards a possible future direction. Separating abstract syntax from concrete can help the appropriate design of language families, since context-specific customizations can be analyzed and described in a more clearer manner.

Aßmann and Zschaler (2006) compare the usage of conventional models and ontologies in the context of model-driven software engineering. The authors point out, ontologies fit better with domain knowledge representation, because of the open-world semantics and its descriptive manner. They also define *upper-level* and *domain ontologies* as the two fundamental types of knowledge representation. This is in line with our approach of establishing a semantic knowledge base, where a *generic* and several *domain ontologies* are merged together in a layered architecture. Despite the fact that there are slight differences between our knowledge base and the proposed solution of Aßmann and Zschaler, the Semantic Knowledge Base proved to be a feasible solution in our framework.

Happel and Seedorf (2006) describe use-cases for applying ontologies in software engineering lifecycle. The most relevant cases, related to our research are: using ontologies as the *object models* of the targeted domain, *integrating domain knowledge with modeling languages* and *component reuse*. All of these ideas are employed in our framework by the Semantic Knowledge Base depicting the domain concepts and facilitating knowledge reuse on language level in different forms, depending on the concrete language of a given language family.

Kalibatiene, Vasilecas and Guizzardi (2010) present a method for transforming ontologies into information processing rules. The approach is closely related with complex event processing, since the latter one is also an implementation of a rule system, thus the solution would fit well with CEP as well.

Decker, Grosskopf and Barros (2007) present a graphical notation for complex event modeling. Despite the graphical approach, BEMN holds numerous drawbacks compared to CEDL: it does not provide any code generation logic in order to be executable on an event processing engine; it focuses on the dynamics of the event patterns, but does not deal with the static structure; because of the previous point, it is not possible to extend the language with full domain-specific knowledge, thus the modeling cannot be aided with this kind of automation; finally, models are built manually, thus in case of a vast problem-space the model could be easily insufficient or incomplete, which defects are not trivial to detect. Nevertheless, the approach could serve as a good starting point for extending CEDL with graphical elements.

CONCLUSION AND FUTURE DIRECTIONS

In this book chapter, a novel technique has been presented for automating the design of domain-specific language families by separating the problem from the context domain. This enables targeting the problem itself with a generic, extensible language. The language is extensible due to its appropriately designed interface architecture. That is, defining general-purpose, common elements for the context domains which can be customized. In the case of CEDL, event source is an example for this.

This technique fits well to the overall model-driven approach, where platform-independent artifacts (here: the generic language) are getting merged with platform-definition models (here: the context-specific domain knowledge from SKB) to yield a platform-specific model (Aßmann & Zschaler, 2006, p. 252).

The chapter also presented proof of concept implementations and case studies as evidence for the theoretical results. The CEPWorkbench is an open source framework for modeling complex events. (See the project Web page: http://home. mit.bme.hu/~gonczy/cep.) Its central element is the Complex Event Description Language, which is an extensible generic language, without context-specific references. The framework allows automated customization of the language for a given context domain by generating context-specific artifacts for the language. Conceptually, the language serves as a modeling layer between the high-level criteria and the executable code. This provides the usual benefits of the model-driven approach, like traceability and elimination of syntactical errors in the executable code.

The generative approach in development of language families is greatly facilitated by the Semantic Knowledge Base (SKB). This layered architecture of ontologies is a feasible solution for depicting domain knowledge. The bottom-level Generic ontology is an appropriate collection of elements driven by the metamodel of the modeling language. This allows the top-level Domain ontologies to be mapped upon and this enables fixed mapping logic to be implemented between the SKB and the language regardless the targeted application domain. Whilst the mapping, the SKB (including of course the Domain ontologies) is traversed using the types defined in the Generic ontology.

This book chapter also provided methodological pointers for building of the SKB and for modeling complex events, e.g. by defining design patterns.

The next few sections designate some of the possible future directions for the domain-specific language design in general as well as for the modeling language presented in this chapter (CEDL).

Comprehensive Mapping of High-Level Criteria for Assessing Traceability

Well-defined model transformations between abstraction levels provide a traceability model for the software lifecycle.

We presented a way for mapping building elements of high level requirements onto the language level of CEDL in the form of atomic event patterns. After the mapping, the developer's task is to compose complex event patterns from the atomic ones. With an appropriate metamodel, the comprehensive mapping of requirements could be handled as well, i.e. mapping the atomic event patterns and automatically composing complex event patterns.

Possible Enhancements for the Complex Event Description Language

As discussed, the UML class diagram is not an efficient graphical notation for modeling complex events. This is because of two reasons: the problematic domain-specific extensibility of the core language (using stereotypes) and the low descriptive power, since despite the domain-specific extensions, the UML class diagram is still a general-purpose notation.

A real *domain-specific graphical interface* would be a useful supplement to the textual notation. There at least two aspects we plan to support with it: (1) the dynamics of complex event patterns; (2) the topology modeling.

To set up realistic measurement thresholds, a *feedback system* could provide valuable data for the SKB. Since only the relevant requirements manifest eventually in event patterns, it is sufficient to deal with feeding back data related to the event patterns. The logging of recognized patterns is easy to establish, since the actions, i.e. right-hand sides of the patterns offer suitable solution for that. The related data can be stored in relational databases and in a later phase intelligent data analysis to be executed on it. This might yield results like for example re-aligning measurement values. *Feeding the data back* can be executed later in batch-mode (offline), or real-time (online), when required. A necessary prerequisite here is to find the appropriate data representation in order to enable intelligent data analysis and automated data mining. This might depend on the domain, thus the data schema shall conciliate with SKB, and moreover, it is desirable to derive it from the domain knowledge.

In earlier sections, a metrics metamodel has been briefly introduced. Although it turned out to be sufficient for the current goals of this chapter, it is desirable to establish an *extended metrics metamodel* in order to meet requirements possibly emerging in large-scale industrial projects.

Such a requirement can be the definition of real measurement points of the underlying system and linking them to the high-level criteria. In this way, the whole "measurement stack" can be controlled, i.e. in a top-down direction hardware-level configuration can be controlled, meanwhile the bottom-up approach can aim model verification, refactoring and deriving domain parameters by projecting the measured data back. This obviously can facilitate utilizing the measured (historical) data and thus keeping the SKB up-to-date. It is also important to design a metamodel to be *extendable*. Expansive metrics- and measurement metamodels, e.g. presented by ASAM-MCD (Weichel, 2001) often prove themselves to be over complicated for the practical problems and hard to be managed on modeling and implementation levels. This problem can be precluded by using model extension methods like the one presented in this chapter, i.e. establishing a generic metrics metamodel and enable its extendibility in accordance with the context domain.

In practice, collecting and storing events is usually accomplished by employing relational databases, e.g. in form of log databases. As a possible future direction, relational databases can be targeted as the execution platform for the approach discussed in this chapter. Event pattern matching in this case is executed on database level. We assume an external (possibly already implemented) service inserting event records into one or multiple tables. Given a fixed schema for storing events, trigger definitions (on insert or update) can be produced from complex event patterns. An inserted or updated event triggers the execution and pattern matching is performed based on the signature of the event (name, id, etc.) and the timestamp of creation, if time windows are specified. As a prerequisite, we already produced insert statements for databases (implemented in MySQL) as action part of events defined in CEDL (According to van Amstel, van den Brand and Engelen [2010] it is not trivial at all, that a newly targeted platform won't require changes in the semantics.).

REFERENCES

Allen, J. F. (1983). Maintaining knowledge about temporal intervals. *Communications of the ACM*, *26*(11), 832–843. doi:10.1145/182.358434.

Aßmann, U., & Zschaler, S. (2006). Ontologies, meta-models, and the model-driven paradigm. In Calero, C., Ruiz, F., & Piattini, M. (Eds.), *Ontologies for Software Engineering and Software Technology* (pp. 249–273). Berlin: Springer-Verlag. doi:10.1007/3-540-34518-3_9.

Bernhardt, T., & Vasseur, A. (2007). Esper: Event stream processing and correlation. *O'Reilly OnJava*. Retrieved from http://onjava.com/pub/a/onjava/2007/03/07/esper-event-stream-processing-and-correlation.html

Carlson, J. (2004). *An intuitive and resource-efficient event detection algebra*. (Unpublished doctoral dissertation). Mälardalen University, Eskilstuna, Sweden.

Dávid, I. (2012). A model-driven approach for processing complex events. In *Proceedings of EDCC 2012 - Fast Abstracts & Student Forum*. Retrieved from http://arxiv.org/abs/1204.4428

Decker, G., Grosskopf, A., & Barros, A. (2007). A graphical notation for modeling complex events in business processes. *Business Information Systems, 4439,* 29–40. doi:10.1007/978-3-540-72035-5_3.

Efftinge, S., & Völter, M. (2006). oAW xText: A framework for textual DSLs. In *Proceedings of the Workshop on Modeling Symposium at Eclipse Summit*. IEEE.

Fowler, M. (2010). *Domain-specific languages*. Boston, MA: Addison-Wesley Professional.

Gönczy, L., Csertán, G., Urbanics, G., Khelil, A., Ghani, H., & Suri, H. (2011). Monitoring and evaluation of semantic rooms. In *Collaborative Financial Infrastructure Protection: Tools, Abstractions, and Middleware* (pp. 99–116). Berlin: Springer.

Gruber, T. R. (1993). Toward principles for the design of ontologies used for knowledge sharing. *International Journal of Human-Computer Studies, 43,* 907–928. doi:10.1006/ijhc.1995.1081.

Happel, H.-J., & Seedorf, S. (2006). Applications of ontologies in software engineering. In *Proceedings of the 2nd International Workshop on Semantic Web Enabled Software Engineering*. IEEE.

Kalibatiene, D., Vasilecas, O., & Guizzardi, G. (2010). Transforming ontology axioms to information processing rules – An MDA based approach. In *Proceedings of the 3rd International Workshop on Ontology, Conceptualization and Epistemology for Information Systems, Software Engineering and Service Science*. Amsterdam, The Netherlands: IEEE.

Ráth, I., Ökrös, A., & Varró, D. (2010). Synchronization of abstract and concrete syntax in domain-specific modeling languages. *Software & Systems Modeling, 9*(4), 453–471. doi:10.1007/s10270-009-0122-7.

Spinellis, D. (2001). Notable design patterns for domain-specific languages. *Journal of Systems and Software, 56*(1), 91–99. doi:10.1016/S0164-1212(00)00089-3.

Szatmári, Z., Izsó, B., Polgár, B., & Majzik, I. (2010). Ontology-based assessment of software models and development processes for safety-critical systems. In *Monographs of System Dependability* (Vol. 2). Wroclaw.

van Amstel, M., van den Brand, M., & Engelen, L. (2010). An excercise in iterative domain-specific language design. In *Proceedings of the Joint ERCIM Workshop on Software Evolution (EVOL) and International Workshop on Principles of Software Evolution (IWPSE)* (pp. 48-57). ERCIM.

Weichel, B. (2001). *Introduction for ASAM-MCD-2MC audience*. Retrieved from http://www.msr-wg.de/medoc/download/msrsw/v222/msrsw-tr-intro/msrsw-tr-intro.pdf

ADDITIONAL READING

Amador, L. (2012). *Drools developer's cookbook. Olton*. Packt Publishing.

Atkinson, C., Gerbig, R., & Kennel, B. (2012). Symbiotic general-purpose and domain-specific languages. In *Proceedings of the Eighth International Conference on Aspect-Oriented Software Development Charlottesville (AOSD'09)*. AOSD.

Atkinson, C., & Kühne, T. (2001). The essence of multilevel metamodeling. In *Proceedings of the 4th International Conference on the Unified Modeling Language, Modeling Languages, Concepts, and Tools*. IEEE.

Calero, C., Ruiz, F., & Piattini, M. (2010). *Ontologies for software engineering and software technology*. Berlin: Springer-Verlag.

Chomsky, N. (2005). Three factors in language design. *Linguistic Inquiry*, *36*(1), 1–22. doi:10.1162/0024389052993655.

Domokos, P., & Majzik, I. (2007). Aspect-oriented modelling and analysis of information systems. *Periodica Polytechnica –. Electrical Engineering*, *51*(1-2), 21–31.

Eckert, M., & Bry, F. (2009). Complex event processing. *Informatik Spektrum*, *32*(2), 163–167. doi:10.1007/s00287-009-0329-6.

Falbo, R. D. A., Ruy, F. B., & Moro, R. D. (2005). Using ontologies to add semantics to a software engineering environment. In *Proceedings of the 17th International Conference on Software Engineering and Knowledge Engineering (SEKE'2005)*. SEKE.

Gamma, E., Helm, R., Johnson, R., & Vlissides, J. (1994). *Design patterns: Elements of reusable object-oriented software*. Reading, MA: Addison-Wesley.

Gasevic, D., Djuric, D., Devedzic, V., & Selic, B. V. (2010). *Model driven engineering and ontology development* (2nd ed.). Berlin: Springer-Verlag.

Hanmer, R. S. (2007). *Patterns for fault tolerant software*. Chichester, UK: Wiley Publishing.

Heer, J., & Bostock, M. (2010). Declarative language design for interactive visualization. *IEEE Transactions on Visualization and Computer Graphics*, *16*(6), 1149–1156. doi:10.1109/TVCG.2010.144 PMID:20975153.

Knublauch, H. (2004). Ontology-driven software development in the context of the semantic web: An example scenario with Protégé/OWL. In *Proceedings of the 1st International Workshop on the Model-Driven Semantic Web (MDSW2004)*. MDSW.

Kocsis, I., Pataricza, A., Micskei, Z., Szombath, I., Kövi, A., & Kocsis, Z. (2012). Cloud based analytics for cloud based applications. In *Proceedings of 1st International IBM Cloud Academy Conference*. IBM.

Levendovszky, T., Karsai, G., Ledeczi, A., & Maroti, M. (2002). Model reuse with metamodel based-transformations. In *Proceedings of the 7th International Conference on Software Reuse: Methods, Techniques, and Tools*, (pp. 166-178). IEEE.

Mernik, M., Heering, J., & Sloane, A. M. (2005). When and how to develop domain-specific languages. *ACM Computing Surveys*, *37*(4), 316–344. doi:10.1145/1118890.1118892.

Meyer, B. (2000). Principles of language design and evolution. In *Proceedings of the 1999 Oxford-Microsoft Symposium in Honour of Sir Tony Hoare*, (pp. 229-246). Microsoft.

Meyer, B. (2012). *Never design a language*. Retrieved from http://bertrandmeyer.com/2012/01/31/never-design-a-language/

Miksa, K. (2011). *MOST – Marrying ontology and software technology*. Retrieved from http://cordis.europa.eu/fp7/ict/ssai/docs/fp7call1a-chievements/most.pdf

Nanz, S. (2011). *The future of software engineering*. Berlin: Springer-Verlag. doi:10.1007/978-3-642-15187-3.

Nolte, T., Hansson, H., & Lo Bello, L. (2005). Automotive communications – Past, current and future. In *Proceedings of the 10th IEEE Conference on Emerging Technologies and Factory Automation* (Vol. 1, pp. 985-992). IEEE.

Peraldi-Frati, M.-A., Blom, H., Karlsson, D., & Kuntz, S. (2012). Timing modeling with AUTOSAR: Current state and future directions. In *Proceedings of the Design, Automation & Test in Europe Conference & Exhibition (DATE),* (pp. 805-809). DATE.

Roebuck, K. (2011). *AUTOSAR – Automotive open system architecture: high-impact strategies - What You need to know: Definitions, adoptions, impact, benefits, maturity, vendors*. Tebbo.

Salatino, M. (2012). *jBPM 5 developer guide*. Olton: Packt Publishing.

van Deursen, A., Klint, P., & Visser, J. (1998). Domain-specific languages. *ACM SIGPLAN Notices, 35*(6), 26–36. doi:10.1145/352029.352035.

Zhou, C., Chia, L.-T., & Lee, B.-S. (2004). DAML-QoS ontology for web services. In *Proceedings of the IEEE International Conference on Web Services,* (pp. 472-479). IEEE.

KEY TERMS AND DEFINITIONS

Complex Event Processing (CEP): A technique which aims at processing high volume of data generated by event-based systems, usually with just a slight processing delay.

Domain-Specific Language (DSL): A modeling formalism designed explicitly for a concrete domain. Since it reuses the concepts of the targeted domain, allows very precise recording of modeled phenomena.

Domain-Specific Language Family: "Bunch of DSLs" closely related to each other because of the shared problem, but in different contexts.

Ontology: A knowledge base shared by a group of people, assuming an open world (i.e. if something is not defined in a model, it *might* exist.)

Problem Domain, Context Domain: The DSL's targeted domain can be characterized by two aspects: the problem domain, which formally defines the problem; and the context domain, in which the problem emerges. For example: modeling complex events (the problem) for stock market trading (the context).

Semantic Knowledge Base (SKB): A formal, descriptive model of the DSL's context domain, implemented as a hierarchical architecture of ontologies.

Chapter 7
The Human Role in Model Synthesis

Steven Gibson
California State University – Northridge, USA

ABSTRACT

This chapter highlights one concept representing the human role in requirements engineering and analysis for model synthesis. The production of design documentation to support model development requires elicitation of user requirements. The process of requirement elicitation plays a primary role in all Model-Driven Software Engineering (MDSE). Issues addressed include how requirements are gathered by the use of surveys, interviews, and questionnaires, and the importance of using validated constructs when gathering user information during requirement elicitation. Survey constructs, as used in requirements engineering, are analogues to the models in the final engineering product. A solution to improving the use of survey methods in the gathering of requirements is introduced. A small application is shown that suggests an example use of this proposed solution. This review of current practices explores areas where challenges are faced in the field with a concluding discussion that points to future trends in this research field.

INTRODUCTION

The use of Model-Driven Software Engineering (MDSE) methods is a valuable approach to address challenges requiring change from traditional software development regimes. One change from older methods to MDSE results in end-users shar-

ing involvement in all aspects of the development process. For this reason and others, the software engineer is a facilitator between the end-user and the working product. Another change impacting software development methods is the likelihood that final software products often result from modifications or transformations of existing

DOI: 10.4018/978-1-4666-4494-6.ch007

systems. Developers seldom have the option of designing a system from the ground up; instead, systems are often syntheses of existing products, uses, user practices, and data structures. The process of building an application can be seen as the transformation and composition of multiple domain models to synthesize a functional system. The process of building a software model delivers a system which is an abstract match with the problem conception of the users. One challenge addressed here is how to standardize verification of the elicitation of the requirements.

Models encode human knowledge, experience, and expectations into graphical or structural forms. The primary concern of this chapter is exploring the methods which best represent the mental models of the users as part of a required system. This model representing the user requirements is described as a construct for this discussion. The resulting MDSE system presents a realization of the requested, required and projected functionality to address the domain challenge. The first focus, during requirements engineering, is on requirements elicitation, analysis, and interaction with the users. The next focus is on the principles of using communication for building understanding between users and engineers around the model systems. The model synthesis step directly follows from the successful requirements engineering process. Model synthesis can be seen as resulting from appropriate formalization of requirements (Desel, 2002).

This chapter focuses on the importance of users and domain specialists during the requirements stage of model development. The techniques discussed include writing and reading requirements, communicating with users about requirements, preparing documentation, synthesizing models, applying data gathering methods, identifying essential domain knowledge, and exploring requirement assumptions. Requirements analysis plays a role in the understanding of problems, communicating with users and driving the system

development process. Requirements engineering serves a key role in model-driven software engineering. Boehm (1981) reported that approximately 60 percent of all errors in system development projects originate during the phase of requirements engineering.

Requirements engineering directs attention on aspects of analysis and design and does not address the full development life cycle of model-driven software engineering. These methods can cover the needs of a wide variety of applications in both large and small projects. Our focus here addresses requirements elicitation and requirements analysis. A definition for requirements has not been formally agreed to by software engineering researchers, although a definition described by Hull, Jackson and Dick (2011, p. 6) includes the ideas of statements which identify characteristics that are unambiguous, and are necessary for system acceptance. Requirements engineering involves developing documents to address the problems as described by users. The analyst and user establish a set of conventions to describe and discuss the problem domain and scope of the system. Analysis can be seen as the methods used to study the concepts, procedures and activities in a problem domain. Each problem domain in a certain modeling situation is dependent on the contexts and goals in the system environment. The first steps of analysis for a MDSE project is carried out in order to increase understanding of the problem domain.

When performing requirements engineering of planned MDSE systems, the requirements documents become the grounding for the entire system. It is important to continually update the set of descriptions which cover the system requirements and the models which generate implementation outcomes. The latest approaches in requirements documentation methods and communication tools, for the purpose of model synthesis, are discussed and explored in this chapter. The philosophy underlying requirements engineering for MDSE should be that gathering information for modeling is appropriate for all domain challenges, and will always deliver some value.

A subtopic of requirements engineering is the importance of analyzing legacy systems for the purposes of requirements documentation. Older legacy systems often need to be interwoven together with the overall system requirements. Model-driven software engineering creates functionality by the transformation of models based on deep knowledge of existing and available systems and tools. The models are derived from requirements of the problem space describing expected inputs and outputs of the system, as obtained from users and system knowledge. Requirements documents are abstractions that portray the essentials of a complex system, with descriptions that are constructed by defining structures and filtering out nonessential details. An early step in domain requirements involves understanding the system at design time.

In order to understand the system under design the engineer and the users must agree on the representations of the system. Constructs and metaphors can help analysts communicate about the interactions between the users and the abstract representations of the models (Rilling, Meng, Charland and Witte, 2008). During communication with the user in requirements elicitation, there is development of constructs specific for the system being modeled. These concepts that are produced are representations for the abstractions of the system behaviors and activities. These representations include the use of constructs which describe the system as sets of communicating objects that behave based on rules. Creating constructs to share with users is a key technique during the requirements elicitation step during the MDSE development process. The requirements derive from information from the users' existing operating environment and expected system needs. The analysis phase often means sketching a rough model of the problem,

while later design phases involve synthesizing a fuller model of the implementation environment. The final steps of model implementation involve translating models into working systems using automated procedures.

Because requirements engineering does not cover implementation details, those later design phases are not covered in this chapter. Formalization in requirements engineering began development even earlier than MDSE approaches, and some methods are shared by practitioners and researchers. While some methods in requirements engineering are mentioned here, this is not a tutorial and as such avoids detailed coverage. Formalisms from other fields used in analysis and design methods, such as state transition diagrams, process diagrams, Petri nets, entity–relationship diagrams, data flow charts, etc., are not discussed here. The focus here is on the latest research directions in requirements elicitation.

The design phase that follows analysis includes some implementation dependencies which are not explored in requirements elicitation. When examining typical requirements through analysis, there is a high level of concern with how to find the abstractions and procedures needed to produce shared understanding of the system with users. Drawing an exact line between analysis and design is somewhat arbitrary, so heuristic methods are often used to discriminate between them. This chapter does not address project management, but there is value in continually assessing and controlling the elements in the software development process. An early step in the requirements process includes the identification and delineation of system boundaries and overlaps. The term analyst is used in this text to represent system engineers, or the systems experts eliciting requirements from the users. Users represent stakeholders of all types who relate to the system. The approach used here is to conceptualize all users as equal potential contributors to the requirements, and consequently labels them all as users.

An analyst often leads the requirements elicitation phase of the MDSE project. This analyst is responsible for determining which information should be documented for modeling the system. The analyst is involved in a number of actions and steps to produce the requirements documents, starting by focusing on high-level descriptions which should help prevent premature design decisions. Requirements engineering involves a series of compromises, with outcomes exhibiting trade-offs between dependency and simultaneous independence from the existing and planned implementation environments. When performing requirements analysis, two different roles must be fulfilled: the user role and the analyst role. Requirements analysis develops a series of structures and constructs in order to communicate the requirements. There often are misunderstandings and misinterpretations between the user and the analyst during the requirements elicitation process, however creating valid constructs and establishing explicit agreement on the requirements documents should decrease the errors of communication. Mismatches between the users' perceptions of the system and the analyst's requirements description can be a leading cause for misunderstandings. Communication mismatches must be confronted directly and continually throughout the requirements process in order to produce valid requirements documents.

The analyst must take steps to increase clarity and correctness during communication with the user. Communication improvements can be encouraged in several ways: better shared vocabulary, better documentation including help systems, and tailored constructs. Requirements elements are expressed in a combination of standard language and model-centric expressions. These abstractions capture the complex behavior of the system that can be communicated through constructs. Both requirements and models must share the same abstraction base in order to avoid divergence. The requirements tell the story of the system, and should have the flexibility and formal structure to

be transformed into usable models. It is important for both to remain seamlessly tied to each other during the development process.

This chapter also reviews current and evolving practices in requirements elicitation to support MDSE. The chapter accomplishes its objectives by conveying the following priorities: encouraging collaborative information gathering at the requirements stage; promoting clear understanding of the preexisting system; keeping users central throughout the process; and focusing on validity of requirements elicitation. Through noting the importance of the human users at the first stages of MDSE, the best use is made of assets and domain resources. Through clearly understanding the existing domain resources, the preexisting elements and new software models can be merged in the requirements documents. Communication thoroughness and accuracy is key to all these priorities in requirements elicitation.

While no in-depth use cases are explored in this text, the outline of an example using a music/media data system at a radio station is used to illustrate some of the methods being discussed. The topics covered in the remaining sections include: user to analyst communication, legacy systems, survey constructs, problem domains, and requirements documentation. The following section explores the background of the field of requirements elicitation.

REQUIREMENTS ENGINEERING FOR MODEL SYNTHESIS

Requirements analysis begins with the gathering of information through user surveys which will obtain a set of details which describe the domains involved in a problem space. Included in these elements of information are vocabulary, concepts, and behaviors. The users and domain experts are tied to the requirements analysis through their operational experience and communication interactions. The systematic requirements analysis starts with identifying needs through the collection and collation of existing resources and likely needed elements. In order to produce an accurate and complete model, multiple domains need to be explored. The goal is the synthesizing of a solution from the available domain resources. Requirements elicitation refers to the specific process of obtaining domain and system information from the users.

In model-driven software engineering, the initial step in requirements analysis reflects risk analysis and problem domain identification. Next, during system development, there follows a series of process tasks and steps, including gathering analysis information, describing the requirement elements, and synthesizing the requirement documents. At the beginning, it is often desirable to assume simplified models of parts of the system, and so reduce initial complexity in order that the overall system is easier to describe. The next steps in requirements analysis involves describing the system, seeking acceptance checks from the user, and producing outcomes that include analysis results and design documentation. The process of requirements analysis can be improved by taking environmental and system risk factors into account. Some details in the requirements documentation should be useful for future costs versus benefit model analysis.

Model synthesis serves to reformulate the needs of users into system outcomes. Modeling involves elements of formal reasoning to help specify properties of a system and transformation process during system development. Designing software is a tradeoff between alternate choices of models. There are always multiple ways to achieve the same working outcomes. For software approaches to system design, the requirements phase is vital for MDSE projects. Software plays an increasingly important role for organizations and individuals, and when software is changed, it changes the organizational culture.

One of the key skill sets for requirement elicitation is the ability to improve communication between the analyst and users. Part of this communication is understanding existing culture, which includes the need to get past assumptions. Users are being recognized as central to the software process and successful software engineering depends on people and clear communication. As software systems become omnipresent in business and society there is increasingly a loss of distinction between the roles of users and software developers (Costabile, Mussio, Piccinno, and Provenza, 2008).

The abstractional power of MDSE analysis can describe organizational entities, and can be understandable to people with various roles in the operation. The requirements phase of MDSE involves analysis and design. Looking at analysis and design from the perspective of uses in other disciplines gives a wider view of their meaning and applicability. Analysis is used to describe taking things apart for examination, while design describes constructing structures or procedures to accomplish tasks; both approaches are needed in software development at different times. However, in MDSE these terms hold somewhat different uses. In MDSE methods, analysis often involves creating models from the user supplied requirements. Agreed upon models of the problem are derived from the requirements process. The design phase relates more to the implementation steps of producing the platform specific models.

During the ongoing MDSE project development, it is possible to extend and modify parts of the requirements to meet changing needs. In addition to analysis and design, there are issues involved in conducting successful requirements engineering, including system modeling of user communication and data transactions. Requirements analysis is the top-level instrument leading to the design of the models and the development of final implementation. Any system consisting of activities exhibiting certain behaviors is suitable for model-driven engineering. User communication and understanding within an organization may be enhanced by constructing models of that organization's concepts, tasks and behaviors.

Abstraction is used to simplify the communication about the requirements. This abstraction is needed in communicating complicated ideas between people, since the human capacity of keeping many things in mind simultaneously is extremely limited (Miller, 1956). Abstracting ideas and information should be undertaken carefully. Concepts are used to group various similarities so that details do not need to be individually enumerated. Moreover, the concepts chosen must be easy to understand in relation to the users and have enough precision to accurately represent the system.

Analysts seek to develop an abstract model of the system under study. Analysts can create a shared language between the user and the analyst. A well developed system presents a clear abstract model of what is going on during execution. This picture can be gathered from users by analysts through surveys. A survey can often be constructed to research the engineering needs and practices of the users (Kitchenham and Pfleeger, 2002). A system model is composed of layers of abstraction, which serve as a specialized language. Surveys can help build a construct that shows congruence with the user's knowledge and practices. Interviews are carried out with potential users, to understand what mental pictures they share regarding their work. Surveys are a valuable addition to the requirements elicitation process.

Requirements engineering proceeds by refining abstract structure representations. MDSE development requires postponing decisions. Refinement involves working in detail and filling in missing parts. As analysis constructs are built they contain abstractions which make it easy to understand the desired system behavior. This may help users identify additional concepts of the system that they had not previously considered. Analysis makes aspects visible which are beyond the common view held by users in the problem domain, since the construct can demonstrate multiple depths and levels. The conceptual construct represents a simplified view of reality, which should directly relate to the user's understanding of the problem domain. The human limitations in memory capacity and precision often prevent complete requirements analysis when depending on a single user. Multiple users are helpful for building valid system constructs, but the construct requires a congruence that represents all the referenced users. The conceptual aspects of the model offered by the system should closely match the user's mental model of the system (Douglas and Moran, 1983).

How a system will be used is an important factor in development. One step is to go through scenarios to check the completeness of the evolving structure of abstractions. The goal is to find abstractions that lead to robust systems. It is necessary to synthesize requirements that are not vague or incomplete and it is important to understand what underlying problem the user is trying to solve. Surveys, focus groups and interviews represent helpful tools to communicate the abstract requirements between users and analysts.

There are many approaches taken to conduct requirements elucidation. As reported by Mulla and Girace (2012), in-person interviews and other direct analyst-to-user contact are the most often used method. The user is the expert in many aspects of the problem space. Interface features and functions are obvious first aspects to be discussed with users. Also the user can best report uses, perceptions and expectations about the existing system. The requirement outcomes most often directly derive from the user input. The outcomes include requirement documents and documentation. Document readability often correlates to reliability of outcomes. Clarity of communication helps as part of addressing requirements elucidation.

Delineating the system is an early step leading to decisions about what aspects of a problem area should be modeled at the highest level and how to describe relationships between elements in this model. Describing the system may include modeling views of external systems, such as database management systems and GUI systems.

Requirements analysis, in this case, means the merging of users and designers in the software development activity. The development process, which includes implementation and maintenance, moves backward and forward through analysis, design, implementation and maintenance.

Table 1 represents the steps available in requirements building. Each element is shown in two modes, the first is a more formal approach in the requirements framework, while the second is a more user oriented mode. By choosing the appropriate mode at different stages in the requirements elicitation process, more complete requirements documentation can be produced. These requirements techniques represent different approaches for eliciting domain and problem information from users. These and other techniques lead to the production of requirements documentation that can be shared with the users.

The documentation of a model-driven software system is the outcome from requirements engineering. It describes the components that define the system behavior and its correctness as a prerequisite for a working system. Useful documentation needs to be abstract, which means that many details of what is described are hidden. This abstract documentation needs to be understandable, and the documentation needs to be precise. This depends on what is being modeled and what are the principal views of the world that is being explained and formalized. Requirements analysis serves to build understanding about the system as requested. This leads to a complete software synthesis process, through design and implementation to maintenance and re-engineering. The key is to ensure a direct mapping between a problem and its software solution, which should result in a high level of quality for the working product.

Current design methods are dominated by hybrid approaches, meaning they contain both the application of model-driven principles with techniques drawn from legacy systems. Requirements engineering for MDSE projects use high levels of

Table 1. Requirements technique pairs (derived from Cheng & Atlee, 2007)

Task/Technique	Strategy	Scope
Elicitation	Model/ Formal	Stakeholders, Metaphors, Context
User Metaphors	User Centered	Stories, Scenarios
Requirement Analysis	Model/ Formal	Ontologies, Risks, Priorities
Domain Information	User Centered	Language, Data
System Analysis	Model/ Formal	Legacy Identification
Existing Environment	User Centered	System Examination
Validation and Verification	Model/ Formal	Checking, Satisfiability
Customer Acceptance Procedure	User Centered	Checking, Approving
Model Synthesis	Model/ Formal	Merging, Synthesis
Documentation and Help Synthesis	User Centered	Synthesis
Requirements Management	Model/ Formal	Traceability
System Evolution	User Centered	Change Management

requirements analysis because of the detail needed for model validity. Requirements engineering develops and explains static and dynamic models synthesized in the MDSE process. The complexity required to solve a problem depends not only on the problem, but equally on the concepts available for reasoning about the problem. The abstractions developed through the process gradually become part of the standard vocabulary for the system.

The requirements analysis process can be viewed as a succession of iterations between system stages, each hopefully representing a better view than the previous one until acceptance is reached. It is important to balance premature abstraction against needed abstraction. Abstraction is strong medicine, which means that the view of the world represented by the set of abstractions available to us can influence the way we think. When introducing new abstractions by encapsulating some behaviors, we protect the users from complexity. Abstraction is the key to clear communication between users and analysts during software development. Abstraction is useful for new systems and legacy systems. There are many industrial software projects that are still using hybrid languages or hybrid methods, or both. The need for backward compatibility means system design must take into account descriptions describing legacy systems as well as newer MDSE developed parts of the system.

The description of the software system can be static or dynamic and covers components, users and dynamic behaviors. Static descriptions document the structure of a system including what the components are and how these components are related to each other. The documentation of the working system uses constructs and descriptions to explain the functioning of the system. Dynamic descriptions document how the system behaves through time and interacts with users. A MDSE developed system can be viewed as a structured collection of models which each demonstrate abstract concepts. These models constitute the blueprints for the system to be constructed. The analysis of a system using requirements engineering methods result in descriptions of the system which are shared with users. Alternative notations used in requirements analysis include graphical descriptions, textual communication and formal notation.

Graphical forms in documentation are intended for use with automatic tools as well as for sketches on paper and whiteboards. Textual forms are intended for communicating model-driven design descriptions between various tools, models and users. Formal notations are expressed with logical or mathematical symbols. In requirements documentation, the language used must be useful for users and analysts.

There are downsides to MDSE approaches. There is not always a clear vision of the working system during the requirement process. There can be valid reasons to not use models in particular cases, such as when the scope of future needs is sure to change and to grow substantially. There is the constant danger that changes in design might not be updated back to the requirements. The essence of modeling is abstraction, and requirements elicitation must be used to obtain the correct information to abstract. When requirements are missing, MDSE methods can become difficult and subject to error. A continuing challenge for MDSE projects is the need to elicit clear, complete requirements.

IMPROVING REQUIREMENTS ELICITATION VALIDITY

Challenges for Requirements Elicitation

There are inherent risks involved when developing software projects. These risks can be attributed to needs for shared data, dispersed control and complex processing. Our society depends increasingly on correctly functioning software systems. We are dependent on the activities of computers and computer software in our daily lives and in every industry and type of technology. In order to reduce risks in software, developers need to improve dependability during development and conceptual phases. Model-driven software engineering is a method computer scientists are currently using to confront the challenges of software design and maintenance.

Risk can be seen as relating to the probability of uncertain future events. For example, according to Jones (2006), risk is the probable frequency and probable magnitude of future loss. In computer science this definition of risk is used by The Open Group (Dobson and Hietala, 2011, p. 49). Fred

Brooks (1996), in the article "No Silver Bullet: Essence and Accidents in Software Engineering," divides the difficulties of building software into essence and accidents. The essence of a piece of software is a construct of interlocking concepts: data sets, relationships among data items, algorithms, and function invocations. Brooks believes the hard part of building software was the requirements, design, and testing of the essential conceptual constructs, as opposed to representing them and testing the fidelity of the representations. Brooks concludes that the building of software will always be hard.

The model-driven engineering method has the potential to turn analysis, design, and implementation of general software systems into a more seamless process. A smooth transition from user requirements through analysis and design into working systems is a goal of software engineering. What makes MDSE attractive is that abstraction mechanisms can be used in all phases of development. The basic concepts needed to model elements representing such external notions as businesses, avionics, and wide area networks are essentially similar (Cabot & Yu, 2008). There is a need to keep the same paradigm from initial feasibility study all the way through to a working system.

A step during requirements analysis involves identifying users and usage patterns, and part of the process involves identifying which user groups have which different needs. Analysts interact with users through requirements communication and later in the process with usage of the working system. Many methods can be used to increase communication between users and analysts. Textual methods and graphical charts are used for communication with non-technical people and users. The first steps in requirements elicitation can be seen as building intellectual guides for finding good system elements and then providing high-level documentation of the system. While requirements engineering should remain divorced from implementation details, automatic reverse generation is helpful for maintaining consistency. Therefore, a balance is needed between avoiding implementation details and matching models with requirements. Depending on the particulars of the type of system, its size, the people involved, and the analysis, the information gathered by the developers may enter the system model through different mappings.

As the requirements are built, some newly developed abstractions can yield insights and new potentials that may even change the initial requirements. Alternative views into a requirements analysis can aid in producing documentation. Detailed requirements are needed for completeness before moving to the model phase. Another important point is that consistency requires the models and requirements to be based on one conceptual view of the world. When specifying the behavior of the objects in a system, we are reasoning about the behavior of the entire system. The information that can be conveyed by an overview of some part of a system should be consistent with any other part. Requirements documentation is structured in order to achieve functionality for MDSE development and outcomes.

Because the challenges for understanding and communicating requirements increase as the systems size increases, abstraction is a vital tool. Abstraction is a key aspect while transforming user input into requirements documents. The difficult task is to apply the appropriate degree of abstraction. The user is likely to accept extreme abstraction which then results in a very simple model. For the model to be validated, a high level of complexity must be included. While an early step involves constructing a high level abstraction, in order to correctly frame the questions, additional steps are required to fill out detail for the model. A contradiction exists for the analyst of building complete requirements while hiding most of the extensive details that exist in all problem domains. Often systems which appear simple to the user are actually complex when fully implemented using computer systems.

The boundaries can be tricky working on requirements for a new project. Knowing where the problem domain begins and ends is often difficult. Any system has a priori structure, which needs to be understood and detailed. The requirements representing the system are not just there for the analyst to discover, and understanding the problem domain requires the imposition of structure on its elements. This begins by examining the set of concepts and views shared by the users. Requirements engineering for MDSE projects entails refining the requirements that can be transformed into models of a solution. Analysis should be performed in a recursive fashion. Any system requires analysis so that requirements can be produced. The continual steps from requirement to documentation is repeated until reaching the desired level. Part of communicating the requirements between users and analyst involves abstractions which describe the system; as the requirements are being synthesized, they take on the form of abstractions that the analyst and user can employ to communicate with each other.

Requirements engineering requires the analyst to investigate the data structure, the user interfaces, the categories of users, communication requirements, security and platforms. Part of the process can involve a re-engineering that involves successive replacement of old parts by new model-driven ones. A challenge exists when attempting to build complete requirements of extremely large systems because of the constant changes which the system is likely to undergo (Walden and Nerson, 1994). While synthesizing large systems, some functions can be minimized in the requirements documents. Often many functions offered by an environment are seldom used.

There is increasing need for formal approaches to requirements engineering. At the early stage of requirements analysis, it is useful to use simple framing to avoid incomplete analysis. An important issue to prioritize is the minimality principle which advises including exactly the functionality needed to solve the problem domain. It is also necessary to explore the problem domain and system borders to produce system constructs. Some researchers are examining techniques to aid in the adoption of formal methods in survey use during requirements elicitation.

Interaction between users and analysts regarding current processes and system state is important to help users detect potential errors. A problem is the inherent conflict between providing just enough information and supplying too much. As the requirements analysis begins, the analyst has formed at least a rough idea about the collective set of models which are initially discussed. Often, parts of the domain are studied in parallel to detect common patterns of needed behavior as early as possible.

A goal in requirements engineering is to strike a balance between the extreme simplicity required by some groups and the functionality needed by all users. Implementation language and MDSE design features need not be represented as such during the requirements phase. Requirements analysis does need to describe behaviors including activities, vocabularies, graphical outcomes and textual explanations. A goal is to strive for high-level abstractions that can be described with fewer elements. Events can be recognized which include system events and selected external events which trigger behavior.

Instead of using defensive programming, software engineers should establish defined responsibilities and a large number of model systems to reduce the cost of error testing and detection. Requirements analysis faces the need for a high-level of detail while communicating with less detail. In the future, analysts will likely use more formal methods to balance between simplicity and power of expression. Formal languages can aid with their expressive power to increase the thoroughness of description, and transformations to simpler forms should be possible. The complexity inherent in requirements may be gradually reduced by successively modeling parts of the system that we do understand. These highly descriptive models will assist in the discovery of additional issues to correct.

When we are working in a very narrow application area, the goal of building system views make the task of finding a good model similar to general human problem solving. The analyst may make the mistake of not fully exploring the desired formalization step. The goal of the requirements process is to gradually build the requirement documents. These represent the most precise requirements of the structure and behavior of the system that exists prior to implementation. Reversibility is another trait linked to formal methods; requirements can be tested formally through successful performance of reverse engineering runs.

Requirements engineering methods are important when connecting existing legacy systems into systems using new models. Model-driven approaches can act as a very efficient gluing mechanism between heterogeneous components (Qiao, Yang, Xu, and Chu, 2003). In this way, a model-driven system can tie together the complex behavior of existing products, and increase access to off-the-shelf components. Encapsulation is a general approach to hiding and incorporating data and information. It can be used to capture abstractions of the system to enable understanding for anyone reading the requirements. Encapsulation can both simplify the problems and make understanding difficult. Software engineering solutions are developed to meet multiple needs, so the concepts used to mold it must be usable for multiple purposes as needed.

When increased information sharing is facilitated, barriers between users and analysts are removed so that the system requirements can be consistently communicated. The elements introduced in the analysis phase will be present in the final system, and tracing the propagation of initial requirements through design and implementation becomes much easier. High-level requirements can

only represent a simplified view of a system, many details and problems ignored at that point will have to be taken care of before the requirements can be made executable. Successive refinements and corrections will flow to and from the implementation, because some resolving of obscurities cannot be addressed by the requirements. When the working system is automatically generated from the models, the latter needs to incorporate and feed back to the requirements.

For abstract system descriptions to retain value beyond the first translation into a working system, changes to the system must repeatedly be reflected back into the requirements. The implementation details not directly mapped to models in the MDSE approach would lead to creeping divergence between requirements and models. It is difficult, but important, to keep the two worlds consistent as the system evolves and continue the matches between difficult-to-comprehend conceptual structures.

The risk of missing requirements is pervasive in software development. In fact, even if efforts are made to keep all requirements complete, the task is difficult because of incomplete communication between users and analysts. The value of the requirements is directly related to the ease in discussing the system with users and the problem solving communication that takes place between analysts and users.

Solutions to Requirements Elicitation Challenges

Several steps are needed to improve the elicitation of accurate and complete information from users. Guidelines for MDSE approaches offer some suggestions for producing requirements documents. Bernand (2011) states that, "in the foreseeable future, the production and verification of such documents will only be accomplished by human means" (p. 120). This encourages prioritization of methods of improving user and analyst interactions. Cabot and Yu (2008) suggest that for model-driven software engineering, "providing a better support for the requirements specification and analysis is still a research challenge" (p. 1).

Bernand also writes that requirements engineering is "the very foundation of systems engineering" (p. 122). Requirements engineering has always been a challenge because of the difficulties with the semantic gap between design intent and computerized models. Analysts need to focus on elicitation of requirements from users that will enable model synthesis. Blanes and Insfran (2012) suggest the need for increased interest in using MDSE techniques in requirements engineering.

Communication between user and analyst is difficult to achieve because users often have trouble identifying and articulating their needs. In addition, those needs often change as a result of system implementation. Davey and Cope (2008) state that poor requirements elicitation continues to be a major problem. Mulla and Girase (2012) state that, "effective requirements elicitation depends upon the ability of users and analysts to understand and appreciate one another's words" (p. 1). Several tools can aid in user to analyst communication. Scenario analysis can aid as a systematic approach to eliciting user requirements. Improved survey production can assist in the elicitation process.

Metaphors and scenarios can aid in producing the requirements. As a starting point, we may select a set of interesting system scenarios to illustrate important aspects of the overall system behavior. A short description of each scenario may then be collected as part of the requirements elicitation. Clear constructs are also vital to describe the requirements from the users perspective. After sharing detailed views of the system with users, some of these scenarios will be produced as requirements documentation.

Table 2 shows different requirements approaches. Some of these approaches were developed in past years while some are more recent. The trend in requirements development seems to be incorporating more formal methods. The table describes

Table 2. Requirements approaches (derived from Mulla & Girase, 2012)

Method	Description
Analytic Hierarchy Process (AHP)	Priorities are relative and based on a ratio scale.
Brainstorming	Stakeholder representatives gather together and rapidly develop a large and broad list of ideas.
Binary Search Tree (BST)	Provides a simple ranking of requirements as no priority values are assigned to the requirements.
Hierarchical cumulative voting (HCV)	Prioritizations to be performed at different levels of a hierarchy.
Interviews	Analyst discusses the desired product with different groups of people and builds up an understanding of their requirements.
Mind mapping	Mind maps are diagrams that are used to capture, visualize and organize ideas.
100-point test	Requirements are then prioritized based on the total points allocated to them.
Pairwise comparison approach	Requirements engineers compare two requirements to determine the more important one.
Prototyping	Prototyping is used to provide a version of the software and which may be functional, but is not final.
Requirements triage method	Relative priorities of requirements depend on the stakeholders who attend the prioritization meeting.
Scenarios, passive storyboards	Include interaction sessions to describe a sequence of actions and events for a specific case of some generic task which the system is intended to accomplish.
Survey/ Questionnaires	Surveys can aid in reaching users and obtained structured input.
Value-oriented prioritization method	Prioritizes requirements based on their contribution to the core business values and their perceived risks.
Win-win approach	Each user ranks the requirements privately before negotiations start.
Workshop, focus groups	Stakeholder representatives gather together for short but intensely focused problem descriptions.

some aspects of each approach and how it serves the requirement tasks of a model-driven development project. An analytic hierarchy process builds priorities based on ranks in a hierarchy. Binary search tree approaches provide a simple ranking of requirements as no priority values are assigned to the requirements. Brainstorming encourages free expression of ideas, but usually does not result in decisions. Hierarchical Cumulative Voting is employed to allow the users to vote together for the ranking. Interviews are useful for gathering deep information about the requirements, but has weaknesses on thoroughness and spanning input from many users. Mind mapping involves graphical representations of the domain space. The 100-point test grants points to users to assign to the priorities.

The pairwise comparison approach compares priorities two at a time. The requirements triage method reduces the number of options. Scenarios and storyboards are helpful in sharing information clearly between users and analysts, but are not fully accurate about the operation of the system. Prototyping offers software examples of the projected model. Surveys and questionnaires can aid in reaching users at a distance. Value-oriented prioritisation methods determine which aspects of the system deserve the highest priority. In the Win-win approach each user assign ranks, then negotiations begin. Workshops and focus groups

build compromises among users, but produce complex information sets.

Figure 1 points out four elements of requirements development. Each of these four elements must be communicated in interaction with the users. This division of the requirements process can be a way of organizing the tasks and development steps in the process.

For requirements elicitation techniques to be effective, they should increase the communication clarity between the system user and the system analyst. When there is lack of communication clarity the result is increased expenditure of intellectual effort by the analysts and the users to take a system through the stages of analysis. Encapsulation is helpful when it hides details of the total system. Other important features include reversibility. Both model and requirements should map completely between each other. The MDSE resultant systems should not be considered independent from the requirements. The completely modeled project should represent a mapping onto the elicited requirements.

Re-engineering of older systems takes place in order to bring about gradual migration from old technologies and uses model-driven encapsulation of existing components written in hybrid system

Figure 1. Elements of requirements (derived from Holt, 2012)

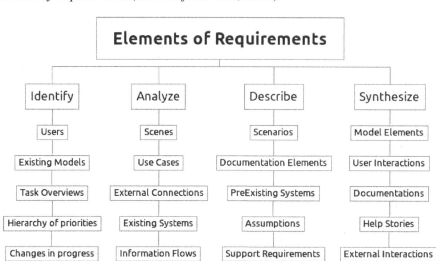

languages. This can include external calls from legacy software and invocation of external software from newer model-driven software.

Employing survey research methods for requirements elicitation involves formulating the goals, selecting data elicitation methods, developing the survey instrument, pretesting the questionnaire with expert judges, pilot testing the questionnaire, collecting data from a sample population and analyzing the data for convergent and discriminant validities, and writing the results (Pfleeger and Kitchenham, 2001).

Table 3 offers a representation of a sample survey which includes previously validated survey questions and newly developed survey questions. A construct is the aspect of the model that is to be verified through input from the users. Each survey item represents refinements questions to build up the correspondence between the suggested model and the users' actual responses. Valid survey results aid in producing requirements documentation that is thorough and accurate. Because the requirements documentation is a key asset of MDSE project development, survey validity will make the difference in achieving successful project outcomes.

Requirements elicitation needs to change during the course of a software project. The first steps will likely be short oral interviews with limited notes. The next stages will include surveys, focus groups and textual and graphic production. Later stages may include automated systems to help to maintain consistency between the requirements, the models and the working system. These requirements should act as intellectual guides for finding

Table 3. Survey validit. (derived from Agarwal, 2011)

Construct Name	Survey Item	Source
Construct 1	Survey item 1	source
	Survey item 2	
Construct 2	Survey item 3	Self-developed
Construct 3	Survey item 4	

models to serve user communication tasks, implementation task and as high-level documentation of the system. Automatic reverse generation is useful for maintaining consistency, in order to maintain high levels of detail and accuracy.

The requirement level of the model-driven analysis is not concerned with implementation, but there should be confidence in the later implementation steps. There is a further system development process which aims to create the working system. Through communication and feedback from the user, the analyst describes the requirements and leads to future development steps.

As an example of a MDSE requirement elicitation task, consider a music and audio file search and distribution system for a radio station. The highest-level requirement may be a construct that includes the ability of users to identify and locate recorded audio segments. Other requirements of the system may be the need to achieve short search times and 100% delivery success on the files that are sought. A starting point in determining the survey instrument should always be the survey's purpose and objectives. A working outcome that fulfills this requirement needs to coordinate between the elements of:

- Methods of access by users, which include desktop, laptop, tablets and phones;
- Authority and responsibilities, regarding who has access to files and who updates them;
- Previous legacy system elements which are retained;
- Documentation and help which details how people get usage questions answered; and
- Storage of files, and the size, speed and dependability of the storage medium.

The analyst will describe the elements of the requirements by formulating a construct. In order to obtain verification that the requirements construct accurately represents the user's needs a survey instrument must be used. A survey consists of multiple constructs that contain one or more questions. Each question will directly relate to the construct being investigated and a thoughtful answer will increase clarity about the opinions of the respondent about the construct being questioned. For example, one construct in this example might relate to the methods of interacting with the system. The construct is a model representing how the system interacts with users. One question might then be: "Are office computers used to access the system?" Another question to assess the same construct might be: "Are home computers used to access the system?"

The validity of surveys can be ensured through a number of steps: first, build constructs based on the likely system design needs; second, reduce analyst bias; third, base questions on likelihood of resolving construct issues. Graphical methods may be useful for achieving construct validity. Forza (2002) describes one method of visualizing constructs when, "the theoretical framework is depicted through a schematic diagram" (p. 5). Forza gives a thorough overview of research directions for improving survey practices.

Several researchers have examined approaches to producing valid survey results. Agarwal (2011) proposes a two-stage sorting procedure to verify survey items. Jansen (2010) explores the basic steps needed for survey methods. Boudreau (2001) explains methods of validation in research. Pfleeger and Kitchenham (2001) offer a series of five articles describing the methods and techniques for effective survey construction and use.

The analyst works to insure complete requirements documentation by expressing the shared understanding of the problem domain using clear communication of the requirements, including environmental conditions, reducing jargon, sequestering as much detail as possible and defining the sources of responsibility. Next steps describe the elements involved in the system. These include the production of the requirements documentation. The analyst explains, communicates, consults and negotiates with the involved users. Once the requirements documentation is complete and accepted, model synthesis can begin.

Surveys are a foundational part of the requirements engineering process. There has been some highlighting of steps which contribute to achieving construct validity and explanation of how these steps can improve requirements elicitation. By integrating the survey method into the elicitation process requirements engineering can be successfully completed. Some of the aspects of the system development will undergo changes away from previously defined procedures during the requirements engineering process. The goal is to deliver a product with all required functionality and performance to the user. When the requirements elicitation steps are completed, the human factors involved in the system will have been properly interrogated and integrated into the system development process.

FUTURE DIRECTIONS IN REQUIREMENTS ELICITATION

The future of requirements elicitation will continue to increase in visibility and importance for the field of software engineering. As assumed in this chapter, the requirements step is arguably the most important for model-driven software engineering. The trend in requirements engineering is likely to mirror the way that model-driven engineering is moving, including more formal methods and more user involvement. While Cheng and Atlee (2007) state that requirements engineering faces challenges which differ from those in other areas of software engineering, some approaches can be shared between these different fields.

Additional research in requirement elicitation for MDSE is needed to address the unanswered challenges facing requirements engineering. Research is needed regarding challenges of communication, documentation formats and end-goal versus requirement abstraction conflict. A serious challenge remains in the difficulties in communication between users and analysts. More efforts are needed to elicit information clearly and successfully in the requirements stage of software development. The actual requirements documentation recording format is an important topic because a key belief is that users and analyst must review the single set of documentation and negotiate agreements based on shared understanding. It is unlikely that users will learn formal mathematical languages for the purpose of software development. Requirements documentation language will need to remain understandable to non-technical users.

Requirements engineers struggle to retain the correct level of abstraction during analysis to avoid being influenced by implementation details. One area where implementation details are difficult to avoid is when analyzing legacy systems, which have existing implementations that can be seen and studied. It is a challenge to build abstract requirement documentation that does not incorporate implementation details from legacy systems. It is important to avoid implementation details in requirements because MDSE projects need to generate the implementation based on the abstract model without predetermined platforms or hardware systems.

Future steps in the development of requirements engineering will involve addressing the issues of communication, documentation and abstraction. Some likely directions of research will center on attention on questionnaires and survey methods. There is a body of research on questionnaires methodology in statistical scholarship and among social studies academics. Some valuable scholarship is available for increasing questionnaire correctness and statistical validity.

Another area where interdisciplinary approaches may be helpful for requirements engineering is the social aspects of users and stakeholders. There are political and social implications and

side-effects involved when eliciting systems details from users in organizations. There are formal and informal power relations and differences of communication competency, experience and expectations. The areas of organization power differences and communication expectations and behaviors are likely subjects for future research.

There is active research taking place which addresses some challenges facing requirements engineering. Several paths that are being explored by researchers include prioritization, survey methods, Bayes methods, user segmentation, and requirements model building. Another approach to requirements analysis tests for form correctness errors, ambiguity errors, inconsistency errors and incompleteness errors (Cheng & Atlee, 2007).

One area of research is software contracting (Meyer 1992), which is an important attempt to improve correctness and robustness in software systems. Designing by contract specifies clear divisions of responsibility between client and supplier regarding the checking of various details of system construction.

Some researchers are exploring using mind mapping to aid in the requirements process. There has been discussion of metamodeling approaches to requirements engineering. One approach has suggested metamodeling to increase flexibility in multiple requirements methodologies, and some possible use can be made of Bayesian graphical and mathematical methods. There is a possibility that use of Bayesian methods will aid in eliciting unspoken assumptions and prioritizing requirements (Tang, Feng, Cooper and Cangussu, 2009). Scenario-based models have been the focus of some recent research.

Bernard and others have discussed Property-Based Requirements (PBR) theory. Bernand (2012) claims the PBR approach will avoid redundancy in requirements efforts. Other practitioners and researchers are using UML for specifying, visualizing, constructing and documenting all aspects of a software system. State diagramming may be useful in special cases, in other circumstances dynamic models can be used to describe behavior through simple message passing between objects in the system. The future of requirements engineering to support MDSE will be successfully developed through continual research, coordination between researchers and practitioners, and incorporating interdisciplinary scholarship.

CONCLUSION

By emphasizing the importance of end-users being involved in development process, this chapter addresses a core issue for requirements engineering. Model-driven software engineering has a strategic dependence upon requirements engineering as a foundational part of the process. Another influence on software development methods is the likelihood that the new system will include modifications or transformations of existing systems.

MDSE creates functionality by the transformation of models based on deep knowledge of existing and available systems and tools. An early step in domain synthesis involves understanding the system through requirements documents. Requirements are abstractions that portray the essentials of a complex system, with descriptions that are constructed by defining structures and filtering out nonessential details. The requirements are derived from specification of the problem space describing expected inputs and outputs of the system, as obtained from users and system knowledge.

This chapter focuses on the centrality of users and domain specialists in the first stage of model development. Communication plays a key role during requirements elicitation, through writing and reading specifications, developing and following requirements, preparing documentation, synthesizing models, sharing graphical descriptions, identifying essential domains, and meshing old systems with new requirements. Communication also plays a role in aiding understanding of

problems, communicating with users and driving the system development process.

The goal of the requirements engineering phase is to successfully lead to MDSE development of a working system. The software engineer serves in the role of the analyst who facilitates between the end-user and the requirements documentation. The requirements are an encoding of human knowledge, experience and expectations into graphical or textual forms. Through the exploration of various tools and methods for eliciting user input as part of the software development, the role of communication has been highlighted.

This chapter presents key concepts which highlight the human role in model synthesis construction, engineering and re-engineering. One key point is the important role in fostering a tight coupling between the system engineers and the users of the system. The users are key because they are the experts in the problem domain and will actually use the implementation. The engineers must cultivate the domain knowledge of the experts in order to communicate using common notations. This chapter shows how users' contributions to domain knowledge are first-order priorities in model development. Another key point conveyed is how multiple domain models need to be integrated to build MDSE systems.

Survey construction and validation plays an important part in the requirements process. A value of construct validation for requirements engineering is to build abstractions of user models in order to fit them within the documentation of the complete system. These user models can then be interwoven together with the overall system requirements. Through this interweaving of different domain resources, the system can use user constructs to produce new software applications. Surveys are required to insure that user needs are being accurately represented by the constructs and the models. Survey validity can be improved through approaches suggested here.

Requirements elicitation supports MDSE as shown through this review of current practices and best-of-breed approaches. Finally, future research trends point to ways in which the field of requirements engineering can continue to develop in support of model-driven software engineering. Remembering the importance of humans at all stages of software development, requirements engineering and Model-Driven Software Engineering will produce valid software systems that fulfill user needs.

REFERENCES

Agarwal, N. K. (2011). Verifying survey items for construct validity: A two-stage sorting procedure for questionnaire design in information behavior research. *Proceedings of the American Society for Information Science and Technology*, *48*(1), 1–8. doi:10.1002/meet.2011.14504801166.

Bernard, Y. (2012). Requirements management within a full model-based engineering approach. *Systems Engineering*, *15*(2), 119–139. doi:10.1002/sys.20198.

Blanes, D., & Insfran, E. (2012). A comparative study on model-driven requirements engineering for software product lines. *Revista de Sistemas e Computação*, *2*(1), 1–11.

Boehm, B. (1981). *Software engineering economics*. Englewood Cliffs, NJ: Prentice Hall.

Boudreau, M., Gefen, D., & Straub, D. W. (2001). Validation in information systems research: A state-of-the-art assessment. *Management Information Systems Quarterly*, *25*(1), 1–16. doi:10.2307/3250956.

Brooks, F. P. Jr. (1995). *The mythical man-month: Essays on software engineering: Anniversary Ed.* Boston, MA: Addison Wesley.

Cabot, J., & Yu, E. (2008). *Improving requirements specifications in model-driven development processes*. Paper presented at the 1st International Workshop on Challenges in Model-Driven Software Engineering (MoDELS'08). Toulouse, France.

Cheng, B., & Atlee, J. (2007). Research directions in requirements engineering. In *Proceedings of FOSE 07 Future of Software Engineering* (pp. 285-303). Los Alamitos, CA: IEEE Computer Society Press.

Costabile, M., Mussio, P., Piccinno, A., & Provenza, L. (2008). End users as unwitting software developers. In *Proceedings of the International Conference on Software Engineering, ISSU*, (pp. 6-10). ISSU.

Davey, B., & Cope, C. (2008). Requirements elicitation - What's missing? *Issues in Informing Science and Information Technology*, *5*(1), 543–551.

Desel, J. (2002). Formalization and validation - An iterative process in model synthesis. In *Proceedings of the Workshop on Foundations for Modeling and Simulation Verification and Validation in the 21st Century* (pp. 1-18). The Society for Modeling and Simulation International.

Dobson, I., & Hietala, J. (Eds.). (2011). *Risk management: The open group guide*. Norfolk, UK: Van Haren Publishing.

Douglas, S., & Moran, T. P. (1983). Learning text-editing semantics by analogy. In *Proceedings of ACM Human Factors in Computing Systems (CHI)*. Boston, MA: ACM.

Forza, C. (2002). Survey research in operations management: A process-based perspective. *International Journal of Operations & Production Management*, *22*(2), 152. doi:10.1108/01443570210414310.

Holt, J. (2012). *Model-based requirements engineering*. London: The Institution of Engineering and Technology.

Hull, E., Jackson, K., & Dick, J. (2011). *Requirements engineering*. Dordrecht, The Netherlands: Springer. doi:10.1007/978-1-84996-405-0.

Jansen, H. (2010). The logic of qualitative survey research and its position in the field of social research methods. *Forum Qualitative Sozial Forschung, 11*(2).

Jones, J. (2006). An introduction to factor analysis of information risk. *Norwich Journal of Information Assurance, 2*(1), 67–76.

Kitchenham, B. A., & Pfleeger, S. L. (2002). Principles of survey research: Part 3: Constructing a survey instrument. *ACM SIGSOFT Software Engineering Notes, 27*(2), 20–24. doi:10.1145/511152.511155.

Meyer, B. (1992). Applying 'design by contract'. *IEEE Computer, 25*(10), 40–51. doi:10.1109/2.161279.

Miller, G. A. (1956). The magical number seven, plus or minus two: Some limits on our capacity for processing information. *Psychological Review, 63*, 81–97. doi:10.1037/h0043158 PMID:13310704.

Mulla, N., & Girase, S. (2012). Comparison of various elicitation techniques and requirement prioritisation techniques. *International Journal of Engineering, 3*(3), 51–60.

Pfleeger, S. L., & Kitchenham, B. A. (2001). Principles of survey research: Part 1: Turning lemons into lemonade. *ACM SIGSOFT Software Engineering Notes, 26*(6), 16–18. doi:10.1145/505532.505535.

Qiao, B., Yang, H., Xu, B., & Chu, W. (2003). Bridging legacy systems to model driven architecture. In IEEE Computer Society Staff (Ed.), *Proceedings 27th Annual International Computer Software and Applications Conference* (pp. 304-309). IEEE Digital Library.

Rilling, J., Meng, W., Charland, P., & Witte, R. (2008). Story-driven approach to software evolution. *IET Software, 2*(4), 304–320. doi:10.1049/iet-sen:20070095.

Tang, Y., Feng, K., Cooper, K., & Cangussu, J. (2009). Requirement engineering techniques selection and modeling an expert system based approach. In M. A. Wani, et al. (Eds.), *International Conference on Machine Learning and Applications, 2009, ICMLA'09* (pp. 705-709). Los Alamitos, CA: IEEE.

Walden, K., & Nerson, J. (1994). *Seamless object-oriented software architecture: Analysis and design of reliable systems*. New York: Prentice Hall.

ADDITIONAL READING

Adzic, G. (2011). *Specification by example: How successful teams deliver the right software*. Westampton, NJ: Manning Publications.

Aiguier, M. (2010). *Complex systems design & management*. Dordrecht, The Netherlands: Springer. doi:10.1007/978-3-642-15654-0.

April, A. (2012). *Software maintenance management: Evaluation and continuous improvement*. Hoboken, NJ: John Wiley & Sons.

Bass, L., Bergey, J., Clements, P., Merson, P., Ozkaya, I., & Sangwan, R. (2006). *A comparison of requirements specification methods from a software architecture perspective* (CMU/SEI-2006-TR-013). Retrieved October 05, 2012, from http://www.sei.cmu.edu/library/abstracts/reports/06tr013.cfm

Cody, K., & Hope, B. (1999). EX-SERVQUAL: An instrument to measure service quality of extranets. In *Proceedings of the 10th Australasian Conference on Information Systems*. IEEE.

Cordova, S. (2012). *Model-driven engineering*. New Delhi: World Technologies.

Deci, E. L., & Ryan, R. M. (1991). A motivational approach to self: Integration in personality. In R. Dienstbier (Ed.), *Nebraska Symposium on Motivation: Vol. 38: Perspectives on motivation* (pp. 237-288). Lincoln, NE: University of Nebraska Press.

Dennis, A., Wixom, B. H., & Tegarden, D. (2012). *Systems analysis and design with UML* (4th ed.). New York, NY: John Wiley & Sons.

Dooley, J. (2011). *Software development and professional practice*. Dordrecht, The Netherlands: Springer. doi:10.1007/978-1-4302-3802-7.

Habrias, H. (2010). *Software specification methods*. Hoboken, NJ: John Wiley & Sons.

Henri, H., & Marc, F. (2010). *Software specification methods: An overview using a case study*. New York, NY: John Wiley & Sons.

Holt, J. (2012). *Model-based requirements engineering*. Stevenage, UK: The Institution of Engineering and Technology.

Kaindl, H., Arnautovic, E., Ertl, D., & Falb, J. (2009). Iterative requirements engineering and architecting in systems engineering. In *Proceedings of the Fourth International Conference on Systems (ICONS 2009)*, (p. 6). Cancun, Mexico: ICONS.

Kamierski, T. J., & Morawiec, A. (n.d.). *System specification and design languages*. Boston: Springer.

Koch, N., & Kozuruba, S. (2005). Requirements models as first class entities in model-driven web engineering. *Lecture Notes in Computer Science, 3579*, 557–575.

Krogstie, J. (2012). *Model-based development and evolution of information systems: A quality approach*. Boston, MA: Springer. doi:10.1007/978-1-4471-2936-3.

Lacy, L. (2012). *Software development process*. New Delhi: World Technologies.

Lahman, H. S. (2011). *Model-based development: applications*. Upper Saddle River, NJ: Addison-Wesley.

Land, S. K. (2012). *Practical support for lean six sigma software process definition: Using IEEE*. Hoboken, NJ: John Wiley & Sons.

Langer, A. M. (2012). *Guide to software development*. Dordrecht, The Netherlands: Springer. doi:10.1007/978-1-4471-2300-2.

Montali, M. (2010). *Specification and verification of declarative open interaction models*. New York, NY: Springer. doi:10.1007/978-3-642-14538-4.

Montali, M. (2010). *Specification and verification of declarative open interaction models*. New York, NY: Springer. doi:10.1007/978-3-642-14538-4.

Monteiro, R., Araújo, J., Amaral, V., Goulão, M., & Patrício, P. (2012). Model-driven development for requirements engineering: The case of goal-oriented approaches. In J. P. Faria, A. Silva, & R. Machado (Eds.), *8th International Conference on the Quality of Information and Communications Technology (QUATIC 2012)* (pp. 75-84). Lisbon, Portugal: IEEE CPS.

Parreiras, F. S. (2012). *Semantic web and model-driven engineering*. Hoboken, NJ: John Wiley & Sons. doi:10.1002/9781118135068.

Pollack, L. (2012). *Software project management*. New Delhi: World Technologies.

Rawlings, A. (2012). *Specification languages in computer science*. Delhi: Research World.

Robertson, S., & Robertson, J. (2012). *Mastering the requirements process: Getting requirements right* (3rd ed.). Boston, MA: Addison-Wesley Professional.

Saris, W. E., & Gallhofer, I. N. (2007). *Design, evaluation, and analysis of questionnaires for survey research*. Hoboken, NJ: Wiley-Interscience. doi:10.1002/9780470165195.

Skroch, O. (2010). *Developing business application systems*. Dordrecht, The Netherlands: Springer. doi:10.1007/978-3-8349-8858-4.

KEY TERMS AND DEFINITIONS

Abstraction: The process of generalizing descriptions of things or activities which remove the individual details in order to produce less bounded conceptual results.

Analyst: A technical adviser who studies systems methodical and gives recommendations or produces documentation.

Construct: A theoretical or abstract conception that has been operationalized through the provision of definitions and descriptions.

Legacy System: Any older method, technology, computer system, or application program that is or was operating and is relevant to a project.

Metaphor: When one detail or concept is represented by a symbolic portrayal or story.

Problem Domain: The scope of topics and information that is directly associated with the system being analyzed or studied.

Requirements Documentation: A set of information that has been recorded, in graphical, textual or other forms, which represents the system that is being analyzed or subjected to systematic study.

Requirements Elicitation: Collecting, gathering, refining or discovery of information about a system that can be recorded in requirements documentation.

Software Engineering: The application of systematic, disciplined approaches to the development and maintenance of software.

Survey: A tool to aid in gathering of data and opinions that represent mental conceptions or models.

Chapter 8
Quality Assurance in Agile Software Development

Iwona Dubielewicz
Wroclaw University of Technology, Poland

Zbigniew Huzar
Wroclaw University of Technology, Poland

Bogumiła Hnatkowska
Wroclaw University of Technology, Poland

Lech Tuzinkiewicz
Wroclaw University of Technology, Poland

ABSTRACT

Agile methodologies have become very popular. They are defined in terms of best practices, which aim at developing good quality software faster and cheaper. Unfortunately, agile methodologies do not refer explicitly to quality assurance, which is understood as a planned set of activities performed to provide adequate confidence that a product conforms to established requirements, and which is performed to evaluate the process by which products are developed. The chapter considers the relations of agile practices with software life cycle processes, especially those connected to quality assurance, and tries to answer the question of which agile practices ensure software quality. Next, agile practices associated with quality assurance are assessed from different perspectives and some recommendations for their usage are given. It is observed that modeling has a particular impact on quality assurance.

INTRODUCTION

Agile methodologies have become very popular in software companies. In spite of that they do not refer directly to model-driven approach, they help develop software faster and more effectively. Many reports confirm the advantages of an agile approach when applied to small and medium-size scale software projects. Unfortunately, agile methodologies do not refer explicitly to quality assurance. Practitioners of agile methodologies have not yet provided a convincing answer to the question "What is the quality of the software product?"

DOI: 10.4018/978-1-4666-4494-6.ch008

ISO/IEC/IEEE 24765:2010 (p. 287) provides some general definitions of quality assurance (QA):

A planned and systematic pattern of all actions necessary to provide adequate confidence that an item or product conforms to established technical requirements.

A set of activities designed to evaluate the process by which products are developed or manufactured.

The definitions show that QA is process oriented and focuses on technical requirements (e.g. conformance to standards, procedures, etc.). From this perspective QA may be thought of as a defect-preventing set of activities.

On the other hand, ISO/IEC 12207:2008 (p. 6) defines QA as:

The planned and systematic activities implemented within the quality system, and demonstrated as needed, to provide adequate confidence that an entity will fulfill requirements for quality.

Similar definition of QA is given in KPI Library (http://kpilibrary.com/store/items/33):

Quality assurance is the systematic monitoring and evaluation of the various aspects of a project, service or facility to maximize the probability that minimum standards of quality are being attained by the production process. QA cannot absolutely guarantee the production of quality products.

Two principles included in QA are: "Fit for purpose" – the product should be suitable for the intended purpose; and "Right first time" – mistakes should be eliminated. QA includes regulation of the quality of raw materials, assemblies, products and components, services related to production, and management, production and inspection processes.

These definitions take into consideration not only the technical requirements a product should fulfill but also the requirements for quality, so they have a broader meaning. From this perspective, QA is also thought of as a defect-detecting set of activities.

Sometimes, the activities that focus on defects detection are called Quality Control (QC): "A set of activities designed to evaluate the quality of developed or manufactured product" (ISO/IEC/IEEE 24765:2010, p. 287)).

In the chapter, software quality assurance (SQA) is understood in the broader sense, so this notion includes QC. There are two different purposes for SQA activities (ISO/IEC/IEEE 24765:2010, p. 287): (1) to convince the customer about the quality of the software which he is going to acquire (external perspective) and (2) to support an organization in developing high-quality software (internal perspective)

The first purpose is fulfilled by agile's approach features e.g. sprint demos, retrospective meetings, acceptation and implementation of proposed changes and so on. In terms of the second aspect, SQA is cited as the one of the software quality management processes (the others two are: quality planning process and quality control process (PMBOK, 2004). All these well-structured processes, together with some additional elements (procedures, quality standards, documentation), constitute the Quality Management System (QMS) inside the organization. QMS is usually adapted to any new project which is undertaken in an organization (i.e. appropriate quality standards, documentation, procedures are chosen). A QMS includes range of techniques and tools for quality management which are used when performing SQA.

In this chapter we are interested in the internal SQA perspective but at the same time we have restricted our considerations only to the execution phase of the project (i.e. software development) and only to the development team perspective. This means that, for example, QMS planning processes are not in the field of our interest (it is assumed that the quality planning process was finished before the development process started). Hence, the quality management system will not be considered further.

The peculiarity of SQA definitions is that the notion of quality is not used directly. SQA abstracts from a concrete definition of software quality, but regardless of the specific definition, emphasizes all the activities that foster quality.

In general, SQA definitions leave room for a range of interpretations. Review of the literature related to SQA does not deliver an unambiguous definition of this notion. For example, one paper (Hongying & Cheng, 2011) proposes a model of SQA and does not define or reference the notion. Some papers identify QA with testing, while the paper written by Janus et al. (2012) identifies it with some arbitrary selected practices. In the paper (Mnkandla & Dwolatzky, 2006) SQA is seen through a list of different quality factors to which some practices are allocated.

The paper (Bhasin, 2012) seems to be most aware of the issue of SQA definition, raising questions such as:

How can we manage the quality of software in Agile?

Which practices of Agile will ensure software quality?

What are the key drivers of quality in Agile?

but without providing answers to them.

As was mentioned above, SQA is process oriented. One of the questions we try to answer is which software development processes may be considered as strongly related to SQA. To do so, we assume that the ISO/IEC 12207:2008 recommendation is the basic source which identifies all the processes. Now, SQA related processes may be considered as a subset of the processes defined by this recommendation.

The set of 43 processes identified by ISO/IEC 12207:2008 is divided into 7 groups. The ISO describes the processes very generally and does not describe how to implement them. There are many possible ways to realize these processes, and in real projects many rules, suggestions and hints are in use – they are sometimes referred to as practices and some of them as best practices. The best practices may be recommended to be used as a whole or a part of a given process. In the case of the use of a practice, the given practice is considered as a fragment of a given process or, more precisely, as a fragment of an instance of the process.

There are numerous practices in use, especially in an agile approach to software development. The questions arise: "Which of them may be applied within processes supporting quality assurance?" and "Which of them are the most influential ones?"

Modeling is a basic paradigm of modern software development. It is considered effective in software development if models may be used to describe an application domain and can serve as a basis for implementing systems. The paradigm has found its application in model-driven engineering (MDE). MDE focuses on creating and exploiting models through the entire software development cycle. The vast majority of MDE approaches are heavyweight methodologies. RUP (Kruchten, 1999) and UP (Scott, 2001) are examples of such methodologies and have proved their usefulness, especially for large software projects. In opposition to heavyweight methodologies there is a group of so-called lightweight software development methodologies which are based on an agile approach. The agile approach, assuming iterative and incremental software development, focuses on the user's requirements and programming code. The approach tends to reject modeling because it entails some form of extra documentation, which conflicts with the Agile Manifesto postulates (http://agilemanifesto.org/).

Lightweight methodologies draw attention to best practices – recommended methods and techniques to perform project tasks. Modeling, although rare, is also considered a best practice. Therefore, it is interesting to consider what the role of modeling is for quality assurance in agile software development. Does the use of modeling have a significant impact on software quality assurance? In particular, what is its impact on software quality assurance in the context of other widely recommended practices? Our findings show that modeling is underestimated by agile methodologies, in spite of the fact that its adoption is less difficult than the adoption of many other commonly used practices.

The aim of the chapter is to provide answers to the above stated questions. We start with a review of the practices applied by the most popular agile methodologies. Next, the practices which may influence quality assurance are identified, and finally, after analysis and assessment, the best practices for quality assurance are recommended.

The chapter is organized as follows. Section 2 provides a short overview of software life-cycle processes defined by ISO/IEC 12207:2008 associated with QA. Section 3 lists the most popular agile methodologies together with their practices. Section 4 checks if there exists any relation between agile practices and ISO processes, in other words if an agile practice can be treated as a part of selected ISO process. Section 5—the core section—presents an assessment of practices supporting ISO quality assurance processes from different angles. Conclusions and recommendations are gathered in Section 6.

ISO PROCESSES SUPPORTING QUALITY ASSURANCE

According to ISO/IEC 12207:2008 the activities that may be performed during the software life cycle contain seven groups of processes:

- Agreement Processes (2 processes)
- Organizational Project-Enabling Processes (5 processes)
- Project Processes (7 processes)

- Technical Processes (11 processes)
- Software Implementation Processes (7 processes)
- Software Reuse Processes (3 processes)
- Software Support Processes (8 processes)

Each of the processes is described in terms of its purpose and desired outcomes, and lists activities and tasks which need to be performed to achieve those outcomes.

Agreement Processes involve activities that serve to establish agreement between two organizations, i.e. the supplier and acquirer of a product or service.

Organizational Project-Enabling Processes contain different management processes that are implemented within an organization for initiation, support, and control of the project. The Quality Management Process is one of them.

Project Processes are those "concerned with planning, assessment, and control" (ISO/IEC 12207:2008, p. 15) regardless of the type of project. They are split into two categories: Project Management Processes ("used to plan, execute, assess and control the progress of a project", (p. 15)), and Project Support Processes (that "support specialized management objectives," p. 15).

Technical Processes are typical software engineering activities which aim to define the requirements, system architecture design, implementation, integration, and testing, as well as software maintenance and keeping the software in use.

Software Implementation Processes aim at production of a specified software item or service.

Software Reuse Processes support reusing software items across project boundaries.

The last group, Software Support Processes (SSP), contain activities which "assist the Software Implementation Process as an integral part with a distinct purpose, contributing to the success and quality of the software project" (ISO/IEC 12207:2008, p. 17). The group is described as one that contributes "to the success and quality of the software project" (ISO/IEC 12207:2008, p. 17), and comprises the following processes:

1. Software Documentation Management Process (SDMP)
2. Software Configuration Management Process (SCMP)
3. Software Quality Assurance Process (SQAP)
4. Software Verification Process (SVerP)
5. Software Validation Process (SValP)
6. Software Review Process (SRP)
7. Software Audit Process (SAP)
8. Software Problem Resolution Process (SPRP)

Below, basing on (ISO/IEC 12207:2008),a short overview of Software Support Processes is given.

The purpose of the Software Documentation Management Process "is to develop and maintain the recorded software information produced by a process" (ISO/IEC 12207:2008, p. 67). Having this process implemented the developers know what, why and how to document (languages, standards). Requirement specifications, models, source code are examples of different documents.

The Software Configuration Management Process is a specialization of the Configuration Management Process which focuses on software items understood as "source code, object code, control code, control data, or a collection of these items" (ISO/IEC 12207:2008, p. 7). Its purpose is to maintain the integrity of software items that is to identify and define them, to control modifications and releases.

The Software Quality Assurance Process aims at assuring "that work products and processes comply with predefined provisions and plans" (ISO/IEC 12207:2008, p. 70). One of the main outcomes of that process is a quality assurance plan which defines, for example, quality standards, procedures, and tools for quality, procedures for maintaining quality records, and employs activities from other, above mentioned processes. The process concerns both the software product (checks if the product together with documentation complies with the contract and adheres to plans) and development process.

Software Verification Process aims at confirming that software products and/or services conform to the requirements. Verification can be done at different levels and for different artifacts, e.g.

*   **Requirement Verification:** Both system, and software requirements are verified for consistency, feasibility, testability; system requirements are verified for their proper allocation to hardware and software items, while software requirements are verified for their traceability to system requirements;
*   **Design Verification:** Verification against correctness, traceability to requirements, fulfilling non-functional demands;
*   **Code Verification:** Verification against testability, correctness, compliance with requirements, and coding standards, as well as implementing non-functional demands;
*   **Integration Verification:** Verification against completeness and correctness of integration of all units (software, and hardware);
*   **Documentation Verification:** Verification against adequacy, completeness, consistency, and being produced on time.

The purpose of the Software Validation Process is to confirm "that the requirements for a specific intended use of the software work product are fulfilled" (ISO/IEC 12207:2008, p. 74). Validation is strongly related to testing, but not only dynamic techniques can be employed by it. Interesting alternatives are analysis, modeling, and simulation.

The purpose of the Software Review Process is to gain a common understanding among all involved parties (developers, customers, users) of progress in relation to the objectives of the project, and "what should be done to help ensure development of a product that satisfies the stakeholders" (ISO/IEC 12207:2008, p. 76). The reviews can be done at different levels – managerial and technical. Further we are interested first of all in technical reviews that check whether products or services are complete, comply with appropriate standards and specifications, and are ready for further development. Any problems identified are recorded and passed to the Problem Resolution Process.

The purpose of the Software Audit Process is to determine the compliance of selected products/processes with requirements, plans, and agreements by independent parties. A specific audit can be done for software to ensure, for example, that a software product reflects the design documentation, test data comply with specifications, and user documentation complies with standards as specified. Any problems identified are recorded and passed to the Problem Resolution Process.

The purpose of the Problem Resolution Process is to ensure that "all discovered problems are identified, analyzed, managed and controlled to resolution" (ISO/IEC 12207:2008, p. 78). Problems should be classified and prioritized to facilitate analysis and problem resolution.

Software Support Processes are considered as processes addressing and supporting SQA activities. The SQAP process relates directly to SQA while the processes SVerP, SValP, SRP, and SAP, are related indirectly through their reference to software quality. The other processes SDMP, SCMP, and SPRP also have a strong influence on software quality.

Although the SQA is associated with the performance of all the above mentioned processes with due diligence, it seems that it is particularly important to the Software Support Process group. It can be observed that some processes from this group are also associated with processes from other groups. For example, the Software Qualification Testing Process as well as the System Qualification Testing Process may be used in the Software Verification Process or the Software Validation Process. For these reasons, the Qualification Testing Process will not be considered separately in this paper. Similarly, the Quality Management Process (QM) belonging to the Organizational Project-Enabling Process group aims at management of activities, tasks, and deliverables of other processes associated with SQA at the level of the organization. Management means defining goals

to achieve, monitoring the status of the managed element, and undertaking appropriate, perhaps corrective actions. Status monitoring should definitely be accompanied by objective measures provided by the Measurement Process (belonging to the Project Processes group). The purpose of the Measurement Process "is to collect, analyze, and report data relating to the products developed and processes implemented within the organization" (ISO/IEC 12207:2008, p. 41).

In summary, we have selected these processes identified by ISO/IEC 12207:2008 that are related to SQA and SQC, which, according to (SQA in Practice, http://www.sqa.net/softwarequalitycontrol.html) are defined as:

SQA – the function of software quality that assures that the standards, processes and procedures are appropriated for the project and are correctly implemented. It also collects software measures that are relevant to the evaluation of the use of the standards, processes, and procedures.

SQC – the function of software quality that checks that the project follows its standards processes, and procedures, and that the project produces the required internal and external (deliverable) products.

AGILE METHODOLOGIES AND APPLIED PRACTICES

The methodology of software development is recognized as an agile one when it allows adaptation to an ever-changing environment, especially unexpected changes, and this adaptation is profitable. Profit is obtained due to the leanness of the methodology i.e. the methodology shortens the development time and cost, simultaneously ensuring the stable or improved quality of the product.

To guarantee continuous adaptation agile methodologies postulate creating an empowered development team which is quick to react to changes and which learns during and after the development process. The essential difference which is observed while comparing agile and traditional approaches is that the former do not define the development process precisely, so there is no need for detailed plans.

Another important feature of the agile approach is its 'people-orientation' after recognition that people and their collaboration are the most important drivers for a project's success. So, software is developed by a collaborative self-organized, cross-functional team. As the team is self-organized, the agile methodologies do not define in details any development process but instead recommend several practices (sometimes called best practices) which allow delivering the desired solution with fewer problems. The next common characteristic of the agile group of software development methodologies is: iterative and incremental development (thus enabling the evaluation of the current solution), and the iterative approach is placed in a well-defined time frame (time-box). Each agile methodology recommends such a set of interrelated practices which adhere to the agile values and principles stated in the Agile manifesto (http://www.agilemanifesto.org/) and Agile principles (www.agilemanifesto.org/principles.html), respectively.

The meaning of a practice is ambiguous; in the literature various definitions can be found of a practice (or best practice). For example, according to (Michaelson, 2006) best practice is defined as *"the most efficient (with least amount of effort) and effective (giving best results) way of accomplishing a task, based on repeatable procedures that have proven themselves over time for large numbers of people."* Closer to our opinion is the practice definition given by (Ambler, 2010) *"a practice is a self-contained, deployable component of a process."*

Since the middle of the nineties many agile methodologies have appeared: Extreme Programming (XP) (Beck & Andres, 2004), Scrum (Schwaber, 2003), Crystal Clear (Cockburn, 2004), Feature Driven Development (FDD)

(Palmer & Felsing, 2002), Dynamic Software Development Methodology (DSDM) (DSDM, 2008), Agile Unified Process AUP (Ambler, 2005), Agile Modeling (Ambler, 2002) are the best known and most frequently applied among them.

The practices recommend by the seven most popular—according to (State of Agile survey, 2011)—methodologies are listed below.

1. **Extreme Programming:** Sit together, Whole team, Informative workspace, Energized work, Pair programming, Stories, Weekly cycle, Quarterly cycle, Slack, Ten-minute build, Continuous integration, Test-first programming, Incremental design, Real customer involvement, Incremental development, Team continuity, Shrinking teams, Root-cause analysis, Shared code, Code and tests, Single code base, Daily deployment, Negotiated scope contract, Pay per-use.

2. **Crystal Clear:** Frequent Delivery, Reflective Improvement, Osmotic Communication, Personal Safety, Focus, Easy Access to Expert Users, A Technical Environment with Automated Tests, Configuration Management and Frequent Integration.

3. **Feature Driven Development:** Domain object modeling, Developing by feature, Individual class ownership, Feature teams, Inspection, Regular builds, Configuration management, Reporting/visibility of results.

4. **Scrum:** Product Backlog, Daily Stand-Up Meeting, Sprint Burndown chart, Sprint Demo, Sprint review, Retrospectives; (this methodology is a collection of practices only for management aspects of software development on the team level).

5. **Dynamic Software Development Methodology:** Facilitating Workshops, Modeling (related to managerial and development aspects) Prototyping Time-boxing Prioritization; it recommends the Scrum and XP practices for the management and development aspects on the team level.

6. **Agile Unified Process:** Architecture Envisioning, Continuous Integration, Database Refactoring, Ranked Work Item List, Requirements Envisioning, Test-Driven Development.

7. **Agile Modeling:** Active Stakeholder Participation, Executable Specifications, Iteration Modeling, Prioritized Requirements, Requirements Envisioning (this methodology is a collection of practices only for lightweight modeling and documentation).

According to (State of Agile Survey, 2011) there exist other agile methodologies, for example: some varieties of Crystal methodology (Crystal Orange, Crystal Red), Lean, Kanban or Scrum/XP or Scrumban. These methodologies are not considered below because they do not essentially recommend practices other than those mentioned above. We have also omitted those methodologies that recommend other practices but which are used very rarely.

AGILE PRACTICES AS PARTS OF ISO PROCESSES SUPPORTING QUALITY ASSURANCE

Now, all the practices of the agile methodologies, mentioned in Section 3, will be compared with the Software Support Processes and Quality Management Process that have been selected in Section 2. The aim of the comparison is to determine whether a given practice may be considered as an instance of an activity within a given process.

The results of the comparison are presented in Table 1. The rows of the table represent only a subset of all the considered practices. To the subset belong these and only these practices that may be considered as an instance of some activity within a process from the set of selected processes. Practices with different names but the same meaning are listed on a single line; different names are separated by a slash. The columns of the table represent selected ISO processes: eight

processes from the group of Software Support Processes, and one Quality Management Process from the group of Organizational Project-Enabling Processes. An 'x' on the intersection of a row and a column indicates that the practice may be considered as a part of the process.

The decision whether to put an 'x' on the item was based on the comparison and analysis of two definitions – the definition of the practice and the process definition (the mark is put if the definitions have something in common).

Now we can perceive a given practice in conjunction with the relevant ISO processes; the practice can be instantiated within the chosen processes, if needed.

Note that none of the practices is associated with the Software Audit Process. Note also that 20 practices are associated with ISO quality assurance processes. The remaining 19 practices that are not associated with these ISO processes are as follows:

Sit together, Whole team, Energized work, Slack, Ten-minute build, Incremental design, Team continuity, Shrinking teams, Shared code, Single code base, Daily deployment, Negotiated scope contract, Pay per-use, Osmotic Communication, Personal Safety, Focus, Sprint planning meeting, Prototyping, Requirements envisioning.

ASSESSMENT OF PRACTICES SUPPORTING ISO QUALITY ASSURANCE PROCESSES

In the previous section, we selected agile practices, which can be applied as a part of ISO quality assurance processes. This section focuses on an assessment of practices. The assessment can be made from different perspectives.

The first assessment perspective is an application perspective. On the basis of existing surveys we analyze to what extent the practices selected in section 4 are perceived as important, and are used in IT companies.

Table 1. Practices supporting quality assurance within ISO processes

Practice	SDMP	SCMP	SQAP	SVerP	SValP	SRP	SAP	SPRP	QM
Informative workspace			x						x
Pair programming				x					
Timeboxing/Quarterly cycle									x
Continuous integration		x	x	x					
Incremental development									x
Root-cause analysis								x	
Code and tests	x								
Product Backlog	x*								
Requirement Prioritization/Prioritized Requirements/ Ranked Work Item List	x*								
Stories	x*								
Daily Stand-Up Meeting			x					x	
Sprint Burndown chart			x						x
Sprint Demo/Weekly cycle/Frequent Delivery					x				
Sprint review /Reflective Improvement			x		x			x	
Retrospectives/Facilitating Workshops			x				x	x	
Product Owner/Real customer involvement/On-site Customer/Easy Access to Expert Users/Active Stakeholder Participation					x				
Modeling/Iteration Modeling/Domain object modeling	x*								
Executable Specifications	x			x					
Database Refactoring	x								
Test-Driven Development/Test-first programming	x			x					

*The practice is optional provided it is required by documentation management plan elaborated within the process.

The second assessment perspective takes into account the difficulty of practice implementation. It appears that an absolute assessment of difficulty is not possible, as it strongly depends on the context of a project. In some cases, a specific practice implementation can meet some impediments or even be impossible.

Implementation of all "quality assurance" practices could be very costly, and—sometimes—not justified. Each practice addresses some problems, which can also be considered as part of a project's context. When these problems occur, the practice can be recommended as a means to solve the problem. Hence, the third perspective of practice assessment is their usefulness in solving specific problems.

In all assessments we consider the selected practices separately. It cannot be ruled out that the practices are mutually related in various ways and it is conceivable that they might be dependent. Unfortunately, the relationships between practices are presented in the literature (e.g. http://www. agilemodeling.com/) without an exact interpretation or semantics of these relationships. As a result, it is difficult to assess the impact of the relationships between practices on the way the practices are implemented. In our opinion, the order of implementation of these practices may be relevant, for example, because of the experience and expertise of the teams. With this in mind we decided that the practices can be adopted and used independently, which is also reflected in practice. However, it should be noticed that the common use of certain practices provides a synergy effect.

Assessment of Practices from Application Perspective

Agile practices have been assessed by IT practitioners through a variety of questionnaires (e.g. Ambler, 2008; Sochova, 2009; Appelo, 2009).

The first one was performed by Ambler in 2008, February. The results obtained, as well as row data, are available on the Internet (http://www.ambysoft.com/surveys/agileFebruary2008.html). The questionnaire had 123 experienced respondents – only 3.3% had less than 2 years work experience, and 60% had used agile practices for more than 3 years.

The Ambler' assessment took plenty of aspects into account, i.e. practice effectiveness, ease of learning, difficulty of learning, practices abandoned, practices a company wants to adopt. Unfortunately, there is no information, as to which practices were being used by respondents when the questionnaire was filled in. What's more, the list of considered practices is shorter than we are interested in. In Table 2 we quoted the results of two aspects: "effectiveness", and "want to adopt".

Continuous integration is perceived as the most effective (64%) and desired practice (16.9%), next to Test-Driven Development (46.9%, 21.3%). Similarly, good results have daily stand-up meeting (46.9%, 6.7%), and refactoring (42.5%, 9%). Pair programming is one of the more controversial practices. On the one side it was rated at 36% (effectiveness), and 19.1% (wanted), but on the other 24.3% of respondents marked this practice as abandoned.

A second survey of agile practices was made by Zuzana Sochova as a part of her MBA dissertation work (Sochova, 2009). The 181 respondents were from different companies, in term of size, business, industry, culture, and geography. Almost 50% of them had used agile methods for more than 2 years. The practices were assessed in ranked scale (from 1 to 5) by their difficulty in adoption, and usefulness. Not all practices we are interested in were covered by the survey. The obtained results (average values calculated by us) for effectiveness are presented in Table 2. The highest marks were given to daily stand-up meeting (4.25), and product backlog (4.24). Next are sprint demo (4.02), and retrospectives (3.82).

The most complete survey of agile practices was carried out by Jurgen Appelo (2009). Opinions were gathered from 341 participants about agility, importance, and applicability of practices. The results for the two later characteristics are given in Table 2. It seems that the survey was filled in by agility fans. The marks for almost all practices are very high. The highest marks in both areas were given to product backlog, requirement prioritization, daily stand-up meeting, and refactoring. Very high or high scores were awarded to time-boxing, continuous integration, stories, sprint demo, frequent delivery, sprint review, retrospectives, and executable specifications.

It should be noticed that some practices were absent in all surveys, i.e. quarterly cycle and incremental development. The former seems not very common, while the later is so obvious that it is even not assessed.

Taking the application perspectives into account, which are encouraged by experiences gathered in the IT industry, we can, first of all, recommend the following practices: daily stand-up meeting, and retrospectives. Subsequently recommended practices are: continuous integration, product backlog, sprint demo, frequent delivery, on-site customer, executable specifications, refactoring, and test-driven development.

Table 2. Summary of survey results related to selected practices

Applicability Dimension	Ambler – 2008		Sochova – 2009	Appelo – 2009	
Practice	Effectiveness	Want to Adopt	Effectiveness	Level of Importance	Level of Appliance
Informative workspace (Information Radiators)				89.9%	70.5%
Pair programming	36.3%	19.1%	2.95	75.2%	52.2%
Weekly cycle (Timeboxing)				92.5%	90.8%
Quarterly cycle					
Continuous integration	64.6%	16.9%		96.7%	86.5%
Incremental development					
Root-cause analysis				84.1%	46%
Code and tests/Coding standards (Ambler)	12.4%	9%			
Product Backlog	23%	3.4%	4.24	98.2%	93.2%
Requirement Prioritization/Ranked Work Item List	23%	3.4%		98.2%	94.4%
Stories				95.3%	83%
Daily Stand-Up Meeting	46.9%	6.7%	4.25	95.6%	90.8%
Sprint Burndown chart	25.7%	10.1%	3.61	83.1%	80%
Sprint Demo	25.7%	7.9%	4.02	96.1%	84.9%
Frequent Delivery/Potentially Shippable Software Each Iteration (Ambler)	28.3%	21.3%		98.3%	83.3%
Sprint review/Reflective Improvement				96.1%	84.9%
Retrospectives/Facilitating Workshops	38.9%	4.5%	3.82	98.0%	84.7%
Product Owner/Real customer involvement/ On-site Customer/Easy Access to Expert Users/ Active Stakeholder Participation	35.4%	13.5%		94.7%	66%
Modeling/Iteration Modeling/Domain object modeling/Domain Driven Design/ JIT Model Storming (Ambler)	3.5%	14.6%		79.2%	46.4%
Executable Specifications	4.4%	18%		95.3%	83%
Refactoring	42.5%	9%		98.9%	91.8%
Test-Driven Development/Test-first programming	46.9%	21.3%	3.49	93.7%	71.4%

Assessment of Difficulties of Practices Application

IT companies can come across different obstacles while implementing agile practices. The impediments can differ according to their nature and cause. According to Ambler's observation, an assessment of the difficulties of a practice's adoption is not absolute, but depends on the context of its use.

To simplify the discussion we assume that the companies we are talking about are at least at the second managed level – of the CMMI maturity model (CMMI® for Development v. 1.2., 2006) so they are able to plan, perform, measure, and control the development process. In this way, we want to eliminate obstacles resulting from lack of experience or applying a disorganized process.

Ambler defines the context of a practice's use through a set of eight factors (Ambler, 2009):

1. Team size,
2. Geographical distribution,
3. Regulatory compliance,
4. Domain complexity,
5. Organizational distribution,
6. Technical complexity,
7. Organizational complexity,
8. Enterprise discipline.

Each of the factors embraces a set of specific situations or properties, which cannot be completely enumerated and ordered. Therefore, we decided to concentrate only on the factors for which it is relatively simple to extract such properties and give them a linear ordering. We concentrated on five factors: team size, geographical distribution, regulatory compliance, domain complexity, and technical complexity. Additionally, we took into account one more factor: people's competency (skills and experience), because we are convinced that it has a fundamental influence on the context of practices used.

To simplify our analysis, we decided to extract only two extreme situations or properties for each selected factors. These are:

- {small, large} for team size,
- {distributed (yes), not distributed (no)} for geographical distribution,
- {with regulation (yes), without regulation (no)} for regulatory compliance,
- {simple, complex} for domain complexity,
- {simple, complex} for technical complexity, and
- {novice, advanced} for people's competency.

The set of these six factors determines the dimensions of the context space for the practices used. Each valuation of the factors defines one, concrete context of practices' use. As a result, we have 2^6 different contexts.

For other omitted factors we have assumed that:

- There is no organizational distribution; development teams are made up of people working for the same organization.
- Organizational complexity is on a level providing a mature, pro-IT culture.
- Enterprise discipline is also on a mature level.

For each given practice, we estimated difficulty of its adoption with respect to a given value (situation or property) attached to a factor.

The difficulty level is one of three values:

- **1:** Practice is difficult to adopt,
- **0.5:** Adoption of the practice may meet some problems,
- **0:** Practice may be adopted without difficulty.

Table 3 summarizes our estimates of the difficulty of adoption for all selected practices for the set of values assigned to the factors. The factors are considered independently, and the levels of difficulty in the table are the upper estimates of the difficulties of adoption. The estimates are based on our professional experience, and also on two years' observation of student teams developing software systems within group projects at the Faculty of informatics and Management, Wroclaw University of Technology.

On the basis of Table 3 some observations were derived.

First, we compare the practices with respect to their difficulty of adoption. Figure 1, in the form of a bar diagram, presents the average and the deviation in difficulty of practice adoption, assuming uniform distribution of possible context usage.

Note, that *Incremental development* practice does not involve any difficulty. Also easy to adopt are (in most cases) *Sprint Burndown chart, Infor-*

Table 3. Difficulty levels for practice adoption

Context Dimension	Team Size		Domain Complexity		Technical Complexity		Geographical Distribution		Regulatory Compliance		People Competency	
Practice	Small	Large	Low	High	Low	High	No	Yes	No	Yes	Novice	Advanced
Informative workspace	0	0	0	0	0	0.5	0	0.5	0	0.5	0.5	0
Pair programming	0	0.5	0	0.5	0	0.5	0	1	0	0	1	0.5
Time-box, e.g. weekly cycle	0	0.5	0	0.5	0	0.5	0	0	0	0.5	0.5	0
Continuous integration	0	0.5	0	0	0	0.5	0	0	0	0.5	1	0.5
Incremental development	0	0	0	0	0	0	0	0	0	0	0	0
Root-cause analysis	0	0.5	0	0.5	0	0.5	0	0.5	0	0.5	1	0.5
Code and tests	0	0.5	0	1	0	0.5	0	0	0	0.5	0.5	0
Product Backlog	0	0.5	0	0.5	0	0.5	0	0	0	0	0.5	0
Requirement Prioritization/Ranked Work Item List	0	0.5	0	0.5	0	0.5	0	0	0	0	1	0.5
Stories	0	0.5	0	0.5	0	0.5	0	0	0	0	0.5	0
Daily Stand-Up Meeting	0	1	0	0	0	0	0	1	0	0	0.5	0
Sprint Burndown chart	0	0	0	0	0	0	0	0	0	0.5	0.5	0
Sprint Demo	0	0.5	0	0.5	0	0.5	0	0	0	0.5	0.5	0
Frequent Delivery	0	0.5	0	0.5	0	0.5	0	0	0	0.5	1	0.5
Sprint review/Reflective Improvement	0	1	0	0	0	0	0	0.5	0	0	1	0.5
Retrospectives/Facilitating Workshops	0	0.5	0	0	0	0	0	0.5	0	0.5	1	0.5
Product Owner/Real customer involvement/ On-site Customer/Easy Access to Expert Users/ Active Stakeholder Participation	0	0.5	0	0.5	0	0	0	0.5	0	0.5	0.5	0
Modeling/Iteration Modeling/Domain object modeling	0	0.5	0	0.5	0	0	0	0	0	0.5	0.5	0
Executable Specifications	0	0	0	0.5	0	0.5	0	0	0	0.5	1	0.5
Refactoring	0	0	0	0.5	0	0.5	0	0	0	0.5	1	0.5
Test-Driven Development/ Test-first programming	0	0	0.5	1	0.5	1	0	0	0	0.5	1	0.5

mative workspace and *Modeling*. Hard to adopt are *Test-Driven Development, Pair programming* and *Root-cause analysis*. *Test-Driven Development, Sprint Review, Daily stand-up meeting* and *Pair programming* have the highest deviation,

which means that these practices strongly depend on the context of their adoption.

Our results are partially confirmed by Ambler's (2009), and Sochova's (2009) surveys, in which the difficulty of adopting practices was

Figure 1. The average and standard deviation of difficulty of practices adoption

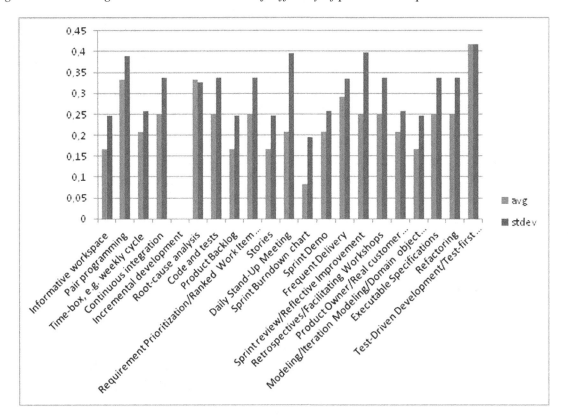

also assessed. These surveys did not consider any specific contexts. Both point out that *Test-driven development* and *Pair programming* are the most difficult to adopt (*Root-cause analysis* was not investigated). According to the surveys *Daily stand-up meeting* and *Sprint demo* should not encounter obstacles during adoption. It should also be noted that—according to Ambler's results—*Modeling* practice belongs to the group of practices with low difficulty level.

Figure 2 shows the range of difficulties of practice adoption for extreme contexts. The higher bars relate to the maximal context, i.e. the context designated by the upper values for each of the context dimensions, while the lower bars relate to minimal context, i.e. the context designated by lower values for each of the dimensions.

To sum up, agile practices should be carefully selected depending on the context of the particular project. There exist some practices that

seem to be context insensitive, e.g. *Sprint burn-down chart*, *Product Backlog*, or *Modeling*. They may be adopted first. On the other side, there are several practices whose difficulty of adoption is strongly influenced by project context (e.g. *Pair programming*) or which are generally hard to adopt (e.g. *Test-driven development*). They should be applied only if the expected benefits overcome the obstacles and if they solve vital problems during software development.

Recommended Practices to Solve Problems

The considered practices were selected as practices supporting quality assurance. The support is reflected in the fact that the practices help to solve some problems. The problems are understood here as some urgent needs or troubles manifested by specific situations or challenges.

Figure 2. The average of difficulty of practices adoption for two extreme contexts

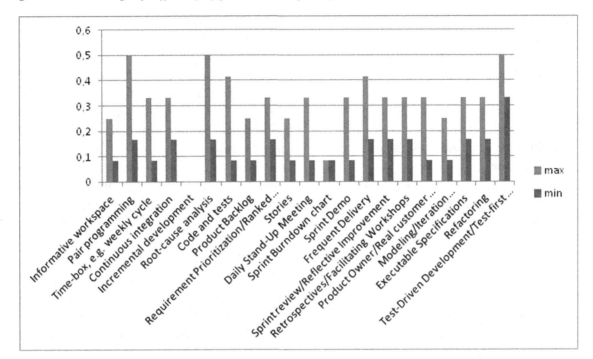

To find out how problems may be assisted by the practices, a survey was carried out among software development teams.

The participants of the survey were four teams from Lower Silesian industry, representing software development and IT companies and one team from a university representing experience resulting from managing many students' collaborative projects.

The starting point to the research was a list of selected problems following the structure of software development processes. The problems were divided into two categories: engineering problems related to software development, and managerial problems related to the management of software development process.

The problems were chosen subjectively based on our personal experience as well as an analysis of the problems contained in the descriptions of different software development methodologies.

The set of identified problems belonging to the first category was structured into the following disciplines: software requirements, software design, software construction, software testing, and

software maintenance. Below, there is a structured list of respective problems.

Software Requirements:

- Elicitation of customer needs
- Unstable requirements
- Inaccurate, incomplete requirements specification
- Incorrect interpretation of the requirements
- Systematic presentation of the requirements
- Key needs identification
- Lack of domain knowledge

Software Design:

- Variety of implementation platforms
- Complex domain

Software Construction:

- Problems with code integration
- Reduction of repeatable programming activities via code generation
- Simple domain

- Complex domain
- Unreadable code structure or database structure
- Different quality of code written by different members of a team

Software Testing:

- Verification of artifacts against their specification
- Desire of acceptable quality of code
- The need for continuous validation of product
- Extensive documentation requirements

Software Maintenance:

- Corrective maintenance
- Adaptive maintenance
- Perfective maintenance
- Preventive maintenance

The set of identified problems belonging to the second category was structured as follows:

Team Communications:

- The need for current information about the state of the project

Risk Analysis:

- Requirement for successful implementation of the medium or large scale project
- Deviation from the plans
- Too late disclosure of problems and their removal
- Working in a multiplayer team

Time Planning:

- Planning of iterations and releases

The results of the survey are presented in Table 4, where the numbers of practices are explained in the list below it.

Figure 3 informs how many times a practice was selected as a remedy for potential problems that can arise during the software life-cycle. Surprisingly, the winner is *Modeling*, which was pointed out 15 times. Next to it come *Executable specifications*, *Product owner*, and *Stories*. On the other end we have practices suggested by single survey participants: *Team velocity*, *Sprint planning* (its result is Sprint Backlog, so we combine it with *Product Backlog*), *System metaphor*, and *Kanban* (more a methodology than a practice).

Figure 4 presents the number of practices selected by survey participants in solving problems from different disciplines. The results confirm the commonly known fact that requirements engineering is a crucial phase in software development. The problems emerging at this phase must be solved effectively, which explains why the number of practices recommended here is so large.

It turns out that the most important problem in the requirements discipline is the 'complex domain' problem. In support of this problem, the practices *Modeling*, *Product Owner*, and *Stories* had four recommendations, and the practices *Code and tests*, *Informative workspace*, *Pair programming*, *Refactoring*, *System metaphor*, and *Test-Driven Development* had two recommendations. The survey shows that the main problem in the context of "Software design", which were recommended for most practices, were "complex domain", which confirms that the understanding of the field of "problem domain" is the key to quality.

CONCLUSION AND FURTHER WORKS

Based on the ISO/IEC standard describing the processes involved in software development, we have indicated the processes that support quality

Table 4. Survey results

Problems	Numbers of Practices Recommended by:				
	Team 1	Team 2	Team 3	Team 4	Team 5
Engineering Problem					
Software Requirements					
Elicitation of customer needs	15,17	6,10,13,19,5,11,8,17,14	11,19,4,10,13,17	7,9,13,19,11,8,4	4,7,10,17,19
Unstable requirements	6	6,13,19,5,11,8,17	6,21,13	6,13,3,7	4,6,13,21
Inaccurate, incomplete requirements specification	11,17	6,10,13,19,5,11,8,17	4	7,21,13,19	4,10,19
Incorrect interpretation of the requirements	11,17	6,19,5,11,8, 4,17	4,22	7,11,19	4,17
Systematic presentation of the requirements	10,19	10,19,8,4,17	10,19,4,17	7,10,17	4,10,19
Key needs identification	13	6,10,13,5,11	13	7,10,11,13	4,10,19
The lack of domain knowledge	11	9,8,25,19,4, 5,11	11,8	4,7,11	4,19
Software Design	Team 1	Team 2	Team 3	Team 4	Team 5
Variety of implementation platforms	8		1,2	1,12	7
Complex domain	8	8,19,11,25	8	8	8
Software Construction	Team 1	Team 2	Team 3	Team 4	Team 5
Problems with the code integration	2	2,18,14,20	1,2	2,20	2,6
Reduction of repeatable programming activities via code generation	8	12,8	20,12	8,12	12,20
Simple domain	1	15,12,1,20		1,12,19,20	1,12,20
Complex domain	8,11	19,12,1,20,9		1,7,8,9,12, 19,20	1,9,12,20
Unreadable code structure or database structure	12	12,9,23	12	12,23	12
Differentiated quality of code written by different members of the team	9	12,9,23,18, 14	9,23	9,12,23	9,12
Software Testing	Team 1	Team 2	Team 3	Team 4	Team 5
Verification of artifacts against their specification	1,2,4,8,9,20	11,19,1,20,4	10	2,4,19	4,19,20
Desire of acceptable quality of code	1,4,8,9,20	23,12,1,18,14	9,23	1,9,12,23	12
The need for continuous validation of product	1,11	2,1,20,17	2	1,2,5,12,20	2
Extensive documentation requirements	8			10	
Software Maintenance	Team 1	Team 2	Team 3	Team 4	Team 5
Corrective maintenance	1,2,20	26			
Perfective maintenance	12				
Preventive maintenance	8,12	26			
Managerial Problem					
Team Communications	Team 1	Team 2	Team 3	Team 4	Team 5
The need for current information about the state of the project	3,7,16	17,3,18,7	7	3,5,10,11,16	3,16

continued on following page

Table 4. Continued

Managerial Problem					
Risk Analysis	**Team 1**	**Team 2**	**Team 3**	**Team 4**	**Team 5**
Requirement for successful implementation of the medium or large scale project	6	6,13,9		1,13,15	5
Deviation from the plans	14,16,21	18,14,3	16	4,10	3
Too late disclosure of problems and their removal	3,16	18,3,6,5	4,5,16	3,18	3,18
Working in a team multiplayer	2,3,7	9	3,14,18		3,9
Time Planning	**Team 1**	**Team 2**	**Team 3**	**Team 4**	**Team 5**
Planning of iterations and releases	10	5,22	10,24	5,22	22

List of practice numbers: 1. Code and tests, 2. Continuous integration, 3. Daily Stand-Up Meeting, 4. Executable Specifications, 5. Frequent Delivery, 6. Incremental development, 7. Informative workspace, 8. Modeling/Iteration Modeling/Domain object modeling, 9. Pair programming, 10. Product Backlog, 11. Product Owner/Real customer involvement/On-site Customer/Easy Access to Expert Users/Active Stakeholder Participation, 12. Refactoring, 13. Requirement Prioritization/Ranked Work Item List, 14. Retrospectives/Facilitating Workshops, 15. Root-cause analysis, 16. Sprint Burndown chart, 17. Sprint Demo, 18. Sprint review/Reflective Improvement, 19. Stories, 20. Test-Driven Development/Test-first programming, 21. Time-box, e.g. weekly cycle, 22. Sprint planning/planning game, 23. Code review, 24. Team velocity, 25. System metaphor, 26. Kanban

The practices numbered from 1 to 22 are those identified previously, while the last three practices (23, 24, 25, and 26) were additionally suggested by participants of the survey.

Figure 3. Number of practices recommendations in the context of all identified problems

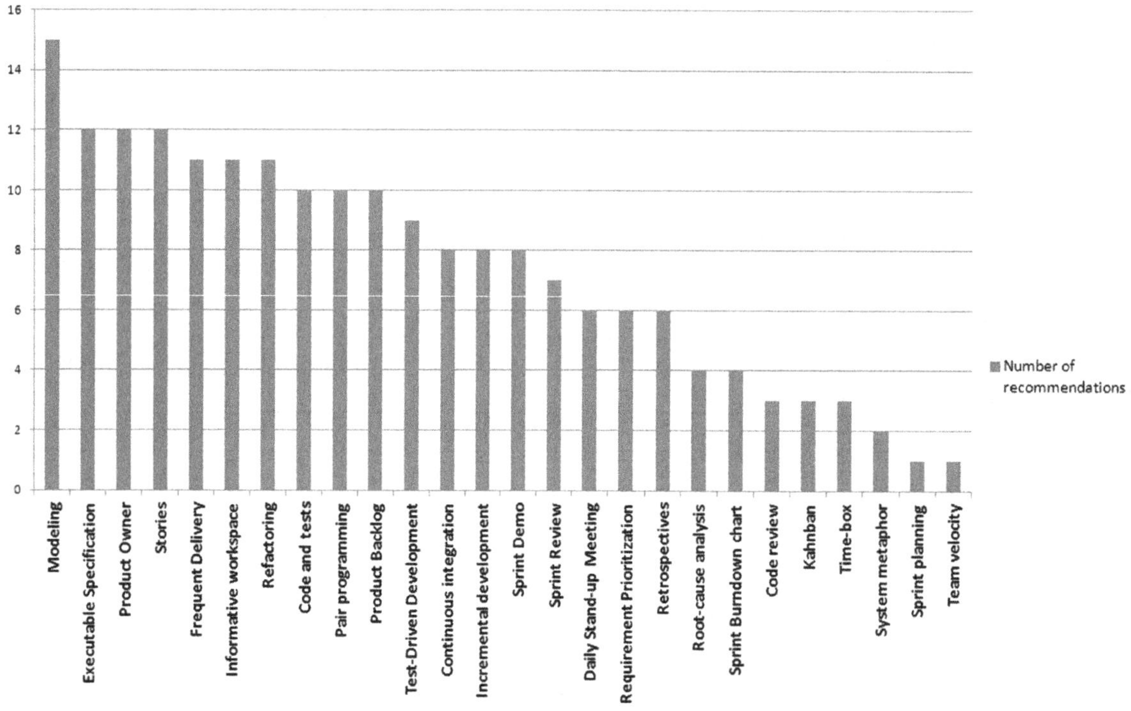

assurance. Then we reviewed the practices of the agile methodologies and chose from among them those that might be treated as fragments of these processes.

Our main, original contribution consists of two elements.

The first one is an analysis of the difficulty of adoption of practices in the software development process. We have observed that it is not reasonable to assess that difficulty in absolute terms. These difficulties of adoption depend on the context of software development. We have proposed an approach to define the context and context-sensitive assessment of difficulties of practices adoption.

The second element is a method of practice recommendation. We have observed that the practices are used to support the solving of problems that may be encountered during software development. Assuming a set of common problems, we asked some software development teams for their

recommended practices. The survey pointed out that the quality assurance point of view indicated that the most important problems are related to the requirements discipline, and that within the disciple modeling is the most important.

Modeling was recognized as a practice with the biggest influence on quality assurance. On the one hand, this is not surprising, as the role of modeling is appreciated and well justified, but on the other hand, it is not reflected in the results of the surveys presented in section 5.2. Modeling seems to be underestimated by agile practitioners.

As we know, the modeling approach enables software developers to better understand an application domain, especially its key notions. An initial domain model forms a good basis for the estimation of project scope as well as its feasibility. For the customer, this model allows for an awareness of his/her needs. In the development process models may be at least partly automatically

Figure 4. Number of practices recommendations in the context of disciplines

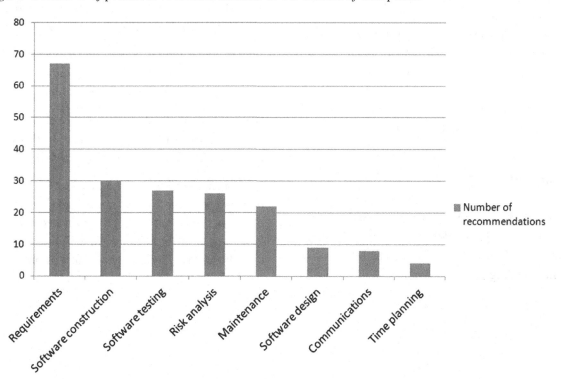

transformed and verified. Finally, the models are the type of documentation that is necessary in the maintenance phase.

Trying to explain this discrepancy, we should first pay attention to the lack of awareness of the importance of modeling. Perhaps this is due to a lack of skills and knowledge about abstract thinking, but more likely this is due to the idea of the agile approach itself. The agile approach focuses on achieving a temporarily determined aim, and tends to avoid additional work that pays off in the future.

So the interesting question arises, what and how (on what level of abstraction) to model in order to benefit from the existence of models during agile development. In the future we plan to propose a set of canonical models (diagrams) that can be recommended especially for data-intensive systems as primary drivers of quality achievement.

The considerations carried out here reveal that any assessment of the difficulties of implementation of practices and their recommendation is not absolute, but must be considered in context. Of course, only the outline of the proposed approach was presented. It should be noted that the recommendations presented are based on objective data. The refinement of assessments of practices and recommendations requires more data gathering. Perhaps a good start would be to create a Web forum gathering experience about adoption of practices and the results of their use.

REFERENCES

Agile Modeling (AM) Home Page. (n.d.). *Effective practices for modeling and documentation.* Retrieved January 13, 2013, from http://www.agilemodeling.com/

Ambler, S. (2002). *Agile modeling: Effective practices for extreme programming and the unified process.* New York: John Wiley & Sons.

Ambler, S. (2005). *AmbySoft agile unified process page.* Retrieved January 13, 2013, from http://www.ambysoft.com/unifiedprocess/agileUP.html

Ambler, S. (2008). *Agile adoption rate survey results.* Retrieved January 13, 2013, from http://www.ambysoft.com/surveys/agileFebruary2008.html

Ambler, S. (2009). *The agile scaling model (ASM), adapting agile methods for complex environments.* Somers, NY: IBM Corporation Software Group. Retrieved from ftp://ftp.software.ibm.com/common/ssi/sa/wh/n/raw14204usen/RAW14204U-SEN.PDF

Ambler, S. (2010). *Context count: Position paper for SOMAT.* Retrieved January 13, 2013, from www.semat.org/pub/.../WorkshopPositions/SEMAT_position_Ambler.doc

Appelo, J. (2009). *A big agile practices survey.* Retrieved January 13, 2013, from http://www.noop.nl/2009/04/the-big-agile-practices-survey.html

Beck, K., & Andres, C. (2004). *Extreme programming explained: Embrace change* (2nd ed.). Reading, MA: Addison-Wesley Professional.

Bhasin, S. (2012). Quality assurance in agile – A study towards achieving excellence. In *Proceedings of AGILE INDIA* (pp. 64-67). AGILE.

Cockburn, A. (2004). *Crystal clear. A human-powered methodology for small teams, including the seven properties of effective software projects.* Retrieved from http://st-www.cs.illinois.edu/users/johnson/427/2004/crystalclearV5d.pdf

DSDM Atern Handbook v.1.0. (2008). Retrieved January 13, 2013, from http://www.dsdm.org/atern-handbook/flash.html

Hongying, G., & Cheng, Y. (2011). A customizable agile software quality assurance model. [NISS]. *Proceedings of Information Science and Service Science, 2*, 382–387.

ISO/IEC 12207:2008. (2008). Software and systems engineering – Software lifecycle processes. Geneva, Switzerland: ISO.

ISO/IEC/IEEE 24765:2010. (2010). Systems and software – Vocabulary. Geneva, Switzerland: ISO.

Janus, A., Schmietendorf, A., Dumke, R., & Jager, J. (2012). The 3C approach for agile quality assurance. In *Proceedings of Emerging Trends in Software Metrics (WETSoM)* (pp. 9-13). WETSoM. doi:10.1109/WETSoM.2012.6226998.

Kruchten, P. (1999). *Rational unified process—An introduction*. Reading, MA: Addison-Wesley.

Michaelson, D. (2006). *Setting best practices in public relations research*. Retrieved January 13, 2013, from http://www.instituteforpr.org/iprwp/wp-content/uploads/Michaelson_09IPRRC.pdf

Mnkandla, E., & Dwolatzky, B. (2006). Defining agile software quality assurance. In *Proceedings of Software Engineering Advances* (p. 36). IEEE.

Palmer, S. R., & Felsing, J. M. (2002). *A practical guide to feature-driven development*. Upper Saddle River, NJ: Prentice Hall PTR.

Project Management Institute. (2004). *PMBOK: A guide to the project management body of knowledge* (3rd ed). Retrieved from http://ebookee.org/A-Guide-to-the-Project-Management-Body-of-Knowledge-Fourth-Ed._282107.html

Schwaber, K. (2003). *Agile project management with scrum*. Seattle, WA: Microsoft Press.

Scott, K. (2001). *The unified process explained*. Reading, MA: Addison-Wesley Professional.

Sochova, Z. (2009). *Agile adoption survey*. Retrieved January 13, 2013, from http://soch.cz/AgileSurvey.pdf

Software Engineering Institute. (2006). *CMMI® for development, version 1.2*. 2006. Retrieved from http://www.sei.cmu.edu/publications/documents/06.reports/06tr008.html

State of Agile Survey. (2011). *Agile methods and practices*. Retrieved January 13, 2013, from http://www.versionone.com/state_of_agile_development_survey/10/page3.asp

KEY TERMS AND DEFINITIONS

Agile Software Development: An iterative and incremental approach to software development based on the Agile Manifesto. It recommends adaptive planning, evolutionary development and delivery, and time-box iterative approach. In addition, this approach is characterized by close collaboration between the team and business representatives, and face-to-face communication.

Best Practice: A component of a life cycle process of a software product which defines how to accomplish a specific goal, for example improve effectiveness of a process or quality of a product.

Context of the Project: A set of dimensions that should be considered, due to they could affect the software development process and should be taken into the account when the project is going to be implemented. Project's context can be defined, among others, by means of: the domain complexity, project size, team size, etc.

Modeling: An approach to software construction based on definition of a set of models which describe software from different perspectives.

Practice Adoption: Applying a practice by putting some parts of it into operation; it is achieved by *practice adaptation* so the practice becomes usable in a given project (better fits to a given project and its context).

Quality Assurance: Overall activities performed within software development processes aimed at ensuring the required quality of a software product.

Software Development Disciplines: A partition of activities performed during software development cycle. In general, the disciplines are classified, similarly to software life cycle processes, in three types: development, for support and enterprise management. For example, USDP methodology distinguishes the following disciplines: business modeling, requirements, analysis and design, implementation, testing, and deployment.

Chapter 9
A Rigorous Approach for Metamodel Evolution

Claudia Pereira
Universidad Nacional del Centro de la Provincia de Buenos Aires, Argentina

Liliana Favre
Universidad Nacional del Centro de la Provincia de Buenos Aires, Argentina & Comisión de Investigaciones Científicas de la Provincia de Buenos Aires, Argentina

Liliana Martinez
Universidad Nacional del Centro de la Provincia de Buenos Aires, Argentina

ABSTRACT

Model-Driven Development (MDD) is an initiative proposed by the Object Management Group (OMG) to model centric software development. It is based on the concepts of models, metamodels, and automatic transformations. Models and metamodels are in constant evolution. Metamodel evolution may cause conforming models to become invalid. Therefore, models must be migrated to maintain conformance to their metamodels. Metamodel evolution and model migration are typically performed manually, which is an error-prone task. In light of this, the authors propose a framework for metamodel evolution and model migration that combine an operator-based approach with refactoring and specification matching techniques. They combine classical metamodeling techniques with formal specifications to reason about transformations, metamodels, and their evolution. The authors describe foundations for MDA-based metamodel evolution that allow extending the functionality of the existing CASE tools in order to improve the MDA-based process quality.

INTRODUCTION

Model-Driven Development (MDD) is a software development approach in which models are considered first-class entities. The goal is to raise the level of abstraction and use concepts closer to the problem and application domain rather than the solution and technical domain. As a result, there is an increasing need for effective techniques and tools to maintain and evolve the models.

Model-Driven Architecture (MDA) (MDA, 2012) is the most known realization of MDD

DOI: 10.4018/978-1-4666-4494-6.ch009

proposed by the Object Management Group (OMG). The key idea behind MDA is to separate the specification of the system functionality from its implementation on specific platforms, managing the software evolution from abstract models to implementations increasing the degree of automation and achieving interoperability with multiple platforms, formal languages and programming languages. Artifacts generated during software development are represented using common metamodeling languages. Metamodeling is an important technique for defining modeling languages. Modeling languages allow transformations can be performed automatically by tools that can understand models. Besides, the transformation rules use the metamodels of the source and target language to define transformations relating elements of the respective languages.

OMG has established different technologies that enable model-driven approach. The object-oriented models used in MDA can be expressed using UML (UML, 2011a; UML, 2011b). Models can be enriched with OCL expressions which specify invariant conditions that must hold for the system being modeled or queries over objects described in a model (OCL, 2012). As a result, models are more accurate and complete. Metamodels in MDA are specified with MOF (MetaObject Facility) (MOF, 2011).

Metamodels, like any software artifact, are subject to constant evolution during its life cycle. This evolution may be due to changes produced during the metamodel definition to improve or correct the abstract syntax of the language, as well as changes in requirements and technological progress in specific application domain. These modifications include changes at all levels, from requirements through architecture and design, to executable models, documentation and test suites. Changes typically affect various kinds of models including data models, behavioral models, domain models, source code models and goal models.

Changes introduced in a metamodel can be classified by their corrupting or non-corrupting effects on existing models (Becker, Gruschko, Goldschmidt, & Koziolek, 2007):

- **Non-Breaking Changes:** Changes which do not break the conformance of models to the corresponding metamodel;
- **Breaking and Resolvable Changes:** Changes which do break the conformance of models, even though they can be automatically resolved;
- **Breaking and Not Resolvable Changes:** Changes which do break the conformance of models which cannot automatically resolved and user intervention is required.

Metamodel evolution can invalidate the models that conform to its previous version. To avoid building models from scratch, they should be migrated in order to conform to the modified metamodel. The activity of changing the metamodel together with its models is called co-evolution.

Model migration, like metamodel evolution, is typically performed manually. This is an error-prone task leading to inconsistencies between the metamodel and related artifacts. Therefore, it is necessary to automate this process using tools that support MDD. This task requires the definition of verification and validation techniques for the effective development of metamodels and the definition of rigorous techniques for migrating models.

Commercial MDD CASE tools provide support for the construction of metamodels and its models as well as mechanisms to check the conformance between models and its corresponding metamodel. However, tools do not support the automation of metamodel evolution. As a result, models become invalid if the metamodel evolution break model conformance. On the other hand, CASE tools do not integrate tools such as automated theorem provers and model checkers that support validation and verification to reason about models.

This chapter describes the state of the art in the area of metamodel evolution and model migration discussing different approaches and the benefits and limitations of each one, as well as strategic directions. We propose a framework for metamodel evolution that assists developer in the construction and editing of metamodels. On the one hand, we propose to combine metamodel matching and operator-based approaches. On the other hand, considering that validation, verification and consistency are crucial activities in metamodel evolution we propose to combine classical metamodeling techniques with formal specification to reason about transformations, metamodels and their evolution. Also, we propose an adaptation of traditional refactoring techniques to support metamodel evolution. Besides, we describe how the proposed metamodel evolution approach can be integrated with related standards.

BACKGROUND

In this section, we start describing the state of the art about metamodel evolution throughout several research approaches analyzing its advantages and weaknesses. Next, valuable techniques that enable to carry out our approach are presented. Finally, we describe commercial and research tools and how they support MDA approach.

Related Work

Models and metamodels are in constant evolution. Metamodel evolution may cause that conforming models become invalid. Hence, models need to be migrated to reflect the changes. There exist different approaches of model migration that vary from proposals that provide languages to manually specify the migration, to approaches that automate the migration process. Several proposals

are described below and categorized according to Rose, Paige, Kolovos, and Polack (2009).

Manual Specification Approaches

The migration strategy is specified manually by the metamodel developer using a general purpose programming language (such as Java) or a model transformation language such as Atlas Transformation Language (ATL, 2012; Jouault & Kurtev, 2005), the Epsilon Transformation Language (Kolovos, Paige, & Polack, 2008) and Flock (Rose, Kolovos, Paige, & Polack, 2010). The transformation languages facilitate the building of the migration strategy by providing mechanisms that are specific for model migration. However, manual specification approaches generally require a great effort on the part of the metamodel developer.

Difference-Based Approaches

The problem of determining model differences can be separated into three phases (Brun & Pierantonio, 2008):

- **Calculation:** Method o algorithm able to compare two distinct models.
- **Representation:** The outcome of the calculation must be represented in some form which is amenable to further manipulations.
- **Visualization:** Model differences are often required to be visualized in a human-readable notation.

Approaches related to the calculation task that is called model matching, are mention below (Kolovos, Di Ruscio, Pierantonio, & Paige, 2009):

- **Static Identity-Based Matching:** In this approach, it is assumed that each model element has a persistent and non-volatile unique identifier that is assigned during its creation. Therefore, a basic approach for matching models is to identify matching model elements based on their corresponding identities. The main advantages of this approach are that it requires no configuration from the user perspective and that it is particularly fast. On the other hand, this approach does not apply to models constructed independently of each other, and to model representation technologies that do not support maintenance of unique identities. Alanen and Porres, (2003) discuss metamodel independent algorithms that calculate the difference between two models, merge a model with the difference of two models and calculate the union of two models. The algorithms are simplified by requiring that each element has a universally unique identifier.

- **Signature-Based Matching:** The identity of each model element is calculated dynamically from the values of its features by means of a user-defined function specified using a model querying language. This approach can be also used to compare models that have been constructed independently of each other. The disadvantages of this approach are that developers need to specify a series of functions that calculate the identities of different types of model elements. In (Reddy, France, Ghosh, Fleurey, & Baudry, 2005) a composition technique for composing aspect and primary models that uses a signature-based approach is presented. A model element is merged with another if their signatures match. A signature consists of some or all properties of an element as is defined in the UML metamodel.

- **Similarity-Based Matching:** This approach treats models as typed attribute graphs and attempts to identify matching elements based on the aggregated similarity of their features. However, not all features of model elements are equally important for model matching. Therefore, similarity-based algorithms typically need to be provided with a configuration that specifies the relative weight of each feature. In (Toulmé, 2006) an approach to compare models based on a total comparison of their elements is presented. The author uses a fixed configuration based on a weight system that computes the importance of the differences between two elements. The comparison algorithm uses the information theory to analyze the model semantic domain. This approach was implemented in EMF Compare tool. In (Falleri, Huchard, Lafourcade, & Nebut, 2008) authors present an approach that produces an alignment between two metamodels in an automated way. It is based on the application on metamodels of the matching algorithm called Similarity Flooding. The main contribution of this paper is the study on the various ways of encoding a given metamodel into a directed labeled graph that can be exploited by the Similarity Flooding algorithm.

- **Custom Language-Specific Matching Algorithms:** The main advantage of this approach is that it can incorporate the semantics of the target language in order to provide more accurate results and reduce the search space. However, developers need to specify the complete matching algorithm manually, which can be a challenging task. In (Xing & Stroulia, 2005) UMLDiff is presented, an algorithm for automatically detecting structural changes between the designs of subsequent versions of object-oriented software based on UML semantic.

There exists some works related to the difference representation. Cicchetti, Di Ruscio, Eramo, and Pierantonio (2008) present a transformational approach to model co-evolution that use higher-order transformations which take a difference model recording the metamodel evolution and produce a model transformation that can migrate the involved models. The previous difference application engine has been improved by adding the supporting knowledge from the weaving model to perform the desired manipulations on arbitrary input models. Such application is realized by coupling a difference model with a weaving model, that is a morphism between models that allows linking the intended modifications to the model elements to be modified (Cicchetti, Di Ruscio, & Pierantonio, 2010).

Operator-Based Approaches

These approaches specify metamodel evolution by a sequence of operator applications. Each operator specifies a metamodel evolution along with a corresponding model migration strategy. These approaches provide a set of reusable operators which work at metamodel level as well as at model level. The usefulness of the operator-based approaches depends heavily on the richness of the operator library and the integration of this library with tools for editing metamodels.

Among the first publications on the area it is worth mentioning the research of (Gruschko, Kolovos, & Paige, 2007). Authors manage model evolution by small elementary transformation steps and use the Epsilon Transformation Language (ETL) to migrate models.

Wachsmuth (2007) combines ideas from object-oriented refactoring and grammar adaptation to provide a basis for automatic metamodel evolution. Author defines several semantics and instance preservation properties in terms of metamodel relations. Besides, a library of QVT relations for the stepwise adaptation of MOF compliant metamodels is presented.

Herrmannsdoerfer, Benz, and Juergens (2008) analyze the occurred changes during the evolution history of two industrial metamodels and classify them according to their level of potential automation. Based on the results, authors present a list of requirements for effective tool to support coupled evolution and implement the operator-based tool COPE that records operator histories on metamodels of the EMF (Herrmannsdoerfer, Benz, & Juergens, 2009). Then, additional functions were added to COPE in order to analyze, refactor and recover the coupled evolution (Herrmannsdoerfer, 2011). In Herrmannsdoerfer, Vermolen, and Wachsmuth (2010), an extensive catalog of coupled operators was presented. This catalog is organized according to a number of criteria that helps developers to evaluate the impact on models as well as to select the right operator for a metamodel change at hand.

Existing approaches to coupled evolution focus on a single, homogeneous domain. Vermolen and Visser (2008) present an architecture to automate coupled evolution on an arbitrary software domain (e.g. programming languages, modeling, or data modeling) and a tool that support the architecture. The tool requires a coupled software evolution scenario and a mapping from software language transformations to software transformations. It generates a Domain Specific Transformation Language (DSTL) for an arbitrary software language domain. It generates an interpreter of transformations defined in the DSTL and it supports generic abstraction from the basic transformations that are defined in the DSTL.

An approach to reconstruct complex evolution traces from difference models is shown in (Vermolen, Wachsmuth, & Visser, 2011). It supports operator dependencies and mixed, overlapping, and incorrectly ordered complex operator components. It also supports interference between operators, where the effect of one operator is partially or completely hidden from the target metamodel by other operators.

Related Techniques

In this section, we describe the main techniques that allow achieving our MDA-based metamodel evolution approach. We use metamodeling technique to specify the transformations applied to metamodels in order to reflect its evolution and to the models in order to maintain its conformance to the new version of the metamodel. Different kinds of metamodel transformations enable its evolution such as introducing or deleting elements in a metamodel and the application of refactoring techniques. Besides, we adapt the foundations of specification matching to the metamodel matching context in order to identify metamodel elements that take part in a transformation.

Metamodeling

The Model Driven Architecture (MDA) is an evolution of OMG standards to support model driven development. A model is a description or specification of the function, structure and/ or behavior of an application or system. Certain models describe the system independently of the technical concepts involved in their implementation on a software platform, whereas other models aim to describe these technical concepts.

A system model is specified using a modeling language. Modeling languages are used to capture, relate and manipulate different aspects of a problem domain. The success of the graphic modeling languages, such as UML, is based on graphic constructions that have an intuitive meaning. These languages are easy to understand and use.

A model specifies which elements may exist in a system. The definition of a modeling language states which elements may be in a model. A language can be described by a model, called metamodel. The language metamodel describes which elements can be used in the language and how they can be connected. As a metamodel is also a model, the metamodel itself should be written in a well-defined language. This language is called a metalanguage.

In order to models and metamodels can be used in a unified way that enable the model driven proposal, OMG has adopted different technologies. Next, the most relevant standards for applying the MDA approach are described.

MOF is a standard that defines the language to define modeling languages allowing interoperability among different languages, platforms, and tools. MOF uses a modeling framework, which is a subset of UML core. Modeling constructs are: classes, modeling metaobjects MOF; associations, modeling binary relations between metaobjects; data types, modeling other data such as primitive types; package, modularizing the models. MOF metamodels are expressed as a combination of UML class diagrams and OCL specifications. The role of MOF within MDA is that it provides the concepts and tools to reason about modeling languages.

OCL language allows writing expressions on each element of a model obtaining more accurate and complete models. This language has a simple syntax, has no side effects for being a pure expression language and its syntactic context is graphically determined. OCL can be effectively used in the definition of transformations. Many transformations can only be applied under certain conditions which can be specified in OCL. All OCL expressions used in a transformation definition are specified on source and target language metamodels.

MOF and OCL are used to define other languages that emerge in MDD. This allows tools to be able to read and write without ambiguity all languages standardized by the OMG and also facilitates interoperability between different tools.

The standard UML is a modeling language to visualize, specify, and document software system component. UML consists of a number of different diagrams that allow different views of a software system to be modeled at different stages of the development lifecycle. Both static and dynamic aspects of a system can be captured. UML can be applied in MDA in two different ways. On the one hand, a developer can use UML to develop models of the system that are precise and consistent enough to be used within MDA. On the other hand, a developer can use the standard to define transformations between models. The developer must understand UML and its use. Additionally, developer needs to be familiar with the UML metamodel in order to define MDA transformations.

Refactoring

Software restructuring refers to transformations applied to a representation form that produce a new software representation form. Source program restructuring is called program transformation but also could be applied at higher levels of abstraction, such as designs or architectures. Refactoring is the equivalent term to restructure in the context of object-oriented programming.

Fowler (1999) states that "the word refactoring has two definitions depending on context:

- **Refactoring (noun):** A change made to the internal structure of software to make it easier to understand and cheaper to modify without changing its observable behavior;
- **Refactor (verb):** To restructure software by applying a series of refactorings without changing its observable behavior" (p. 46).

According to Fowler, refactoring is an important technique that should be used for several purposes: it improves software design, it makes software easier to understand, it helps to find bugs and to program faster.

Refactorings are typically applied at program level, but may also be applied as model transformations. France and Bieman (2001) categorize model transformations according to two dimensions:

- Vertical transformations occur when a source model is transformed into a target model at a different level of abstraction. Refining a model and realizing a model

in a target programming language are instances of vertical transformations. In the context of MDA, vertical transformations are useful when transforming a Platform-Independent Model (PIM) to a Platform-Specific Model (PSM) (MDA, 2012).

- Horizontal transformation involves transforming a source model into a target model that is at the same level of abstraction as the source model. Horizontal transformations are carried out to support model evolution. Three types of model evolution can be distinguished:
  - Perfective evolution is concerned with modifying a design in order to improve model characteristic.
  - Corrective evolution is concerned with correcting errors in the design.
  - Adaptive evolution is concerned with modifying a design model to accommodate changes in requirements and design constraints.

Model refactoring is defined as the process to restructure object-oriented model by applying a sequence of transformations that preserve its functionality in order to improve some quality factor. It is a transformational proposal for iterative software development. It is based on the idea of make changes in a model by small steps that support perfective evolution of models to improve some aspect of quality such as robustness, adaptability, reusability, compatibility, ease of use, portability and understandability.

Model Driven Development is carried out through a sequence of model transformations which include the following steps: the building of a computation independent model, the transformation to a independent model of implementation details, the transformation of this model to one or more dependent models of any implementation platform and the code generation from models. MDD proposes the use of models for addressing the course of each stage in the process of developing systems. Therefore, it is important to apply refactorings as a tool for restructuring models which are the starting point in the sequence of transformations. The objective of refactorings is the perfective evolution of the models to improve quality aspects without changing the observable behavior of the system.

Specification Matching

In our approach of metamodel evolution, we adapt the specification matching technique described by Zaremski and Wing (1997) to identify metamodel elements. The transformation rules, corresponding to the operators of metamodel evolution and model migration, need to detect metamodel elements that match with certain constraints established by the rule. These metamodel elements can be packages, classes, operations, properties and associations.

Specification matching is a process of determining if two software components are related. Zaremski and Wing (1997) state the foundations for different kinds of semantic matching of software component such as modules and operations. Next, we briefly describe matching of operations in particular.

Associated with each operation O, there is a Signature, O_{sig} and a specification of its behavior, O_{spec}. The signature describes type information of the component and the specification describes its dynamic behavior.

Given two operations, $O = (O_{sig}, O_{spec})$ and $O' = (O'_{sig}, O'_{spec})$, a generic component match predicate, Match, is defined:

$$\text{Match : Operation, Operation} \rightarrow \text{Boolean}$$
$$\text{Match}\left(O, O'\right) = \text{matchsig}\left(O_{sig}, O'_{sig}\right) \wedge$$
$$\text{matchspec}\left(O_{spec}, O'_{spec}\right)$$

Two operations O y O' match if:

- Their signatures match, given some definition of signature match and
- Their specifications match, given some definition of specification match.

Signature match is used as a filter discarding the obvious no matches before trying the more expensive specification matches.

Operation Signature Matching: Operation matching based on just signature information result in type matching. Given the operation signatures (domain and codomain) Sig and Sig', a generic form of operation signature match, Match (Sig, Sig'), is defined:

$$\text{Match}\left(\text{Sig, Sig'}\right) = \exists \text{ a transformation pair,}$$
$$T = \left(T_s, T_{s'}\right), \text{such that } T_s\left(\text{Sig}\right) R T_{s'}\left(\text{Sig'}\right)$$

where R is some relationship between signatures (e.g., equality) and T_s and $T_{s'}$ are transformations that are applied to Sig and Sig', respectively. A transformation is a function from signature to signature.

Given two operation signatures Sig and Sig', the operation signature equality, is defined:

$$\text{Sig} = T \text{ Sig'}, \text{if they are syntactically identical.}$$

Two operations exactly match if renaming domains and codomain of one operation, syntactically match to the domains and codomain of the another one:

$$\text{match}_E\left(\text{Sig, Sig'}\right) = \exists \text{ a sequence of renaming}$$
$$\text{of types, V, such that V Sig} = T \text{Sig'}$$

Exact match is a useful starting point, but it may miss useful operations whose signatures are close but do not exactly match. One kind of relaxed match transforms a signature to achieve a match, for example, change the order of operations domain (for operations that take more than one argument).

Other kind of partial matches relax the relation R to be a partial order on signatures. Type substitution is used to define the partial ordering, based on the "generality" of the types.

Operation Specification Matching: Two operations can have the same signature, but completely opposite behavior. The specification match takes into consideration the behavior of the components, it allows increasing the precision with which we determine when two components match.

An operation specification S is denoted in terms of its precondition and postcondition as S_{pre} and S_{post}, respectively. The precondition states the properties that must hold whenever the operation is called; the postcondition states the properties that the operation guarantees when it returns. S_{pred} defines the interpretation of the operation specification as an implication between the two: $S_{pred} = S_{pre} => S_{post}$. This interpretation means that if S_{pre} holds when the operation specified by S is calls, the S_{post} will hold after the operation has executed. If S_{pre} does not hold, there are no guarantees about the behavior of the operation. There exist different kind of pre/post matches and predicate matches defined on operation specifications.

Case Tools and MDA

All of the MDA CASE tools are partially compliant to MDA features with regard to metamodel evolution. A study of tools on the basis of a set of specific characteristics of the MDA approach is presented in (Fernández Sáez, 2009) where the tools studied were: IBM Rational Software Architect, Borland Together and Sparx Systems Enterprise Architect. The main MDA Case tools and their facilities are described in (CASE MDA, 2012). On the one hand, they provide good support

for modeling and limited support for metamodel evolution and model migration. The Eclipse Modeling Framework (EMF) (Eclipse, 2012) was created for facilitating system modeling and the automatic generation of Java code. EMF started as an implementation of MOF resulting Ecore, the EMF metamodel comparable to EMOF. EMF has evolved starting from the experience of the Eclipse community to implement a variety of tools and to date is highly related to Model Driven Engineering (MDE). Commercial tools such as IBM Rational Software Architect, Spark System Enterprise Architect or Together are integrated with Eclipse-EMF (CASE MDA, 2012). A comparison of research tools that support model migration is presented in (Rose et al., 2010). The comparison is based on practical application of the tools to the common migration examples. The compared tools are:

- **AtlanMod Matching Language (AML):** A model matching migration tool
- **COPE:** An operator-based migration tool
- **Ecore2Ecore:** A manual specification migration tool
- **Epsilon Flock:** A manual specification migration tool

On the other hand, CASE tools provide little support for validating models in the design stages. Modern metamodeling frameworks work as follow: a metamodel captures the key assumptions and constraints to a problem domain, then instance models are constructed against a metamodel and must adhere to its regulations. Metamodeling tools generate authoring environments, enable code generation, and reporting constraint violations. Reasoning about models of systems is well supported by automated theorem provers and model checkers, however these tools are not integrated into CASE tools environments. Only research tools provide support for formal specification and deductive verification. As an example, we can mention Use 3.0 that is a system for specification of information systems in OCL. Use allows snapshots of running systems can be created and manipulated during an animation, checking OCL constraints to validate the specification against non-formal requirements (Use, 2011). Besides, we can mention Formula that is a new formal specification language and toolset for describing, transforming, and analyzing metamodels and instance model. Formula finds instance models satisfying all the rules of a given metamodel (Formula, 2012).

A FRAMEWORK FOR METAMODEL EVOLUTION

We propose an architectural framework for metamodel evolution structured at different levels of abstraction linked to models, metamodels and formal specifications (Figure 1).

At model level, models need to be adapted to reflect the changes during its evolution. This level represents the model-to-model transforma-

tions that support both the metamodels during its evolution and also the models to maintain its conformance with its corresponding metamodel. These transformations are called metamodel evolution and model migration respectively.

At metamodel level, the model-to-model transformations are defined by metamodeling technique. Source and target MOF metamodels are defined for each transformation. Source metamodel describes a model family to which the transformation may be applied. Target metamodel characterizes the models that are generated. The models to be transformed and the resulting models of the transformations will be instances of the corresponding metamodel.

The transformations between models are described relating each element of the source model to one or more elements of the target model at metamodel level. In other words, relating the metaclass of the element of the source model with the metaclasses of the elements of the target model.

Figure 1. A framework for MDA software evolution

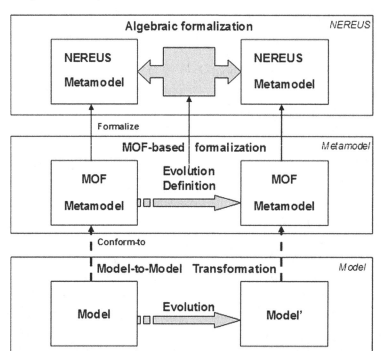

The level of formal specification links MOF metamodels and metamodel evolution to formal specifications. We propose to formalize MOF metamodels and metamodel evolution by using the NEREUS language. NEREUS can be used as a common specification language and is connected with different semiformal, formal and programming languages. A detailed description of NEREUS may be found at (Favre, 2009).

This article focuses on the specification of metamodel evolution at metamodel level. As an example, a metamodel refactoring is specified as metamodel-based transformation.

Specifying Metamodel Evolution in MDA

We present an operator-based approach to specify metamodel evolution. We define a library of operators that works at metamodel level as well as at model level. Metamodel evolution and model migration is carry out by a sequence of operator applications.

At metamodel level, each operator defines a metamodel transformation. There exist different kinds of metamodel transformations to enable its evolution:

- Transformations that introduce elements in a metamodel, such as adding a class, relation or attribute;
- Transformations that eliminate elements in a metamodel such as deletion a class, relation or attribute;
- Refactorings, which were detailed in Background section, describe sequences of transitions that are made according to precise rules based on the redistribution of classes, variables, operations and associations across models. Examples of refactorings are renaming a class, relation, or attribute, pulling up an attribute or relation from subclasses into a superclass, pushing down an attribute or relation from a su-

perclass into the subclasses, extracting a superclass.

At model level, a coupled operator defines a model transformation capturing the corresponding migration in order to maintain the conformance between models and the corresponding metamodels. Figure 2 shows this approach.

Operators can be classified according to model preservation property, which indicate when the migration of instances of a modified metamodel is needed:

- **Model-Preserving Operator:** When all models conforming to an original metamodel also conform to the evolved metamodel and hence, these operators do not require model migration.
- **Model-Migrating Operator:** When models conforming to an original metamodel might need to be migrated in order to conform to the evolved metamodel.
- **Safely Model-Migrating Operator:** When migration preserves distinguishability, that is to say, different models (conforming to the original metamodel) are migrated to different models (conforming to the evolved metamodel).
- **Unsafely Model-Migration Operator:** When migration might yield the same model when migrating two different models.

The classification is related to a classification of changes described in a previous section. Model-preserving operators perform non-breaking changes, whereas model-migrating operators perform breaking and resolvable changes. However, there is no correspondence for breaking and not resolvable changes, since coupled operators always provide a migration to resolve the breaking change.

Our approach to migrate models presents the following features:

Figure 2. Software evolution

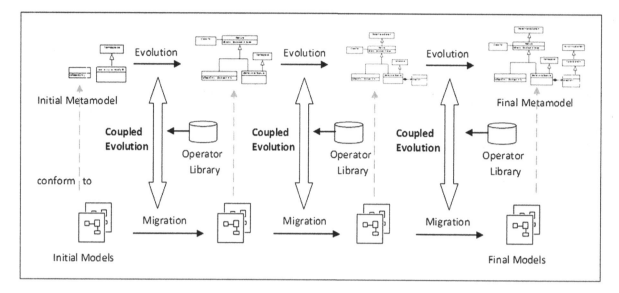

- It describes sequences of transitions that are made according to precise rules based on the redistribution of classes, variables, operations and associations across models.
- The catalog of operators enables the reuse of recurring coupled operators reducing the effort of building a new model migration.
- Operators can be combined in order to specify more complex ones.
- The operator-based approach allows recording the history of the applied transformations.
- The approach allows identifying elements in the model to which a transformation will be applied. For example, the operator for extracting a super class introduces a new class, makes it the super class of a set of classes and pulls up features from these classes. In existing model migration approaches, the classes from which the super class is extracted and the features to be extracted should be identified by the developer as a parameter of the transformation. In our approach, the elements of the model to

be transformed are identified by specification matching. If the transformation needs to detect operations with the same signature or functionally equivalent, to obtain the features to be pull up to the super class, these operations are matched through the operation signature matching or operation specification matching. This is possible because models are described in UML and enriched with OCL restrictions. To carry out this, we adapt the ideas of specification matching described in Background section (Zaremski & Wing, 1997).

An Example: The Extract Composite Refactoring

During a metamodel construction, we might need to restructure the metamodel applying a sequence of operators in order to enable its evolution. To carry out this task, we need to specify two transformations:

- At metamodel level, a transformation to restructure the metamodel and
- At model level, a transformation to migrate the instances of the original metamodel in order to conform the evolved metamodel.

Figure 3 depicts source and target pattern of the *Extract Composite* refactoring that was adapted of the Kerievsky definition (Kerievsky, 2004, p. 214). The source pattern depicts subclasses in a hierarchy that store children from the hierarchy and have functionally equivalent operations (duplicated operations) that operate on the children. The transformation extracts a *Composite* superclass and moves the duplicated fields and operations from the subclasses to the *Composite* as it is shown in target pattern. The main steps in the proposed transformation are:

- Create a composite, a class that will become a *Composite*,
- Make each child container (a class in the hierarchy that contains duplicated methods) a subclass of the *Composite* and
- Identify duplicated operations across the subclasses of a Composite and move them to the *Composite*.

Figure 4.a shows an example of a simplified hierarchy of the HyperText Markup Language (HTML) tags that correspond to a metamodel under construction specified in UML/OCL. The HTML tags can be form, link and image tag. The form and link tags are child containers; for example, a link tag can contain an image tag. The HTMLLinkTag and HTMLFormTag metaclasses have equivalent operations. As a consequence, we need to refactor the metamodel to improve the design. The evolved metamodel (Figure 4.b contains an abstract metaclass CompositeHTMLTag that has the common operations. In order to carry out this transformation, the metamodels should be enriched with OCL constraints; in this example the methods have pre- and post-conditions that enables the identification of functionally equivalent operations.

We propose to adapt the ideas of specification matching described by Zaremski and Wing (1997) to identify functionally equivalent operations. These ideas were exposed in section Related Techniques. Next, we only detail those specification matches to be used in the example where S is an operation specification denoted in terms of its precondition and postcondition as S_{pre} and S'_{post}, and R is some relationship between specifications.

Figure 3. The extract composite refactoring

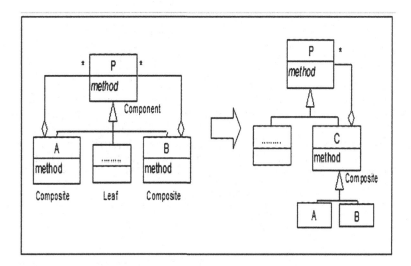

Figure 4. An example of HTML metamodel

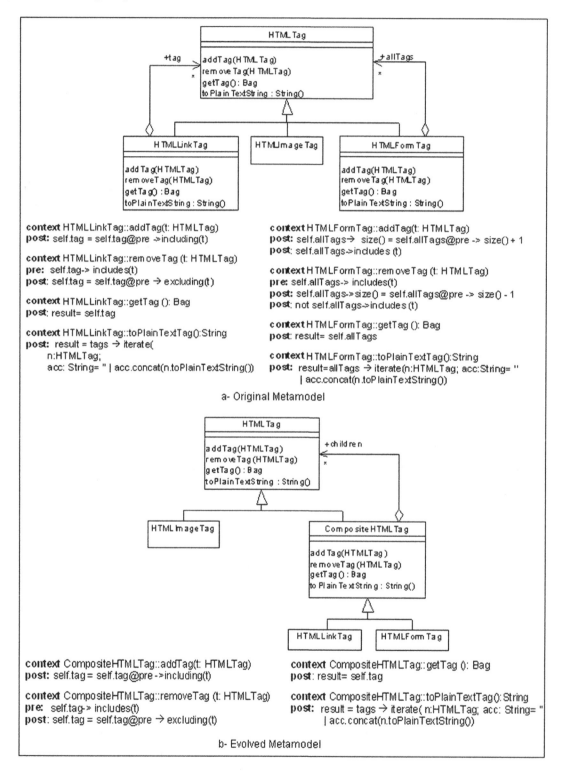

Pre/Post Matches: Pre/post matches on specifications S and S' relate S_{pre} to S'_{pre} and S_{post} to S'_{post}. Each match is an instantiation of the generic pre/post match:

$$\text{match}_{pre/post}\left(S, S'\right) = \left(S'_{pre} R_1 S_{pre}\right) \wedge \left(S_{post} R_2 S'_{post}\right)$$

Two kinds of pre/post matches are detailed as follows.

Exact Pre/Post Match: It establishes if two components are essentially equivalent and thus completely interchangeable. Two specifications satisfy the exact pre/post match if their preconditions are equivalent and their postconditions are equivalent:

$$\text{match}_{E-pre/post}\left(S, S'\right) = \left(S'_{pre} <=> S_{pre}\right) \wedge \left(S_{post} <=> S'_{post}\right)$$

Plug-In Match: Under this match, S' is matched by any specifications S whose precondition is weaker (to allow at least all of the conditions that S' allows) and whose postcondition is stronger (to provide a guarantee at least as strong as S').

$$\text{match}_{plug-in}\left(S, S'\right) = \left(S'_{pre} => S_{pre}\right) \wedge \left(S_{post} => S'_{post}\right)$$

S is behaviorally equivalent to S', since we can plug in S for S' and have the same observable behavior, but this is not a true equivalence because it is not symmetric.

We apply these definitions to the example of HTML metamodel in order to identify functionally equivalent operations. Under signature matching, the addTag operation of HTMLLinkTag metaclass is matched by both addTag and removeTag operations of the HTMLFormTag metaclass. However, under Plug-In Match, addTag operation of HTMLLinkTag metaclass is only matched by addTag operation of HTMLFormTag metaclass. Let S be the specification of the addTag operation of the HTMLLinkTag metaclass and let S' be the specification of the addTag operation of the HTMLFormTag metaclass with allTag renamed to tag. The precondition requirement ($S'_{pre} => S_{pre}$) holds, since $S'_{pre} = S_{pre}$ = true, thus the match plug-in (S, S') reduces to proving ($S_{post} => S'_{post}$), in OCL (Box 1):

As a result, S is behaviorally equivalent to S', since we can plug in S for S' and have the same observable behavior, however this is not a true equivalence because it is not symmetric. Therefore, addTag operation of the HTMLLinkTag class is moved to the new *Composite* metaclass generated in the target model and addTag operation of the HTMLFormTag metaclass is removed.

Similarly, matchings are applied to the remaining operations. The operation removeTag of HTMLLinkTag metaclass is matched by removeTag operation of HTMLFormTag metaclass under Plug-In Match. The operations getTag and toPlainTextTag of HTMLLinkTag metaclass are matched by getTag and toPlainTextTag of HTMLFormTag metaclass respectively under Exact Pre/Post Match.

Specifying Metamodel Evolution

A metamodeling technique is used to specify software evolution in MDA. MOF metamodels are used to describe the transformations. For each

Box 1.

```
(self.tag = self.tag@pre->including (t)) implies
(self.tag→size () = self.tag@pre->size ()+1 and self.tag->includes (t))
```

transformation, source and target metamodels are specified. A source metamodel defines the family of source models to which transformation can be applied. A target metamodel characterizes the generated models.

Figure 5 partially shows the specialized UML metamodel of the Extract Composite refactoring corresponding to source and target metamodel. These metamodels include metaclasses linked to the essential participants in the models of Figure 3: Composite, Component and Leaf and, three relationships: Composite-Component-Assoc, Component-Leaf-Generalization and Component-Composite-Generalization. The metamodel also shows metaclasses linked to properties such as AssEndComposite and AssEndComponent and, shaded metaclasses that correspond to the UML metamodel.

We can remark the following difference between the source and target metamodel. On the one hand, in the source metamodel, an instance of Component has two or more instances of Component-Composite-Generalization (compositeSpecialization) and two or more association-ends (associationEnd). On the other hand, in the target metamodel, an instance of Component has exactly one instance of Component-Composite-Generalization and one association-end.

The metamodel evolution is expressed as OCL contracts that consist of a name, a set of parameters, preconditions, postconditions, and declarations of local operations that are used in preconditions and postconditions. Each parameter is a metamodel element. Preconditions, which deal with the state of the model before the transformation, state relations at the metamodel level between the elements of the source model. Postconditions, which deal with the state of the model after the transformation, state relations at metamodel level between the elements of the source model and a target model. Figure 6 partially depicts the specification of the Extract Composite transformation, which is explained by comments.

Specifying Model Migration

The application of Extract Composite refactoring can invalidate the instances of the original metamodel.

Considering the original metamodel (Figure 4a), the HTMLLinkTag and HTMLFormTag metaclasses have an association with HTMLTag metaclass where one of its association ends is an aggregation and the other one is a navigable association end that has a role name. There are two possible situations:

- If role names are the same, the transformation keeps the same name for the evolved metamodel, as a result, the instances of the original metamodel remain valid.
- If role names are different (as in the example), the transformation changes the role name in the evolved metamodel, as a result, the instances of the original metamodel become invalid. In this case, it is necessary migrate the instances of the original metamodel in order to conform the evolved metamodel. The migration is specified in the same way as the transformation for metamodel evolution.

Implementation Alternatives

In this work, we present the foundations to specify metamodel evolution by means of metamodeling technique and formal specification. We define a library of operators that works at metamodel level as well as at model level. At this moment, transformations corresponding to basic operators have been implemented in the Eclipse Modeling Framework using ATL. However, to implement complex operators we face up with some challenges.

In the example presented in this chapter, the transformation at metamodel level needs to identify functionally equivalent operations by signature matching. Transformation languages such

Figure 5. MOF-metamodel for the extract composite refactoring

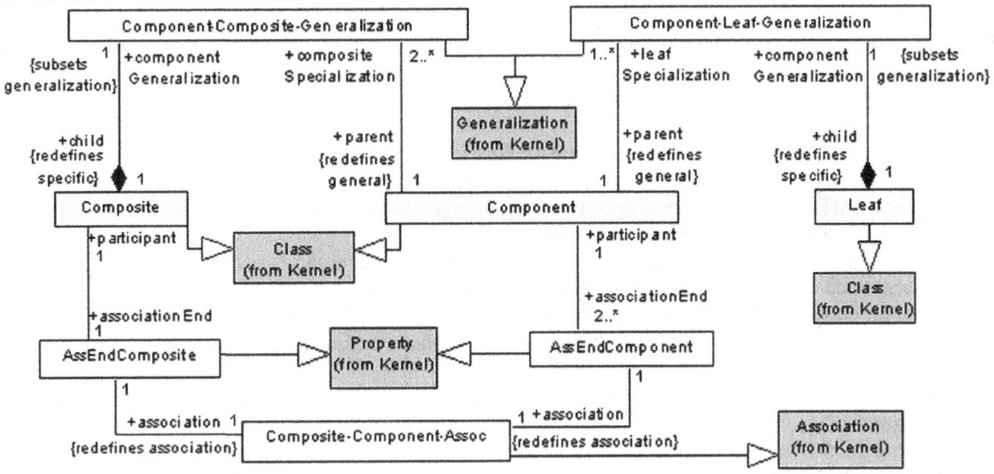

context AssEndComposite inv:
    self.aggregation = #shared or self.aggregation = #composite

context Component inv:
-- The associations between Composite and Component are equivalent.
    self.associationEnd.association → forAll (a1, a2 | a1 = a2 or
    a1.isEquivalentTo(a2))

context Component inv:
-- For each class Composite there is an operation
    self.compositeSpecialization.child → forAll (class |
        class.ownedOperation → exists (op |
            -- that is equivalent to operations of the others classes Composite
            self.compositeSpecialization.child → excluding (class) → forAll (c |
                c.ownedOperation → exists (o | op.isEquivalentTo(o)))))
...

a- Source Metamodel

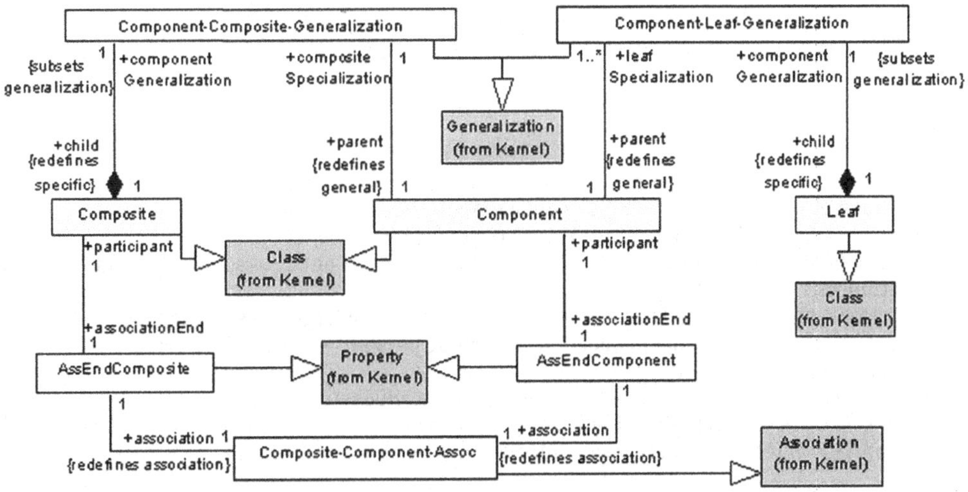

context AssEndComposite inv:
    self.aggregation = #shared or self.aggregation = #composite
...

b- Target Metamodel

Figure 6. Extract composite refactoring in OCL

```
Transformation Extract Composite {
parameters
    source: Extract Composite Source Metamodel:: Package
    target:  Extract Composite Target Metamodel:: Package
local operations
Operation::isEquivalentTo(op:Operation): Boolean;
--checks if the argument is equivalent to Operation.
isEquivalentTo = self.hasSameSignature(op)- signature matching of functions and
    -- specification matching: Exact pre/post match:
    (self.precondition.AsOclExp implies op.precondition.AsOclExp and
    op.precondition.AsOclExp implies self.precondition.AsOclExp and
    op.postcondition.AsOclExp implies self.postcondition.AsOclExp and
    self.postcondition.AsOclExp implies op.postcondition.AsOclExp)    or
    -- specification matching: plug-in match ....
post:
-- For each class Component 'sourceClass' in the source package,
source.ownedMember → select(oclIsTypeOf(Component))→ forAll ( sourceClass |
--there is a class Component 'targetClass' in the target package so that
target.ownedMember →  select(oclIsTypeOf(Component)) → exists ( targetClass |

--targetClass has one Component-Composite generalization,
targetClass.oclAsType(Component).compositeSpecialization → size() =1  and

-- targetClass has one association end that is associated with Composite,
targetClass.oclAsType(Component).associationEnd → size() =1  and

-- targetClass and sourceClass have the same Leaf classes, ...
post:
-- For each class Composite 'sourceClass' in the source package,
source.ownedMember → select(oclIsTypeOf(Composite)) → forAll ( sourceClass |
-- there is a class 'targetClass' in the target package so that
target.ownedMember → select(oclIsTypeOf(Class)) → exists ( targetClass |

-- targetClass is subclass of a class whose type is Composite,
targetClass.oclAsType(Class).superClass.oclIsTypeOf(Composite)  and

-- targetClass and sourceClass have the same name,
targetClass.name = sourceClass.name  and

-- for each equivalent operation of Composite classes in source model
sourceClass.oclAsType(Composite).ownedOperation → forAll( op |
(source.ownedMember →  select(oclIsTypeOf(Composite))
→excluding(sourceClass))→collect (oclAsType(Composite).ownedOperation) → forAll ( o |
    if o.isEquivlalentTo(op) then
        -- in target package there is an equivalent operation in the targetClass superclass,
        targetClass.oclAsType(Class).superClass.ownedOperation → exists ( targetOp |
        op.isEquivalentTo (targetOp)  ) and
        targetClass.oclAsType(Class).ownedOperation→ excludes(op)
    else -- otherwise, the operation is owned by targetClass.
        targetClass.oclAsType(Class).ownedOperation→ includes(op)  endif

-- for each equivalent property of Composite classes in source model
-- in target package there is an equivalent property in the targetClass superclass
...)) }
```

as ATL and QVT do not support the evaluation of OCL preconditions and postconditions during the transformation. A possible solution to this problem would be to modify the transformation language in order to call the OCL evaluator on demand to evaluate the pre and postconditions. However, this alternative could only applied to different snapshots of a system. Another alternative would be translate OCL constraints into an input for a theorem prover and prove that a match holds between two specifications. Although there is a formal basis that has been validated, formal languages are not integrated into metamodeling tools. Our challenge is integrate formal techniques with metamodeling tools.

Once components have been identified to apply the transformation at metamodel level, information about these components is used by the transformation that enables the migration of the evolved metamodel instances. Transformation languages do not support parameterized modules. Hence, a transformation could be built by instantiation of a transformation scheme with the required information.

Formalizing Metamodel Evolution

Most known software failures are due to the tradition of applying trial-and-error. Formal specification addresses this problem advising against error before the trial begins. In our context, a central problem is how to define metamodels correct and aligned with MOF. Inconsistencies in a MOF specification will affect to models and its implementations.

Formal and semiformal techniques have complementary roles in the software development process based on MDA. Both techniques provide benefits.

On the one hand, semiformal techniques lack precise semantic, however, these techniques have the advantage of visualizing language constructions allowing high productivity in the specification processes especially when they are supported by tools. The UML specification is defined using a metamodeling approach but "it is important to note that the specification of UML as a metamodel does not preclude it from being specified via a mathematically formal language at a later time" (UML, 2011a, p.11). OCL is widely used by MOF-based metamodels to constrain and perform evaluations on them. OCL has a denotational semantics that has been implemented in tools allowing dynamic validation of snapshots. However, it can not be considered strictly a formal specification language due to does not support logic deductions in the style of solid formal languages; for instance, it does not provide the notions of refinement, implementation correctness, observable equivalences and behavioral equivalences that play an essential role in model-to-model transformations.

On the other hand, a formal specification technique must at least provide syntax, some semantics and an inference system. The inference system allows defining deductions that can be made from a formal specification. These deductions allow new formulas to be derived and checked. So, the inference system can help to automate testing, prototyping or verification. However, formal techniques require some familiarity with formal notation that most designers do not have and the learning of these techniques requires considerable time.

A combination of MOF metamodeling and formal specification can help us to address MDA processes, software evolution in particular. Current metamodeling tools enable code generation and detect invalid constraints; however, they do not find instance models. Formal specification such as algebraic specification allows automatically generate instance models satisfying all the semantics of a given metamodel and automate proofs of transformations.

In light of this, we define a special-purpose language called NEREUS to provide extra support for metamodeling. The semantics of MOF metamodels (that is specified in OCL) can be enriched and refined by integrating it with NEREUS.

This integration facilitates proofs and tests of models and model transformations via the formal specification of metamodels. In addition, to define strictly the type system, NEREUS, like algebraic languages, allows finding instance models that satisfy metamodel specification. Some properties can be deduced from the formal specification and could be re-injected into the MOF specification without wasting the advantages of semi-formal languages of being more intuitive and pragmatic for most implementers and practitioners.

Our approach has two main advantages linked to automation and interoperability. On the one hand, we propose to generate automatically formal specifications from MOF metamodels. Due to scalability problems, this is an essential requisite. On the other hand, our approach focuses on interoperability of formal languages. Considering that there exist many formal algebraic languages, NEREUS allows any number of source languages such as different Domain Specific Languages (DSLs) and target languages (different formal language) could be connected without having to define explicit metamodel transformations for each language pair. Such as MOF is a DSL to define semi-formal metamodels, NEREUS can be viewed as a DSL for defining formal metamodels.

We define the NEREUS semantics by giving a precise formal meaning to each of its constructs in terms of the CASL language (Bidoit & Mosses, 2004). CASL was selected due to it is at the center of a family of specification languages. It has restrictions to various sublanguages, and extensions to higher-order, state-based, concurrent, and other languages. CASL is supported by tools and facilitates interoperability of prototyping and verification tools. A detailed description of this integration may be found at (Favre, 2009). NEREUS allows specifying metamodels such as the Ecore metamodel, the specific metamodel for defining models in EMF (Eclipse Modeling Framework) (Eclipse, 2010). Today, we are integrating NEREUS in EMF.

FUTURE RESEARCH DIRECTIONS

MDA is a software development approach based on the concepts of models as a representation of a system. Models are expressed in languages called metamodels. Model-to-model transformations play a crucial role in this approach. The artifacts used in model driven approaches are often tightly coupled. Models are bound to metamodels by a conformance relation. When metamodels change during its development or due to requirement changes, models need to be adapted. This becomes a problem when faced with evolution: as metamodels are extended or adapted, the dependencies may break until all dependent artifacts are adapted accordingly. This issue has been addressed by several research approaches. However, existing commercial MDA-based tools provide little support for metamodel evolution. Next, we describe some aspects worthwhile further research to provide support for metamodel evolution.

- Constraints and transformation rules also evolve triggered by metamodel evolution. Constraint and transformation rules might be adapted. There are proposals that identify the impact of a change, made at the metamodel, on its associated constraints and suggest constraint adaptations to remain consistent. Modeling framework should include these facilities.
- Metamodeling is the basis of many modeling framework and it is important to formalize and reason about it. Formal methods should be included in modeling frameworks to provide support for evolving specifications and incremental verification approaches.
- Refactoring is an important technique for evolving metamodels. CASE tools should provide an extensive catalogue of refactorings in order to assist the developer in the construction and editing of metamodels.

- Combine metamodel evolution techniques with specification matching that enable identify automatically metamodel elements on which a transformation could be applied. Existing proposals provide facilities for transformations through an explicit selection made for the designer.

- Metamodel evolution allows improving quality factors such as extensibility, modularity, reusability, complexity and readability. Metamodel evolution can be classified according to the quality attributes that improve, allowing applied relevant transformations where is needed and thus, increase the quality of the models. Software metrics can be used to identify problem areas and to evaluate the resulting improvements after applying a transformation.

CONCLUSION

There is an increasing adoption of MDA initiative for software development. MDA can help to develop and support a common application framework for software evolution that raises issues such as common exchange formats, tool integration and interoperability. When the system evolves, MDA maintains the interrelation between software entities accommodating the evolution of higher-level artifacts together with the code in a consistent way. Metamodels are crucial artifacts in MDA-based processes and a challenge on MDA software evolution is metamodel evolution. MDA tools should assist the developer in the metamodel evolution process. Besides, they should incorporate model migration in order to maintain the model conformance to its previous metamodel.

This article presents a framework for metamodel evolution, which combines metamodeling technique to specify the transformations and algebraic formalization.

We present an operator-based approach that works in two levels. At metamodel level, operators define transformations that correspond to metamodel evolution in order to assist the developer in its construction and edition. To define these transformations we incorporate refactoring technique and signature matching to identify metamodel elements. At model level, coupled operators define transformations capturing the corresponding model migration to maintain the conformance between models and the modified metamodel.

We propose formalize our approach in terms of the notation NEREUS, a formal metamodeling language that allows us interoperability with different formal languages. We define bridges that allow automating transformations from MOF-metamodels to NEREUS and from NEREUS to formal languages too. The formalization should be used to ensure that each transformation maintains the consistency between models.

We define foundations for MDA metamodel evolution. We foresee to integrate our results in the existing MDA-based CASE tools experimenting with different platforms.

REFERENCES

Alanen, M., & Porres, I. (2003). Difference and union of models. InStevens, , Whittle, , & Booch, (Eds.), *UML 2003 - The Unified Modeling Language. Modeling Languages and Applications*LNCS), (Vol. *2863*, pp. 2-17). Berlin: Springer-Verlag. doi:10.1007/978-3-540-45221-8_2.

ATL. (2012). *Atlas transformation language (ATL) documentation*. Retrieved September 24, 2012, from http://www.eclipse.org/atl/documentation/

Becker, S., Gruschko, B., Goldschmidt, T., & Koziolek, H. (2007). A process model and classification scheme for semi-automatic meta-model evolution. In GiTO-Verlag (Ed.), *Proceeding of the 1st Workshop MDD, SOA and IT-Management (MSI'07)*, (pp. 35-46). MSI.

Bidoit, M., & Mosses, P. (2004). *CASL user manual- Introduction to using the common algebraic specification language (LNCS) (Vol. 2900)*. Berlin: Springer-Verlag.

Brun, C., & Pierantonio, A. (2008). Model differences in the eclipse modeling framework. *UPGRADE: The European Journal for the Informatics Professional, 9*(2), 29–32.

CASE MDA. (2012). *Committed companies and their products*. Retrieved September 24, 2012 from www.omg.org/mda/committed-products.htm

Cicchetti, A., Di Ruscio, D., Eramo, R., & Pierantonio, A. (2008). Automating co-evolution in model-driven engineering. In IEEE Computer Society (Ed.), *Proceedings of the 2008 12th International IEEE Enterprise Distributed Object Computing Conference (EDOC 2008)* (pp. 222-231). Washington, DC: IEEE.

Cicchetti, A., Di Ruscio, D., & Pierantonio, A. (2010). Model patches in model-driven engineering. [LNCS]. *Proceedings of Models in Software Engineering, 6002*, 190–204. doi:10.1007/978-3-642-12261-3_19.

Eclipse. (2012). *The eclipse modeling framework*. Retrieved September 24, 2012 from http://www.eclipse.org/emf/

Falleri, J. R., Huchard, M., Lafourcade, M., & Nebut, C. (2008). Metamodel matching for automatic model transformation generation. In Busch, Ober, Bruel, Uhl, & Völter (Eds.), *Proceedings of the 11th International Conference on Model Driven Engineering Languages and Systems (MoDELS '08)* (LNCS), (Vol. 5301, pp. 326–340). Berlin: Springer-Verlag.

Favre, L. (2009). A formal foundation for metamodeling. In *Proceedings of the 14th ADA-Europe International Conference on Reliable Software Technologies* (LNCS), (Vol. 5570, pp. 177-191). Berlin: Springer-Verlag.

Fernández Sáez, A. (2009). *Un análisis crítico de la aproximación model-driven architecture. (Máster en Investigación)*. Madrid, Spain: Informática Facultad de Informática, Universidad Complutense de Madrid.

Formula. (2012). *Formula - Modeling foundation*. Retrieved from http://research.microsoft.com/en-us/projects/formula/

Fowler, M. (1999). *Refactoring: Improving the design of existing programs*. Reading, MA: Addison-Wesley.

France, R., & Bieman, J. M. (2001). Multi-view software evolution: A UML-based framework for evolving object-oriented software. In *Proceedings of the IEEE International Conference on Software Maintenance* (ICSM 2001), (pp. 386-395). IEEE.

Gruschko, B., Kolovos, D., & Paige, R. (2007). Towards synchronizing models with evolving metamodels. In *Proceeding of Workshop on Model-Driven Software Evolution (MODSE 2007)*. Amsterdam, The Netherlands: MODSE.

Herrmannsdoerfer, M. (2011). COPE – A workbench for the coupled evolution of metamodels and models. In Malloy, Staab, & van den Brand (Eds.), *Software Language Engineering (SLE 2010)* (LNCS), (Vol. 6563, pp. 286-295). Berlin: Springer-Verlag.

Herrmannsdoerfer, M., Benz, S., & Juergens, E. (2008). Automatability of coupled evolution of metamodels and models in practice. In Czarnecki, Ober, Bruel, Uhl, & Völter, (Eds.), *Model Driven Engineering Languages and Systems (MoDELS'08)* LNCS), (vol. *5301*, pp. 645-659). Berlin: Springer-Verlag. doi:10.1007/978-3-540-87875-9_45.

Herrmannsdoerfer, M., Benz, S., & Juergens, E. (2009). COPE - Automating coupled evolution of metamodels and models. In *Proceedings of the ECOOP 2009 – Object-Oriented Programming* (LNCS), (Vol. 5653, pp. 52-76). Berlin: Springer-Verlag.

Herrmannsdoerfer, M., Vermolen, S., & Wachsmuth, G. (2010). An extensive catalog of operators for the coupled evolution of metamodels and models. In *Proceedings of the 3rd International Conference on Software Language Engineering (SLE' 10),* (pp. 163-182). Berlin: Springer-Verlag.

Jouault, F., & Kurtev, I. (2005). Transforming models with ATL. InBruel, (Ed.), *Model Driven Engineering Languages and Systems (MoDELS 2005) LNCS),* (Vol. *3844,* pp. 128-138). Berlin: Springer-Verlag.

Kerievsky, J. (2004). *Refactoring to patterns.* Reading, MA: Addison-Wesley.

Kolovos, D., Di Ruscio, D., Paige, R., & Pierantonio, A. (2009). Different models for model matching: An analysis of approaches to support model differencing. In *Proceedings of the 2009 ICSE Workshop on Comparison and Versioning of Software Models (CVSM'09),* (pp. 1-6). ICSE.

Kolovos, D., Paige, R., & Polack, F. A. (2008). The epsilon transformation language. In Vallecillo, Gray, & Pierantonio (Eds.), *Theory and Practice of Model Transformations, First International Conference, ICMT 2008* (LNCS), (Vol. 5063, pp. 46-60). Berlin: Springer.

MDA. (2012). *The model-driven architecture.* Retrieved September 24, 2012, from http://www. omg.org/mda/

MOF. (2011). *Meta object facility (MOF) core specification version 2.4.1* (OMG Document Number: formal/2011-08-07). Retrieved September 24, 2012, from http://www.omg.org/spec/MOF/2.4.1

OCL. (2012). *OCL: Object constraint language, version 2.3.1* (OMG Document Number: formal/2012-01-01). Retrieved September 24, 2012, from http://www.omg.org/spec/OCL/2.3.1/

Reddy, R., France, R., Ghosh, S., Fleurey, F., & Baudry, B. (2005). Model composition - A signature-based approach. In *Proceedings of Aspect Oriented Modeling (AOM).* Montego Bay, Jamaica: MoDELS.

Rose, L., Paige, R., Kolovos, D., & Polack, F. (2009). An analysis of approaches to model migration. In *Proceeding of the 1st International Workshop on Model Co-Evolution and Consistency Management* (pp. 6-15). IEEE.

Rose, L. M., Herrmannsdoerfer, M., Williams, J., Kolovos, D., Garces, K., Paige, R., & Polack, F. (2010). A comparison of model migration tools. In Petriu, Rouquette, & Haugen (Eds.), *Proceedings of the 13th International Conference on Model Driven Engineering Languages and Systems (MoDELS)* (LNCS), (Vol. 6394, pp. 61-75). Berlin: Springer.

Rose, L. M., Kolovos, D. S., Paige, R. F., & Polack, F. A. C. (2010). Model migration with epsilon flock. In *Theory and Practice of Model Transformation (LNCS) (Vol. 6142,* pp. 184–198). Berlin: Springer. doi:10.1007/978-3-642-13688-7_13.

Toulmé, A. (2006). *Presentation of EMF compare utility.* Paper presented at 10th Eclipse Modeling Symposium. New York, NY.

UML. (2011a). *Unified modeling language: Infrastructure, version 2.4.1* (OMG Specification formal/2011-08-05). Retrieved September 24, 2012, from http://www.omg.org/spec/UML/2.4.1/

UML. (2011b). *Unified modeling language: Superstructure, version 2.4.1* (OMG Specification: formal/2011-08-06). Retrieved September 24, 2012, from http://www.omg.org/spec/UML/2.4.1/

Use. (2011). *A UML-based specification environment*. Retrieved September 24, 2012, from http://sourceforge.net/apps/mediawiki/useocl/

Vermolen, S., & Visser, E. (2008). Heterogeneous coupled evolution of software languages. In *Proceeding of the 11th International Conference on Model Driven Engineering Languages and Systems (MODELS'08)* (LNCS), (Vol. 5301, pp. 630-644). Berlin: Springer.

Vermolen, S., Wachsmuth, G., & Visser, E. (2011). Reconstructing complex metamodel evolution. In Sloane & Aßmann (Eds.), *Proceeding of the 4th International Conference on Software Language Engineering (SLE 2011)* (LNCS), (Vol. 6940, pp. 201-221). Berlin: Springer.

Wachsmuth, G. (2007). Metamodel adaptation and model co-adaptation. In *Proceeding of ECOOP 2007-Object Oriented Programming* (LNCS), (Vol. 4609, pp. 600-624). Berlin: Springer.

Xing, Z., & Stroulia, E. (2005). UMLDiff: An algorithm for object-oriented design differencing. In *Proceedings of the 20th IEEE/ACM International Conference on Automated Software Engineering (ASE'05)*, (pp. 54-65). ACM.

Zaremski, A., & Wing, J. (1997). Specification matching of software components. *ACM Transactions on Software Engineering and Methodology*, 6(4), 333–369. doi:10.1145/261640.261641.

KEY TERMS AND DEFINITIONS

Atlas Transformation Language (ATL): A model transformation language and toolkit developed on top of the Eclipse platform that provides ways to produce target models from source models.

Eclipse Modeling Framework (EMF): An open source modeling framework that enables developing applications based on models. It provides the foundation for interoperability with other tools and applications based on EMF.

Model Driven Architecture (MDA): An initiative of the Object Management Group (OMG) for the development of software systems based on the separation of business and application logic from underlying platform technologies. It is an evolving conceptual architecture to achieve cohesive model-driven technology specifications.

Metamodel Evolution: Is the process of gradual change and development from source metamodels to higher, more complex or better ones.

Model Migration: The process to transform a model in order to maintain the conformance with its corresponding metamodel that has been modified.

Meta-Object Facility (MOF): A meta-metamodel from the Object Management Group (OMG) that defines a common way for capturing the diversity of modeling standards and interchange constructs involved in MDA.

Object Constraint Language (OCL): a notational language for analysis and design of software systems that allows software developers to write constraints and queries over object models such as UML models.

Query, View, Transformation (QVT): A metamodel from the Object Management Group for expressing transformations in MDA-based processes.

Refactoring: A change to a system that leaves its behavior unchanged but enhances some non-functional quality factors such as simplicity, flexibility, understanding, and performance.

Specification Matching: Is the process to determine if two software component matches.

Chapter 10
Rosetta Composition Semantics

Megan Peck
University of Kansas, USA

Perry Alexander
University of Kansas, USA

ABSTRACT

The Rosetta specification language aims to enable system designers to abstractly design complex heterogeneous systems. To this end, Rosetta allows for compositional design to facilitate modularity, separation of concerns, and specification reuse. The behavior of Rosetta components and facets can be viewed as systems, which are well suited for coalgebraic denotation. The previous semantics of Rosetta lacked detail in the denotational work, and had no firm semantic basis for the composition operators. This thesis refreshes previous work on the coalgebraic denotation of Rosetta. It then goes on to define the denotation of the composition operators. Several Rosetta examples using all types of composition serve as a demonstration of the power of composition as well as the clean modular abstraction it affords the designer.

INTRODUCTION

System-level design is characterized by the need to assemble information from multiple, heterogeneous domains when making design decisions. Languages intending to support system-level design must therefore support defining and composing multiple models representing heterogeneous information. To achieve this, such languages must eschew single semantics approaches and instead provide support for modeling and composition across multiple semantics. Following this approach one can reason about smaller pieces within an appropriate domain vocabulary and semantics, rather than trying to apply automated reasoning to a monolithic specification, then compose results to make a system-wide correctness assessment (Frisby et al., 2011).

DOI: 10.4018/978-1-4666-4494-6.ch010

The Rosetta specification language (Alexander, 2006; Alexander et al., 2000) provides a language and semantics in support of system-level design. Rosetta focuses on heterogeneous model composition in its domain-based specification system and model composition system. Like traditional hardware description languages, it allows the designer to compose systems from components, or decompose a system into its parts. Unlike traditional hardware description languages, Rosetta supports concurrent engineering by allowing multiple models to describe a system simultaneously. Rosetta's built-in composition operators give the designer the ability to compose components and systems from domain-specific pieces describing different system aspects or components.

The basic specification building block in a Rosetta specification is the *component* that collects a set of declarations and states assumptions, definitions, and implications about those declarations within a single domain. In Rosetta, components are first-class structures and can be manipulated as data. Here we describe three operation classes defined for component composition: structural composition constructs a component that includes the operand facets as components; conjunctive composition defines a component that satisfies all given operand components; and disjunctive composition defines a component that satisfies one or more of the given operand facets.

Formalizing the semantics of a specification language gives us assurances as to the validity of the specifications we write with the language. This work refreshes previous work on the coalgebraic denotation of components and facets. While the ideas of the Rosetta composition operators are not new, they have yet to be formally denoted. Here we fill that semantic void by defining the denotation of the composition operators.

BACKGROUND

By their nature, complex systems must be viewed from many different perspectives. A building provides comfort (HVAC, plumbing, heating, lighting and aesthetics), shelter (structure and safety), and consumes resources (power, water, maintenance) in addition to supporting its business function. Similarly, an embedded system has weight, must be manufactured, must operate in real-time, generates electromagnetic noise, and consumes power in addition to performing its basic function. These systems requirements are as important as a systems basic function. For example, no one buys a cell phone because it makes phone calls – battery life, weight, interoperability, and smart functions play a huge role in such purchases. Given the importance of these system-level issues, design languages should support their inclusion as first-class citizens in the design process.

Rosetta accommodates these multiple specification views by providing a framework for heterogeneous specification, called *domains* (Streb et al., 2006; Streb and Alexander, 2006). Each domain defines a modeling vocabulary and semantics for representing information related to some specification viewpoint. The basic building blocks of Rosetta are *components* and *facets* use domains to represent system models. A *component* defines assumptions, definitions, and implications over a set of declarations specifying not only behavior, but assumptions on that behavior and secondary behaviors that follow from the specification. A *facet* is a component with no assumptions or implications – a basic, ideal model. Every facet or component models a specific system aspect by extending a domain with specific definitions for the system being modeled. For example, an encryption device might be specified in the discrete_time domain while a filter might be written

Box 1.

```
facet halfAdder(x,y::input bit;s,c::output bit)::state_based is
begin
     s'=x xor y;
     c'=x and y;
end facet halfAdder ;
```

in the frequency domain. In these cases, the domain is serves as the *type* of the facet.

Rosetta includes a pre-defined domain hierarchy (Streb et al., 2006; Streb and Alexander, 2006) that includes domains for modeling various types of predominantly electronic systems and the constraints they operation under. New domains extend existing domains in the hierarchy to include more specific knowledge or to write models using a new semantics. These extensions can be a simple set of declarations and or as complicated as a complete new semantics or timing models. The Rosetta domain hierarchy forms a complete lattice ordered by theory homomorphism and relationships defined in the lattice form a Galois connection (Streb et al., 2006; Streb and Alexander, 2006) between adjacent domains. This relationship is critical when viewing facet transformation as moving among domains in the lattice.

Because specifications are written in multiple domains, a mechanism is needed for understanding combined specifications and their interactions. In Rosetta, *interactions* provide language level support for specifying how specifications from different domains impact each other. The interaction construct describes how information flows between two domains – as the name suggests, how two domains interact with each other. Within the domain lattice, standard interactions are defined that allow information move among domains in the canonical style. Just as users can extend domains to form new domains, they can define new interactions between domains. In this way, Rosetta does not attempt to support all conceiv-

able domains using a single semantics, but instead offers a framework where the domains and their interactions may be constructed.

As noted previously, the basic specification unit in Rosetta is the *component* or *facet*. A component definition extends a particular domain with definitions, assumptions, assertions, and implications. Components may have inputs and outputs that allow them to be parameterized and to communicate with other specification constructs. The terms defined by a component may either be Boolean expressions written in Rosetta's expression language or may instantiate other components to define structural specifications. Since it is common to not need the assumptions or implications of a component, facets are more commonly seen in Rosetta than the more general components, as facets are simply components with no assumptions or implications.

As an example of Rosetta specification, consider the half-adder definition in Box 1.

The example halfAdder facet shown has two input bits, x and y, and sum and carryout outputs, s and c, respectively. The domain of the facet is state_based, as the sum and carry are computed for the next state, given the current inputs. The next states of outputs, denoted by a "ticked" symbol in the classical styel, are constrained by Boolean expressions that equate them with values calculated from current state variables. This intentionally follows closely the convention used by hardware designers writing VHDL (IEEE, 1994) or Verilog (IEEE, 1995).

COMPOSITION OPERATORS

Rosetta specification constructs—components, facets, and domains—are first-class entities in the language allowing models to be treated as values in definitions. Thus, composition operators are literally Rosetta operations—usually binary or unary—over models. Following, we explore three such composition operations formally: *structural; conjunctive;* and *disjunctive* composition. Each composition type is motivated, described, and a Rosetta example of each type demonstrates its use.

Structural, or *hierarchical composition* allows the designer to specify a system as a collection of communicating components. Structural composition is the dominant method of system design where components are composed into systems and is define as component instantiation and inclusion. It is extensively used by hardware design languages such as VHDL, Verilog, SystemVerilog (Acc, 2002) and SystemC (Grötker et al., 2002) or software architecture specification languages (Allen and Garlan, 1997, Nuseibeh et al., 2003).

Structural composition allows the designer to reuse component models to construct systems. The concept of structural composition provides the semantic tool for specifying systems in a manner that reflects how they are already designed, allowing direct representation of the structure already inherent in the system being specified. Thus, the reusable units in a specification will mirror the recurring units throughout the actual structure of the system. This is common for composable elements in a hierarchical specification that is particularly popular in hardware design.

Continuing with the halfAdder example from the previous section, we now create the fullAdder facet by instantiating two halfAdders and appropriately interconnecting their input and output parameters in Box 2.

The fullAdder uses three internal variables, s1, c1 and c2 to share information between the instantiated halfAdder facets. The fullAdder uses structural composition to create the traditional implementation of a full adder using two half adders. The output, s, of fullAdder is constrained through the constraints given by halfAdder, while co is constrained through a new Boolean expression. As noted, structural composition in this style is exceptionally common in hardware and systems design, but less so in software engineering.

By *instantiating* the halfAdder facet, we get a distinct copy of the original facet. Those copies are no longer halfAdder, but copies named ha1 and ha2 respectively. Their instantiations are separate, and the only shared information between them are the parameters s1 and c1. There is no depth limit to hierarchical composition – we could use fullAdder instantiations to create ripple carry adders for inclusion in an ALU, and could then instantiate an ALU in a CPU design, instantiate the CPU in an embedded systems design, continuing as far as necessary. It should also be noted that the instantiated half adders are first-class in the language. Facets and components are both first-class in that they may be referenced by the

Box 2.

```
facet fullAdder(x,y, ci::input bit;s,co::output bit):: state_based is
        s1,c1,c2::bit;
begin
        ha1: halfAdder(x,y,s1,c1);
        ha2: halfAdder(s1,ci,s,c2);
        co' = c1 or c2;
end facet fullAdder ;
```

Box 3.

```
facet controlUnit (..)::HW ...
facet alu (..)::HW ..
facet regFile (..)::HW ...
facet memory (..)::HW ...
facet cpu(clk):: HW is
        enable:: bit ;
        instruction,A,B,C::word;
        addressA,addressB,addressC,aluOP::nibble;
        memControl::bitVector;
begin
        c: controlUnit(instruction,memControl,addressA,addressB,addressC,
                            enable,aluOP);
        a: alu(aluOp,A,B,C);
        rf: regFile(clk,enable,addressA,addressB,addressC,A,B,C);
        m: memory (clk,memControl,C);
end facet cpu ;
```

language as any other data. This is in principle the same as first-class functions in functional programming languages.

This composition is available at any level of the specification from simple combinational circuits through entire processors, embedded systems, and systems-of-systems. As a more complex example, Box 3 shows how a simple structural CPU model is constructed from components in the canonical fashion.

Limiting ourselves to just hierarchical composition has drawbacks, however. We are required to define all behavior and constraints in place. Any non-functional behaviors would necessarily be involved in the same specifications that defined the behavior. Conjunctive composition alleviates this problem by adding a 'horizontal' composition that allows two specifications to represent the *same* unit, both constraining what that system is and how that system behaves.

Specification conjunction allows the designer to specify multiple views of a single component and compose them into a single model. Specification composition provides language level support for separation of concerns by defining models of different system aspects and composing the

result. Conjunctive composition is done using the product operator, *. We can define a facet f3 as the product of facets f1 and f2 as follows:

```
facet f3 = f1 * f2;
```

In the new facet f3 simultaneously exhibits behaviors specified by f1 and f2 as well as emergent behaviors defined by their interaction.

Using specification conjunction a system designer specifies functional requirements that define what a system does separately from physical constraints such as resource limitations, available implementation fabrics, and usage assumptions. Conjoining the resulting specifications allows concurrent design, modeling all aspects simultaneously directly supporting what is traditionally called concurrent engineering – simultaneously considering multiple aspects during design. Using specification conjunction, a designer specifies system behavior separately from implementation architecture specifics. This feature supports co-design applications where a system designer defines the functional requirements of a component without reference to particular hardware or software architecture details (Peck, 2011).

Box 4.

```
facet qamAESArch(i::input word(2);o::output real;f::input frequencyType;
                        length:: design; keyLengthType; k::input
                        word(length))::static is
        ho::bit ; aesi:: word(16);
                        mi:: word (2);)
begin
        code: huffEncoder(i, ho);
        buff1: buffer(ho,aesi);
        enc: aesEncryptor(aesi, aeso, length, k); buff2: buffer(aeso,mi);
        modulate: qamModulator(mi,o,f);
end facet qamAESArch;
facet structure1 ()::fabric is begin
        code: hardware (fpga);
        buff1: hardware (fpga);
        enc: hardware(crypto);
        buff2:
        hardware (fpga);
        modulate: hardware (fpga);
end facet structure1 ;
facet structure2 ()::fabric is begin
        code: hardware (fpga);
        buff1: hardware (fpga);
        enc: software(proc1);
        buff2: software(proc2);
        modulate: software (proc2);
end facet structure2 ;
```

As another example of how Rosetta supports specification at the language level, consider the functional behavior definition for a QAM modulator with encryption (Kimmell et al., 2008) in the qamAESArch facet below. In the structure1 and structure2 facets, we define two alternative non-functional views of the same system that differ by requiring the sub-components to be implemented in hardware or software in different configurations (Box 4).

The two models are initially independent models. The facet product operator defines the conjunctive composition of the qamAESArch facet with each non-functional requirements facet to describe two separate implementations (Box 5).

Box 5.

```
facet implementation1:: static is qamAESArch * structure1;
facet implementation2:: static is qamAESArch * structure2;
```

Composition requires any resulting implementation to satisfy both facet specifications. Thus, implementation1 is constrained by domains static and fabric, and by the definitions of code, buff1, enc, buf2 and modulate of behavior, and the code, buff1, enc and modulate as defined in structure1. This composition searches for terms within the two facets with the same names, that are then similarly conjoined. The overall composed entity therefore must have within it one term for each shared name that satisfies both sets of requirements. An interesting and useful side effect of this composition style is that when a component like code is referenced in either of the implementation definitions, both the code models from architecture and structure are included. Specifically, referencing code in the product results in the product of code definitions used in the product.

Both structural composition and conjunctive composition are still limited in that a facet must meet all definitions and constraints of each facet involved in the construction. The designer may desire the ability to compose separate aspects, requiring only one set of definitions and constraints to hold. *Specification disjunction* allows the designer to specify alternate views of a system. Disjunctive composition is a means of separately defining alternative. Disjunctive composition uses the sum operator, +. We can define a facet f3 as the sum of facets f1 and f2 by saying:

```
facet f3 = f1 + f2;
```

The disjunction of two facets is itself a facet where at least one definition needs to be valid at all times. However, facet sum does not necessitate

mutual exclusivity. All alternate definitions might hold in a valid sum. In the running example, we might use disjunction to define possible structures:

```
facet anyStructure:: fabric is struc-
ture1 + structure2;
```

anyStructure is a composed facet that must satisfy either structure1's behavior and constraints or structure2's behavior and constraints. Often, in practice the terms of the facets being composed will themselves define mutually exclusive sets of requirements, but nothing about the Rosetta mechanism guarantees this. Consider two semantically equivalent specifications from software defined radio design defining alternative implementations (Box 6).

In these implementations, we require behavior to be satisfied, and we require *either* structure1 or structure2 to be satisfied. This implementation style allows us to consider a larger system containing this implementation without having to select one structure and exclude the other. We can define the behavior once, and have reuse with respect to the different possible implementation structures and details of the system.

Disjunction allows the designer to specify multiple pieces of a component and com- pose them to make the whole. In this way, the designer can isolate functionalities. For instance, if an instruction or command in a system can be one of many options, the designer can specify each separately and compose them to create the entire instruction. This approach is also used in the specification language Z, via disjoints (Woodcock, 1995).

Box 6.

```
facet implementation3  = qamAESArch * anyStructure ;
facet implementation4  = qamAESArch * (structure1 + structure2);
```

Box 7.

```
domain processor:: state_based is
        registers:: array(16, word);
        pc:: word;
        instruction:: word is memFetch (pc);
begin
end domain
facet plus:: processor is
        src1:: nibble is decodeSrc1(instruction);
        src2:: nibble is decodeSrc2(instruction);
        dest:: nibble is decodeDest(instruction);
begin
        op = plusOp ;
        registers '= replace(registers,dest,registers(src1)+registers(src2));
        pc' = pc + x"0002";
end facet plus;
facet jmp:: processor is
begin
        op = jmpOp;
        registers ' = registers ;
        pc' = newPC(instruction);
end facet jmp ;
facet processorBeh:: processor is plus + ... + jmp;
```

As an example, consider a microprocessor specification for a commonly used architecture (Box 7)

The processor domain extends the state_based domain with declarations for a 16-register register file, registers, and a program counter, pc. Using a function memFetch::word->word, it constrains the instruction to be the value fetched from memory at address pc. Similarly, using a function to decode the instruction, decodeOp::word->nibble, the domain constrains the op.

The behavior for each operator can now be written in its own facet with this new processor domain. For instance, the plus facet enforces the constraint that operator be the operation for addition. Since the domain defines registers and pc, the plus facet must provide the next state constraints for these. The register file is updated using the replace function that replaces the given index parameter with the new given value, and leaves

the rest alone. In this way, we correctly define the framing rules by updating the destination register while leaving the rest unchanged. The facet also updates the pc's next state simply by adding 2 to the current state. Similarly, the jmp facet constrains that op must be the jmpOp, explicitly states that registers does not change, and updates pc with the newly calculated program counter value given by the function newPC::word->word.

Other processor instructions would be written in the same fashion. Given the individually written facets for each instruction, we can sum them to create a new facet, processorBeh, that defines the behavior of the processor for all possible operations. Since each individual facet has the domain processor, the facet sum will as well.

Disjunction in this example illustrates two major benefits. The first is the idea that a designer can separate the concerns of the different operations and focus on the behavior of one instruction at a time. The second is the ease of extensibility of the processor design. New instructions can be written and added in a clear way by adding one more facet to the disjunction. This allows for the modularity a programmer is accustomed to utilizing, at a per-behavior level, and ensures that every facet in the disjunction is constraining the needed pieces of the facet — in this case, the registers and pc.

SEMANTICS

With Rosetta's composition operators identified, we define their formal semantics by building on previous work (Kong et al., 2003). The previous work sets up two parts for denoting a facet – defining the coalgebraic system structure, and denoting the syntactic pieces of a facet. The definition of coalgebraic system structure lacked a framework for the general case of denoting any facet. It describes the structure and denotes some specific facets. Here we expand upon that, giving the framework for the general case. The previous work thoroughly describes the general case of denoting the syntactic pieces of a facet. However, we have updated these with newer Rosetta requirements. The previous work also lacked discussion of components, defining denotation for only facets only.

Facets are denoted in two parts. One defines its behavior as a coalgebraic structure. The abstract state of a facet is its observable behavior that itself can be viewed as a system and is thus a coalgebra. The other part denotes the facet pieces going inside the facet definitions, giving the semantic details of the facet.

First consider the denotation of the syntactic pieces of a facet. Recall the general syntactic parts of a typical facet f are:

```
facet f (#parameters#)::#domain# is
        #variables#
begin
```

```
    #terms#
end facet f;
```

Rosetta facets consist of observers defined from parameters and variables and any variables that come from the domain definition, a domain, and sets of definitions, or terms. Every facet has only one set of observers. These observers are essentially the interface of the facet. There can be multiple sets of terms, though it is most typical to only have one set.

We now define facets as 3-tuples, $<O,D,T>$ where:

- O contains the observers (parameters and variables, including all domain variables) of f. Thus. $O = (O1,O2,...,On)$ where each Oi is an observer of f.
- D is the domain of the components.
- $T = (T1,T2,...Tn)$, where each Tj is a set of denotations of f's terms.

Any specification of a facet is consistent if at least one of its sets of terms is consistent. Note that the majority of facets will only have one term set. The denotations for the terms themselves has already been done (Kong et al., 2003) using three valuation functions, E, O, and V, for expressions, operators, and values, respectively,

In previous Rosetta work, a component was denoted as three facets – one for assumptions, one for definitions, and one for implications. The component has now become the primitive building block. A facet is now a special case of a component with no assumptions or implications. Therefore, the denotation of components has not previously been addressed. The observers and domain details of components and facets are identical, so the denotations only differ in that the bodies of the component denotation must also include the assumptions and implications.

Recall the general syntactic parts of a typical component as seen in Box 8.

We will consider a component a 3-tuple. $<O,Dom,Bodies>$ and where:

- O contains the observers (parameters and variables, including all domain variables) of c. So, $O = (O_1,O_2,...,O_n)$ where each O_i is an observer of c.
- *Dom* is the domain of the components.
- *Bodies*$=((A_1,D_1,I_1),(A_2,D_2,I_2),...,(A_p,D_p,I_p))$, *a list of triples containing* each of the following:
  - Each *Ai* is the set of denotations of the assumptions in the ith triple in c
  - Each D_i is the set of denotations of the definitions in the ith triple in c
  - Each I_i is the set of denotations of the implications in the ith triple in c

Rather than just sets of terms, there are now sets of triples containing assumptions, definitions, and implications. A consistent component is one in which every term within one set of (assumptions, definitions, implications) triples in the component is consistent.

The coalgebraic structure of facets was initially described by Kong et al. (2003), but lacked a denotation for the most general case. Their denotation was never defined as a function with only specific examples demonstrated. These techniques were

Box 8.

```
component c (#parameters#)::#domain# is
#variables#
        begin assumptions
                #assumptions#
        end assumptions ;
        begin definitions
                #terms#
        end definitions ;
        begin implications
                #implications#
        end implications ;
end c;
```

used to denote the coalgebraic structure of specific facets, but there was no general denotation to apply to any facet.

The behavior, or *abstract state*, of a component or facet is defined by its observers. This imples that there is no distinction between the coalgebra denoted by a component and a facet. Consider the abstract state, S, of the facet f. S is defined by all possible observations of f, meaning S is the same as ON. We will see that S is the coalgebra denoted by f. The behavior of a facet is what we observe of that facet over transitions. So, the system can be thought of as all possible observations. Given a transition function, ξ, we take a step, which results in the observations from that transition as well as the rest of the system behavior. So we can define the structure of f, $O^N \rightarrow^\xi O \times O^N$, where we describe the behavior of the facet as a sequence of observations of the facet.

Facets are described via their observations over transitions. We can abstract these observations to observations within their domains, or their domain coalgebra. Every facet with a given domain observes the variables defined by that domain. When we abstract a facet's coalgebra to its domain coalgebra, we get the behavior from observing only the domain variables. Essentially, we start with the abstract state of the facet. We abstract to get the behavior of the facet, or facet coalgebra. We can abstract once more to get the behavior from only observing domain variables. The commuting diagram shows that domain coalgebras are final, meaning they are complete in the sense that any facet in that domain can be uniquely mapped/abstracted to that domain. Essentially, this reiterates that every facet with a given domain extends that domain, and therefore observes the variables of that domain.

Consider the counter facet as an example denoted to a coalgebra in Box 9.

Let S be the set of states, or all the observations, of the counter facet. The system defined by counter is a stream from S to S, exhibiting observations. The observations are the 3 input bits, the

Box 9.

```
facet counter(en,clk,rst::input bit;out::output word(3))::state_based is
        internal:: word(3);
begin
        t1: if rst=1 then internal ' = "000"
            elseif (en=1 and clk=1 and clk'event) then
               internal ' = case internal is b"000" -> b"001" |
                                             b"001" -> b"010" |
                                             b"010" -> b"011" |
                                             b"011" -> b"100" |
                                             b"100" -> b"101" |
                                             b"101" -> b"110" |
                                             b"110" -> b"111" |
                                             b"111" -> b"000"
               end case ;
            else internal '=internal ;
            end if;
        t2: out' = internal ';
end facet counter ;
```

output 3-bit bit vector (word(3)), and the internal 3-bit bit vector of the facet, as well as the set of states used in the specification of counter. The coalgebraic structure shows the observations over taking one transition with the rest of the possible transitions in Box 10.

The first part of the denotation gives all observers of counter, which includes all parameters and variables of the counter. Next is the domain, which is state_based. Then is the list of denoted term lists. As with most facets, this facet only has one term body. Therefore, there is only one list of terms. That list contains the denotations for t1, and t2, shown above. The denotation of counter follows the two parts of a facet denotation. First, the observers form the coalgebraic structure of

the denotation. This consists of all possible states of counter's behavior, and the observations over the possible transitions. Second, the syntactic pieces of counter make up the rest of the denotation. These pieces are the observers, domain, and denotations of all terms within counter.

Components and facets may have more than one body, but only one body needs to be consistent. Therefore, a consistent component is one in which every term within one set of (*assumptions, definitions, implications*) triples, corresponding to one body, in the component is consistent. Similarly a consistent facet is one in which every term within at least one set of terms, corresponding to one body, in the facet is consistent. A Boolean term, or assertion, is consistent if it is true. In other words,

Box 10.

```
<(en:input bit, clk:input bit, rst:input bit, out:output word(3),
   internal:word(3)),
   State_based,
   ((T[t1],T[t2]))  >
```

if no term is false, then false has not been asserted, so the component is consistent. In components, the assumptions and definitions are typically used in the implications, i.e. (*assumptions* ∧ *definitions* ⇒ *implications*). All instantiated facets within the instantiating facet must themselves be consistent given their formal parameters replaced with the actual parameters. Essentially, consistency is a structurally recursive or inductive concept, in that something is consistent if all of its parts are consistent (the base case being true Boolean assertions). So a facet with only assertion terms is consistent if all of its assertions hold true. Once that facet is instantiated in another facet, the instantiating facet is consistent if its assertions hold and if the instantiated facet holds under the instantiation.

Using our counter example, an implementation of counter is consistent if the assertions t1 and t2 hold true. Any facet that instantiates counter is only consistent if that instantiation of counter, which replaces counter's formal parameters with actual parameters, is consistent. An inconsistent component is invalid, in that we can say nothing about it. There is no basis for reasoning about an inconsistent component.

Having defined the denotation for components and facets. This section will show how we denote composed components and facets. First, we'll look at sum and product. For each, we give the denotation of the resulting sum or product, respectively. We explain how the result is constructed, as well as the validity of the domain of the result. We also describe how the result is still a valid component/facet, and exhibits the desired behavior appropriate to sum and product. Each section gives an example to illustrate the construction and denotation of the sum or product. We then move on to instantiation and inclusion, noting the subtle distinction between the two. We give an example of each, followed by each denotation. We end the chapter with a discussion of homomorphisms. While not a true composition operator, it is an important

factor in relating and reasoning about multiple components and facets.

Product and sum are binary operators. As such, we will define two components, *c1* and *c2* as operands in future discussion. Say *c1* and *c2* are components denoted as:

$< O_1, Dom_1, Bodies_1 >$ and $< O_2, Dom_2, Bodies_2 >$, where

- O_1 and O_2 are the observers (parameters and variables, including all domain variables) of *c1* and *c2*, respectively. So, $O_1 = (O_{11}, O_{12}, ..., O_{1n})$ and $O_2 = (O_{21}, O_{22}, ..., O_{2m})$, where each O_{1i} is an observer of *c1* and each O_{2j} is an observer of *c2*.
- Dom_1 and Dom_2 are the domains of the components.
- $Bodies_1 = ((A_{11}, D_{11}, I_{11}), (A_{12}, D_{12}, I_{12}), ..., (A_{1p}, D_{1p}, I_{1p}))$, a list of triples containing each of the following:
  ○ Each A_{1i} is the set of denotations of the assumptions in the ith triple in *c1*.
  ○ Each D_{1i} is the set of denotations of the definitions in the ith triple in *c1*.
  ○ Each I_{1i} is the set of denotations of the implications in the ith triple in *c1*.
- $Bodies_2 = ((A_{21}, D_{21}, I_{21}), (A_{22}, D_{22}, I_{22}), ..., (A_{2q}, D_{2q}, I_{2q}))$, like in $Bodies_1$

Similarly, when needed, we will define two facets, *f1* and *f2*, as the operands in future facet composition discussion. Say *f1* and *f2* are facets denoted as $< O_1, D_1, Terms_1 >$ and $< O_2, D_2, Terms_2 >$ where

O_1 and O_2 are the observers (parameters and variables, including all domain variables) of *f1* and *f2*, respectively. In this case $O_1 = (O_{11}, O_{12}, ..., O_{1n})$ and $O_2 = (O_{21}, O_{22}, ..., O_{2m})$, where each O_{1i} is an observer of *f1* and each O_{2j} is an observer of *f2*. D_1 and D_2 are the domains of the facets. $Terms_1 = (T_{11}, T_{12}, ... T_{1m})$, where each T_{jk} is a set of terms (like sets of definitions in a component) in *f1*, likewise for $Terms_2 = (T_{21}, T_{22}, ... T_{2n})$ in *f2*.

Sum Operator

The sum operator allows for disjunctive composition of components and facets. It gives us the ability to define multiple views of a system, where only one must hold. As described earlier, the specifier can separate alternative functionalities and sum them to create the entire functionality. This section explains how the sum is constructed. We look at the denotation, explain the validity of the pieces of the denotation and the resulting component or facet, and give a full example.

Using the definitions of *c1* and *c2* from above, when we take the sum of two components, we get the following:

Say *c3 = c1+c2*, then *c3* is denoted as $<O_1 ++ O_2, Dom_1 \sqcap Dom_2, Bodies_1 ++ Bodies_2>$ where

- $O_1 ++ O_2 = (O_{11}, O_{12}, ..., O_{1n}, O_{21}, O_{22}, ... O_{2m})$. Note that duplicates (i.e. some $O_{1i} = O_{2j}$) are excluded. Also note that it is common that $O_1 \equiv O_2 \equiv O_1 ++ O_2$. All parameters and variables of both operands are included in the sum.
- $Dom_1 \sqcap Dom_2$ is the least common domain of Dom_1 and Dom_2. It is also common for $Dom_1 \equiv Dom_2 \equiv Dom_1 \sqcap Dom_2$.
- $Bodies_1 ++ Bodies_2$ is a simple append yielding $((A_{11}, D_{11}, I_{11}), (A_{12}, D_{12}, I_{12}), ..., (A_{1p}, D_{1p}, I_{1p}), (A_{21}, D_{21}, I_{21}), (A_{22}, D_{22}, I_{22}), ..., (A_{2q}, D_{2q}, I_{2q}))$.

Rosetta domains and the transformations between domains form the Rosetta domain lattice (Lohoefener, 2011). As domains are extended (down the lattice), constraints are added. Since the domains form a lattice, any two domains on the lattice have a *least common domain,* or *meet,* above them in the hierarchy. So if *c1* has domain Dom_1 and *c2* has domain Dom_2 as described above, when we disjoin them to produce *c3 = c1 + c2,* we can safely say that *c3* has the domain Dom_3

$= Dom_1 \sqcap Dom2$. Because any domain lower in the lattice is more constrained, all definitions in *c1* meet all the constraints of Dom_3 and likewise, the definitions of *c2* meet all constraints of Dom_3. So no work is necessary to transform the terms in either component of the sum into the new domain – they are already in that domain.

It should be noted that the specifier can safely explicitly abstract (moving up the lattice) or concretize (moving down the lattice) a component into a desired domain prior to taking the sum to control the domain of the resulting component. For instance, the designer may want the specificity provided by a more constrained domain. They may safely transform one of the components into the domain of the other component prior to taking their sum to gain that specificity.

We can think of sum as a disjunction of components. We have two components that we sum together, and we know the result is either the first part of the sum or the second part of the sum. We need a way of separating the disjoint parts within a summed facet. Essentially, we need a way of separating entire sets of assumptions, definitions, and implications, and enforcing that only one of those sets needs to be consistent. This need prompted the addition of multiple bodies within a component, which was not previously supported in Rosetta, and is the reason that only one body needs to be consistent for the component to be consistent.

Say you know your system will behave in one of two ways. With the addition of multiple bodies, you have two choices for specifying this system. You could explicitly write a facet with two bodies – one for each behavior.

```
facet system(#parameters#)::#domain#
begin

    #body describing behavior 1#
begin

    #body describing behavior 2#
end system ;
```

Alternatively, you could define two components, behavior1 and behavior2. This way you are able to separate the assumptions, definitions, and assumptions of these two behaviors into their own components. When behavior1 and behavior2 are summed to describe the entire system, their assumptions, definitions, and assumptions need to be reflected in the entire system, though, in a way that allows for the situation that only one behavior at a time need be enforced. To that end, we have chosen to append each set of (*assumptions,definitions,implications*) triples onto the list of possible behaviors. One or more of these triples needs to hold in a consistent component. Appending these (*assumptions,definitions,implications*) triples gives us exactly the notion of disjunction we need. Note that it is not always possible to know which body (or bodies) is active within a specification. There is no notion of tagging that identifies what set of (*assumptions,definitions,implications*) is consistent at that time. This is intentional, as some specifications may not be constrained enough to determine which option is currently active. We opt for a strategy that allows for expressivity and under-constrained specifications.

Facets are simply components without sets of assumptions or implications. So the facet sum is a simplified version of the component sum. We will use the previously defined facets, *f1* and *f2* as the operands of the sum. Say *f3=f1+f2*. Then it is denoted as $<O_1++O_2, D_1 \sqcap D_2, Terms_1++Terms_2>$ where:

- $O_1 ++O_2 = (O_{11}, O_{12},...,O_{1n}, O_{21}, O_{22},...O_{2m})$ as in component sum. Note that duplicates (i.e. some $O1i = O2j$) are excluded. Also note that it is common that $O_1 \equiv O_2 \equiv O_1++O_2$. All observers from each operand are included in the result.
- $D_1 \sqcap D_2$ is the least common domain of D_1 and D_2 as in component sum. It is also common for $D_1 \equiv D_2 \equiv D_1 \sqcap D_2$.

- $Terms_1++Terms_2$ would be $(T_{11}, T_{12},...,T_{1m}, T_{21}, T_{22},...,T_{2n})$, where each T_{ij} is a set of terms.

Recall our earlier processor example where instructions are fetched and decoded, and based on the decoding, different operations are executed. We modularly define the execution for each possible instruction. The processor is defined as the sum of each instruction execution.

Note, this is an example where the different parts of the sum are in fact mutually exclusive. The first line of each facet asserts that the *op* is equal to that particular instruction. That assertion will hold in only one of the facets. For instance, if the op is decoded as the plusOp, then the assertion op = plusOp will hold in the facet plus, but in any other facets those assertions will fail. In the jmp facet, the assertion op = jmpOp will hold, etc. The facet processorBeh is consistent if any of the facets in the sum are satisfied. The denotation of this facet is located in Box 11.

All observers are combined. The domain for all of the operands was *processor,* so the domain of the result is still processor. The list of all body terms in the result contains separate lists of terms from each operand – the lists of terms are all appended in the result.

In sum, sharing clauses give us a similar power as domain definitions. In domain definitions, we add constraints based on our knowledge of that domain. For example, in state_based, we add the constraints of having a current state and next state. When we do a facet sum, we have the potential of losing domain information since we may have to go up the domain lattice to find the least common domain of the summed facets. However, we may have knowledge of certain constraints that should still be part of each of those facets. We can add that information to the sharing clause of the conjunction to enforce those constraints. Note that with facet sum, nothing is automatically

Box 11.

```
< (registers::array(16,word),pc::word,instruction::word is
   memFetch(pc), op::nibble is decodeOp(instruction)),
processor,
((T[op= plusOp],
   T[registers'=replace(registers,dest,registers(src1)+registers(src2))],
   T[pc' = pc + x"0002"],
 ..., (...), ...
 (T[op = jumpOp],
   T[registers'=registers],
   T[pc' = newPC(instruction)]))
>
```

shared to avoid name capture issues. Anything that should be shared must be explicitly added to the sharing clause.

It should be noted that the result of summing two components (or facets) is a valid component. A valid component/facet would contain valid observers, a valid domain, and a list of valid bodies/terms. We have appended all observers from each operand to form the observers of the sum. Since those operand observers were all the parameters and variables of the operands, the appending of the observers gives us valid parameters and variables for the sum. We have already discussed that the domain of the new facet exists and is valid. Since we have appended sets of bodies/terms from the operands together for the sum, we get a list of valid sets of bodies/terms. These are the three parts of a component's (or facet's) denotation.

Also, since the structure of the component coalgebra is defined over the observers,

we still have a valid coalgebra as the observers, $(O_1 + {+}O_2)$, are valid. Therefore, the behavior of the summed component still denotes a coalgebra. The structure of this coalgebra is

$$(O_1 {++}O_2)^n \to^\varsigma (O_1 {++}O_2) \times (O_1 {++}O_2)^n$$

Product Operator

The product operator allows for conjunctive composition of components and facets. It gives us the ability to define multiple views of a system, where all views must hold. As described earlier, the specifier can separate concurrent requirements, often in different domain vocabularies, and take the product to address all requirements. This section explains how the product is constructed. We look at the denotation, explain the validity of the pieces of the denotation and the resulting component or facet and give an example.

Using the definitions of *c1* and *c2* from above, when we take the product of two components, we get the following: Say $c3=c1*c2$, then c3 is denoted as $<O_1{++}O_2, Dom_1 \sqcap Dom_2, Bodies_1 {**} Bodies_2>$ where:

- $O_1 + {+}O_2 = (O_{11}, O_{12}, ..., O_{1n}, O_{21}, O_{22}, ... O_{2m})$. Note that duplicates (i.e. some $O_{1i} = O_{2j}$) are excluded.
- $Dom_1 \sqcap Dom_2$ is the least common domain of Dom_1 and Dom_2
- $Bodies_1 * {*}Bodies_2$ is essentially the cross product of $Bodies_1$ and $Bodies_2$, however there are some intricacies in combining

shared items. In the case of no shared items,
$Bodies_1 * *Bodies_2 = ((A_{11} ++A_{21},D_{11} ++D_{21},I_{11} ++I_{21}),...,(A_{1p} ++A_{21},D_{1p} ++D_{21},I_{1p} ++I_{21}), (A_{11+}+A_{22},D_{11} ++D_{22},I_{11} ++I_{22}),...,(A_{1p} ++A_{22},D_{1p} ++D_{22},I_{1p} ++I_{22}), ...,(A_{11} ++A_{2q},D_{11} ++D_{2q},I_{11} ++I_{2q}),...,(A_{1p} ++A_{2q},D_{1p} ++D_{2q},I_{1p} ++I_{2q})).$

For the same reasons as explained for component sum, any two components have a least common domain and it is safe to use this domain as the domain for the product, with no additional work necessary as all terms in the product will be in the least common domain.

Regardless of shared items in the definitions sections, there is no sharing in the assumptions or implications. Therefore, we always use the simple append operator for these sets, as done above. Within the sets of definitions, we *can* have sharing. Since there is no sharing in the assumptions or implications, we will describe the details of sharing within the confines of the simpler case of a facet product as facets have no assumptions or implications. The same notion is applied to the appending of definitions in the case of component product.

Facets are simply components without sets of assumptions or implications. So the facet product is a simplified version of the component product. We'll use the previously defined facets, $f1$ and $f2$, as the operands of the product. Say $f3 = f1 * f2$. Then $f3$ is denoted as $<O_1 ++O_2, D_1 \sqcap D_2, Terms_1 ++Terms_2>$, where:

- $O_1 ++O_2 = (O_{11},O_{12},...,O_{1n},O_{21},O_{22},...O_{2m})$. Note that duplicates are excluded (e.g. O_{2j} is excluded if some $O_{1i} = O_{2j}$).
- $D_1 \sqcap D_2$ is the least common domain of D_1 and D_2.
- $Terms1**Terms2 = (T_{11} ++T_{21},T_{12} ++T_{21},...,T_{1m} ++T_{21}, T_{11} ++T_{22},T_{12} ++T_{22},...,T_{1m} ++T_{22}, ..., T_{11} ++T_{2n},T_{12} ++T_{2n},...,T_{1m} ++T_{2n})$ if there is no sharing. However, for any shared items, their definitions must be conjoined.

If there are no shared definitions in $Terms_1$ and $Terms_2$, then $Terms_1 ** Terms_2$ is the cross product of all of the sets of terms (each T_{jk}) in $Terms_1$ and $Terms_2$, where all of the terms in each part of the cross product are appended, as done above. We are constraining $f3$ with all of the constraints of $f1$ and all of the constraints of $f2$. Often, these definitions are constraining the same item. Any terms that have the same labels are considered shared items. Consider one cross product, say $T_{ab} ++T_{cd}$ within $T_1 ** T_2$. If T_{ab} and T_{cd} each contain a term with label sharedItem, then those items are conjoined into one item within $T_{ab} ++T_{cd}$. Let's look at some examples of the kinds of shared items encountered.

Rosetta terms are either Boolean assertions or are instantiated facets. When we conjoin facets, we must essentially conjoin their terms. The conjunction of two Boolean assertions $a1$ and $a2$ would then instinctively be $a1$ and $a2$. For conformity, we can use the * operator in Rosetta, which subsumes *and*. Consider two simple facets $g1$ and $g2$ that both have an item *sum,* where $g1$ defines:

```
sum: z'=input+z;
```

and g2 defines:

```
sum: power'=power+loss;
```

Note that these definitions of *sum* are Boolean assertions. Then $g3 = g1 * g2$ would contain the item with the Boolean assertion:

```
sum:(z'=input+z)*(power'=power+loss);
```

The conjunction of two facet instantiations is done using facet product. Shared facet instantiations can be seen in the following example of the denotation of a facet product. Note that it is invalid to have a shared item where one is an assertion and the other is a facet instantiation.

Recall the prior QAM example where we defined the behavior in qamAESArch, and the implementation details in structure1 and structure2.

We then composed the behavior and structure to get fully constrained implementation details in implementation1, and implementation2. The denotation for qamAESArch is located in Box 12.

Note each of these facets have only one set of terms. We are taking the cross product of two sets with cardinality one, which yields a set with cardinality one. The shared items in our QAM example are code, buff1, enc, buff2, and modulate. These terms are facet instantiations. Conjoining these terms involves taking their facet products. So, for the code item, we have the new code item code:huffEnconder(i,ho)* hardware(fpga). The denotation of implementation1 is located in Box 13.

Assertion terms have type Boolean, while instantiations have the type of the instantiated facet. Rosetta does not support heterogeneity with respect to types. Thus, it is considered invalid to take the product of a term that is an assertion and a term that is a facet instantiation. The previous example has many shared items, but they are implicitly shared. Sharing clauses explicitly force facets in a facet product to each define every item in the sharing clause. In other words, a valid specification of $f3 = f1 * f2$ *sharing* $x_1, x_2, \ldots x_n$ explicitly forces f_1 and f_2 to each define items $x_1, x_2, \ldots,$ and x_n.

It should be noted that the product of two components (or facets) is a valid component. We have appended all observers, giving the resulting component a valid set of observers. We've already discussed that the domain of the new facet exists and is valid. We have appended the assumptions, implications, and unshared definitions of each cross-product of bodies, which gives valid new assumptions, implications and definitions. Any shared definitions are either Boolean assertions that are multiplied to give a valid Boolean assertion, or are facet instantiations, where we take the product of these instantiations. Using structural induction, we can assume that we start with the product of any instantiated facets being valid. With that assumption, we can show that for all

Box 12.

```
< (i::input word(2),o::output real, f::input frequencyType; length::design
    keyLengthType, k::input word(length), ho::bit,aesi::word(16),
    mi::word(2)),
    static,
    ((T[code: = huffEncoder(i,ho)],
       T[buff1: buffer(ho,aesi)],
       T[enc: aesEncryptor(aesi,aeso,length,k)],
       T[buff2: buffer(aeso,mi)],
       T[modulate: qamModulator(mi,o,f)]))
>
and the denotation of structure is
< (),
    fabric,
    ((T[code:=hardware(fpga)],
       T[buff1:=hardware(fpga)],
       T[enc:=hardware(crypto)],
       T[buff2:=hardware(fpga)],
       T[modulate:=hardware(fpga))])
>
```

Box 13.

```
< (i::input word(2), o::output real, f::input frequencyType; length::design
    keyLengthType, k::input word(length), ho::bit,aesi::word(16),mi::word(2)),
    static,
  ((T[code: huffEncoder(i,ho) * hardware(fpga)],
    T[buff1: buffer(ho,aesi) * hardware(fpga)],
    T[enc: aesEncryptor(aesi,aeso,length,k) * hardware(crypto)],
    T[buff2: buffer(aeso,mi) * hardware(fpga)],
    T[modulate: qamModulator(mi,o,f) * hardware(fpga)]))
>
```

products of two components, the result is a valid components.

Also, since the structure of the component coalgebra is defined over the observers, we still have a valid coalgebra as the observers, $(O_1 + +O_2)$, are valid. Therefore, the behavior of the product still denotes a coalgebra. The structure of this coalgebra is $(O_1 + +O_2)^N \rightarrow^\xi (O_1 + +O_2) \times (O_1 + +O_2)^N$

Instantiation

We can specify systems structurally or hierarchically using facet instantiation. Facet instantiation happens within a body of another facet, by replacing the formal parameters of the instantiated facet with actual parameters. A facet declaration, say *fd*, is similar to a class. An instantiation of a facet is a value whose type is the facet it instantiates. If *f* is an instance of *fd*, *f* is a value with type *fd*.

Let's look at an example of facet instantiation. Recall the halfAdder facet definition and that we can now structurally define a fullAdder facet that instantiates two halfAdders. We did this by connecting the correct inputs to the correct outputs,

i.e. by assigning the correct actual parameters to the formal parameters of each halfAdder.

The denotations of both the instantiated facets and the instantiating facet are the same as for any facet. We'll illustrate this using our example. The denotation of halfAdder is located in Box 14.

The denotation of each term in fullAdder goes in its denotation. So, the above denotation of halfAdder will appear twice in the denotation of fullAdder, with the formal parameters replaced with the actual parameters. So, the denotation of fullAdder is located in Box 15.

Note that since the instantiation occurs in the body of the instantiating facet, it has no direct bearing on the observers of the instantiating facet. It therefore has no bearing on the coalgebraic structure of the instantiating facet. Rather the denotation of the instantiated facet just becomes part of the denoted terms within the body of the instantiating facet.

Inclusion

Recall that the earlier fullAdder example instantiates two halfAdders to construct a full adder.

Box 14.

```
< (x: input bit,y: input bit,s: output bit,c: output bit),
    state_based,
    ((T[s' = x xor y],(T[c' = x and y]))
>
```

Box 15.

```
< (x: input bit,y: input bit,ci: input bit,s: output bit,co: output bit,
   s1: bit,c1: bit,c2: bit),
   state_based,
((ha1:< (x: input bit,y: input bit,s1: output bit,c1: output bit),
        state_based,
        ((T[s' = x xor y],T[c' = x and y]))
     >,
  ha2:< (s1: input bit,ci: input bit,s: output bit,c2: output bit),
        state_based,
        ((T[s' = x xor y],T[c' = x and y])) >,
          T[co = c1 or c2]))
```

Notice that each instantiation is given an item label, namely ha1 and ha2. These items labels are essentially the facet inclusion. So ha1 is a facet inclusion, and halfAdder(x,y,s1,c1) is the facet instantiation. What this does is allow the observable behavior of what is included to be observable by the facet inclusion. For instance, any observable behaviors of the halfAdder(x,y,s1,c1) instantiation are observable by ha1. The inclusions ha1 and ha2 in effect rename the instance allowing for multiple instances of the same facet.

Box 16 is a more detailed example of facet inclusion, that will help illustrate the distinction between instantiation and inclusion, as well as what is in scope within the including facet (Ros, 2008).

We've essentially put a box around the pf instantiation and called it f1 (and f2). Anything observable from pf is now observable in the f1 and f2 inclusions. Therefore, f1.power is in scope. And while the facet definition pf is in scope in the example facet, pf. power is not in scope, because pf is a facet definition, or class, and not a value.

Box 16.

```
facet pf(x::input integer; y::output integer)::static is
export power;
       power:: real;
begin
       power = 0.2;
       y=x+3;
end facet pf;
facet example(x1,x2::input integer;y1,y2::input integer;z::output integer)
     :: static is
export power ;
       power:: real;
begin
       power = f1.power + f2.power + 0.2; z = y1 + y2;
       f1: pf(x1, y1);
       f2: pf(x2, y2);
end facet example ;
```

Labels are left alone in the denotation, whereas the actual facet instantiation is what is denoted. So in our example, we left ha1 and ha2 alone, and then denoted the two instantiations of halfAdder. So, within the terms of the denotation of fullAdder we had what is seen in Box 17.

In essence, there is no real denotation of the inclusion, but rather, the instantiation of what is included.

Homomorphism

Facet and component *homomorphism* is not technically a composition operator, but as a first-class relation defined on facets used heavily in specification, merits discussion. A homomorphism $A => B$ exists between two facets A and B when properties of B can be derived from A. This can be thought of in several ways, but the best is that the closure of B under logical inference is completely contained in the closure of A. In the Rosetta literature, facet homomorphism is frequently called facet implication because all the behaviors of one facet are implied by another. An *isomorphism* between two facets A and B id defined in the classical way when both $A => B$ and $B => A$. Facet isomorphism is referred to as facet equivalence because A and B have the same properties and are otherwise indistinguishable.

Homomorphism and isomorphism are used to define correctness conditions and express inheritance relationships among facets. If A represents a system and B represents a minimal set of properties that system must exhibit, then $A => B$ formally defines a correctness condition on A

that would be checked with a theorem prover or through testing. Similarly, if B represents a system and A represents a maximal set of properties the system is allowed to exhibit, then $A => B$ formally defines a correctness condition on B that would be checked using model checking techniques. In essence, we are able to express both algebraic and coalgebraic correctness conditions.

Homomorphism plays an exceptionally important role in the domain lattice by defining the partial ordering used to define lattice. Recall that for a set to represent a lattice, a partial order on the set, a minimum element, and a maximum element must be defined. In Rosetta, homomorphism is the partial order while the static and bottom domains represent the minimum and maximum elements respectively.

ANNOTATED EXAMPLE

Here we will define and denote a complete system using facet composition as a case study on system design using the composition operators and their semantics. Our system, a small MIPS style CPU will contain a dual port RAM and a CPU. We will specify both the functional/behavioral design as well as implementation de- tails, and use conjunction to combine these different design aspects to fully specify our entire system. This example will use instantiation and inclusion, product and sum. The functional design of our system is shown in Figure 1, with all descriptions of instructions in Table 1.

Box 17.

```
ha1:<
        (x: input bit,y: input bit,s1: output bit,c1: output bit),
        state_based,
        ((T[s' = x xor y□,T[c' = x and y□))
>
```

Controller

The controller that decodes instructions into control signals lends itself well to modular design using facet sum. The controller's function is to configure the CPU to execute each instruction in its instruction set. Each instruction can be modeled independently specifying a constant configuration of control signals. Then a particular model selected based on the input opcode. Thus, we write a facet for each instruction plus one for resetting the CPU and disjoin them to create the entire controller.

We define the facet definitions for the *Reset*, *AddOp*, *LW*, *SW*, *LI*, *BLT*, and *Jmp* facets following. The *SubOp*, *AndOp*, and *OrOp* facets only differ from *AddOp* in their *ops* and the *aluOp*. In this example, the disjoined facets are mutually exclusive. Only one facet will be consistent when there is a reset, and one for when there is no reset and its instruction is specified by the opcode. Each facet starts with an assertion that the inputs to the facets have the values associated with that facet. For the *reset* facet, there is an assertion that *rst* = 1. This is only a consistent facet under

Figure 1. System block diagram for a simple microprocessor

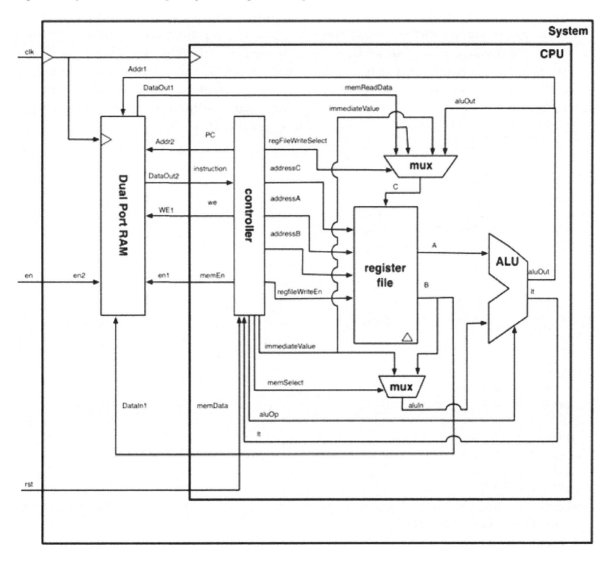

Table 1. Instruction set for a simple microprocessor

Instruction	Description	Op	Source1	Source2	Source3
Add R_s,R_t,R_d	$R_d = R_s + R_t$	0000	0-15	0-15	0-15
Sub R_s,R_t,R_d	$R_d = R_s - R_t$	0001	0-15	0-15	0-15
And R_s,R_t,R_d	$R_d = R_s$ AND R_t	0010	0-15	0-15	0-15
Or R_s,R_t,R_d	$R_d = R_s$ OR R_t	0011	0-15	0-15	0-15
LW R_s,R_t,off	$R_t = M(R_s+off)$	0100	0-15	0-15	0-15
LI $R_s,immed$	$R_s = extend(immed)$	0101	0-15	$immed_{7-4}$	$immed_{3-0}$
SW R_s,R_t,off	$M(R_s + off) = R_t$	0110	0-15	0-15	0-15
BLT R_s,R_t,off	$if(R_s < R_t)$ $PC = PC + off$	0111	0-15	0-15	0-15
Jmp *addr*	$PC= extend(addr)$	1000	$addr_{11-8}$	$addr_{7-4}$	$addr_{3-0}$

the condition of a reset. Similarly, each facet for a particular instruction operator has an assertion that *rst* = 0 and *instruction = opcode for the associated instruction*. Following the assertions that select each model are assignments appropriate for each instruction.

In the *reset* facet below following the assertion that a reset is present, each output is reset to 0's as required by the reset state. The assertion *rst=1* assures that the reset signal is set and is implicitly conjuncted with the remaining terms. Thus, this facet cannot be consistent unless the reset signal is high (Box 18).

Box 18.

```
facet Reset(rst::input bit; we,memEn,memSelect,we,memEn,
            regFileWriteEn:: output bit;
            aluOp,refFileWriteSelect:: output word(2);
            addressA,addressB,addressC::output word(4);
            PC,immediateValue:: output word(16)
        ):: State_based is
begin
  rst =1;
  we'=0; memEn'=0;
  memSelect '=0;
  regFileWriteEn'=0;
  regFileWriteSelect'=b"00";
  aluOp'= b"00";
  addressA'=x"0";
  addressB'=x"0";
  addressC'=x"0";
  PC' = x"0000";
  memAddr'=x"0000";
  immediateValue'=x"0000";
end facet Reset;
```

In this fashion, we write each facet separately with the assertions and assignments appropriate to the desired functionality. When we sum them, only one will be consistent for each instruction. Thus, there is a facet that will be consistent and give the appropriate assignments for that instruction.

The *AddOp* facet implements functionality for the add operation similar to the reset operation above. In this case, the assertion *rst=0* ensures that there is no reset while the assertion that the opcode is binary 0000 assures that we are executing an add instruction. The remaining outputs are set so that the CPU elements properly execute an add instruction (Box 19).

Operations that access memory are no different than arithmetic operations except that they must configure memory to perform read/write operations. The racets for load word, load immediate, and store word are all constructed like the add operation facet. Specifically, reset must not be active and the opcode specified must be the opcode associated with the instruction (Box 20).

Finally our redirection operators execute conditional branches and absolute jumps respectively. The distinction here is that memory is disabled and outputs remain invariant while the PC value is manipulated to implement a branch or jump. The branch differs from the jump only in that the change to the PC is conditional on the comparison of two values (Box 21).

The facet representing the system controller is simplest while in many ways being the most interesting of the specifications. It says only that the controller is the sum of facets for each instruction (Box 22).

Recall that each instruction facet has two assertions – that reset is or is not asserted and that a particular opcode is specified by the current instruction. The disjunction says that at least one of those facets must be consistent. If each opcode has an associated facet and only one asserts that reset is present, the *controller* facet will have only

Box 19.

```
facet AddOp(rst::input bit; we,memEn,memSelect,we,memEn,
            regFileWriteEn:: output bit;
            aluOp,refFileWriteSelect:: output word(2);
            addressA,addressB,addressC::output word(4);
            PC,immediateValue:: output word(16)
         ):: State_based is
begin
  rst =0;
  instruction(15 downto 12)=b"0000";
  we'=0;
  memEn'=0;
  memSelect '=0;
  regFileWriteEn'=1;
  regFileWriteSelect'=b"00";
  aluOp'- b"00";
  addressA'=instruction(11 downto 8);
  addressB'=instruction(7 downto 4);
  addressC'=instruction(3 downto 0);
  PC' = PC+x"0002";
  immediateValue'=x"0000";
end facet AddOp;
```

Box 20.

```
facet LW(rst::input bit; we,memEn,memSelect,we,memEn,
            regFileWriteEn:: output bit;
            aluOp,refFileWriteSelect:: output word(2);
            addressA,addressB,addressC::output word(4);
            PC,immediateValue:: output word(16)
         ):: State_based is
begin
  rst =0;
  instruction(15 downto 12)=b"0100";
  we'=0;
  memEn'=0;
  memSelect '=0;
  regFileWriteEn'=1;
  regFileWriteSelect'=b"00";
  aluOp'= b"00";
  addressA'=instruction(11 downto 8);
  addressB'=x"0";
  addressC'=instruction(7 downto 4);
  PC' = PC+x"0002";
  immediateValue '=signExtend(instruction(3 downto 0));
end facet LW;
facet LI(rst::input bit; we,memEn,memSelect,we,memEn,
            regFileWriteEn:: output bit;
            aluOp,refFileWriteSelect:: output word(2);
            addressA,addressB,addressC::output word(4);
            PC,immediateValue:: output word(16)
         ):: State_based is
begin
  rst =0;
  instruction(15 downto 12)=b"0101";
  we'=0;
  memEn'=0;
  memSelect '=0;
  regFileWriteEn'=1;
  regFileWriteSelect'=b"00";
  aluOp'= b"00";
  addressA'=x"0";
  addressB'=x"0";
  addressC'=instruction(11 downto 8);
  PC' = PC+x"0002";
  immediateValue'=signExtend(instruction(7 downto 0));
end facet LI;
facet SW(rst::input bit; we,memEn,memSelect,we,memEn,
            regFileWriteEn:: output bit;
            aluOp,refFileWriteSelect:: output word(2);
```

continued on following page

Box 20. Continued

```
            addressA,addressB,addressC::output word(4);
            PC,immediateValue:: output word(16)
         ):: State_based is
begin
  rst =0;
  instruction(15 downto 12)=b"0110";
  we'=1;
  memEn'=1;
  memSelect '=1;
  regFileWriteEn'=0;
  regFileWriteSelect'=b"11";
  aluOp'= b"00";
  addressA'=instruction(7 downto 4);
  addressB'=x"0";
  addressC'=x"0";
  PC'  = PC+x"0002";
  immediateValue'=signExtend(instruction(3 downto 0));
end facet SW;
```

one possible configuration for any given input. Although this is not always what we want, for this CPU model it is precisely what is desired.

Looking at our denotation function defined earlier, the *controller* facet's coalgebraic structure is located in Box 23.

For analysis, the several facets making up *controller* would be expanded appropriately. We have not done this hear as it adds nothing to the exposition. However, as an example (#*denotations of all terms in Reset*#) would be seen in Box 24.

All observers from every part of this sum are now part of the observers for the *controller*, meaning all parameters and variables of each facet. The domain of each piece is *State_based*, so the domain of the *controller* is still *State_based*. Because we are in doing a sum, the terms from each piece of the sum are kept separate in their own list of terms. The separate lists of terms are appended in the *controller*'s list of term lists. In this way, only one instruction (or reset) is addressed at a time.

Full CPU

The full CPU is defined by composing facets, each of which is defined like the controller. Instead of using facet disjunction, the CPU is constructed structurally by composing various components in parallel. Like traditional hardware description languages, Rosetta specifications use share signals to move data among components. In the *cpu* facet, terms represent instantiated facets rather than Boolean assertions. Thus, the *cpu* facet is instantiating and including definitions structurally in a manner that is common in system design Box 25).

The CPU itself consists of a controller, two MUXes that control devices, a simple ALU and a register file. Each of these components is easily identifiable in the *cpu* facet providing a syntax that is consistent with what traditional designers expect. Using our denotation function, the *cpu* facet's coalgebraic structure is located in Box 26.

Box 21.

```
facet BLT(rst::input bit; we,memEn,memSelect,we,memEn,
            regFileWriteEn:: output bit;
            aluOp,refFileWriteSelect:: output word(2);
            addressA,addressB,addressC::output word(4);
            PC,immediateValue:: output word(16)
         ):: State_based is
begin
  rst =0;
  instruction(15 downto 12)=b"0111";
  we'=0;
  memEn'=0;
  memSelect '=0;
  regFileWriteEn'=1;
  regFileWriteSelect'=b"00";
  aluOp'= b"00";
  addressA'=x"0";
  addressB'=x"0";
  addressC'=x"0";
  PC' = if lt=1 THEN PC+instruction(3 downto 0);
                ELSE PC+x"0002"
        endif;
  immediateValue'=x"0000";
end facet BLT;
facet Jmp(rst::input bit; we,memEn,memSelect,we,memEn,
            regFileWriteEn:: output bit;
            aluOp,refFileWriteSelect:: output word(2);
            addressA,addressB,addressC::output word(4);
            PC,immediateValue:: output word(16)
         ):: State_based is
begin
  rst =0;
  instruction(15 downto 12)=b"1000";
  we'=0;
  memEn'=0;
  memSelect '=0;
  regFileWriteEn'=1;
  regFileWriteSelect'=b"00";
  aluOp'= b"00";
  addressA'=x"0";
  addressB'=x"0";
  addressC'=x"0";
  PC' = signExtend(instruction(11 downto 0));
  immediateValue'=x"0000";
end facet Jmp;
```

Box 22.

```
facet controller:: State_based = Reset + AddOp + SubOp
                                + AndOp + OrOp + LW + LI
                                + SW + BLT + Jmp
```

Box 23.

```
(bit,word(16),bit,bit,bit,bit,word(2),word(2),word(4),word(4),word(4),word(16),word(16))
ᴺ↦ᶠ (bit,word(16),bit,bit,bit,bit,word(2),word(2),word(4),word(4),word(4),word(16),wo
rd(16)) × (bit,word(16),bit,bit,bit,bit,word(2),word(2),word(4),word(4),word(4),word(16)
,word(16))ᴺ
while its denotation is:
< (rst:: input bit,lt:: input bit, instruction:: input word(16),
    we:: output bit,memEn:: output bit,memSelect:: output bit,
    regFileWriteEn:: output
    bit,aluOp:: output word(2),refFileWriteSelect:: output word(2),
    addressA:: output word(4),addressB:: output word(4), addressC:: output word(4),
   PC:: output word(16),immediateValue:: output word(16)),
  State_based,
  ((# denotations of all terms in Reset #),
   (# denotations of all terms in AddOp #),
   (# denotations of all terms in SubOp #),
   (# denotations of all terms in AndOp #),
   (# denotations of all terms in OrOp #),
   (# denotations of all terms in LW #),
   (# denotations of all terms in LI #),
   (# denotations of all terms in SW #),
   (# denotations of all terms in BLT #),
   (# denotations of all terms in Jmp #))>
```

Box 24.

```
(rst = 1;,
we(next(a)) = 0;,
memEn(next(a)) = 0;,
memSelect(next(a)) = 0;,
regFileWriteEn(next(a)) = 0;,
regFileWriteSelect(next(a)) = b"00";,
aluOp(next(a)) = b"0)";,
addressA(next(a)) = x"0";,
addressB(next(a)) = x"0";, addressC(next(a)) = x"0";, PC(next(a)) = x"0000";,
memAddr(next(a)) = x"0000";, immediateValue(next(a)) = x"0000";,
)
```

This denotation involves the simple case of instantiations. The observers are the parameters and variables from the *cpu* itself. Its domain is HW. And there is one list of terms, which contains the five facet instantiations in *cpu*.

System Constraints

System-level design languages must support composition of heterogeneous specifications. Here we show how a designer might us specification conjunction to specify implementation details such as power constraints of a system. We can write facets to model these constraints like we did instructions where each specification is viewed independently. We can do this at a course-grained or fine-grained level as desired. For example, below we define facets to model the power constraint corresponding with each block in Figure 1 and their composition. These facet definitions model power consumption the ALU, the CPU, the RAM, and the entire composite system (Box 27).

The implementation details of the power consumption facets model the same general strategy of the behavioral model. The power consumption of the system is defined in a block-diagram fashion using instantiation.

System Under Constraint

We now have a model for the behavior and the power consumption of our system. We can conjoin them to create the desired specification of the system that defines a desired functional behavior and power consumption. A consistent implementation would be one in which the behavior were consistent and the power consumption were consistent. In this case, given the appropriate design parameters, the power consumption for each clock cycle must remain under the given threshold (Box 28).

The denotation of the *completeSystem* facet differs from both the facets using structural composition and disjunction in that multiple models define the behavior of the same component si-

Box 25.

```
facet cpu(clk,rst::input bit;
        memReadData,instruction::input word(16);
        we, memEn:: output bit;
        pc, aluOut,memData:: output word(16)):: HW is
  memLoadSelect, regFileWriteEn, immediateSelect:: bit;
  aluOp, aluStatus:: word (2);
  addressA, addressB, address:: word(4);
  immediateValue, A,B,C, aluIn::word(16);
begin
  c: controlUnit (rst,lt,instruction,
                  we,memEn,memSelect,regFileWriteEn,
                  aluOp,refFileWriteSelect,
                  addressA,addressB,addressC,
                  PC,immediateValue);
  a: alu(aluOp,A,aluIn,aluStatus,aluOut);
  rf: regFile(clk,regFileWriteEn,addressA,
              addressB,addressC,A,B,C);
  mux1: mux(memReadData,AluOut,memLoadSelect,C);
  mux2: mux(B,immediateValue,immediateSelect,aluIn);
end facet cpu;
```

Box 26.

```
(bit,bit,word(16),word(16),bit,bit,word(16),word(16),word(16)
        bit,bit,bit,word(2),word(2),word(4),word(4),word(16),word(16),
        word(16),word(16),word(16))ᴺ
  →ᶠ (bit,bit,word(16),word(16),bit,bit,word(16),word(16),word(16)
        bit,bit,bit,word(2),word(2),word(4),word(4),word(16),word(16),
        word(16),word(16),word(16))
   × (bit,bit,word(16),word(16),bit,bit,word(16),word(16),word(16)
        bit,bit,bit,word(2),word(2),word(4),word(4),word(16),word(16),
        word(16),word(16),word(16))ᴺ
and its denotation is:
< (clk:: input bit,rst:: input bit,
        memReadData:: input word(16),instruction:: input word(16),
        we:: output bit,memEn:: output bit,
        pc:: output word(16),aluOut:: output word(16),memData:: output
word(16),
        memLoadSelect:: bit,regFileWriteEn:: bit,immediateSelect:: bit,
        aluOp:: word(2),aluStatus:: word(2),
        addressA:: word(4),addressB:: word(4),addressC:: word(4),
        immediateValue:: word(16),A:: word(16),B:: word(16),
        C:: word(16),aluIn:: word(16))
  HW,
  ((#denotation of controlUnit instantiation#,
    #denotation of alu instantiation#,#denotation of regFile instantiation#,
    #denotation of mux instantiation#,#denotation of mux instantiation#)
  )
>
```

multaneously. It is not sufficient for the functional design and power constraints to be satisfied by an implementation – they must be jointly satisfied for a aystem to be correct. The *completeSystem* facet's coalgebraic structure is located in Box 29.

These denotations follow the details of facet product. The observers consist of all observers from each piece of the product–so the observers of *completeSystem* include the inputs from *system*, as well as the inputs, outputs, and design parameters of *systemPower*, along with all variables of both. The domain is the least common domain of *HW* and *State_based*, which is *State_based*.

There are several important things to note in constructing the list of term lists of *completeSystem*. The *system* facet has one list element in its

list of term lists. The *systemPower* also has one list element in its list of term lists. Therefore, *completeSystem* will have one element in its list of term lists since it is the cross product of two lists with one element The *system* and *systemPower* facets have two shared items, *processor* and *mem*. They are facet instantiations, so *completeSystem* will contain the facet products of these items. The denotations of one of these facet products is given above, following the denotation of *completeSystem*. The non-shared items in *system* and *systemPower– np*, *ot*, and *p*–will also be part of *completeSystem*.

Box 27.

```
facet aluPower(rst::input bit;
                wordSize,operandSize,statusSize:: input natural;
                calculation,switch,leakage:: design real ;
                power::output real):: is
begin
 power'= if rst=1 then leakage
                  else wordSize*(switch + operandSize*calculation)
                       + switch*statusSize
        endif;
end facet aluPower ;
facet cpuPower(rst:input bit;
                calculation,switch,leakage:: design real;
                power::output real):: state_based is
  aluP, mux1P,mux2P, regFileP, controllerP:: real;
  controllerOutputBits:: natural;
begin
  cob: controllerOutputBits = 1*4 + 2*2 + 4*3 + 16*2;
  a: aluPower(rst, 16,2,1,calculation, switch, aluP);
  mux1: muxPower(rst, 2,16,switch, mux1P);
  mux2: muxPower(rst, 4,16,switch, mux2P);
  rf: regFilePower (rst, 16, 16, switch, leakage, regFileP);
  c: controllerPower(rst, 10,16,controllerOutputBits, switch, leakage);
  p: power'= if rst=1 then leakage
                      else aluP + mux1P + mux2P + regFileP + controllerP
                      endif;
end facet cpuPower;
facet systemPower(rst:input bit;
                   calculation,switch,leakage,threshold::design
                   power::output real,
                   overThreshold:: output boolean):: state_based is
  cpuP,memP,nextPower::real;
begin
  processor: cpuPower(rst,calculation,switch,leakage,cpuP);
  mem: memPower(rst,switch,leakage,memP);
  mp: nextPower = if rst=1
                  then leakage
                  else cpuP + memP
                  endif;
  ot: overThreshold' = nextPower > threshold;
  p: power' = nextPower ;
end facet systemPower ;
```

Box 28.

```
facet completeSystem = system(clk,en,rst)
                        * systemPower(rst,calculation,switch,
                        leakage,threshold,power,overThreshold);
```

Evaluation

Our resulting facet is only useful if it gives us all of the behavior desired from the system design. Furthermore, a useful specification is one that can be designed modularly. We argue that facet composition gives us the usefulness of modular design, while maintaining the correct behavior.

The denotation of the *completeSystem* facet gives evidence of its desired behavior. The denotations's observers are seen in Box 30.

Box 30 shows the entire state necessary to describe the complete system. Beyond the *bit* in-puts and *design* inputs, all of the internal state (for instance, *cpuP* or *nextPower*) of the *completeSystem* is present in its denotation. Note that in the facet definition of *completeSystem* there are no explicit observers listed. Rather, these come from each piece of the facet product used to construct *completeSystem*. Furthermore, inside *completeSystem*'s list of denotation bodies as seen in Box 31.

Box 31 shows the denotation of any facet instantiation or facet composition at all levels within *completeSystem*. As we push into the denotation of *completeSystem*, we get the constraints at every level of the modular design. Pushing in one level, we've shown that the denotation of *cpu*cpuPower* is part of the denotation of *completeSys- tem*. This continues for each level of vertical composition in the design.

The denotation of every instantiated component throughout the system is included in the denotation body of the facet that instantiates it. The constraints of each instan- tiated component are included in the instantiating component. So, *alu*'s denotation is included in the denotation body of *cpu*, *cpu*'s denotation is included in the denotation body of *system*, etc.

When two facets are summed, their constraints are added in separate denotation bodies in the resulting facet, indicating that only one of the bodies must hold. So the *controller* is modularly built up. The denotations for each instruction in the *controller* are in their own denotation body (Box 32).

This means that only one instruction path must be satisfied for each instruction that comes through the *cpu*.

The product of two facets results in all of the constraints from both facets. So, by taking the product of *cpu* and *cpuPower*, we are requiring a valid implementation to meet all constraints of each. The denotation reflects that all of the terms and facet instantiations of both facets are in the same term denotation body in the resulting facet (Box 33).

So, we have enforced that a valid implementation meets all of the behavioral constraints of *system* as well as the power constraints of *systemPower*.

The simplicity of the *completeDesign* facet speaks to the simplicity of design that facet composition enables. The *completeDesign* facet is at the same high level as the block design. However, the denotation of *completeDesign* shows how intricate and detailed that facet is under the hood, so to speak. We have the power to fully specify a complex system with the simplicity and elegance of a high-level "black box" feel. Modular design is

Box 29.

$(bit, bit, word(16), word(16), bit, bit, word(16), word(16), word(16),$
      $bit, bit, bit, word(2), word(2), word(4), word(4), word(4),$
      $word(16), word(16), word(16), word(16), word(16),$
      $real, real, real, real, real, real, real, real, real)^N$
$\rightarrow^\xi$ $(bit, bit, word(16), word(16), bit, bit, word(16), word(16), word(16),$
       $bit, bit, bit, word(2), word(2), word(4), word(4), word(4),$
       $word(16), word(16), word(16), word(16), word(16),$
       $real, real, real, real, real, real, real, real, real)$
    $\times$ $(bit, bit, word(16), word(16), bit, bit, word(16), word(16), word(16),$
        $bit, bit, bit, word(2), word(2), word(4), word(4), word(4),$
        $word(16), word(16), word(16), word(16), word(16),$
        $real, real, real, real, real, real, real, real, real)^N$

```
and its denotation is
<(clk,en,rst::input bit; we1,en1:: bit;
     dataOut1,dataOut2,dataIn,Addr1,Addr2::word(16);
     calculation,switch,leakage,threshold:: design real;
     power:: output real; overThreshold:: output boolean;
     cpuP,memP,nextPower:: real),
  State_based,
  ((#denotation of facet product of cpu and cpuPower#,
    #denotation of facet product of memory and memPower#,
    #denotation of np term#,
    #denotation of ot term#
    #denotation of p term#))
>
where the denotation of the facet product of cpu and cpuPower (and similarly
for the product of memory and memPower) is
<(clk::input bit,rst::input bit,
    memReadData::input word(16),instruction::input word(16),
    we::output bit,memEn::output bit,
    pc::output word(16),aluOut::output word(16),
    memData::output word(16),
    memLoadSelect::bit,regFileWriteEn::bit,immediateSelect::bit,
    aluOp::word(2),aluStatus::word(2),
    addressA::word(4),addressB::word(4),addressC::word(4),
    immediateValue::word(16),A::word(16),B::word(16),C::word(16),
    aluIn::word(16),
    calculation::design real,switch::design real,
    leakage::design real;
    power::output real,
    aluP::real,mux1P::real,mux2P::real,regFileP::real,
    controllerP::real)
  State_based,
  ((#denotation of facet product of controlUnit and controllerPower#,
    #denotation of facet product of controlUnit and controller#,
```

continued on following page

Box 29. Continued

```
    #denotation of facet product of alu and aluPower#,
    #denotation of facet product of mux and muxPower#,
    #denotation of facet product of mux and muxPower#,
    #denotation of facet product of regFile and regFilePower#,
    #denotation of cob term#,
    #denotation of p term#))
>
```

Box 30.

```
(clk::input bit,rst::input bit,
    memReadData::input word(16),instruction::input word(16),
    we::output bit,memEn::output bit,
    pc::output word(16),aluOut::output word(16),
    memData::output word(16),
    memLoadSelect::bit,regFileWriteEn::bit,immediateSelect::bit,
    aluOp::word(2),aluStatus::word(2),
    addressA::word(4),addressB::word(4),addressC::word(4),
    immediateValue::word(16),A::word(16),B::word(16),C::word(16),
    aluIn::word(16),
    calculation::design real,switch::design real,
    leakage::design real;
    power::output real,
    aluP::real,mux1P::real,mux2P::real,regFileP::real,
    controllerP::real)
```

Box 31.

```
((#denotation of facet product of controlUnit and controllerPower#,
    #denotation of facet product of controlUnit and controller#,
    #denotation of facet product of alu and aluPower#,
    #denotation of facet product of mux and muxPower#,
    #denotation of facet product of mux and muxPower#,
    #denotation of facet product of regFile and regFilePower#,
    #denotation of cob term#,
    #denotation of p term#))
```

Box 32.

```
((#denotations of all terms in Reset#),
  (#denotations of all terms in AddOp#),
  (#denotations of all terms in SubOp#),
  (#denotations of all terms in AndOp#),
  (#denotations of all terms in OrOp#),
  (#denotations of all terms in LW#),
  (#denotations of all terms in LI#),
  (#denotations of all terms in SW#),
  (#denotations of all terms in BLT#)
  (#denotations of all terms in Jmp#))
```

important for complex systems in that it gives us the benefits of reuse and flexibility. Our coalgebraic semantics gives us the benefits of modularity, while maintaining the power of a detailed design capability. We could have written a course-grained model of power consumption to conjunct with the system design and could easily compare the two. This kind of flexibility allows quick design space exploration. At the same time, the cumbersome nature of the coalgebraic specifications – most of which are not fully expanded – justifies language support. Using a coalgebra directly would prove difficult if not impossible to manage.

RELATED WORK

There are several works on applications that are particularly well suited to coalgebras. Two major categories that motivate our use of coalgebras

for Rosetta specifications are *modal logics* and *systems,* especially state-based or reactive systems. We describe the uses of coalgebras in these categories and describe why the techniques used are appropriate for Rosetta.

Modal logics are the family of logics whose operators conditionalize formulas to hold under certain criteria such as "in the future," "normally," "necessarily." There are several works that address coalgebras' suitability for modal logics. Cirstea et al. (2011) discuss that the more common non-normal modal logics are not amenable to the standard Kripke semantics. Rather, since these modal logics are essentially reactive systems, they are much better suited to coalgebraic semantics.

Cirstea et al. (2011) argue that the first major advantage of using coalgebraic semantics is the generality. A coalgebraic framework is constructed per application, and is therefore applicable to a larger class of modal logics. The second major advantage is the compositionality. The coalgebraic framework allows for integration of different requirements, and many different logics co-exist in the same framework. This allows for the "modular combination of reasoning principles" (Cirstea et al., 2011). Lastly, the coalgebraic framework lends itself to adaptability. The previous two traits allow for new requirements to be easily added to existing requirements.

Rosetta components are logics in that they express assumptions, definitions – which are essentially assertions – and implications that must hold for the specification to be valid. Furthermore,

Box 33.

```
((#denotation of facet product of controlUnit and controllerPower#,
   #denotation of facet product of controlUnit and controller#,
   #denotation of facet product of alu and aluPower#,
   #denotation of facet product of mux and muxPower#,
   #denotation of facet product of mux and muxPower#,
   #denotation of facet product of regFile and regFilePower#,
   #denotation of cob term#,
   #denotation of p term#))
```

Rosetta has the ability of expressing temporal concepts, such as current state and next state transitions. For instance, some assertions must hold for the next state of a state-based specification. This makes Rosetta modal. Therefore we can apply the principles of using coalgebras for modal logics to Rosetta domains, components, and facets.

Kurz (2001) describes the theory of systems, and lays out how coalgebras are a natural model of these theories. Systems are understood by their interfaces – how they interact and communicate with other systems. Essentially, they are a set of states and the observable transitions on those states. Systems are reactive in nature, and we look at them as "black boxes." Jacobs and Rutten (1997) also give a thorough tutorial of algebras versus coalgebras, and induction versus coinduction and bisimulation.

With an algebraic definition, the initiality gives us a base to stand on. For instance, when describing a list, we have the base case of an empty list, and all lists can be thought of as the pieces constructed to the base case of an empty list – we have a base case and constructors that build any list. With a coalgebraic definition, we have the dual, finality. Think of a stream. We don't think of streams constructively, but rather we have the entire stream, and we take observations, destructing the stream.

Rather than an inductive principle from the initiality of an algebra, we can utilize a coinductive principle from the finality of a coalgebra. Along with coinduction, we can also exploit bisimulation. Informally, a bisimulation exists between two systems or coalgebras if all of their observations match. So, if A and B are two state machines, then A and B are bisimilar if upon each state transition, their outputs or actions match.

When we specify or design a system in Rosetta, we in a sense do so constructively, i.e. we build up a system using the components that make up that system. However, since we still wish to view the systems we're specifying or designing as black boxes, reasoning about systems is done by observing their behavior. We look at the system as a whole, and the only things we need to know about it are what we are able to observe from its interface. This notion is exactly what coalgebras give us. We start with the structure and take observations of all transitions. To this end, coalgebras are the suitable choice to describe Rosetta specifications. We can then use the notions of coinduction and bisimilarity to reason about and compare behavioral equivalence of systems. This work shows how we can still build models constructively using composition, while maintaining the coalgebraic structure that is suitable for systems.

The second area of related works motivating this work is in using coalgebras specifically in semantics. We describe the use of coalgebraic denotation in process calculi, the use of coalgebras in the semantics of Java, as well as address previous semantic work of Rosetta.

Hausmann et al. (2006) describe the use of a coalgebraic denotation for process calculi. The paper claims coalgebraic semantics add clarity to the calculi as well as allow for comparison and unification of process calculi. It gives the formal denotation of the ambient calculus (for mobile computing) using CoCASL (Mossakowski et al., 2003), an extension of the CASL specification language (Astesiano et al., 2001) that adds built-in coalgebraic structures. Process calculi model concurrent systems. Rosetta is for system specification, and we have shown that behaviors of Rosetta specifications can themselves be considered systems. As such, the approaches used for denoting process calculi with coalgebraic semantics can be applied to our denoting Rosetta with coalgebraic semantics.

Hausmann et al. (2006) first lay out the signature functor for the design of the coalgebraic model of the ambient calculus. This sets up the structure, or type of transition system. Then the paper lays out the transition rules. Essentially, this is the structure and observations of the calculus. The coalgebraic framework gives more structure than the typical approach of a labeled transition system, but also gives the generality to be able to relate it to (via bisimulation, etc.) and combine it with other calculi.

Jacobs and Poll (2002) explore the use for a combination of monads and coalgebras in the semantics of sequential Java. The monadic approach gives a clean model of the computational structure of Java, while coalgebras give the program logic based on that monadic structure. The coalgebraic view provides reasoning principles via modal operations and bisimulation. This approach is able to cleanly deal with complexities such as multiple termination patterns in the model of computation.

While Rosetta is significantly different from sequential Java, this work still supports our use of coalgebras in the denotation semantics of Rosetta. We show that the structure of Rosetta is well suited for Rosetta components, and, as for Jacobs and Poll (2002), the use of coalgebras gives us valuable reasoning principles, such as modal logic and bisimulation.

Kong et al. (2003) lay down the foundation for the coalgebraic semantics for Rosetta. The work gives the general approach to denoting facets, giving the two-part denotation. It then goes on to give the denotation details within a few specific domains. Kong et al. (2003)'s work is the basis for this continued work. The denotation of components and facets is necessary in developing the denotation of the composition of components and facets. We have refreshed the denotation with necessary updates and details.

A number of specification approaches similar to the Rosetta approach appear in the literature. Nuseibeh et al. (2003) discuss a software engineering approach based on multiple system viewpoints. The approach is similar to Rosetta in that multiple views simultaneously describe a system. View-Points are quite software centric and do not address many issues specific to system-level design. However, there is little question that the approach could be extended to work at the system-level and would represent a complimentary approach. SysML (Vanderperren et al., 2005) and work specifically by Balmelli (2006) in model-driven

design use an UML-centered approach comparable to Rosetta using structural composition to perform system-level design in an object-oriented fashion. During the early stages of SysML and Rosetta we worked with SysML designers to denote SysML in Rosetta. Although initially successful, that work has largely ended. Also worth mentioning is the Z specification language (Woodcock et al., 1996) where specification conjunction and disjunction arise. Z uses first-order logic and is not coalgebraic, but it did motivate Rosetta designers to think about specifications as first-class objects. Their use of specification and disjunction parallel the approach presented here.

FUTURE RESEARCH DIRECTIONS

We believe strongly that a future direction in language design – particularly specification language design – is the treatment of modules as first-class language structures. Recall that when a structure as first-class in a language it is data in that language with all associated functionality including transformation, typing, and reflection. At its essence, Rosetta tries to be a specification with first-class theories. We say 'tries to be' because it has some limitations in its reflection system, kernel language and type system that make treatment of theories as first class objects difficult, however we have made an important first step.

Beyond Rosetta, making modules first-class is an enabler for many functions currently performed by specialized languages. When modules are first-class design activities such as testing, configuration and compilation, and revision control can be done within the host language. A major addition to SystemVerilog when moving from Verilog is the addition of a verification language supporting writing test harnesses and executing tests. This is great first step. However, in a first-class module system, this operation may be done with the language itself without need for additional constructs or language extensions. Similarly for the functions performed by the venerable make system C preprocessing system used for decades to build and maintain software across multiple platforms. With a first-class module system, the host language would be used to instantiate components and process build commands. Even revision control systems that currently exist outside all languages could become a part of the languages themselves, manipulating modules directly in their associated languages.

Much remains to be done in developing higher-order module systems. Dependent types, an enabling technology for first-class module systems, is a hot research area in language design with significant advancements. Languages such as Agda (Norell 2009) Epigram (McBride 2005) and Cayenne (Xi & Pfenning 1999) provide significant advancements in the use of dependent types in practical programming applications. The Coq proof assistant (Bertot & Castéran 2004) also provides a mixed programming and verification environment with dependent type support. We continue to explore local type inference in Rosetta as well as new reflection capabilities.

CONCLUSION

System-level design is characterized by a need to consider multiple perspectives simultaneously to predict the behavior of complex systems. Specification composition in Rosetta provides an important mechanism for composing heterogeneous specifications in support of system-level design. After introducing the three foundational Rosetta specification composition operations – inclusion, conjunction and disjunction – we have presented the first semantics of those operators in the context of the overall Rosetta semantics. We first refreshed the original Rosetta denotation to be more general than the original denotation and include support for component constructs. We then defined semantics for and provided examples of composition using product for concurrent engineering; sum for expressing alternative behaviors;

and instantiation and inclusion for expressing structural architectures. Finally, we presented a potential generalization of this approach using first-class module systems as a direction for further research. We plan to continue our work on model composition as we develop a standard for the Rosetta language.

REFERENCES

Accellera. (2002). *SystemVerilog 3.0: Accellera's extensions to verilog*. San Francisco, CA: Accellera.

Alexander, P. (2006). *System level design with rosetta*. San Francisco, CA: Morgan Kaufmann Publishers Inc..

Alexander, P., Barton, D., & Kong, C. (2000). *Rosetta usage guide*. Lawrence, KS: The University of Kansas.

Allen, R., & Garlan, D. (1997). A formal basis for architectural connection. *ACM Transactions on Software Engineering and Methodology*, *6*(3), 213–249. doi:10.1145/258077.258078.

Astesiano, E., Bidoit, M., Kirchner, H., Krieg-Brückner, B., Mosses, P. D., Sannella, D., & Tarlecki, A. (2002). CASL: The common algebraic specification language. *Theoretical Computer Science*, *286*(2), 153–196. doi:10.1016/S0304-3975(01)00368-1.

Balmelli, L., Brown, D., Cantor, M., & Mott, M. (2006). Model-driven systems development. *IBM Systems Journal*, *45*(3), 569–585. doi:10.1147/sj.453.0569.

Bertot, Y., & Castéran, P. (2004). *Interactive theorem proving and program development: Coq'Art: The calculus of inductive constructions*. New York: Springer. doi:10.1007/978-3-662-07964-5.

Cirstea, C., Kurz, A., Pattinson, D., Schröder, L., & Venema, Y. (n.d.). Modal logics are coalgebraic. *The Computer Journal*, *54*(1), 31–41.

Frisby, N., Peck, M., Snyder, M., & Alexander, P. (2011). Model composition in rosetta. In *Proceedings of the IEEE Conference and Workshops on the Engineering of Computer Based Systems* (pp. 140–148). IEEE Computer Society.

Grötker, T., Liao, S., Martin, G., & Swan, S. (2002). *System design with SystemC*. New York: Springer.

Hausmann, D., Mossakowski, T., & Schröder, L. (2006). A coalgebraic approach to the semantics of the ambient calculus. *Theoretical Computer Science*, *366*(1), 121–143. doi:10.1016/j.tcs.2006.07.006.

IEEE. (1994). *VHDL language reference manual*. New York: IEEE.

IEEE. (1995). *Standard verilog hardware description language reference manual*. New York: IEEE.

Jacobs, B., & Poll, E. (2003). Coalgebras and monads in the semantics of Java. *Theoretical Computer Science*, *291*(3), 329–349. doi:10.1016/S0304-3975(02)00366-3.

Jacobs, B., & Rutten, J. (1997). A tutorial on (co) algebras and (co) induction. *Bulletin-European Association for Theoretical Computer Science*, *62*, 222–259.

Kimmell, G., Komp, E., Minden, G., Evans, J., & Alexander, P. (2008). Synthesizing software defined radio components from rosetta. In *Proceedings of the Forum on Specification, Verification and Design Languages, 2008, FDL 2008* (pp. 148-153). New York: IEEE.

Kong, C., Alexander, P., & Menon, C. (2003). Defining a formal coalgebraic semantics for the rosetta specification language. *Journal of Universal Computer Science*, *9*(11), 1322–1349.

Kurz, A. (2001). *Coalgebras and modal logic*. Retrieved from http://www.cs.le.ac.uk/people/akurz/CWI/public_html/cml.ps.gz

Lohoefener, J. (2011). *A methodology for automated verification of rosetta specification transformations*. (PhD thesis). University of Kansas, Lawrence, KS.

Mossakowski, T., Roggenbach, M., & Schröder, L. (2003). CoCasl at work—Modelling process algebra. *Electronic Notes in Theoretical Computer Science*, *82*(1), 206–220. doi:10.1016/S1571-0661(04)80640-6.

Norell, U. (2009). Dependently typed programming in Agda. *Advanced Functional Programming*, 230-266.

Nuseibeh, B., Kramer, J., & Finkelstein, A. (2003). ViewPoints: Meaningful relationships are difficult! In *Proceedings of the 25th International Conference on Software Engineering* (pp. 676-681). New York: IEEE.

Peck, W. (2011). *Hardware/software co-design via specification refinement*. (PhD thesis). University of Kansas, Lawrence, KS.

SRI FormalWare. (2011). *PVS specification and verification system*. Retrieved from http://pvs.csl.sri.com/

Streb, J., & Alexander, P. (2006). Using a lattice of coalgebras for heterogeneous model composition. In *Proceedings of the MoDELS Workshop on Multi-Paradigm Modeling* (pp. 27-38). MoDELS.

Streb, J., Kimmell, G., Frisby, N., & Alexander, P. (2006). Domain specific model composition using a lattice of coalgebras. In *Proceedings of the OOPSLA* (Vol. 6). OOPSLA.

Vanderperren, Y., & Dehaene, W. (2005). SysML and systems engineering applied to UML-based SoC design. In *Proceedings of DAC UML-SoC Workshop*. DAC UML-SoC.

Woodcock, J., & Davies, J. (1996). *Using Z: Specification, refinement, and proof (Vol. 1)*. Englewood Cliffs, NJ: Prentice Hall.

Xi, H., & Pfenning, F. (1999). Dependent types in practical programming. In *Proceedings of the 26th ACM SIGPLAN-SIGACT Symposium on Principles of Programming Languages* (pp. 214-227). ACM.

ADDITIONAL READING

Augustsson, L. (1999). Cayenne—A language with dependent types. *Advanced Functional Programming*, 240-267.

Aydemir, B., Charguéraud, A., Pierce, B. C., Pollack, R., & Weirich, S. (2008). Engineering formal metatheory. *ACM SIGPLAN Notices*, *43*(1), 3–15. doi:10.1145/1328897.1328443.

Compagnoni, A. (2004). Higher-order subtyping and its decidability. *Information and Computation*, *191*(1), 41–103. doi:10.1016/j.ic.2004.01.001.

Dreyer, D., & Rossberg, A. (2008). Mixin' up the ML module system. *ACM Sigplan Notices*, *43*(9), 307–320. doi:10.1145/1411203.1411248.

Harper, R., & Lillibridge, M. (1994). A typetheoretic approach to higher-order modules with sharing. In *Proceedings of the 21st ACM SIGPLAN-SIGACT Symposium on Principles of Programming Languages* (pp. 123-137). ACM.

Leroy, X. (1994). Manifest types, modules, and separate compilation. In *Proceedings of the 21st ACM SIGPLAN-SIGACT Symposium on Principles of Programming Languages* (pp. 109-122). ACM.

Leroy, X. (2000). A modular module system. *Journal of Functional Programming*, *10*(3), 269–303. doi:10.1017/S0956796800003683.

Maharaj, S., & Gunter, E. (1994). Studying the ML module system in HOL. *Higher Order Logic Theorem Proving and its Applications*, 346-361.

McBride, C. (2005). Epigram: Practical programming with dependent types. *Advanced Functional Programming*, 130-170.

Pierce, B. C., & Turner, D. N. (2000). Local type inference. *ACM Transactions on Programming Languages and Systems*, *22*(1), 1–44. doi:10.1145/345099.345100.

Rajamani, S., & Rehof, J. (2001). A behavioral module system for the pi-calculus. *Static Analysis*, 375-394.

Sheldon, M. A., & Gifford, D. K. (1990). Static dependent types for first class modules. In *Proceedings of the 1990 ACM Conference on LISP and Functional Programming* (pp. 20-29). ACM.

Wells, J., & Vestergaard, R. (2000). Equational reasoning for linking with first-class primitive modules. *Programming Languages and Systems*, 412-428.

Chapter 11
Model–Driven Data Warehouse Automation:
A Dependent–Concept Learning Approach

Moez Essaidi
Université Paris-Nord, France

Aomar Osmani
Université Paris-Nord, France

Céline Rouveirol
Université Paris-Nord, France

ABSTRACT

Transformation design is a key step in model-driven engineering, and it is a very challenging task, particularly in context of the model-driven data warehouse. Currently, this process is ensured by human experts. The authors propose a new methodology using machine learning techniques to automatically derive these transformation rules. The main goal is to automatically derive the transformation rules to be applied in the model-driven data warehouse process. The proposed solution allows for a simple design of the decision support systems and the reduction of time and costs of development. The authors use the inductive logic programming framework to learn these transformation rules from examples of previous projects. Then, they find that in model-driven data warehouse application, dependencies exist between transformations. Therefore, the authors investigate a new machine learning methodology, learning dependent-concepts, that is suitable to solve this kind of problem. The experimental evaluation shows that the dependent-concept learning approach gives significantly better results.

DOI: 10.4018/978-1-4666-4494-6.ch011

1. INTRODUCTION

The *decision support systems* and *business intelligence systems* (Turban et al., 2010; Poe et al., 1997) are the areas of the information systems discipline that is focused on supporting and improving decision-making across the enterprise. The decision-making process is a strategic asset that helps companies to differentiate themselves from competitors, improve service, and optimize performance results. The data warehouse (Kimball and Ross, 2002, 2010) is the central component of current decision support and business intelligence systems and is responsible for collecting and storing useful information to improve decision making process in organization. Several data warehouse design frameworks and engineering processes have been proposed during the last few years. However, the *framework-oriented* approaches (Luján-Mora et al., 2006; Prat et al., 2006; Simitsis, 2005) fail to provide an integrated and a standard framework that is designed for all layers of the data warehousing architecture. The *process-oriented* approaches (Westerman, 2001; List et al., 2000; Kaldeich and Sá, 2004) fail, also, to define an engineering process that handles the whole development cycle of data warehouse with an iterative and incremental manner while considering both the business and the technical requirements. In addition, not much effort was devoted to unify the framework and the process into a single integrated approach. Moreover, no intelligent and automatic data warehouse engineering method is provided.

The *model-driven data warehouse* gathers approaches that align the development of the data warehouse with a general *model-driven engineering* paradigm (Bézivin, 2006). The model-driven engineering is mainly based on models, meta-models, and transformation design. Indeed, model-driven strategy encourages the use of models as a central element of development.

The models are conforming to metamodels and the transformation rules are applied to refine them. Therefore, transformations are the central components of the each model-driven process. However, transformation development is a very hard task that makes the model-driven approach more complex and entails additional costs. So, designers or programmers must have high skills in the corresponding metamodels and the transformation languages as the *query-view-transformation* (Object Management Group/QVT, 2010). In addition, data warehousing projects require more knowledge about the underlying business domain and requirements. This raises many risks and challenges during the transformations design. One of the main challenges is to automatically learn these transformations from existing project traces. In this context, the *model transformation by-example* (introduced by Varró, 2006a) is an active research area in model-driven software engineering that uses artificial intelligence techniques and proposes to automatically derive transformation rules. It provides assistance to designers in order to simplify the development of model transformations and it reduces complexity, costs and time of development.

In the framework of *model-driven data warehousing*, several steps are needed to automatically learn the transformation rules. The first step (the *modelling* step), which has been addressed in previous papers (Essaidi and Osmani, 2009, 2010a), consists in isolating stages where it is necessary to induce transformation rules; in identifying the metamodels used to define the input/output models of these transformations and in designing a conceptual framework for transformations learning expressed in an adequate representation language. We have focused on effective modelling of the *model-driven data warehouse* architecture in order to simplify machine learning framework integration. This architecture allows also for a flexible deployment of the application, with respect to standards and data warehousing requirements in organisations. Then, we propose to express the model transformation problem as an *inductive logic programming* one (Muggleton and Raedt, 1994) and to use existing project traces to find the best transformation rules. To the best of our knowledge, this work is the only one effort that has been developed for automating model-driven data warehousing with relational learning that provides experimentations in this context.

In a *model-driven data warehouse* application, dependencies exist between transformations. We investigate a new machine learning methodology stemming from the application needs: learning dependent-concepts. Following work about layered learning (Stone and Veloso, 2000; Nguyen et al., 2004), context learning (Turney, 1993; Bieszczad and Bieszczad, 2006), predicate invention (Muggleton and Road, 1994; Stahl, 1994), and cascade learning (Gama, 1998; Xie, 2006), we propose a *Dependent-Concept Learning (DCL)* approach where the objective is to use a pre-order set of concepts on this dependency relationship: first learn non dependent concepts, then, at each step, add the learned concept as background knowledge for next concepts to be learned according to the pre-order. This DCL methodology is implemented and applied to our transformation learning problem. Experimental evaluation shows that the DCL system gives significantly better results.

This chapter is organised as follows: Section 2 presents the terminology used and outlines the related research fields. In Section 3, the learning aspects of the solution are detailed. The section starts by the formalisation of key concepts used in our approach. Then, it studies the proposed machine learning approach (dependent-concepts) to learn transformation rules. Section 4 gives experimental results and discussion. Our main perspectives and the future research challenges are presented in Section 5. Section 6 summarizes our contributions and gives our final conclusions and remarks.

2. BACKGROUND

In the problem we deal with, several concepts and research fields are considered. This section investigates the definition of these fields and the associated terminologies. It brings together the elements that are necessary to understand the context of our work. It provides also a review of works in various areas (i.e., *model-driven data warehouse* approaches, *model transformation by-example* framework and *concept learning strategies*) related to the provided methodology.

2.1. Model-Driven Data Warehouse

The *model-driven engineering* represents a promising approach to support software development practices (Kent, 2003; Bézivin, 2006; Kulkarni et al., 2010). The *Model-Driven Architecture (MDA)* standard (Miller and Mukerji, 2003) represents the Object Management Group implementation to support the model driven approach. The MDA starts with the well-known and long established idea of separating the specification of the operation of a system from the details of the way that system uses the capabilities of its platform. The three primary goals of the MDA are portability, interoperability and reusability. The *model-driven architecture* standard base also includes many specifications. These include the *Unified Modelling Language (UML)*, the *Meta-Object Facility (MOF)*, specific platforms models (i.e., CORBA, JEE), and the *Common Warehouse Metamodel (CWM)* to design data warehouse components. The transformations are essential for each model-driven process. And, a simple model transformation consists in defining the mapping between elements of a source model (i.e., the input parameter of the transformation) and a target model (i.e., the resulted output of the transformation execution). In this context, the *Query-View-Transformation (QVT)* standard plays a central role, since it allows for the specification of model transformation rules (Figure 1 is an example).

Figure 1. Graphical notation of EntityToCube relation

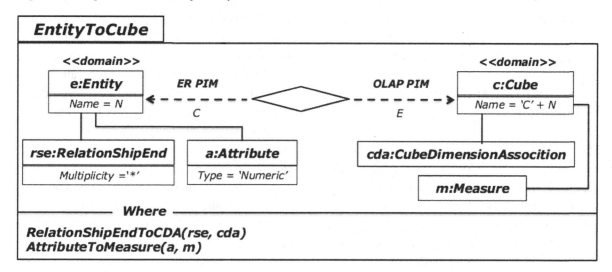

The *model-driven data warehouse* represents approaches that align the development of the data warehouse with a general *model-driven engineering* paradigm. Related work (Mazón and Trujillo, 2008; Zepeda et al., 2008) have tried in 2008 to adapt the model-driven approach for the development of data warehouses using the *model-driven architecture* and the *query-view-transformation*. For example, the approach presented in (Zepeda et al., 2008) describes derivation of *Online Analytical Processing (OLAP)* schemas from *Entity-Relationship (ER)* schemas. The source and target models respectively conform to ER and OLAP metamodels from the *common warehouse metamodel*. Authors describe how an ER schema is mapped to an OLAP schema and also provide a set of *query-view-transformation* rules (e.g., *EntityToCube, AttributeToMeasure, RelationShip-ToDimension*, etc.) to ensure this.

The designed transformation rule (Figure 1) shows a candidate *Entity* that gets transformed to a corresponding *Cube*. The generated *Cube* has the same name of the *Entity*, but prefixed with a "C". Also, the transformation rules *Relation-ShipEndToCDA* and *AttributeToMeasure* must be done as post-conditions. The left part of the rule check the data-source elements (i.e., *Entity, Attri-*

bute, etc) while the right part defines the derived multidimensional elements (i.e., *Cube, Measure*, etc). By the transformation *AttributeToMeasure*, the numeric attributes of the candidate *Entity*, gets transformed to a corresponding measures of the *Cube*. Also, through the transformation rule *RelationShipEndToCDA*, each *RelationShipEnd* role with multiplicity equal to many is matching with a *CubeDimensionAssosiation*.

Through this simple example we show that the designed transformations require an expert who knows the business domain, the principles of model-driven approach, and the transformation languages (Jouault et al., 2008; Eclipse-M2M, 2010). Thus, the actual *model-driven data warehouse* automation processes are partial because the transformation design remains manual. This could increase the time and cost of developing the decision support information system. Also, there is no guarantee that the proposed transformations are used for any given data-model, and that elements defining the mapping are consistent with the initial (business and technical) requirements. The basic idea is that the dependencies between metamodels concepts (e.g., *Entity, Attribute, Cube*, and *Measure*) create a post-conditions dependency within the definition of transformations (e.g.,

EntityToCube and *AttributeToMeasure*). And so, this kind of dependencies may change the way of transformations design and thus enable better finding elements involved in the mapping.

2.2. Model Transformation By-Example

The *model transformation by-example* is related to several others by-example based approaches: *query-by-example, programming-by-example, and XSLT generation by-example*. The *query-by-example* approach (Zloof, 1975) aims at proposing a language for querying relational data constructed from sample tables filled with example rows and constraints. The *programming-by-example* (Repenning and Perrone, 2000; Cypher et al., 1993), where the programmer (often the end-user) demonstrates actions on example data, and the computer records and possibly generalizes these actions, has also proven quite successful. The by-example approach has also been proposed in the XML world to derive XML schema transformers (Erwig, 2003; Ono et al., 2002; Yan et al., 2001), which generate XSLT code to carry out transformations between XML documents. In (Varró and Balogh, 2007), authors present an automated *model transformation by-example* approach using the *inductive logic programming*, an improvement of the initial proposal introduced in (Varró, 2006a). The proposed method (based on Aleph ILP implementation) aims at the inductive construction of first-order clausal theories from examples and background knowledge (restricted to Prolog clauses). A running example is provided where a source class diagram (based on the *unified modelling language*) is mapped into a target relational database diagram.

Authors in (Wimmer et al., 2007), present a conceptual framework for *model transformation by-example* to derive ATL (Atlas Transformation Language [Jouault and Kurtev, 2005]) rules. The approach uses the inter-model mappings representing semantic correspondences between concrete domain models which is more user-friendly then directly specifying model transformation rules or mappings based on the abstract syntax. The inter-model mappings between domain models can be used to generate the model transformation rules, by-example, taking into account the already defined mapping between abstract and concrete syntax elements. Then, in (Strommer et al., 2007), authors extend the *model transformation by-example* approach to the domain of business process modelling languages. The definition of requirements for *model transformation by-example* in the context of business process modelling and the specification of proper mapping operators comprise the main contribution of authors in this paper.

In (Kessentini et al., 2008), authors present a by-example approach, named *Model Transformation as Optimization by Examples (MOTOE)* which combines transformation blocks extracted from examples to generate a target model. Authors use an adapted version of *Particle Swarm Optimization (PSO)* where transformation solutions are modelled as particles that exchange transformation blocks to converge towards the optimal transformation solution. In a second paper (Kessentini et al., 2009) authors use the *Simulated Annealing (SA)* to improve the performances of the approach. In (Dolques et al., 2009), authors study the generation of transformation rules form transformations traces (transformations examples) using an extension of the *Formal Concept Analysis (FCA)*. FCA is based on the philosophical understanding that a concept is constituted by two parts: its extension which consists of all objects belonging to the concept, and its intention which comprises all attributes shared by those objects. Authors use the *Relational Concept Analysis (RCA)*, one of the extensions of FCA that considers links between objects in the concept construction. Then, lattices allow rules classification and help navigation among the generated results to choose the relevant transformation rule. The experimental evaluations are provided using LATEX to HTML transformation examples.

Authors in (Sun et al., 2009), discuss the limitations of above approaches and introduce a new approach called *model transformation by-demonstration* instead of the *model transformation by-example* approach. The *model transformation by-example* idea is about inferring the model transformation rules from a prototypical set of mappings. However, the *model transformation by-demonstration* approach asks users to demonstrate how the model transformation should be done by directly editing (e.g., add, delete, connect, update) the model instance to simulate the model transformation process step-by-step. Finally, ontology-based approaches allow semantic reasoning techniques for metamodels alignment or matching. For example, in (Roser and Bauer, 2006), metamodels are mapped to a pivot ontology, then an ontology-based reasoning is used to generate a Relational-QVT transformation. In (Kapsammer et al., 2006), authors apply refactoring to metamodels in order to make explicit hidden concepts of metamodels and obtain an ontology where all concepts are reified before mapping. The *similarity flooding* (Melnik et al., 2002) algorithm allows similarity values propagation in a labelled graph whose vertices are potential mappings, authors in (Falleri et al., 2008) adapt it for metamodel alignment.

2.3. Related Machine Learning Approaches

The goal of machine learning is to design algorithms that use example data or past experience to improve their performance for solving a given problem (Alpaydin, 2010). Many successful applications of machine learning exist already, including systems that analyze past sales data to predict customer behaviour, recognize faces or spoken speech, and optimize robot behaviour so that a task can be completed using minimum resources, and extract knowledge from bioinformatics data. Machine learning algorithms and techniques have long been used for various purposes in software engineering (testing, validation, security, etc.). For instance (Zhang and Tsai, 2007; Djuric et al., 2007) have studied the advances and perspectives in applications of such approaches in the software and data engineering fields. In our *model-driven data warehouse* framework, we propose to discover transformation rules from previous project experiences using supervised learning techniques.

The *inductive logic programming* (Nienhuys-Cheng and de Wolf, 1997) is an active research subfield of machine learning that addresses relational learning and uses a first-order representation of the problem domain and examples. Its objective is to provide practical algorithms for inductively learning hypotheses, expressed as logical rules (Lavrac and Dzeroski, 1994). An ILP learning task is defined by four aspects: model theory defines the semantic constraints on hypotheses, i.e., what to search for; proof theory describes the strategy to perform the search; declarative bias explicitly defines the hypothesis space, i.e., where to search; and preference bias is concerned with the generalisation performance of ILP. For a hypothesis to become a solution, semantic requirements set by an ILP learning task have to be satisfied. The ILP learning tasks can adopt two important settings: the descriptive or the predictive setting. In our approach, we are concerned by the predictive setting using the well known Aleph framework (Srinivasan, 2006). The predictive setting is formally defined as below:

Definition 1 (Predictive Setting): Given background knowledge B, some positive examples E^+ and negative examples E^-, the predictive setting is to learn a hypothesis H, such that H covers all positive examples and none of the negative examples, with respect to B.

We will investigate a new machine learning methodology stemming from the application needs: *learning dependent-concepts*. Following work about *layered learning, context learning, predicate invention and cascade learning,* we propose a new methodology that automatically updates the background knowledge of the concepts to be learned (i.e., the learned child-concepts are used to update the background knowledge of parent-concepts).

Authors in (Stone and Veloso, 2000), introduce the *layered learning* machine learning paradigm. In (Nguyen et al., 2004) authors study the problem of constructing the approximation of higher level concepts by composing the approximation of lower level concepts. Authors in (Gustafson and Hsu, 2001; Jackson and Gibbons, 2007) present an alternative to standard genetic programming that applies layered learning techniques to decompose a problem. The *layered learning* approach presented by Muggleton in (Muggleton, 1993) aims at the construction of a large theory in small pieces. Compared to *layered learning*, the *Dependent-Concept Learning (DCL)* approach aims to find all concepts theory using the theories of concepts on which they depend. Then, while the layered learning approach exploits a bottom-up, hierarchical task decomposition, the DCL algorithm exploits the dependency relationships between specific concepts of the given dependency-graph. The dependency structure in (Stone and Veloso, 2000) is a hierarchy, whereas our dependency structure is a directed acyclic graph. A breadth-first search algorithm is used to explore the dependency-graph.

Within the field of *inductive logic programming*, the term *predicate invention* has been introduced in (Muggleton, 1991) and involves the decomposition of predicates being learned into useful sub-concepts. In (Muggleton and Road, 1994), authors define *predicate invention* as the augmentation of a given theoretical vocabulary to allow finite axiomatisation of the observational predicates. Stahl in (Stahl, 1994; Stahl, 1995), studies the utility of Predicate Invention task in ILP and its capabilities as a bias shift operation. Authors in (Rios and Matwin, 1998), investigate a specification language extension when no examples are explicitly given of the invented predicate. The proposed *dependent-concept learning* and *predicate invention* approaches share the fact that they correspond to the process of introducing

new theoretical relationships. However, in the case of *predicate invention*, the approach is usually based on decomposition of the theory to learn on simple sub-theories and the DCL approach is based on the composition of a theory from the learned theories.

In (Gama and Brazdil, 2000), authors introduce the *cascade generalization* method. This approach is compared to other approaches that generate and combine different classifiers like the Stacked Generalization approach (Wolpert, 1992; Ting and Witten, 1997; Ting and Witten, 1999). In (Xie, 2006), author proposes several speed-up variants of the original *cascade generalization* and show that the proposed variants are much faster than the original one. As the *cascade generalization*, the *dependent-concept learning* approach extends the background knowledge at each level by the information on concepts of the sublevel (according to the dependency-graph). But, within the DCL approach, we use the same classifiers for all iterations. In our experiments, we report the results of the extension of the background knowledge by instances (as first setting) and the learned theory/rules (as second setting). The first setting is named DCLI and the second setting is named DCLR.

Machine learning approaches that exploit context to synthesize concepts are proposed in (Turney, 1993; Bieszczad and Bieszczad, 2006). The *model transformation by-example* approach also aims to find *contextual patterns* in the source-model that map *contextual patterns* in target-model. This task is defined as *context analysis* in (Varró, 2006a). In (Turney, 1993) author provides a precise formal definition of context and list four general strategies for exploiting contextual information. Authors in (Bieszczad and Bieszczad, 2006) introduce an enhanced architecture that enables contextual learning in a problem solving system. Nevertheless, the notion of context is different in the *dependent-concept learning* approach. In fact, in the DCL, contextual information is the result of the learning process (which will form the transformation rule); while within the *contextual learning* strategy the context is part of input information that improves the performance of the learner.

3. LEARNING-APPROACH

The main goal is to automatically derive the transformation rules to be applied in the Model-Driven Data Warehouse process. This aims to reduce the contribution of transformations designer and thereby reducing the time and the cost of development of data warehouses. We use the *inductive logic programming* framework to express the model transformation and we find a new methodology (the *dependent-concept learning*) that is suitable to solve this kind of problem. In this section, a formalisation of model-driven concepts is provided and the problem statement in a relational learning setting is expressed. The DCL problem is defined and the learning approach of model transformations is provided.

3.1. Key Concepts Formalisation

First, we define the notion of model. Then, we recall the definition of a metamodel and the relation between models and metamodels. Finally, the definition of a model transformation is given.

Definition 2 (Model): A model $M = (G, MM, \mu)$ is a tuple where:

$G = \left(N_G, E_G, \Gamma_G\right)$ is a directed multi-graph[1], MM is itself a model called the reference model of M (i.e., the metamodel) associated to a graph $G_{MM} = (N_{MM}, E_{MM}, "_{MM})$, and

$\mu : N_G \cup E_G \rightarrow N_{MM}$ is a function associating elements (nodes and edges) of G to nodes of G_{MM}.

The relation between a model and its reference model (metamodel) is called *conformance* and is noted *conformsTo*. The elements of *MM* are called meta-elements (or meta-concepts). μ is neither in-

jective (several model elements may be associated to the same meta-element) nor subjective (not all meta-elements need to be associated to a model element). The relation between elements and meta-elements is an instantiation relation. For example, the *Invoice (respectively InvoiceFact)* element in a DSPIM (respectively a MDPIM) is an instance of *Class* (respectively of *Cube*) meta-class in the UML CORE (CWM OLAP) metamodel.

Definition 3 (Metamodel and Meta-Metamodel): A meta-metamodel is a model that is its own reference model (i.e., it conforms to itself). A metamodel is a model such that its reference model is a meta-metamodel (Jouault and Bézivin, 2006).

In (Czarnecki and Helsen, 2006), authors provide a classification of models transformation approaches (template-based, graph-based, relational and so on). In our case, we are interested in Relational Approaches that can be seen as a form of constraint solving. The basic idea is to specify the relations among source and target element types using constraints. Declarative constraints can be given executable semantics, such as in logic programming. In fact, logic programming with its unification-based matching, search, and backtracking seems a natural choice to implement the relational approach, where predicates can be used to describe the relations (Czarnecki and Helsen, 2006). For example, in (Gerber et al., 2002), authors explore the application of logic programming. In particular, Mercury a typed dialect of Prolog, and F-logic an object-oriented logic paradigm, are used to implement transformations. In (Rutle et al., 2008) authors discuss a formalization of modelling and model transformation using a generic formalism, the *Diagrammatic Predicate Logic (DPL)*. The DPL (Diskin and Wolter, 2008; Rutle et al., 2009) is a graph-based specification format that takes its main ideas from both categorical and first-order logic, and adapts them to software engineering needs.

Definition 4 (Model Transformation): A model transformation is defined as the generation of a target model from a source model. Then formally, a model transformation consists of a set of transformation rules which are defined by input and output patterns (denoted by $\mathbb{P}$) specified at the M2 level (the metamodel level) and are applied to instances of these meta- models. Thus, a model transformation is associated to a relation $R\left(MM,MN\right) \subseteq \mathbb{P}\left(MM\right) \times \mathbb{P}\left(MN\right)$ defined between two metamodels which allows obtaining a target model N conforming to MN from a source model M that conforms to metamodel MM (Stevens, 2010).

3.2. Relational Learning Setting

We consider the machine learning problem as defined in (Mitchell, 1982). A (single) concept learning problem is defined as follows. Given 1) a training set $E = E^+ \cup E^-$ of positive and negative examples drawn from an example language $\mathcal{L}_e$ 2) a hypothesis language $\mathcal{L}_h$, 3) background knowledge B described in a relational language $\mathcal{L}_b$, 4) a generality relation $\geq$ relating formulas of $\mathcal{L}_e$ and $\mathcal{L}_h$, learning is defined as search in $\mathcal{L}_h$ for a hypothesis h such that h is consistent with E. A hypothesis h is consistent with a training set E if and only if it is both complete $(\forall e^+ \in E^+, h, B \geq e^+)$ and correct $(\forall e^- \in E^-, h, B \geq e^-)$. In an ILP setting, $\mathcal{L}_e$, $\mathcal{L}_b$ and $\mathcal{L}_h$ are Datalog languages, most often, examples are ground facts or clauses, background knowledge is a set of ground facts or clauses and the generality relation is a restriction of deduction.

We used in our experiments the well-known Aleph system, because of its ability to handle rich background knowledge, made of both facts and rules. Aleph follows a top-down generate-and-test approach. It takes as input a set of examples, represented as a set of Prolog facts and background knowledge as a Datalog program. It also enables the user to express additional constraints C on the admissible hypotheses. Aleph tries to find a hy-

pothesis $h \in \mathcal{L}_h$, such that h satisfying the constraints C and which is complete and partially correct. We used Aleph default mode: in this mode, Aleph uses a simple greedy set cover procedure and construct a theory H step by step, one clause at a time. To add a clause to the current target concept, Aleph selects an uncovered example as a seed, builds a most specific clause as the lowest bound of its search space and then performs an admissible search over the space of clauses that subsume this lower bound according the user clause length bound. In the next section, we show the reduction of the source-model, the target-model, and the mapping between them into an ILP problem.

3.3. Problem Statement in a Relational Learning Setting

In the ILP framework (regarding the background knowledge and examples), a model M_i is characterized by its description MD_i, i.e., a set of predicates that correspond to the contained elements. The predicates used to represent Mi as logic programs are extracted from its metamodel MM_i. For example, consider a data model used to manage customers and invoices. The classes *Customer* and *Invoice* are defined respectively by *class(customer)* and *class(invoice)*. The one-to-many association that relates Customer to Invoice is mainly defined by *association(customer-invoice, customer, invoice)*. Then, the logic description of models from project's traces (i.e., previous projects) constitutes the generated background knowledge program in ILP.

Definition 5 (Transformation Example): A *transformation example* (or trace model) $R\left(M,N\right) = \left\{r_1,...,r_k\right\} \subseteq \mathbb{P}\left(M\right) \times \mathbb{P}\left(N\right)$ specifies how the elements of M and N are consistently related by R. A training set is a set of transformation examples.

The transformation examples are project's traces or they can be collected from different experts (Kessentini et al., 2010). For instance, we are interested in the transformation of the Data-Source PIM (DSPIM) to the Multidimensional PIM (MDPIM). The DSPIM represents a conceptual view of a data-source repository and it *conformsTo* the UML CORE metamodel (part of the *unified modelling language*). The MDPIM represents a conceptual view of a target data warehouse repository and its *conformsTo* the CWM OLAP metamodel (part of the *common warehouse metamodel*). The predicates extracted from the UML CORE metamodel to translate source models into logic program are: *type(name), multiplicity(bound), class(name), property(name, type, lower, upper), association(name, source, target), associationOwnedAttribute(class, property),* and *associationMemberEnds(association, property)*. Then, according to the CWM OLAP metamodel, the predicates defined to describe target models are: *cube(Name), measure(Name, Type, Cube), dimension(Name, isTime, isMeasure), cubeDimensionAssociation(Cube, Dimension), level(Name), levelBasedHierarchy(Name, Dimension),* and *hierarchyLevelAssociation(LevelBasedHierarchy, Level)*.

So, by analysing the source and target models, we observe that structural relationships (like aggregation and composition relations, semantic dependency, etc.) define a restrictive context for some transformations. For instance, let us consider the concept *PropertyToMeasure*. We know that there is a composition relation between Class and Property and there is also a composition relation between Cube and Measure in the metamodels. This implies that the concept *PropertyToMeasure* must be considered only after the concept *ClassToCube* is learned. Therefore, the *ClassToCube* concept must be added as background knowledge in order to learn the *PropertyToMeasure* concept.

3.4. Dependent-Concept Learning Problem

The model-driven domain specificity induces a pre-order on the concept to be learned and defines a *dependent-concept learning* problem. Therefore, in our approach, concepts are organized to define a structure called dependency-graph. In (Esposito et al., 2000), authors use the notion of dependency graph to deal with hierarchical theories. The dependency-graph in this context is defined as a directed acyclic graph of concepts, in which parent nodes are assumed to be dependent on their offspring.

Definition 6 (Dependency Graph after (Esposito et al., 2000)): A *dependency graph* is a directed acyclic graph of predicate symbols, where an edge $(p \rightarrow q)$ indicates that atoms of predicate symbol q are allowed to occur in the hypotheses defining the concept denoted by p.

Let $\{c_1, c_2, \ldots, c_n\}$ be a set of concepts to be learned in our problem. If we consider all the concepts independently, each concept c_i defines an independent ILP problem, i.e., all concepts has independent training sets E_i and share the same hypothesis language $\mathcal{L}_h$ and the same background knowledge B. We refer to this framework as the Independent-Concept Learning (ICL). It takes into account a pre-order relation[2] $\preceq$ between concepts to be learned such that $c_i \preceq c_j$ if the concept c_j depends on the concept c_i or in other term, if c_i is used to define c_j. So, it is denoted $c_j \rightarrow c_i$ as the dependency-graph definition. More formally, a concept c_j is called parent of the concept c_i (or c_i is the child of c_j) if and only if $c_i \preceq c_j$ and there exists no concept c_k such that $c_i \preceq c_k \preceq c_j$. $c_i \preceq c_j$ denotes that c_j depends on c_i for its definition. A concept c_i is called *root* concept iff there exists no concept c_k such that

$c_k \preceq c_i$ (in other words, a *root* concept c_i does not depend on any concept c_k, for $k \neq i$).

The DCL framework uses the idea of decomposing a complex learning problem into a number of simpler ones. Then, it adapts this idea to the context of ILP multi-predicate learning. A *dependent-concept learning* algorithm accepts a pre-ordered set of concepts, starts with learning root concepts, then children concepts and propagates the learned rules to the background knowledge of their parent concepts and continues recursively the learning process until all dependent concepts have been learned. Within this approach, we benchmark two settings: (1) the background knowledge B_j of a dependent-concept (parent) c_j is extended with the child concept instances (as a set of facts - this framework is referred to as DCLI) and (2) B_j is extended with child concept intensional definitions: all children concepts are learned as sets of rules and are added to B_j – this frameworks is referred to as DCLR in the following sections. In both cases, DCLI or DCLR, all predicates representing children of c_j can be used in the body of c_j's definition. Our claim here is that the quality of the $c_j{'}^s$ theory substantially improves if all its children concepts are known in B_j, extensionnally or intensionnally. In the next Section (i.e., Evaluation), we provide results concerning the impact of child concepts' representation (extensional vs. intensional) on the quality of the c_j.

Finally, the task of empirical *dependent-concept learning* of model-driven context in ILP can be formulated as follows: Given a dependency graph $G_d = (C_d, E_d)$ where $C_d = \{c_1, c_2, ..., c_n\}$ the set of concepts to learn such that $\forall c_i \in C_d$: set of transformation examples (i.e., examples) $E = \{E_1, E_2, ..., E_n\}$ is given; and defined as (where $|TM|$ is the number of training models):

$$E_i = \left\{ R_i^j\left(M^j, N^j\right) \mid R_i^j\left(M^j, N^j\right) \subseteq \mathbb{P}\left(M^j\right) \times \mathbb{P}\left(N^j\right), j \leq |TM| \right\}$$

and a background knowledge B which provide additional information about the examples and defined as:

$$B = \left\{ \mathbb{P}\left(M^j\right) \bigcup \mathbb{P}\left(N^j\right) \mid M^j \, conformsTo \, MM, N^j \, conformsTo \, MN \right\}$$

Find: $\forall c_i \in C_d$, based on E_d and following a *BFS* strategy[3], learn a set of rules $R_i\left(MM, MN\right) \subseteq \mathbb{P}\left(MM\right) \times \mathbb{P}\left(MN\right)$; where *MM* is the reference source-metamodel and *MN* is the reference target-metamodel.

4. EVALUATION

4.1. Materials and Methods

We use a set of real-world data models for experimentations. The models represent projects' traces such as the example presented in Figure 2. In each project trace, we find the source-model(s), the target-model(s) and the transformations. A source-model description mainly includes the definition of classes, associations, and properties. In a target-model, we find elements like cubes, measures, dimensions and levels. From each model, we extract a set of positive and negative examples that define respectively positive and negative transformations as explained below.

We define the language bias using the metamodel level (M2 level) of the meta-modelling architecture. This gives the advantage to define a clear set of predicates with an optimal level of abstraction. Predicates obtained from the M2 level will ensure obtaining understandable transformation rules, equivalent to transformation designed manually. In the proposed Model-Driven Data Warehouse framework, this process will extract predicates from UML CORE and CWM OLAP metamodels. UML CORE defines the predicates used for the representation source-models (denoted as DSPIMs). CWM OLAP defines the predicates used for representing target-

Figure 2. Mapping UML CORE instance (DSPIM) to CWM OLAP instance (MDPIM)

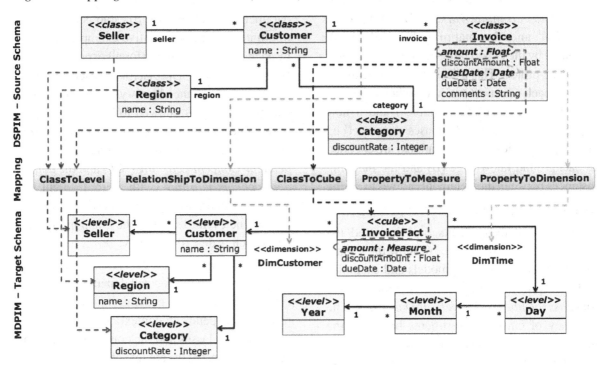

models (denoted as MDPIMs). As an example of trace model, consider a data schema used to manage customers and invoices (see Figure 2). With respect to the defined UML CORE and CWM OLAP predicates the following code is example of the generated background knowledge program of DSPIM: *type(integer); type(float); class(invoice); class(customer); property(amount, float, 1, 1)*. And, *cube(invoiceFact); measure(amount, float, invoiceFact); dimension(customerDim, false, false)* as part of the generated background knowledge program of MDPIM.

The mapping model describes all transformations of the source elements to the target elements. The transformation predicates are of the form: *transformation(sourceElement, targetElement)* where *sourceElement* and *targetElement* represent, respectively, input and output of the transformation rule. Given project traces, we extract the situation where the *Invoice* class is translated into a cube *InvoiceFact*, each such situation defines a positive

example. Similarly, the situation where a class is not transformed into a cube defines a negative example. In the example, *classtocube(invoice)* is a positive example and *classtocube(customer), classtocube(seller), classtocube(region)* are negative examples.

For the experiments presented in (Essaidi et al., 2011), we have selected 10 model instances of database schemas, provided by an industrial partner, and describing several application domains (invoices, sales, e-commerce, banking-investment, and so on). Concerning the experimentations of this paper, we use the *AdventureWorks 2008R2* reference databases (Microsoft, 2011). The Microsoft *AdventureWorks* reference databases are: The *Sample OLTP Database (AdventureWorksOLTP)* and the *Sample Data Warehouse (AdventureWorksDW)*. The AdventureWorksOLTP is a sample operational database used to define the source-model (i.e., the data-source schema - DSPIM). The *AdventureWorksDW* is a sample data warehouse schema used as target-model (i.e., the multidimensional schema - MD-PIM). The *AdventureWorksOLTP*, *Adventure-WorksDW* and the mapping between them, evaluated by the expert, are considered as a reference trace-project. This will allow us to benchmark our approach on a new extended schema (extend the number of learning examples) and a new dependency-graph. The databases elements (i.e., classes, properties and associations) are encoded as background knowledge (*B*) and the mapping instances between their elements allows to define positive $\left(E^+\right)$ and negative $\left(E^-\right)$ examples.

In the first experiments, we examined the accuracy of the learned rules to show the impact of the number of training models and examples and we report the obtained test accuracy curves for *ClassToCube* and *PropertyToDimension*. The accuracy of current experiments based on the new dataset (of *AdventureWorks*) confirms the results reported in (Essaidi et al., 2011). Then, considering the second dependency graph, we study also the performances of the DCL approach (with the two settings DCLI and DCLR) compared to the ICL approach. We report, in this section, the ROC curves of the tested approaches (ICL, DCLI and DCLR) based on the new dataset and the new enhanced dependency-graph. Accuracy is defined, based on the contingency table (or confusion matrix), as: *Accuracy=TP+TN / P+N*, Where *P(N)* is the number of examples classified as *positive (negative)*, *TP(TN)* is the number of examples classified as *positive (negative)* that are indeed *positive (negative)* (see Table 1).

As input, Aleph takes: (1) background information in the form of predicates and rules, (2) a list of modes declaring how these predicates can be chained together, (3) a designation of one predicate as the "head" predicate to be learned, and (4) a lists of positive and negative facts of the head predicate are also required. The learned logical clauses give the relationship between the transformations and the contextual information (elements) in the models. We run Aleph in the

default mode, except for the *minpos* and *noise* parameters; *:-set(minpos, p)* establishes as *p* the minimum number of positive examples covered by each rule in the theory (for all experiments we fix *p = 2*); and *:-set(noise, n)* is used to report learning performance by varying the number of negative examples allowed to be covered by an acceptable clause (we use two setting *n = 5 and n = 10*). We propose also to compare the *Independent-Concept Learning (ICL)* and the *dependent-concept learning* (with more different settings) approaches. We use the concept dependencies illustrated by the graph in Figure 3 and which is described below:

ClassToCube $\preceq$ *PropertyToMeasure:* The *PropertyToMeasure* concept depends on the concept *ClassToCube*. In general, transformation of properties depends on contextual information of transformed classes and the context of obtaining measures is part of the context of obtaining cubes. In fact, *properties* that become *measures* are numeric properties of *classes* that become *cubes*. So, we need information about the context of *ClassToCube* transformation in order to find the context of *PropertyToMeasure*.

Table 1. The number of examples per-concept used for learning

Concept	Number of Positive examples $\left(E^+\right)$	Number of negative examples $\left(E^-\right)$	Total number of examples $\left(E\right)$
ClassToCube	*27*	*44*	*71*
PropertyToMeasure	*47*	*202*	*249*
PropertyToDimension	*38*	*207*	*245*
RelationShipToDimension	*33*	*60*	*93*
ElementToHierarchyPath	*115*	*223*	*338*
ElementToDimensionLevel	*109*	*229*	*338*

Figure 3. The considered dependency graph of second experiments

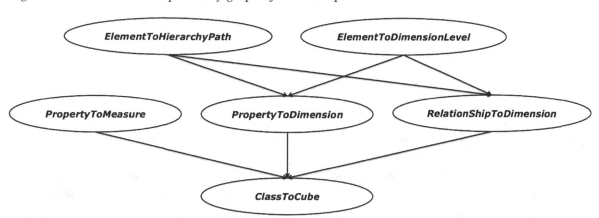

ClassToCube ≼ PropertyToDimension: This defines dependency between classes transformed into cubes and their properties that can be transformed into dimensions. Regarding the UML CORE metamodel, we find a structural dependency between *Class* and *Property* elements (a *Class* includes attributes, represented by the *ownedAttribute* role that defines a set of properties). Then, regarding the CWM OLAP metamodel, we have a structural dependency between *Cube* and *Dimension* elements. Current experiments confirm that structural dependencies in the metamodel act on the ways to perform learning.

ClassToCube ≼ RelationShipToDimension: Indeed, *dimensions* are, also, obtained from relationships of the *Class* that is transformed into Cube. The *CubeDimensionAssociation* meta-class relates a *Cube* to its defining dimensions as showed by the CWM OLAP metamodel in (Essaidi et al., 2011). These relationships define the axes of analysis in the target multidimensional schema (Wrembel and Koncilia, 2007).

(PropertyToDimension, RelationShipToDimension) ≼ ElementToHierarchyPath: A *Dimension* has zero or more hierarchies. A Hierarchy is an organizational structure that describes a traversal pattern through a *Dimension*, based on parent/child relationships between members of a *Dimension*. Then, elements that are transformed into dimensions (properties and relationships)

extend the background knowledge used to find hierarchy paths.

(PropertyToDimension, RelationShipToDimension) ≼ ElementToDimensionLevel: The *LevelBasedHierarchy* describes hierarchical relationships between specific levels of a *Dimension* (*e.g., Day, Month, Quarter and Year levels for the Time dimension*). So, rules of transforming elements into *Dimension* are used to find rules of obtaining the levels.

4.2. Results and Discussion

For *ClassToCube*, Aleph induces the following rules with the best score: a class (from the source-model) is transformed to a cube (in the generated target-model) where it has properties (with type float or integer, *i.e., associationOwnedAttribute(A,B), property(B,float/integer,1,1)*.) and it participate is an association that can define a hierarchy-path of a dimension (*i.e., association(C,A,D)*. where *A* as a second parameter represent the source element of the association, Box 1).

We note that the learned rules for *ClassToCube* are close to the rules designed manually. We compare the resulted rules with those provided by related work. For example in (Zepeda et al., 2008), the source-model context of the proposed *EntityToCube* is formed by Entity, *RelationShipEnd* (with *multiplicity* = '*') and Attribute

Box 1.

```
classtocube(A):  associationOwnedAttribute(A,B),
                              property(B,float,1,1), association(C,A,D).
classtocube(A):  associationOwnedAttribute(A,B),
                              property(B,integer,1,1), association(C,A,D).
```

(with numeric types) relations. Indeed, as it is shown above in the resulted *ClassToCube* transformation rule, we find also the relations: (*classtocube(A):- associationOwnedAttribute(A,B), property(B,numeric,1,1), association(C,A,D).*). Where *numeric* represents *float* and *integer* numbers. Regarding the resulted rules, *associationOwnedAttribute(A,B)* and *association(C,A,D)* atoms are associated with *RelationShipEnd* relation; then *property(B, float,1,1)* and *property(B,integer,1,1)* to *Attribute* (numeric types) relation. We have proposed a good language bias and a good modelling bias based on the same domain metamodel used by experts (UML and CWM) and this explains the obtained rules. Also, all of the resulted rules are found in a reasonable time. Concerning concepts *PropertyToMeasure*, *RelationshipToDimension* and *ClassToLevel*, Aleph induces also the following rules with the best score. For each concept, the obtained rules include predicates of child-concepts in the dependency-graph, when learned

in the *dependent-concept learning* framework as seen in Box 2.

In order to assess the impact of a child concept rules quality on the learning performances of a parent concept, we experiment the case where the child concept is noisy. This experiment is made within the DCL approach, we add noise to the non-dependent concept (i.e., *ClassToCube*) and we observe results of learning dependent-concepts with different acceptable noise setting (n = 5 and n = 10). We report the cases where 10% (denoted N-DCLI and N-DCLR) and 20% (denoted N2-DCLI and N2-DCLR) of the examples are noisy. To add noise, we swap positives and negatives examples.

We use also, the *area under the ROC curve (AUC)* as a common measure to compare the tested methods. As example of obtained graphs within current experiments, learning results for *PropertyToMeasure* and *RelationshipToDimension* are reported by Figures 4 and 5). Figures show that *n = 10* setting (right part of each figure) gives best performances compared to *n = 5*. Indeed, data quality and conceptual models quality (Mehmood

Box 2.

```
% PropertyToMeasure
propertytomeasure(B): associationOwnedAttribute(A,B),
                              classtocube(A), property(B,float,1,1).
propertytomeasure(B): associationOwnedAttribute(A,B),
                              classtocube(A), property(B,integer,1,1).
% RelationshipToDimension
relationshiptodimension(C): association(C,A,D), classtocube(A).
% ClassToLevel
classtolevel(D): association(C,A,D), classtocube(A), relationshiptodimension(C).
```

Figure 4. Learning PropertyToMeasure (n=5 for left) and (n=10 for right)

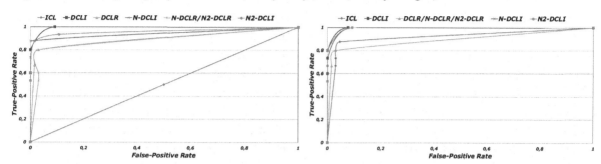

et al., 2009; Kersulec et al., 2009) play an important role in the design of information systems, and in particular decision support systems. Then, comparing ICL, DCLI and DCLR approaches, results show that the DCLI has greater AUC than other tested methods. The DCLI curves follow almost the upper-left border of the ROC space. Therefore, it has better average performance compared to the DCLR and ICL ($AUC_{DCLI} > AUC_{DCLR} > AUC_{ICL}$). The ICL curves almost follow to the 45-degree diagonal of the ROC space, which represents a random classifier. The DCLR setting exhibits good results with respect to the ICL approach, which are nevertheless slightly worse than results of the DCLI setting.

This result is expected, because the DCLI configuration, when learning a parent concept, uses in its background knowledge child-concepts as set of facts (extensional definition), as opposed to DCLR, which previously learns as sets of rules definition for offspring concepts. In case lower level concepts (i.e., child-concepts) are not per-

fectly identified, the errors for offspring concepts propagate to parent concepts. We assume here that examples are noise-free, which explains why DCLI has a better behaviour than DCLR. Thus, for *PropertyToMeasure* and *RelationshipToDimension*, results integrate the error rate from *ClassToCube* learned rules. Another remarkable point concerning curves is that the gap between ICL and DCL becomes more important more when we learn top-level concepts (i.e., parent-concepts) in the dependency-graph. So in this case, the contribution of DCL becomes more significant. This is explained by the fact that when finding top-level concepts using ICL, the learning configuration will be deprived of much more information on all intermediate concepts.

In this evaluation, a sensitivity analysis of classifiers is performed. Indeed, we can tell how robust a classifier is, by noting the classification accuracy of learning approaches using noisy data of different noise levels. We use two different percentages of noise, *10%* of the original data set (approaches

Figure 5. Learning RelationshipToDimension (n=5 for left) and (n=10 for right)

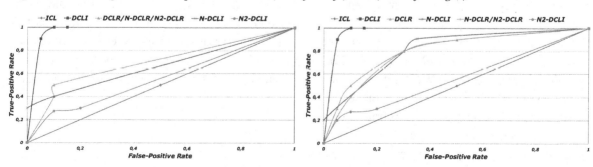

working on the obtained noisy dataset are denoted N-DCLI and N-DCLR) and *20%* (the obtained datasets are denoted N2-DCLI and N2-DCLR). Similar experiments on all datasets are performed and the resulted behaviours are compared to the ICL and DCL results using original datasets. Figures show that, in the presence of noise, and for most concepts, the classification accuracy of the DCLs settings drops less than those of the ICL approach. The performance degradation measures effect of noise on the classifiers. A classifier is more tolerant and resistant to the noise when it shows a smaller performance deviation. Therefore, considering the N-DCLI, N2-DCLI, N-DCLR and N2-DCLR settings, we have mainly: $AUC_{N\text{-}DCLI} > AUC_{N2\text{-}DCLI}$ and $AUC_{N\text{-}DCLR} > AUC_{N2\text{-}DCLR}$. Curves show that the obtained performances depend on the concept to learn and its *degree-of-dependence* on *ClassToCube* (the noisy non-dependent concept of this configuration). For instance, in Figure 5, *RelationshipToDimension* is more impacted than *PropertyToMeasure* (in Figure 4). In fact, based on the dataset, we observe that the *RelationshiptoDimension* concept is highly dependent on *ClassToCube*. This can be observed on most schemas (remarks provided in first experiments) and it is confirmed by the expert point-of-view. For example, in the case of *RelationshipToDimension*, the N2-DCLI curve seems to reach the 45-degree diagonal. This gives us an idea of the noise that we can accept when learning specific dependency relationships.

The model-driven process, based on *two track unified process* is one of the components of the architecture that we propose (Essaidi and Osmani, 2010a). This process offers a comprehensive partitioning by layer (or component) and a local partitioning (by design-level). The proposed frameworks, largely based on *Unified Modelling Language (UML) and Common Warehouse Metamodel (CWM)* define the second component of the architecture. These frameworks are used to define the representation language of transformations to learn. The choice of these

industry standards, recognized by experts, allows ensuring a good level of system integrity and also provides an optimal representation language (for understandable rules). The transformations are bound by the execution-dependency (the where relationship, or the post-condition). The proposed approach, based on the dependency-graph is consistent with this definition of transformations. The execution-dependencies are transformed (or reduced) into search-dependencies. This reduction problem creates the best environment for defining parent-concepts and improves the quality of the obtained rules; thus ensuring an effective assistance to experts.

The DCL approach is better, because adding a child-concept description allows the definition of new information (i.e., a new context) to consider when learning the parent-concept. This adds dependency information considered as an informational context that enriches the search space of the parent-concept and helps finding expected relations. This plays as an additional language bias, but also a search bias, allowing for good learning performances and good rules quality. The learned theory of the child-concept extends the background knowledge with a specific theory simple to learn, but at the same time it defines a sub-context of the parent-concept theory. So, this learning strategy, find first relations that are simple to learn with a minimum number of model elements. Then, the resulted sub-theories are used to set-up the learning context of parents-concepts. This approach is suitable to solve model-driven based problem because: (i) in metamodels definitions, we find dependencies between model elements (class, attribute, cube, measure, etc.), and (ii) in the manually designed rules, the "where" part of the transformation defines rules that must be activated (or executed) as post-condition. This post-condition information is a form of dependency.

5. FUTURE RESEARCH DIRECTIONS

Regarding *model-driven data warehouse* automation, our future work will experiments the case when a business goals model is considered during transformations. For example, the derivation of the MDPIM from the pair models (DSPIM, MDCIM), where MDCIM defines the organisation requirements/goals. We plan also to extend the approach to new application domains that provide a large dependency-graph (e.g., the *Extraction, Transformation, and Loading – ETL* process in the data warehousing architecture). Then we plan for an extension of the proposed *Model-Driven Architecture (MDA)* and the conceptual transformation learning framework to knowledge engineering seems also an interesting and a challenging future work. For example, a recent work (Prat et al., 2012) proposes an MDA approach to knowledge engineering that addresses the problem the mapping between CommonKADS knowledge models and *Production Rule Representation (PRR)*. Below we discuss others important directions in the fields of *data warehouse performance management* and *semantic model-driven policy management* that we consider interesting.

The book entitled DW 2.0: *The Architecture for the Next Generation of Data Warehousing* (Inmon et al., 2008) describes an architecture of the second generation data warehouses. It presents also the differences between DW 2.0 (introduced as the new generation) and the first generation data warehouses. Authors start by an overall architecture of DW 2.0 and give its key characteristics. Then, they present the DW 2.0 components and the role of each component in the architecture. The proposed architecture focuses on three key features: (1) the data warehouse repository structure (organization on four sectors: interactive, integrated, near line, and archival); (2) unstructured-data integration and organization; and (3) unified meta-data management. We confirm that unstructured data and Web-data integration constitutes a future challenge. Thus, we support semantic based approaches (Nebot and Llavori, 2010) for Web-data integration and data warehouses contextualization with documents (Pérez et al., 2009). This kind of approaches will probably represent the essential part of what we call the "DW 3.0 architecture." The DW 3.0 concept (or content data warehouse) is a unified architecture that includes the data warehouse, the *document warehouse,* and the *Web warehouse.* According to authors (Inmon et al., 2008), DW 2.0 represents the way corporate data needs to be structured in support of Web access and *Service Oriented Architecture (SOA)*. For this purpose, an effort is provided by (Wu et al., 2007). So, we believe that *business intelligence-as-a-service* platforms need a more efficient, personalized and intelligent Web-services discovery and orchestration engines. The perfect marriage of SOA/SaaS infrastructures is a key issue to design future on-demand business intelligence services.

In (Russom, 2009), the author studies the evolving state of data warehousing platforms and gives options available for next generation data warehousing. The options include concepts presented in (Essaidi and Osmani, 2012): SaaS and open-source business intelligence tools. It presents also many important features such as: real-time data warehousing, data management practices and advanced analytics. In (Pedersen, 2007), the author discusses other remaining challenges to extend traditional data warehouse architecture. The focus is mainly given for the data warehouse full-scale problem (world warehouse) and the privacy in data warehousing systems.

Metadata management for *business intelligence-as-a-service* infrastructures and cloud-based databases will be an interesting research direction. Indeed, current standards and models should be extended in this new architectural context. Finally, we believe that our proposal for *model-driven data warehouse- as-a-service* is a key characteristic to provide future data warehouse design in the cloud. The purpose of the *Common Warehouse Metamodel (CWM)* specification is to define a common interchange specification for metadata in a data warehouse. This definition provides a common language and metamodel definitions for the objects in the data warehouse. CWM describes a format to interchange metadata, but lacks the knowledge to describe any particular type of interchange. The need to define the context of a CWM interchange was discovered when the CWM co-submitting companies produced the CWM interoperability showcase. In order to make an effective demonstration of CWM technology, the participants needed to agree upon the set of metadata to be interchanged. In this context, the *object management group* proposes the *CWM Metadata Interchange Patterns (CWM MIP)* specification in order to address the limitations of the CWM.

The purpose of CWM MIP specification is to add a semantic context to the interchange of metadata in terms of recognized sets of objects or object patterns. We will introduce the term *Unit Of Interchange (UOI)* to define a valid, recognizable CWM interchange. From this information, a user of CWM, working in conjunction with CWM MIP, should be able to produce truly interoperable tools. CWM MIP augments the current CWM metamodel definitions by adding a new metamodel package. This new metamodel will provide the structural framework to identify both a UOI and an associated model of a pattern, and providing the necessary object definitions to describe both. In our future work, we will study in detail the CWM MIP in order to define new features that can improve the proposed architecture. Also, the *Ontology Definition Metamodel (ODM)* has been used as a basis for ontology development as well as for generation of OWL ontologies. The specification defines a family of independent metamodels, related profiles, and mappings among the metamodels corresponding to several international standards for ontology and *topic maps* definition, as well as capabilities supporting conventional modelling paradigms for capturing conceptual knowledge, such as entity-

relationship modelling. The ODM is used for ontology development and analysis on research in context-aware systems. As part of the OMG metamodeling architecture (ODM is a MOF-compliant metamodel), the ODM enables using *model driven architecture* standards in ontological engineering. The ODM is applicable to knowledge representation, conceptual modelling, formal taxonomy development and ontology definition, and enables the use of a variety of enterprise models as starting points for ontology development through mappings to UML and MOF.

The software engineering community is beginning to realize that security is an important requirement for software systems, and that it should be considered from the first stages of its development. Unfortunately, current approaches which take security into consideration from the early stages of software development do not take advantage of *model-driven development*. Security should definitely be integrated as a further element of the high-level software system models undergoing transformation until the final code generation (Fernández-Medina et al., 2009). Thus, *model-driven development* for secure information systems, *model-driven security,* and dynamic refinement of security policy are a new promising research direction. Another important aspect related to this, is semantics. In fact, dynamic refinement of security requires applying innovative semantic reasoning techniques to security metrics and contextual information. We consider this as an interesting problem for model-driven approach and its application. Our ideas and perspectives around these topics are discussed below. We seek to answer the question of how to provide intelligent methods and techniques to dynamically refine security policy using the contextual information? This work addresses different new issues such as: advanced semantic reasoning, recent security standards integration and deployment of the approach in several application domains. The project covers also the extension and the improvement of several existing approaches such as: related-policies management, context-aware and smart nodes, and the improvement of policies refinement techniques. The *modularity, adaptability, and consistency* will be the main features of the proposed architectures, methods, and standards. Thus, the dynamic policy refinement life-cycle proposed in this document aims to ensure these important aspects.

This proposal addresses the use of the *model-driven development* for security policies derivation and standards recommendation for policies definition and representation. The general research area related to this proposal is called *model-driven security*. The idea is derived from the research challenges discussed in (Fernández-Medina et al., 2009; Villarroel et al., 2005), where authors brie y explore some of the important related works (Basin et al., 2006; Hafner et al., 2006; Seehusen and Stølen, 2008; Seehusen and Stølen, 2007; Moebius et al., 2009; Kallel et al.,) and standards (OMG, 2008, 2009) to this context. So, an adapted architecture using a model-driven approach and that provide our vision of the problem is described. Regarding the problem description, the "*Semantics and Reasoning*" component will represent the core workflow element of the proposed architecture. This component is responsible of semantic policy adaptation (or reactions generation) based on policy changes and users activity. The project description also implies that the overall workflow contains a human approval step.

When we talk about automatic derivation, the application of the *model driven architecture* approach is directly possible. In the case of the proposed framework, the terms are around *model-driven security or model-driven policy*. The aim of our proposal is to take advantages of the semantic-driven approaches and the model-driven approaches. So, considering the "*Semantics & Reasoning*" component definition and the model-driven engineering definition, several questions arise: (1) how allow interoperability between the reasoning process (reasoning mechanisms) and the MDA process (policies generation mechanisms)?

(2) Which representation languages are available to define policy in order to ensure the integrity of the entire process? And (3) in more general, how provide a unified semantic model-driven and reasoning approach? To address this problem, we propose the use of several industry standards covering the semantics, security, and the MDA aspects. In addition, based on our experience on MDA-compliant architectures, a common approach showing the use of these standards is defined. The ontology is the central concept of any semantic-driven development. The *ontology definition metamodel* specification (Object Management Group/ODM, 2009) is an OMG standard (MDA-compliant and extensible metamodel) that allows *model-driven ontology engineering*. It provides standard profiles for ontology development in UML and enables consistency checking, reasoning, and validation of models in general.

The ODM include five main packages: At the core are two metamodels that represent formal logic languages: *DL (Description Logics)* which, although it is non-normative, is included as informative for those unfamiliar with description logics and *CL (Common Logic)*, a declarative first-order predicate language. There are three metamodels that represent more structural or descriptive representations that are somewhat less expressive in nature than CL and some DLs. These include metamodels of the abstract syntax for RDFS, the *ontology Web language (OWL)* and *topic maps (TM)*. Thus, the ODM standard is highly recommended in an MDA-based process because it is conform to MOF metamodeling architecture. In this case, is important to have a generic ontologies representation in order to: (1) facilitate the transformation of semantic models using the MDA-enabled frameworks; and (2) ensure interoperability between the different components/tools (including the reasoning engine and the transformation engine).

Security policy definition is very important in organization because it should cover many aspects. Several security specifications are proposed.

However, the OMG security specification (Object Management Group/Security, 2008) remains the most comprehensive and extensible. The OMG catalogue mainly contains: (1) the *Authorization Token Layer Acquisition Service (ATLAS)* specification which describes the service needed to acquire authorization tokens to access a target system using the CSIv2 protocol; (2) The *Common Secure Interoperability Specification, version 2 (CSIv2)* which defines the Security Attribute Service that enables interoperable authentication, delegation, and privileges; (3) the CORBA security service which provides a security architecture that can support a variety of security policies to meet different needs (identification and authentication of principals, authorization and infrastructure based access control, security auditing, etc.); (4) the *Public Key Interface (PKI)* specification which provides interfaces and operations in CORBA IDL to support the functionality of a PKI (issuance, management, and revocation of digital certificates); and (5) the *Resource Access Decision facility (RAD)* specification which provides a uniform way for application systems to enforce resource-oriented access control policies. The integration of these standards in the final architecture allows more quality in policy representation and a more unified, interoperable *model-driven security* approach with MDA. Note also that the proposed approach is open for integration of others security specifications and profiles.

In the proposed approach we focus on the "*Semantics & Reasoning*" module. Thus, we discuss some details of this module based on our main objectives (i.e., adaptability, consistency) and the model-driven support that we add. Figure 6 illustrate the workflow that we explain below.

Semantic Reasoning: The policies adaptation flow starts by a semantic analysis step. This step considers at the input several semantic conceptions: context, user activities/profiles, auditing, etc. This information is defined in general by ontologies and/or rules. As discussed above, the ontologies models are conform to the *ontology definition metamodel.*

Policy Adaptation: This step corresponds to the *model-driven architecture* transformation process. Based on the results of semantic reasoning step, it selects the appropriate transformation from transformations repository and applies it on current policy model. Thus, information given by the reasoning engine is automatically projected on policy by the transformation engine. In a model-driven context this supposes that policy models are conforming to the specifications cited above (ATLAS, RAD, etc.) or other profiles already recognized by the transformation. This allows for more interoperability between reasoning and transformation engines.

Consistency Verification: It is a machine approval foregoing the human approval. During this step the system checks if the main security constraints are violated or not. If a constraint is violated, the transformation process must be re-executed with new parameters (information about the violation is also audited), else the system waits for human approval.

Human Approval: It can be a very important step in the approval work flow mechanisms of some organizations. Indeed, in the detailed requirements we stress this sentence: automatic reactions on a policy might not be accepted by security officers, or even more simply might not be suitable in real-life.

Finally, in this proposal we discuss brie y the application of the *model-driven engineering* for policy adaptation in the respect of the problem description. So, based on our contributions to improve model-driven methodologies, an adapted *model-driven security* approach is defined. Then, we provide also recommendations for specification to define policy (i.e., perspectives to integrate new policy languages). So, future work will provide a detailed analysis, more research and improvements around the proposed approach.

Figure 6. Semantic model-driven policy adaptation

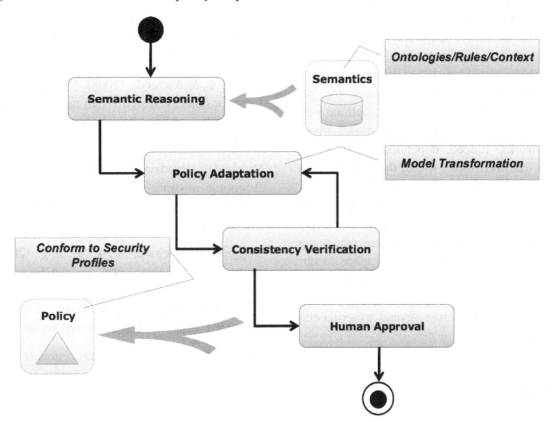

6. CONCLUSION

This chapter studies a complex problem at the crossroad of several research fields. We study the *model-driven data warehouse* engineering and its automation using machine learning techniques. The automation of information systems engineering and, in particular, *decision support systems* remains a difficult and a challenging task. Indeed, the *model-driven data warehouses* require a transformation phase of the models that necessitate a high level of expertise and a large time span to develop. The main goal is to automatically derive the transformation rules to be applied in the model-driven data warehouse process. This process of transformation concerns for instance the creation of an OLAP cube in business intelligence from an UML diagram of the considered application. The proposed solution allows the simplicity of decision support systems design

and the reduction of time and costs of development. First, we use the *model-driven engineering* paradigm for data warehouse components development (as first level of automation). Then, we use *inductive relational learning* techniques for automatic model transformation generation (as second level for automation).

The modelling step (Essaidi and Osmani, 2010a), considered as modelling bias (or architecture bias) is important to manage these risks and make efficient the task of transformations learning. In this step, the *model-driven data warehouse* framework is extended by *inductive logic programming* capabilities in order to support the expert in the transformation process. The ILP offers a powerful representation language and the given results (i.e., transformation rules) are easy to understand. We have focused on providing an optimized representation of the language bias (or declarative bias) based on *Unified Modelling*

Language (UML) and the *Common Warehouse Metamodel (CWM)* standards. This declarative bias addresses the reduction of CWM-UML problem into ILP and aims to restrict the representation to clauses that define best the transformation rules. Through the learning approach step, we contribute to the definition of the optimal way to learn transformation rules in model-driven frameworks. Indeed, dependencies exist between transformations within the model-driven warehousing architecture. We investigate a new machine learning methodology stemming from the application needs: *Learning Dependent-Concepts (DCL)*. This DCL method is implemented using the Aleph ILP system and it is applied to our transformation-learning problem. We show that the DCL approach gives significantly better results and performances across several experimental settings.

REFERENCES

Alpaydin, E. (2010). *Introduction to machine learning* (2nd ed.). Cambridge, MA: The MIT Press.

Basin, D. A., Doser, J., & Lodderstedt, T. (2006). Model driven security: From UML models to access control infrastructures. *ACM Transactions on Software Engineering and Methodology*, *15*(1), 39–91. doi:10.1145/1125808.1125810.

Bézivin, J. (2006). Model driven engineering: An emerging technical space. In *Proceedings of GTTSE*, (pp. 36-64). Berlin: Springer.

Bieszczad, A., & Bieszczad, K. (2006). Contextual learning in the neurosolver. In *Proceedings of ICANN*, (pp. 474-484). Berlin: Springer.

Cypher, A., Halbert, D. C., Kurlander, D., Lieberman, H., Maulsby, D., Myers, B. A., & Turransky, A. (Eds.). (1993). *Watch what I do: Programming by demonstration.* Cambridge, MA: MIT Press.

Czarnecki, K., & Helsen, S. (2006). Feature-based survey of model transformation approaches. *IBM Systems Journal*, *45*, 621–645. doi:10.1147/sj.453.0621.

Diskin, Z., & Wolter, U. (2008). A diagrammatic logic for object-oriented visual modeling. *Electronic Notes in Theoretical Computer Science*, *203*(6), 19–41. doi:10.1016/j.entcs.2008.10.041.

Djuric, D., Devedzic, V., & Gasevic, D. (2007). Adopting software engineering trends in AI. *IEEE Intelligent Systems*, *22*, 59–66. doi:10.1109/MIS.2007.2.

Dolques, X., Huchard, M., & Nebut, C. (2009). From transformation traces to transformation rules: Assisting model driven engineering approach with formal concept analysis. In *Proceedings of ICCS*, (pp. 93-106). Moscow, Russia: ICCS.

Eclipse-M2M. (2010). *The model to model (M2M) transformation framework.* Retrieved June 22, 2010, from http://www.eclipse.org/m2m/

Erwig, M. (2003). Toward the automatic derivation of XML transformations. In *Proceedings of XSDM*, (pp. 342-354). Berlin: Springer.

Esposito, F., Semeraro, G., Fanizzi, N., & Ferilli, S. (2000). Multistrategy theory revision: Induction and abduction in inthelex. *Machine Learning*, *38*(1-2), 133–156. doi:10.1023/A:1007638124237.

Essaidi, M., & Osmani, A. (2009). Data warehouse development using MDA and 2TUP. In *Proceedings of SEDE*, (pp. 138-143). ISCA.

Essaidi, M., & Osmani, A. (2010a). Model driven data warehouse using MDA and 2TUP. *Journal of Computational Methods in Sciences and Engineering*, *10*, 119–134.

Essaidi, M., & Osmani, A. (2010b). Towards model driven data warehouse automation using machine learning. In *Proceedings of IJCCI*, (pp. 380-383). Valencia, Spain: SciTePress.

Essaidi, M., & Osmani, A. (2012). Business intelligence-as-a-service: Studying the functional and the technical architectures. In *Business Intelligence Applications and the Web: Models, Systems and Technologies* (pp. 199–221). Hershey, PA: IGI Global.

Essaidi, M., Osmani, A., & Rouveirol, C. (2011). Transformation learning in the context of model-driven data warehouse: An experimental design based on inductive logic programming. In *Proceedings of ICTAI*, (pp. 693-700). IEEE.

Falleri, J.-R., Huchard, M., Lafourcade, M., & Nebut, C. (2008). Metamodel matching for automatic model transformation generation. In *Proceedings of MoDELS*, (pp. 326-340). Berlin: Springer-Verlag.

Fawcett, T. (2004). *Roc graphs: Notes and practical considerations for researchers (Technical report)*. HP Laboratories.

Fernández-Medina, E., Jürjens, J., Trujillo, J., & Jajodia, S. (2009). Model-driven development for secure information systems. *Information and Software Technology, 51*(5), 809–814. doi:10.1016/j.infsof.2008.05.010.

Gama, J. (1998). Combining classifiers by constructive induction. In *Proceedings of ECML*, (pp. 178-189). Springer.

Gama, J. A., & Brazdil, P. (2000). Cascade generalization. *Machine Learning, 41*, 315–343. doi:10.1023/A:1007652114878.

Gerber, A., Lawley, M., Raymond, K., Steel, J., & Wood, A. (2002). Transformation: The missing link of MDA. In *Proceedings of ICGT*, (pp. 90-105). Springer.

Gustafson, S. M., & Hsu, W. H. (2001). Layered learning in genetic programming for a cooperative robot soccer problem. In *Proceedings of EuroGP* (pp. 291-301). London, UK: Springer-Verlag. doi:10.1007/3-540-45355-5_23.

Hafner, M., Breu, R., Agreiter, B., & Nowak, A. (2006). Sectet: An extensible framework for the realization of secure inter-organizational workflows. *Internet Research, 16*(5), 491–506. doi:10.1108/10662240610710978.

Inmon, W., Strauss, D., & Neushloss, G. (2008). *DW 2.0: The architecture for the next generation of data warehousing*. San Francisco, CA: Morgan Kaufmann Publishers Inc..

Jackson, D., & Gibbons, A. P. (2007). Layered learning in boolean GP problems. In *Proceedings of EuroGP* (pp. 148–159). Berlin: Springer-Verlag.

Jouault, F., Allilaire, F., Bézivin, J., & Kurtev, I. (2008). Atl: A model transformation tool. *Science of Computer Programming, 72*(1-2), 31–39. doi:10.1016/j.scico.2007.08.002.

Jouault, F., & Bézivin, J. (2006). KM3: A DSL for metamodel specification. In *Proceedings of FMOODS*, (pp. 171-185). Springer.

Jouault, F., & Kurtev, I. (2005). Transforming models with ATL. In *Proceedings of MoDELS Satellite Events*, (pp. 128-138). Springer.

Kaldeich, C., & Sá, J. O. (2004). Data warehouse methodology: A process driven approach. In *Proceedings of CAiSE*, (pp. 536-549). Springer.

Kallel, S., Charfi, A., Mezini, M., Jmaiel, M., & Sewe, A. (2009). A holistic approach for access control policies: From formal specification to aspect-based enforcement. *International Journal of Information and Computer Security, 3*(3-4), 337–354. doi:10.1504/IJICS.2009.031044.

Kapsammer, E., Kargl, H., Kramler, G., Reiter, T., Retschitzegger, W., & Wimmer, M. (2006). Lifting metamodels to ontologies - A step to the semantic integration of modeling languages. In *Proceedings of MoDELS/UML*, (pp. 528-542). Springer.

Kent, S. (2003). Model driven language engineering. *Electronic Notes in Theoretical Computer Science, 72*(4). doi:10.1016/S1571-0661(04)80621-2.

Kersulec, G., Cherfi, S. S.-S., Comyn-Wattiau, I., & Akoka, J. (2009). Un environnement pour l'évaluation et l'amélioration de la qualité des modèles de systèmes d'information. In *Proceedings of INFORSID*, (pp. 329-344). INFORSID.

Kessentini, M., Sahraoui, H., & Boukadoum, M. (2008). Model transformation as an optimization problem. In *Proceedings of MoDELS*, (pp. 159-173). Berlin: Springer-Verlag.

Kessentini, M., Sahraoui, H., & Boukadoum, M. (2009). *Méta-modélisation de la transformation de modèles par l'exemple: Approche par méta-heuristiques*. LMO.

Kessentini, M., Wimmer, M., Sahraoui, H., & Boukadoum, M. (2010). Generating transformation rules from examples for behavioral models. In *Proceedings of BM-FA*, (pp. 2:1-2:7). ACM.

Kimball, R., & Ross, M. (2002). *The data warehouse toolkit: The complete guide to dimensional modeling*. New York: John Wiley & Sons, Inc..

Kimball, R., & Ross, M. (2010). *The Kimball group reader: Relentlessly practical tools for data warehousing and business intelligence*. New York: John Wiley & Sons, Inc..

Kulkarni, V., Reddy, S., & Rajbhoj, A. (2010). Scaling up model driven engineering - Experience and lessons learnt. In *Proceedings of MoDELS*, (pp. 331-345). Springer.

Lavrac, N., & Dzeroski, S. (1994). *Inductive logic programming: Techniques and applications*. New York: Ellis Horwood.

List, B., Schiefer, J., & Tjoa, A. M. (2000). Process-oriented requirement analysis supporting the data warehouse design process - A use case driven approach. In *Proceedings of DEXA*, (pp. 593-603). Springer.

Luján-Mora, S., Trujillo, J., & Song, I.-Y. (2006). A UML profile for multidimensional modeling in data warehouses. *Data & Knowledge Engineering, 59*(3), 725–769. doi:10.1016/j.datak.2005.11.004.

Mazón, J.-N., & Trujillo, J. (2008). An MDA approach for the development of data warehouses. *Decision Support Systems, 45*, 41–58. doi:10.1016/j.dss.2006.12.003.

Mehmood, K., Cherfi, S. S.-S., & Comyn-Wattiau, I. (2009). Data quality through conceptual model quality - Reconciling researchers and practitioners through a customizable quality model. In *Proceedings of ICIQ*, (pp. 61-74). HPI/MIT.

Melnik, S., Garcia-Molina, H., & Rahm, E. (2002). Similarity flooding: A versatile graph matching algorithm and its application to schema matching. In *Proceedings of ICDE*, (pp. 117-128). IEEE Computer Society.

Microsoft. (2011). *Microsoft AdventureWorks 2008R2*. Retrieved September 8, 2011, from http://msftdbprodsamples.codeplex.com/

Miller, J., & Mukerji, J. (2003). *MDA guide version 1.0.1 (Technical report). Object Management Group*. OMG.

Mitchell, T. M. (1982). Generalization as search. *Artificial Intelligence, 18*, 203–226. doi:10.1016/0004-3702(82)90040-6.

Mitchell, T. M. (1997). *Machine learning*. New York: McGraw-Hill.

Moebius, N., Stenzel, K., & Reif, W. (2009). Generating formal specifications for security-critical applications - A model-driven approach. In *Proceedings of IWSESS*. IWSESS.

Muggleton, S. (1991). Inductive logic programming. *New Generation Computing, 8*, 295–318. doi:10.1007/BF03037089.

Muggleton, S. (1993). Optimal layered learning: A PAC approach to incremental sampling. In *Proceedings of ALT*, (pp. 37-44). London, UK: Springer-Verlag.

Muggleton, S., & Raedt, L. D. (1994). Inductive logic programming: Theory and methods. *The Journal of Logic Programming, 19/20*, 629–679. doi:10.1016/0743-1066(94)90035-3.

Muggleton, S., & Road, K. (1994). Predicate invention and utilisation. *Journal of Experimental & Theoretical Artificial Intelligence, 6*, 6–1. doi:10.1080/09528139408953784.

Nebot, V., & Llavori, R. B. (2010). Building data warehouses with semantic data. In *Proceedings of EDBT/ICDT Workshops*. ACM.

Nguyen, S. H., Bazan, J. G., Skowron, A., & Nguyen, H. S. (2004). Layered learning for concept synthesis. *Transactions on Rough Sets, 3100*, 187–208.

Nienhuys-Cheng, S.-H., & de Wolf, R. (1997). *Foundations of inductive logic programming*. Berlin: Springer. doi:10.1007/3-540-62927-0.

Object Management Group/ODM. (2009). *The ontology definition metamodel (ODM) specification*. Retrieved April 19, 2009, from http://www.omg.org/spec/ODM/

Object Management Group/QVT. (2010). *The query/view/transformation (QVT) specification*. Retrieved June 14, 2010, from http://www.omg.org/spec/QVT/

Object Management Group/Security. (2008). *The OMG security specifications catalog*. Retrieved December 13, 2008, from http://www.omg.org/technology/documents/formal/omg_security.htm

Ono, K., Koyanagi, T., Abe, M., & Hori, M. (2002). Xslt stylesheet generation by example with wysiwyg editing. In *Proceedings of SAINT*, (pp. 150-161). Washington, DC: IEEE Computer Society.

Pedersen, T. B. (2007). Warehousing the world: A few remaining challenges. In *Proceedings of DOLAP*, (pp. 101-102). New York, NY: ACM.

Pérez, J. M., Berlanga, R., & Aramburu, M. J. (2009). A relevance model for a data warehouse contextualized with documents. *Information Processing & Management, 45*(3), 356–367. doi:10.1016/j.ipm.2008.11.001.

Poe, V., Brobst, S., & Klauer, P. (1997). *Building a data warehouse for decision support*. Upper Saddle River, NJ: Prentice-Hall, Inc..

Prat, N., Akoka, J., & Comyn-Wattiau, I. (2006). A UML-based data warehouse design method. *Decision Support Systems, 42*(3), 1449–1473. doi:10.1016/j.dss.2005.12.001.

Prat, N., Akoka, J., & Comyn-Wattiau, I. (2012). An MDA approach to knowledge engineering. *Expert Systems Application, 39*(12), 10420–10437. doi:10.1016/j.eswa.2012.02.010.

Repenning, A., & Perrone, C. (2000). Programming by example: programming by analogous examples. *Communications of the ACM, 43*(3), 90–97. doi:10.1145/330534.330546.

Rios, R., & Matwin, S. (1998). Predicate invention from a few examples. In *Proceedings of AI*, (pp. 455-466). London, UK: Springer-Verlag.

Roser, S., & Bauer, B. (2006). An approach to automatically generated model transformations using ontology engineering space. In *Proceedings of SWESE*. SWESE.

Russom, P. (2009). *Next generation data warehouse platforms*. Retrieved October 26, 2009, from http://www.oracle.com/database/docs/tdwi-nextgen-platforms.pdf

Rutle, A., Rossini, A., Lamo, Y., & Wolter, U. (2009). A diagrammatic formalisation of mof-based modelling languages. In *Proceedings of TOOLS*, (pp. 37-56). Springer.

Rutle, A., Wolter, U., & Lamo, Y. (2008). A diagrammatic approach to model transformations. In *Proceedings of EATIS*. EATIS.

Seehusen, F., & Stølen, K. (2007). Maintaining information flow security under refinement and transformation. In *Proceedings of FAST*. FAST.

Seehusen, F., & Stølen, K. (2008). A transformational approach to facilitate monitoring of high-level policies. In *Proceedings of POLICY*, (pp. 70-73). IEEE Computer Society.

Simitsis, A. (2005). Mapping conceptual to logical models for ETL processes. In *Proceedings of DOLAP*, (pp. 67-76). ACM.

Srinivasan, A. (2006). *A learning engine for proposing hypotheses (aleph)*. Retrieved from http://web.comlab.ox.ac.uk/oucl/research/areas/machlearn/Aleph

Stahl, I. (1994). On the utility of predicate invention in inductive logic programming. In *Proceedings of ECML*, (pp. 272-286). Springer.

Stahl, I. (1995). The appropriateness of predicate invention as bias shift operation in ILP. *Machine Learning, 20*, 95–117. doi:10.1007/BF00993476.

Stevens, P. (2010). Bidirectional model transformations in QVT: Semantic issues and open questions. *Software & Systems Modeling, 9*(1), 7–20. doi:10.1007/s10270-008-0109-9.

Stone, P., & Veloso, M. M. (2000). Layered learning. In *Proceedings of ECML*, (pp. 369-381). Springer.

Strommer, M., Murzek, M., & Wimmer, M. (2007). Applying model transformation by-example on business process modeling languages. In *Proceedings of ER*, (pp. 116-125). Berlin: Springer-Verlag.

Sun, Y., White, J., & Gray, J. (2009). Model transformation by demonstration. In *Proceedings of MoDELS*, (pp. 712-726). Springer.

Ting, K. M., & Witten, I. H. (1997). Stacked generalization: When does it work? In *Proceedings of IJCAI*, (pp. 866-871). San Francisco, CA: Morgan Kaufmann.

Ting, K. M., & Witten, I. H. (1999). Issues in stacked generalization. *Journal of Artificial Intelligence Research, 10*, 271–289.

Turban, E., Sharda, R., & Delen, D. (2010). *Decision support and business intelligence systems*. Englewood Cliffs, NJ: Prentice Hall.

Turney, P. D. (1993). Exploiting context when learning to classify. In *Proceedings of ECML*, (pp. 402-407). London, UK: Springer-Verlag.

Varró, D. (2006a). Model transformation by example. In *Proceedings of MoDELS*, (pp. 410-424). Genova, Italy: Springer.

Varró, D., & Balogh, Z. (2007). Automating model transformation by example using inductive logic programming. In *Proceedings of SAC*, (pp. 978-984). New York, NY: ACM.

Villarroel, R., Fernández-Medina, E., & Piattini, M. (2005). Secure information systems development - A survey and comparison. *Computers & Security, 24*(4), 308–321. doi:10.1016/j.cose.2004.09.011.

Westerman, P. (2001). *Data warehousing: Using the Wal-Mart model*. San Francisco, CA: Morgan Kaufmann Publishers Inc..

Wimmer, M., Strommer, M., Kargl, H., & Kramler, G. (2007). Towards model transformation generation by-example. In *Proceedings of HICSS*, (p. 285b). Washington, DC: IEEE Computer Society.

Wolpert, D. H. (1992). Stacked generalization. *Neural Networks*, *5*, 241–259. doi:10.1016/S0893-6080(05)80023-1.

Wrembel, R., & Koncilia, C. (2007). *Data warehouses and OLAP: Concepts, architectures and solutions*. Hershey, PA: IGI Global.

Wu, L., Barash, G., & Bartolini, C. (2007). A service-oriented architecture for business intelligence. *Service Oriented Computing and Applications*, 279–285.

Xie, Z. (2006). Several speed-up variants of cascade generalization. In *Proceedings of FSKD*, (pp. 536-540). Xi'an, China: Springer.

Yan, L. L., Miller, R. J., Haas, L. M., & Fagin, R. (2001). Data-driven understanding and refinement of schema mappings. In *Proceedings of SIGMOD*, (pp. 485-496). New York, NY: ACM.

Zepeda, L., Celma, M., & Zatarain, R. (2008). A mixed approach for data warehouse conceptual design with MDA. In *Proceedings of ICCSA*, (pp. 1204-1217). Perugia, Italy: Springer-Verlag.

Zhang, D., & Tsai, J. J. P. (2007). *Advances in machine learning applications in software engineering*. Hershey, PA: IGI Global.

Zloof, M. M. (1975). Query-by-example: The invocation and definition of tables and forms. In *Proceedings of VLDB*, (pp. 1-24). New York, NY: ACM.

ENDNOTES

[1] A directed multi-graph $G = \left(N_G, E_G, "_G \right)$ consists of a finite set of nodes N_G, a finite set of edges E_G, and a function $"_G : E_G \rightarrow N_G \times N_G$ mapping edges to their source and target nodes (Jouault & Bézivin, 2006).

[2] A pre-order is a binary relationship reflexive and transitive.

[3] A BFS strategy: Start by an offspring and non-dependent concept (i.e., a root concept), then follow its parents dependent-concepts.

Chapter 12
Parameterized Transformation Schema for a Non-Functional Properties Model in the Context of MDE

Gustavo Millán García
Pontifical University of Salamanca, Spain

Rubén González Crespo
Pontifical University of Salamanca, Spain

Oscar Sanjuán Martínez
Carlos III University, Spain

ABSTRACT

The integration between design models of software systems and analytical models of non-functional properties is an ideal framework on which lay the foundation for a deep understanding of the architectures present in software systems and their properties. In order to reach this integration, this chapter proposes a parameterized transformation for a model of performance properties derived from a system model in the MDE context. The idea behind a parameterized term is to leave open the transformation framework to adopt future improvements and make the approach reusable. The authors believe that this kind of integration permits the addition of analysis capabilities to the software development process and permits an early evaluation of design decisions.

DOI: 10.4018/978-1-4666-4494-6.ch012

INTRODUCTION

In 1987 F. Brooks published a famous paper called "No Silver Bullet: Essence and Accident in Software Engineering" (Brooks, 1986) as a clear allusion to the fact that there are not magic solutions to the fundamental problems affecting the overrall development of software systems. This paper focused on the inherent complexity of the software and its invisibility . Therefore, the field of software engineering is trying to address this inherent complexity by using models to better understand the characteristics of a software systems increasingly complex and bigger. The use of models allow us to see a problem more clearly and help us to visualize properties that can not easily seen from the encrypted form of the system.

The approach proposed by MDE (Model-Driven Engeneering) (Atkinson, 2003) is a clear recognition of the concept of model as an key artifact within the construction and development activity of a software system. MDE discusses the software development process by promoting the use of models as first-class artifacts. But not only promotes the use of different models but fundamental proper integration and consistency allowing an integrated view of different properties and dimensions of system software at different levels of abstraction.

One of the major advances in the use of models in software engineering is the integration of design tools and semiautomatic code generation (Czarnecki, 2000) from these models. Thereby many potential errors resulting from manual coding could be avoided . A clear example of these design and integration tools is the Eclipse development environment and the Eclipse Modeling Project (Fundation, 2001) .

There is however a class of models that have nothing to do directly with the primary objective of generating system code, but rather to allow quantitative analysis of non-functional properties of system, such analysis would be based on performance models, reliability or security models.

This chapter discusses MDE software development view, focusing on the transformation for non-functional properties models, more precisely a model based on queue theory (Lazowska, 1984). The chapter provides a study of the transformation issues of a system model to a performance model based on queuing theory. As a solution to the gap problem between models this chapter provides a parameterized transformation schema applied to this kind of context. At the end of the chapter, we describe a simple example for proving our approach

THE MDE VIEW OF SOFTWARE DEVELOPMENT PROCESS

The model-driven engineering software view of software development has its roots in a general method to represent the details of the system in some kind of formalism (model) focused mainly on software and hardware issues and then through a chains of transformations to obtain a software system encoded in some execution platform.

The concept of model as a key artifact in the software development is widely accepted. However this term is often overused and misused. A definition of model term could be found in (ModelWare, 2007).

Def. Model: "formal representation of entities and relationships (abstraction) with some correspondence (isomorphism) to the real world for some purpose (pragmatism)."

Many models have been proposed in software development, but not all are formal. This variety of informal models has contributed to make more difficult tracking of consistence between models.

The transformation process promoting by MDE could be shown in the following idealized sequence $M1 \xrightarrow{tM1M2} M2..Mn \xrightarrow{tMncode} code.$

In this scheme of transformations chain *M1, M2, ..., Mn* are design models of software system at several details or levels (formalization) of abstraction and *t* represents any correspondence or association between elements of different models implied in the process. Ideally, this transformation is hierarchical by the different levels of abstraction (isomorphism).

Given two models called *Mi* and *Mf*, with *Mi* as initial model of transformation and *Mf* as final model, if we consider these models as two sets of elements, then informally we could describe the transformation *t* as follows:

$$t: Mi \rightarrow Mf \quad (1)$$

where *t* represents any correspondence or relationship between an item *ei ∈ Mi* and an item *ef ∈ Mf*, then *t(ei)=ef* defines the transformation between elements that have some sort of correspondence defined. It may be that not all elements are related through this relation .For this reason we define the set *G*, which represents the graph of the set (or relation) of elements that has a correspondence rule within the set of pairs defined by the Cartesian product *Mi x Mf*, more formally $G = \{<ei,ef> \in Mi \times Mf\}$.

It is important highlight, that there are several applicable transformations. The overall consistency of a software development approach based on MDE therefore it will be directly related to eliminating gaps between models and the precise definition of mapping rules between model elements at different levels of abstraction.

The ultimate goal of a development approach based on MDE is to obtain an operational system implementation from an abstract specification at different levels of abstraction and that there is consistency between models used. However it is not always possible to find a transformation with completely defined mapping rules. This issue is related to the gaps between models $M1 \xrightarrow{\ ?\ } M2..Mn \xrightarrow{tMncode} code$. In this scenario the transformation between M1 and M2 has a gap of information, and it could be carry out only partially.

Figure 1 shows four different solutions to this problem:

a. Adapting the scope and the conventions of the initial model (*Mi→ Mi'*), so this model can reflect the features required in the final model (*Mf*) of the transformation.
b. Adding the required information to the initial model (*Mi*) without changing its conventions *Mi→Mi extended.*
c. Using a specific model (*Mb*) to provide additional information and combine with the initial model *Mi*
d. Leaving open the transformation using a set of parameters called in this paper C where $C = \{p \mid p \in Mf\}$ and get each parameter value via external procedures *Pext(p)*

Solution a) has the drawback to pervert the meaning of the initial model (*Mi*) making change "ad-hoc" with the consequent loss of effectiveness of *Mi* essence. The solution b) seems less intrusive (no changes of conventions) but can cause an overload of information in the initial model (*Mi*). The solution c) allows an effective combination of models (e.g. architecture model components deployment model) for the transformation. The solution d) is called parameterized transformation in this paper and may be the most appropriate when you do not have sufficient experience to define a deterministic mapping rule in the transformation.

Figure 1. Different solutions to tackle the gap problem between models

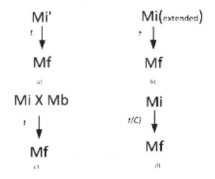

MDA (Model-Driven Architecture): Standard Specification for Model Transformation

Within the context of model transformations worth a special mention the specification MDA (Model-Driven Architecture) of OMG (Object Management Group) (OMG, 2006). Mainly MDA defines an approach of software development with special emphasis on the separation of the abstract functional specification and the actual system specification based on a particular technology platform.

MDA is one of the possible scenarios in the process of model driven development. The main transformation takes place between a PIM (Platform Independence Model) and a PSM (Platform Specific Model).

MDA defines an essential process of transformation. The first model to be considered in the transformation chain is an abstract model defined solely on the problem domain; this model is called CIM (Computation Independent Model). Software developers using this model to create a system model without using technical details related to the final implementation platform. The other model is called PIM (Platform Independence Model). It can be used an automatic transformation between models (M2M) to refine the model and to add PIM implementation details considering a specific platform. The term platform has a very broad sense in this context. For example, you can define the type of realization of a PIM item (database, workflow manager, etc.) or a specific implementation within the different types of existing abstract components in the industry through their platforms (.NET, CORBA, and Java EE). Besides, for example a platform could also refer to implementation details as configuration files. Definitely a model that depends on these details is called PSM (Platform Specific Model). Finally, a code generator could interpret a PSM to generate the necessary system code.

TRANSFORMATION TO NON-FUNCTIONAL PROPERTIES MODEL IN THE MDE CONTEXT

A more ambitious and comprehensive MDE approach is reached by the transformation of models in order to analyze various non-functional properties of software system, such as performance, security or reliability properties. Schematically we can see this kind of transformation like a horizontal transformation, in the sense that it aims is not to descend into the detail level-to-level coding system for execution. Figure 2 shows this idea and an overview of transformations (functional and non-functional elements) within MDE context, which uses model transformation, not only to project functional design decisions at different levels of abstraction, but also promotes the possibility of transforming a given design model to a Model of Analysis (MA) to analyze various non-functional system properties.

The direct benefits obtained from this approach are the analytical capabilities in the development process, making it a more robust process.

Figure 2. Functional models and non-functional models in the context of model transformations

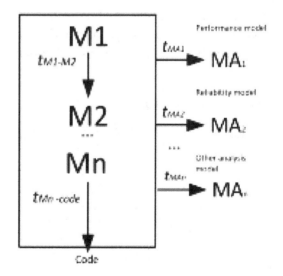

STUDY OF THE TRANSFORMATION TO A NON-FUNCTIONAL PROPERTIES MODEL

In this section we analyze the transformation of a system model based on component architecture and an analytical model of performance properties, such as a queuing model. This will allow the definition of a set of mapping rules in the context of non-functional models transformation.

Queuing Model

The simplest queuing system is the shown in Figure 3. The core of the system is a server that provides some service. Individuals of a certain population reach the system to be served. If the server is free, the individual is served immediately. Otherwise, the individual is added to a queue. When the server has finished serving an individual, this leaves the service. If there are individuals waiting in line, one is selected to the server immediately.

Figure 3 shows the parameters associated with the model. Individuals come to the service with some average rate (e.g. $\lambda = individuals / second$). At a given time, in the queue will have a certain number of individuals (zero or more), the average number of individuals waiting is W, and the average time an individual must wait is tw. The server manages individuals with a service time s. Server utilization p is the fraction of time the server is busy. Finally the parameter Q is the average number of individuals in the system,

including the individual who is in service (if any) and individuals who are waiting (if any), and tq as the average time an individual spends in the system, both waiting and service. Assuming that the system is not saturated ($\lambda < \mu$) the maximum rate of entries that can handle the system is $\lambda = 1 / s$.

Fix this model requires complete knowledge of the probability distribution of the arrivals rate and service time. This is a complex issue to resolve when you do not have a reference implementation. To make the problem more tractable some assumptions can be made to simplify the problem. The most important of these assumptions is that the arrival rate follows a Poisson distribution, which is to say that the time between arrivals is exponential. This assumption is common; with it you can get many useful results knowing only the mean value of the arrival rate and service time.

Table 1 shows a summary of the most important elements of a service queue model with a single server single A/B/1/N/Z in Kendall's notation in terms of its parameters.

Figure 3. Single queuing system

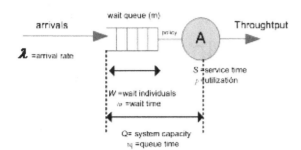

Table 1. Elements and parameters of a model of service queue

Element	Properties
Waiting queue	m:length W:Mean waiting time
Service	n:Number of Servers S:Service Time (S) Queue policy (FIFO,LIFO, etc) R:Mean Response Time X:Throughput p:Utilization
Arrival Time distribution (A)	Exponential (M), Deterministic (D), Erlang (Ek), General (G) Idenpendientes/dependientes
Server Time Distribution (B)	Exponential (M), Deterministic (D) Erlang (Ek), General (G)
Population (type,classes)	Type =open/closed (N) classes

Figure 4. Queuing network

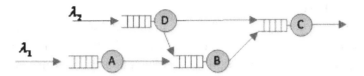

Queuing Network Model

In an environment of distributed processing system, queues are not isolated. The two elements are complicating this scenario are (Stallings, 2008).

- The split and mix of traffic between servers
- The existence of queues in series

Figure 4 shows an example of division and traffic mix in a network model of service nodes.

Jackson's theorem (Jackson, 1957) is used to analyze a queuing network like above figures. This theorem states that in a queuing network under certain conditions, each service node is considered an independent system. Thus, each node can be analyzed independently of the others by a model M/M/1 or M/M/N and the results can be combined using classical statistical methods. Average delays of each node can be added to deduce the total system delay of network, but one cannot say anything about the instants of greatest delay (for example, standard deviation).

Another key element in queuing models is whether the number of individuals is infinite or finite. For finite model, we assume a fixed arrival rate. In the case of endless population, the arrival rate depends on the number of sources. The infinite models are easier to solve and this assumption is reasonable when the number of sources is at least 5 to 10 times the capacity of the system.

The most important properties of a network of service nodes are shown in Table 2.

The parameters of a model generally constitute the set of information necessary to assess a particular situation, in this case a model performance. Therefore, the goal of a transformation approach of a software system model to a performance model is to identify the most amount of information that needs the transformation by the first model (i.e. the system model). The problem is that many times can not accurately determine the value of some of these parameters. In this case we must assume certain values for these parameters as we have seen before. For example a typical system with a series of assumptions about certain parameters for studying the performance is as follows in Table 3.

This can be expressed more conveniently in Kendall notation as follows M/M/1 / ∞ / ∞ / FIFO or M/M/1.

Table 2. Properties of a network queues model

Element	Properties
service nodes network (K) Queuing Network class	K:Total number of queuing nodes in the system class: Jackson,Gordon-Newell, Kelly,BCMP, networks with blocking
Routing Matrix (Pij)	M=‖Pij‖ Pij= routing probability from node-i to nodo-j
Population(N)	N/infinite population

Table 3. Default values of single server

Element	Properties
Waiting queue	M:infinite Policy=FIFO
Service	n=1 Queue policy =FIFO
Arrival time distribution (A)	Markov(M) with λ parameter
Service Time distribution (B)	Markov(M) with μ parameter

Obtaining Parameter Values for Performance Analytical Model

The estimation of certain parameters of the analytical model can be accomplished by testing over generated reference implementations of the system under study, if it has an implementation, but if not the situation is more complicated. These tests are based primarily on use a population with an arrival rate, specific profile, and a particular implementation to derive several metrics such as service time, response time, utilization factor, and so on. Figure 5a shows this scenario for a component testing and runtime environment.

One of the most difficult parameters to measure in a software system is the service time. The diversity of types of resources used (hardware and software) and the complex relationships between these resources make it difficult to obtain a direct measure accurately. As shown consider an example of the various common layers of control that might exist in a runtime environment based on the Java EE platform (see Figure 5b).

In Figure 5b *Tt* represents the total time attributable to the software component consider-

ing control layers of the Java EE platform. This response time is derived from the interaction of the underlying infrastructure control layers used as support for the execution of the component. This interaction can be seen as a process time required for the component can operate. In this publication this time unit it is called response time due to intrinsic factors. Whenever the component is operating will be associated these delays to control stages. This layers structure essentially models the characteristics of a service (implemented by a software component) that are relevant in defining the response time of the service layer (context).

One important idea in the above approach is that the delay associated with the interaction between underlying layers is a determinant factor in the final performance of the component. This assumption is based on works such as (Zhu, 2007; Liu, 2005). So the overall performance of a software system or software component based on a complex software platform is directly related to the number of interactions between layers control underlying and communications between the various components of the system under study. Or put another way, the service time S of a software component depend on

Figure 5. a) Test environment, b) software component underlying control layers

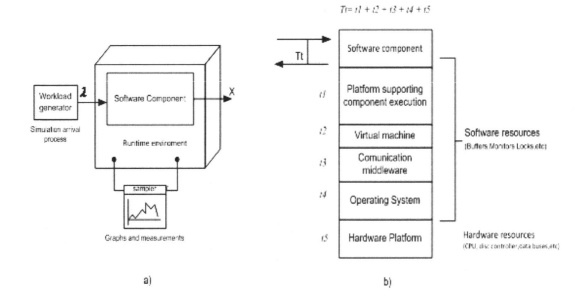

a)   b)

interaction layers necessary for execution, among other factors. In this scenario, a call to a software component involves the amount of time derived from the use of resources (mainly CPU) in each layer until the own control layer of the component. Unfortunately, it is not always possible to determine the accumulated times from the different layers of control (inability to include a probe).

One approach to obtain the service time of a component or system is to use the so-called Little´s general law of service demand formula, whose expression is:

$$\rho = \lambda * S \qquad (2)$$

This expression applied to a software component relates the utilization factor ρ of the component, the rate requests λ and the mean service time S of component. Determining the utilization factor ρ can be carried out by sampling system activity under different loads (Figure 5a), an arrival rate can be assumed and then simulate different scenarios and finally derive the service time S estimate.

The parameters required usually for these test fall into two categories:

- **Workload Parameters:** This type of load measuring parameters through different access profiles. This can be measured in terms of arrival rates (e.g., requests/sec) or by the number of concurrent entities (e.g., clients) requesting work along with the rate at which each client requests work.
- **Service Demand Parameters**: These parameters indicate the amount of each resource used on average by each type of request. Service demands can be measured in time units or in other units from which time can be derived. For example, the service demand on a communication link can be expressed by the size of a message in bits, provided we know the link capacity in bps.

The System Model: Component-Based Software System Architecture

The view of a software system can be understood in terms of the range of services it provides. We can understand a service as a well-defined and limited part of the overall functionality offered by the system. In the domain of information process a service is an operation that it is accessed through an interface using a type of request and response interaction.

A system architecture based on software components ideally is a combination of these components which allow to implement such services. These software components thus provide implementations for various services. In order to provide these services use different resources. An important issue is that a resource can be virtually anything that may be required to develop a software service. There are various types of resources; some will be more influential than others in performance. The resources can be classified into software or logical resources (buffers, locks, semaphores) and physical hardware resources (CPU, bus, disk controller, network, etc.). The most important feature of a resource from point of view of performance is that it has a limited capacity and submits time delays in its use.

In the context of a software component there are two types of essential properties (Zschaler, 2008):

- **Intrinsic Properties:** They are those that can be expressed entirely independently of the use of the component. Describe the implementation of the component itself.
- **Extrinsic Properties:** They are the result of using the component with certain intrinsic properties.

Software components usually develop a well-defined service and require a runtime environment. In the literature this environment is referred to as execution environment or container (Zschaler, 2008). The container among other things manages and controls the life cycle aspects of software components, use of resources and the service requests that come from the outside. All this must be considered to a greater or lesser extent in the performance of the component model.

Different views can be used to define and design a component. Figure 6 shows a simple example of a software component whose most relevant internal functional modules are A1, A2 and A3. Besides, the figure shows two design views: internal structural view (static view) and behavior view (dynamic view) for a usage scenario described in the UML language (OMG, 2007).

The static view of a software component identifies the relevant elements (key subcomponents or modules), and the means used for interaction (connectors). A complete classification of connectors could be found in (Mehta, 2000). It should be noted that each kind of connector could affect performance in a different way.

The dynamic view of software component identifies interaction in the context of the operation (usage scenario).From this view present demand for resources could be derived.

We can appreciate that these models alone there are not sufficient to obtain all the information related to the performance model. From the point of view of the performance model we want to know what will be expected resource demand or defined by the component given an execution

context (usage scenario). A fundamental problem concerning this issue is that a performance model of a software system depends primarily on the characteristics of the execution environment of each component used in the system design.

In view of the above we can deduce that a software architecture model is far from providing this information. A solution is to provide a resource allocation model for architectural description by a deployment model (OMG, 2007).It is worthy to consider that we only contemplate predesigned component architectures, such as those in platforms such as Java EE (Oracle, 2009), which allow us obtain a reference implementation available.

Regardless of how you will specify the behavior in a scenario of use, a general model of resource demand service can be expressed with the following expression:

$$D=V*S \qquad (3)$$

where D represents the demand (in units of time) that the component present in a stage of execution. The term V is the number of times the component interacts within the usage scenario and S is the service time of the component in terms of the resources it uses in its execution environment. Obviously if the component has a single interaction within the scenario of use, the expression (3) is transformed to $D = S$.

Figure 6. Analysis views of a software component

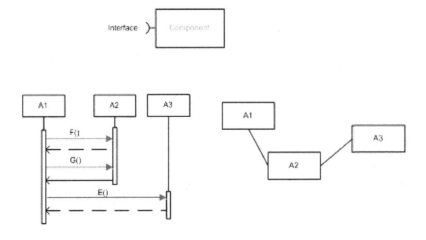

Table 4. Service demand model derived from the component and its internal behavior

Sub-module	Use relative frequency (based on use scenario)	Demand model $Di=Vi*Si$
A1	4/7	$DA1=4 *SA1$
A2	2/7	$DA2=2 *SA2$
A3	1/7	$DA3=1*SA3$
Expected total demand (D)		$D=DA1+DA2+DA3$

This demand model applied to the component architecture and considering recursively for each sub-module in the previous example (see Figure 6) results in Table 4's expected aggregated demands.

Thus, Table 4 shows the total demand D (time units) derived from the interaction of internal component parts in a scenario of use of the component, or what is the same the total estimated service time. This idea will be applied at the level of system´s components.

From the above analysis derives the fact that you need a way to identify the resources that will be in the demand as a result of the interaction of the key components of the architecture. This could be done by a deployment model, so one can obtain a partial characterization of an execution environment of resources mainly hardware. However, the execution environment of a software component (predesigned) further comprises logic resources as seen previously.

CORRESPONDENCE RULES OF A TRANSFORMATION SCHEMA FOR NON-FUNCTIONAL PROPERTIES MODEL

From the analysis of the previous section we can derive a set of general rules of correspondence (see Table 5) or a kind of relationship (R) between the information provided by a system model and the information required to evaluate an analytical model of performance.

Table 5. Relevant information from initial model from final model point of view

System model (Msys)	Information *(view point system model)*	Information *(view point performance system model)*
M1: Architecture model based on software components	-Key software components -Key Conectors	-Key Components of Service -Type of communication -Interaction mechanism -Network topology of service nodes
M2:Model Behavior	- Control sequence and interaction between components - usage scenario	- Service demands of each software component based on the frequency of use on the usage scenario
M3: Deployment model	- Allocation of physical resources to software components -Physical resources system -Logical-Resources -Resources-physical connection topology	-Topology network service nodes -Hardware resources -Software resources
M4: Architecture model (layers service pattern)	Logic control bound -Key service control layer	-Interpretation of service time and response time of the system or components -Logical resources used in each layer
M5:Workload model	-Number of requests - arrival rate of request -classes of requests	-Characterization of the population and system workload
M6: Platform component model	Pre-built software components	-Performance references

Table 5 shows key information present in both models considered in this chapter.

The system model based on components architecture (see Table 5) is defined informally as a set named *Msys* where *Msys =M1 x M2 x M3 x M4 x M5 x M6*.

As discussed in Section IV an analytical model requires solving the model by the parameters values. For the present example case it will primarily determine the distribution of arrivals (A) and service time (B), the type of queue discipline and the capacity of each queue service in the system. Figure 7 shows schematically the idea of this transformation between models considering a software component as a system.

In this particular case, the type of transformation *t: Mi →Mf* where initial model *Mi (Mi= Msys)* may not be able to identify the value of all the parameters required by the target analytical model (*Mf*) of the transformation (see section I). One approach to solving this problem is define the t transformation based on a set of parameters (*C*), which they cannot be solved directly by the processing therefore, so it is necessary use additional procedures to determine its value. This fact can be expressed by changing the general expression of the transformation t to the following expression:

$$t: Mi(C) \rightarrow Mf \qquad (4)$$

where *C* is the set of parameters that need to be resolved or determined to make the transformation *t: Mi→ Mf* and to evaluate quantitatively the final model (*Mf*). In this case is where we speak of a parameterized transformation between two models.

Our study focuses on a parameterized transformation respect to service time(S) of each component of the system architecture. This parameter has the main feature that is highly dependent on the underlying platform. The parameterized choice of this factor allows a general and reusable approach, but the final model cannot be resolved until the value of certain parameters has been obtained. These parameters should be obtained by additional procedures outside the own transformation t. Another different approach would be to perform

Figure 7. General scheme of transformation based on a correspondence relationship

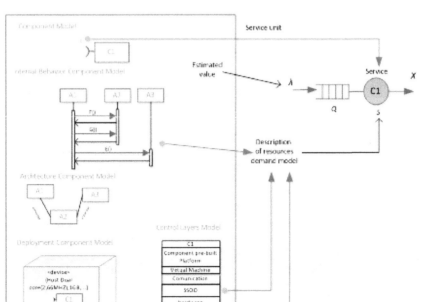

several valid or acceptable assumptions on these parameters directly at the time of performing the transformation.

PARAMETERIZED TRANSFORMATION SCHEMA FOR A NON-FUNCTIONAL PROPERTIES MODEL

This section will define a parameterized transformation between a software system model based on component architecture and an analytical model based on queue theory to evaluate service performance features of software system. In short this transformation is considered a transformation to a non-functional properties model of the system.

Given the transformation named t (see expression 3) and defined as *t: Mi (C) →Mf* where *Mi* is system model based on a component architecture and Mf as a queuing network model. Table 6 shows a mapping relation of *Mi* on *Mf*, where *Mi* will play the role of system model (*Msys*) and Mf will play the role of performance model (*Mperf*).

The information of the transformation has four sections: *Mapping rules, valid assumptions, information of approach and Transformation parameters:*

- **Mapping Rules:** The relation between an element and an element.
- **Assumptions:** The set of valid assumptions that can be made in the context of the transformation.
- **Information of Approach:** The information relevant to know how to interpret metrics taken values
- **Transformation Parameters:** Set of parameters C where $C = \{p \mid p \in Mf\}$ that their value must be obtained by external methods called generally here *Pext (p)*.

Table 6. Transformation schema rules

Correspondence Relationship between Msys And Mperf Models
Conventions
comp-i= ith component of the system architecture A(comp-i)→ component distribution time arrival B(comp-i)→ component distribution time service Fr(comp-i,scenario-k)→ Relative frequency of component on usage k-th scenario q(node-i,node-j)→ routing probability from node-i to node-j for a request
Correspondence rules
t(comp-i)=queue(m,service-i)
t((connector (comp-i,comp-j)=true))=(q(node-i, node-j)=1)
t(Interfacecomp-i)= description(service-i)
t(connector(deploynode-i, deploynode -j)=true)=(q(node-i,node-j)=1)
t(layer(comp-i))= description(s(service-i))
t(deployment (comp-i))= queue(m,service-i)
t(A(comp-i))= A(service-i)
t(B(comp-i))=B(service-i)
t(Fr(comp-i,scenario-k))= D(service-i) with Di=Fr*Si

Assumptions About Model Parameters	
Parameter	*Assumption*
Service time	**S1:** service time distribution is exponential **S2:** service time distribution is independent for each request
Arrival process	**S3:** time between arrivals is exponential.
Population	**S4:** Infinite population

Information Of Approach	
Parameter	*descripción*
Service time	To interpret the service time, consider only the layer service model and intrinsic factors to each component

Transformation Parameters	
Parameter	*description*
Service time	Service time of each key component in the architecture of the system

Table 6 shows a general schema with the previously mentioned sections.

Figure 8. Simple platform independent model (PIM)

The set of metrics that can be obtained as a result of evaluation of *Mperf* model are: system response time (R), component response time (Ri), component utilization (Ui) and throughput (Th) . In this paper we focus on the response time only.

Parameterized Transformation Schema for a Simple Component-Based System

To test the overall parameterized transformation approach described in this paper, we will consider the design of a cartographic maps system and on which we want to analyze their performance properties. The maps size varies between 1 and 4 Mbytes.

The maps access service is designed architecturally by two key software components whose execution occurs in two different execution environments (containers) called E1 and E2. E1 is based on a platform Java EE v5 and E2 is based on commercial data base system (MySQl) running in Windows Vista. Both environments running on a hardware platform Intel Core Duo (2.65 GHz). The maps size range is 1 to 4 Mbytes.

We consider that the only relevant functional property is that service time of the system cannot be greater than 3000 ms in the worst cases (max size) and further we assume that the only relevant resource in each component implementing is the CPU resource.

A simple Platform Independent Model (PIM) regarding with this system is shown in Figure 8 by a component diagram in UML notation.

This model consist of two software components (hereinafter abbreviated C1 and C2), each providing a single service ideally described by their respective interfaces (called IC1 and IC2). The previous design reveals that component named C1 requires component interface service of C2 and in turn provides an external interface called Ic1

To test the ideas under this paper we will solve the parameterized transformation *t: Msys (C) →Mperf*. Where *t* is a "partial" relation between the model (*Msys)* and a performance analysis model based on queueing theory named (*Mperf)* which it has been explained briefly here. *Msys* will consist of as seen in this article is essentially composed of a component architecture model, behavior model, a model of deployment of these components and a service layers model. In addition, the approach will use MDA approach to transform the Platform Independent Model (PIM) to a simple Platform Specific Model (PSM).

Figure 9 shows the general context of MDE transformations applied to this example.

In this scheme, C represents the set of parameters that cannot be solved in the transformation *t: Msys→Mperf* properly and should be resolved with additional procedures. For our example case, these parameters are the service times for each component of the architecture proposed. The set *C* would be defined by the following expression $C = \{S (E1 (c1)), S (E2 (c2))\}$, where $S (E1 (c1))$ and $S (E1 (c2))$ represents he service time estimated for the components (c1 and c2) in the runtime environment E1 and E2 respectively. Additionally Pext1 $(S (E1 (c1))$ is defined how the procedure used to estimate the parameter *S (E1 (c1))* and *Pext2 (S (E2 (c2)))* as the procedure used to estimate the parameter *S (E1 (c1))*.

Within the MDE context and following MDA guidelines we can define a transformation from a

Figure 9. a) General schema of transformation, b) set of parameters and foreign procedures

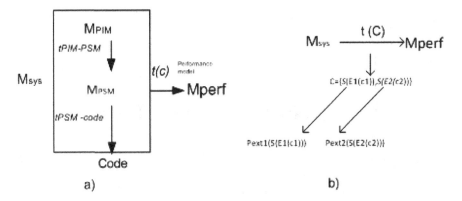

a)   b)

PIM to a PSM. A simple PSM taking account E1 and E2 environments will consist of infrastructure data persistence, connectors and a platform component as a Stateful EJB of Java EE. Figure 10a shows this simple transformation $v: PIM \rightarrow PSM$.

The Figure 10b shows the simplified interaction diagram of the system for the usage scenario called for example "maps access," at bottom of this diagram is shown the expected demand model based on the relative usage frequency of each component in this usage scenario.

Figure 11 shows the matching rule used in t transformation, which assigns a logical resource of *Msys* model to a service node of *Mperf* model with the assumption that the arrival process and service are both Markov processes. The transformation is a correspondence between the software component and a single service node that is accessed with a distribution with parameter (μ) and random arrivals with parameter (λ).

The service layers architectural model of *Msys*, shows the different layers of control considered

Figure 10. a) PIM-PSM transformation, b) internal demand model of system based on a use case

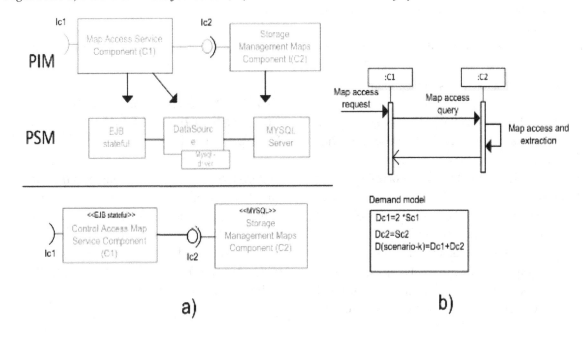

a)   b)

Figure 11. Correspondence between software components and service node M/M/1

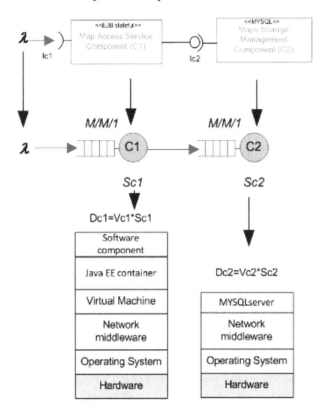

in the interpretation of service time due to intrinsic properties of C1 and C2 .These layers will model service demand *D* of each component. Figure 11 shows the system architecture in terms of the logical control layers most significant in our case.DC1 and DC2 will define demand associated with the component 1 and 2 respectively and SC1 and SC2 is the estimated service time for each service.

The deployment model of the system consists of two separate machines. This model specifies the mapping of logical components to physical resources. Figure 12 shows the correspondence of services nodes and physical resources. The connectors are defining the network topology of service nodes in the queues model:

The final step is to determine how to resolve the parameters values of transformation *t*. This issue is taken in our case using two different procedures. The first is to estimate the service

time of C1 in the environment named E1.This procedure is called *Pext1 (S (E1 (C1)))* and be used primarily Little´s laws [see expression 1]. The second procedure *Pext2 (S (E2 (C2)))* is to estimate the service time of component called C2 running in the environment E2, in this case it will be used a regression technique this procedure.

Pext1: Procedure for Service Time Estimation of Platform Component (C1)

The service time parameter value in this case is based on only intrinsic characteristics of component. Thus, a Stateful EJB service time is obtained using benchmark and measurement on a reference implementation. The procedure consist of a test with a duration T (seconds) and N samples. Table 7 shows the results obtained.

Figure 12. Deployment architecture model

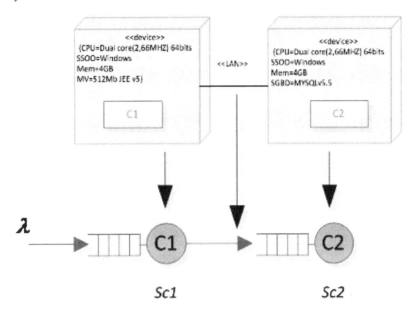

It should be noted that this procedure may be prior to any design and reused in different designs and systems based on the same kind of components.

Pext2: Procedure for Service Time Estimation of Database Server (C2)

The service time parameter of component C2 will be fixed by linear regression techniques taking into account following factors: *size of map (tmap) and response time (R).*Here *tmap* is an independent variable with values in the range [1-4 MBytes] representing map size and *R* is a the dependent variable and represents response time of component during a map access.

Figure 13 shows the results obtained and the lineal relation obtained by the experiment.

The above procedure allows gets the function to get the value of the component response time depending on the map size. This function will be defined as:

$$fa(tmap) = 1,4095tmap + 2477,8. \qquad (5)$$

where *tmap* is the size (MBytes) of the map.

Finally, we will solve the performance model (*Mperf*) by the procedures above to obtain response time index. Table 8 shows all these data schematized.

The general expression of the response time for the previous queuing model is determined by:

$$R = \sum_{i=1}^{k} R_i \qquad (6)$$

Table 7. Results benchmark

N=100 samples T=100 seconds		
Hardware layer	Intel i5 dual core (2,54 GHz)	
Component platform layer	Java EE v5	
Virtual machine layer	JVM(512 KB)	
Measurement results		
Mean utilization (p) %100	Mean Service time(S)	Est.Dev
6,0	144,76 ms	96,7

Figure 13. Response time graph based on the size of the map

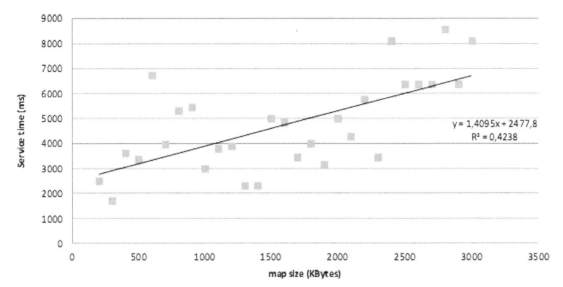

where *R* is the response time of the system and Ri is the response time indexed for each component. For our example and considering the guidelines shown in this paper we will use the following expression for R:

$$R=(S(C1)+fa(tmap))+k \qquad (7)$$

where *S(C1)* represents the service time estimation of C1 (see Table 7), *fa(tmap)* function could be evaluated for different map sizes and *k* is a constant delay associated to transmission over LAN network. As discussed in this publication the response time (R) of the system is determined by the expression $Ri = Di\ (i=1..2)$. Thus the demand on this basis Figure 14 shows the response time and size of the map.

It should be noted that these results are based on average values. But as a first approximation it could be a valuable reference to performance evaluation. For example, we can see that according the graph obtained, in the worst case for accessing over a map of 1.4 Mbytes, the response time is less than 3000 (ms), so the present design of this system will complies with the requirements mentioned.

FUTURE RESEARCH DIRECTIONS

Advanced software development paradigms such as Model Driven Engineering (MDE) represents forward steps on the path to the industrialization of software development and a trend consolidated in software engineering. Models capture the essence of the software at hand while transformations

Table 8. General assumptions about model

Element	Properties
Waiting queue-1 Waiting queue -2	m1=m2=infinite
Service -1 Service-2	N1=N2=1 Mean Sevice time Sc1=144,6 ms Sc2= S(map size) Queue policy Service -1 policy = FIFO Service-2 policy =FIFO
Arrival time diistribution(A)	A(service-1)=M A(service-2)=M
Service time distribution(B)	A(service-1)=M A(service-2)=M
Population	N= infinite Classes =1 Open network

Figure 14. Mperf model evaluation over different map sizes

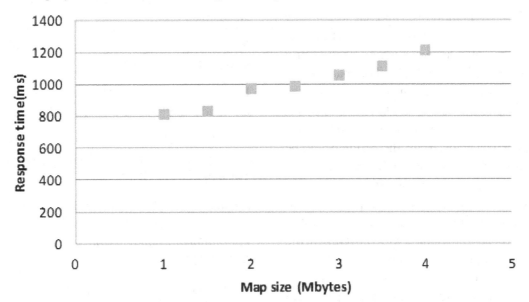

are the means for reuse, as they encode reusable mappings of models to lower abstraction levels. In MDE, coding is substituted by modeling and transforming. Hence, the software development process could become a pipeline of model transformations that eventually leads to an end product. In these scenario tools support is critical for ensuring consistent of system models.

The major challenges that researchers face when attempting to realize the MDE vision are summarized into the following categories (France & Rumpe, 2007):

- **Modeling Language Challenges:** These challenges arise from concerns associated with providing support for creating and using problem level abstractions in modeling languages, and for rigorously analyzing models.
- **Separation of Concerns Challenges:** These challenges arise from problems associated with modeling systems using multiple, overlapping viewpoints that utilize possibly heterogeneous languages.
- **Model Manipulation and Management Challenges:** These challenges arise from

problems associated with (1) defining, analyzing, and using model transformations, (2) maintaining traceability links among model elements to support model evolution and roundtrip engineering, (3) maintaining consistency among viewpoints, (4) tracking versions, and (5) using models during runtime.

CONCLUSION AND FUTURE WORK

This publication has developed the concept of parameterized transformation to reach an easy integration between analytical and design model. The transformation scheme we have presented will depend on many factors such as software system architectural approach, the conventions used, the type of system under study, the analysis and design technique used and the performance model chosen to describe the system. Considering these facts we have tried to provide a comprehensive, reusable and open method to include certain kind of analysis in the software development process. The profile of systems considered are component-based systems and they can provide a reference implementation on which to base load tests like Java EE or .NET platforms. This issue we consider is critical due to the lack of common references.

The presented approach allows us to offer a framework for integrated a kind of analysis that can be used in early design stages, because the approach presented here could be used for semi-automatic generation prototype systems for further analytical study. Thus, the design would be guided by quantitative analysis by eliminating the subjectivity of many current evaluation criteria.

We are now continuing the experiments for augmenting the empirical evidence of the viability of our approach, providing a wider coverage of the possible alternatives of component based distributed systems. Our main objective in the future is integrating more and more number of analytical models to constitute an integrated framework to reach diverse analysis capabilities. These capabilities will be applied to a system design model in order to reach an early evaluation approach.

REFERENCES

Atkinson, T. K. (2003, September). Model-driven development: A metamodeling foundation. *IEEE Software*. doi:10.1109/MS.2003.1231149.

Brooks, F. (1986). No silver bullet — Essence and accident in software engineering. In *Proceedings of the IFIP Tenth World Computing Conference*, (pp. 1069–1076). IFIP.

Czarnecki, U. (2000). *Generative programming*. Reading, MA: Addison-Wesley.

France, R., & Rumpe, B. (2007). Model-driven development of complex software: A research roadmap. In *Proceedings of the Workshop on the Future of Software Engineering (FOSE 2007), at the 29th International Conference on Software Engineering (ICSE 2007)*, (pp. 37–54). FOSE.

Fundation, E. (2001). *Eclipse*. Retrieved from http://www.eclipse.org.

Jackson, J. (1957). *Networks of waiting lines*. Academic Press.

Kent, S. (2002). *Model driven enegnieering*. Berlin: IFM Springe-Verlag.

Lazowska, J. Z. (1984). *Quantitative system performance*. Englewood Cliffs, NJ: Prentice Hall.

Liu, A. F. (2005). *Design-level performance prediction of component-based applications*. Washington, DC: IEEE.

Mehta, N. R. (2000). Towards a taxonomy of software connectors. In *Proceedings of the 22nd International Conference on Software Engineering*, (pp. 178-187). IEEE.

ModelWare. (2007). *ModelWare information society technologies (IST) sixth framework programme glossary*. Retrieved 2008-01-06, from http://www.modelware-ist.org/index.php?option=com_rd_

OMG. (2006). *Modeldriven architecture- Specifications*. Retrieved June 2, 2012, from http://www.omg.org/mda/mda_files/Model-Driven_Architecture.pdf

OMG. (2007). *Unified modeling language: Infrastructure version 2.1.1*. Retrieved January 13, 2013, from http://www.omg.org/cgi-bin/doc?formal/07-02-06

Oracle. (2009). *Oracle*. Retrieved January 20, 2013, from http://www.oracle.com

Stallings, W. (2008). *Operating systems internals & design*. Upper Saddle River, NJ: Prentice Hall.

Zhu, Y. L. (2007). Revelor: Model driven capacity planning tool suite. In *Proceedings of the 29th International Conference on Software Engineering*, (pp. 797–800). IEEE.

Zschaler, S. (2008). Formal specification of non-functional properties of component-based software systems -- A semantic framework and some applications thereof. In *Software and Systems Modeling* (pp. 161–201). Academic Press.

ADDITIONAL READING

Architecture Board MDA Drafting Team. (2001). *Model driven architecture: A technical perspective (Document Number ab/2001-01-01)*. Author.

Barry, T., & Barry, B. M. (2003). Model-driven development: The case for domain-oriented programming. In *Proceedings of the 18th Annual ACM SIGPLAN Conference on Object-Oriented Programming, Systems, Languages and Applications, OOPSLA,* (pp. 2-7). New York: ACM.

Becker, S., Koziolek, H., & Reussner, R. (2007). Model-based performance prediction with the palladio component model. In *Proceedings of the 6ᵗʰ International Workshop on Software and Performance (WOSP'07),* (pp. 54–65). New York: ACM.

Chen, S., Liu, L., Gorton, I., & Liu, A. (2005). Performance prediction of component-based applications. *Journal of Systems and Software, 74*(1), 35–43. doi:10.1016/j.jss.2003.05.005.

Ciancone, A., Filieri, A., Drago, M. L., Mirandola, R., & Grassi, V. (2011). KlaperSuite: An integrated model-driven environment for reliability and performance analysis of component-based systems. In Bishop, J., & Vallecillo, A. (Eds.), *TOOLS 2011 (LNCS) (Vol. 6705,* pp. 99–114). Berlin: Springer. doi:10.1007/978-3-642-21952-8_9.

Gaily, M. (2007). *What is MDD/MDA and where it leads the software development in the future?* Zurich, Switzerland: Department of Informatics at the University of Zurich.

Gorton, I., Liu, A., & Brebner, P. (2003). Rigorous evaluation of COTS middleware technology. *Computer, 36*(3), 50–55. doi:10.1109/MC.2003.1185217.

Jackson, E. K., Kang, E., Dahlweid, M., Seifert, D., & Santen, T. (2010). Components, platforms and possibilities: Towards generic automation for mda. In *Proceedings of EMSOFT.* ACM.

Jurack, S., & Taentzer, G. (2009). Towards composite model transformations using distributed graph transformation concepts. In Schurr, A., & Selic, B. (Eds.), *MoDELS (LNCS) (Vol. 5795,* pp. 226–240). Berlin: Springer. doi:10.1007/978-3-642-04425-0_17.

Koziolek, H. (2010). *Performance evaluation of component-based software systems: A survey.* Ladenburg, Germany: ABB Corporate Research. doi:10.1016/j.peva.2009.07.007.

Lavenberg, S. S. (1983). *Computer performance modeling handbook.* New York: Academic Press.

Lawley, M., & Steel, J. (2006). Practical declarative model transformation with tefkat. In Bruel, J.-M. (Ed.), *MoDELS 2005 (LNCS) (Vol. 3844,* pp. 139–150). Berlin: Springer.

Lazowska, E. D., Zahorjan, J., Graham, G. S., & Sevcik, K. C. (1984). *Quantitative system performance: Computer system analysis using queueing network models.* Englewood Cliffs, NJ: Prentice Hall.

Mens, T., Czarnecki, K., & Gorp, P. V. (2005). A texonomy of model transformations. In *Proceedings of International Workshop on Graph and Model Transformation (GraMoT).* GraMoT.

Meservy, T. O., & Fenstermacher, K. D. (2005). Transforming software development: An MDA road map. *IEEE Computer, 38*(9), 52–58. doi:10.1109/MC.2005.316.

Object Management Group. (2008). *MOF query / views / transformations specification 1.0.* Retrieved from http://www.omg.org/docs/formal/08-04-03.pdf

Raistrick, C., & Carter, C. (2004). *Model driven architecture with executable UML.* Cambridge, UK: Cambridge University Press.

Richters, M., & Gogolla, M. (2000). Validating UML models and OCL constraints. In A. Evans & S. Kent (Eds.), *The Third International Conference on the Unified Modeling Language* (LNCS). York, UK: Springer.

Rutle, A., Rossini, A., Lamo, Y., & Wolter, U. (2012). A formal approach to the specification and transformation of constraints in MDE. *Journal of Logistical Algebraic Programmming, 81*(4), 422–457. doi:10.1016/j.jlap.2012.03.006.

Saxena, T., & Karsai, G. (2010). MDE-based approach for generalizing design space exploration. In Petriu, D. C., Rouquette, N., & Haugen, Ø. (Eds.), *MODELS 2010 (LNCS)* (*Vol. 6394,* pp. 46–60). Berlin: Springer. doi:10.1007/978-3-642-16145-2_4.

Sendall, S., Hauser, R., Koehler, J., Kuster, J., & Wahler, M. (2004). Understanding model transformation by classification and formalization. In *Proceedings of Workshop on Software Transformation Systems*. Vancouver, Canada: IEEE.

Smith, C. U., & Williams, L. G. (2002). *Performance solutions: A practical guide to creating responsive, scalable software*. Reading, MA: Addison Wesley.

Szyperski, C. (1998). *Component software: Beyond object-oriented programming*. Reading, MA: Addison Wesley.

Whittle, J., Clark, T., & Kuhne, T. (Eds.). (2011). Model driven engineering languages and systems. In *Proceedings of the 14th International Conference, MODELS 2011,* (LNCS), (vol. 6981). Wellington, New Zealand: Springer.

Woodside, M., Vetland, V., Courtois, M., & Bayarov, S. (2001). Resource function capture for performance aspects of software components and subsystems. In *Performance Engineering, State of the Art and Current Trends* (pp. 239–256). London, UK: SpringerVerlag.

Xiong, Y., Liu, D., Hu, Z., Zhao, H., Takeichi, M., & Mei, H. (2007). Towards automatic model synchronization from model transformations. In *Proceedings of the Twenty-Second IEEE/ACM International Conference on Automated Software Engineering, ASE '07,* (pp. 164-173). New York: ACM. doi:10.1145/1321631.1321657

KEY TERMS AND DEFINITIONS

Analytical Model: Analytical models are mathematical models that have a closed form solution.

Architecture Model: Set of main structures needed to reason about the software system, which comprise main software elements, the relations between them, and the properties of both elements and relations.

Model-Driven Engineering (MDE): Software engineering paradigm where models play a key role in all engineering activities. Usually, in MDE, the implementation is (semi)automatically generated from the models.

Model: Formal representation of entities and relationships (abstraction) with some correspondence (isomorphism) to the real world for some purpose (pragmatism).

Parameterized Transformation: Parameterized transformation consists on any model transformation based on parameters.

Performance Model: A model created to define the significant aspects of the way in which a system operates in terms of resources consumed, contention for resources, and delays introduced by processing or physical limitations (such as speed, bandwidth of communications, access latency, etc.).

Software Component: Unit of composition with contractually specified interfaces and explicit context dependencies only. A software component can be deployed independently and is subject to composition by third parties.

Chapter 13
Object Model Development/ Engineering:
Proven New Approach to Enterprise Software Building

David Chassels
Procession PLC, UK

ABSTRACT

Solving the business "software problem" of inflexibility with poor user experience was at the heart of the original R&D that started over 20 years ago. Any solution had to acknowledge how people work and most importantly remove the interpretation gap between users and "IT," thus removing the need for programmers in the build process. It was quickly recognised that in reality business logic does not really change, and it was therefore important to separate from the ever-changing technologies, such as operating systems and browsers, to "deliver." After 20+ years of research and development and working with early adopters, this approach has achieved the objectives set. As is explained, generic task objects and the important links were built and displayed in a Graphical Model where the building of custom applications takes place with no change to the core code and no code generation or compiling. This approach has opened a new perspective on capability in software yet does it by adopting simplicity in the core design. Indeed the driving philosophy for all development was to produce enabling technology that was business friendly, simple to use, and generic in application.

INTRODUCTION AND BACKGROUND

The thinking behind this development has been driven by business thinking with a "clean sheet." All the developer coders were selected direct from universities and colleges none of whom where "contaminated" by existing commercial development practices. This "journey" started out in the late 1980s at a time many were expecting the removal of coders as articulated by Naomi Bloom (2013):

DOI: 10.4018/978-1-4666-4494-6.ch013

Writing less code to achieve great business applications was my focus in that 1984 article, and it remains so today. Being able to do this is critical if we're going to realize the full potential of information technology

....how those models can become applications without any code being written or even generated.

The founder was an international banker who shared these views from the perspective of a user of information. He was a business executive but understood enough of "IT" to impose his business ideas into finding a solution. The core thinking was based upon once only entry of data to flow in a horizontal manner (as opposed to vertical silos) addressing all business requirements yet remove the need to use coders to build working applications.

It was established that people at work need to be supported by "generic" task types and this simple reality opened the door on removing the need to program code applications. The early research was carried out using the dos operating system and the Paradox database to create a complete data centric environment utilising the latent power of a relational database. Against then current thinking, it was proven that the identified generic task types and required links could be stored as data in the database ready to be configured.

By 1995, the concept was proven, and it was decided to re-write using Windows 95 and the Oracle database was chosen. Once this new version was created it was decided to create a Graphical Process Designer as an interface over the code that would allow easy click drag and open the task types to allow configuration by business analysts to build the required application..

What follows explains the detail core thinking and how build takes place.

KEY ISSUES

Before a sustainable approach was adopted it was important to understand some basic fundamentals on how organisations actually work.

1. First people are the source creators of all information and second there is the need to recognise that people work in relatively small teams to achieve individual and collective outcomes that make any business or indeed an efficient government. Worthy of note is that book keeping is just that it records history and generally keeps the accountant happy; it does not "run" the business.

2. The other key issue that was established early on was that business logic has never changed since commerce started; indeed business is actually quite "simple" if you focus on supporting people at work. This includes "rules" simple or complex (e.g. means testing) and just reflects business logic. The communication technologies to deliver are both complex and challenging but do not change the fundamentals of business; it is about people, internal and external to the organisation and IT should there to support and contribute to efficiency.

3. When you look at how people work irrespective of the required function there are relatively few work task types, human and system, including the user interface that address all business driven issues? So why repeatedly recode for every function in a business when such standardisation allows unlimited flexibility to build on any business requirement

With these basic fundamentals and recognising the latent power in relational database technology to "connect" the different parameters to deliver a working application work started by identifying the "generic" task types as follows.

- The Normal task "halts" the process for an off-line activity. It is a very useful development aid in a process, but should be used sparingly in a "live" environment.
- The Form task is the task that the user will be mostly concerned about. This is where data is entered into and extracted out of the database. It can be a "simple" display form or a complex interactive form. This superseded by the Web report/form task below but used for quick first cut/ prototype of the application.
- The Program task allows you to "call" applications such as Word, Excel and specific program used in a process.
- The Pending task places the process into the "Pending" tray of the user concerned. This is a very useful task that is used alongside deadlines and delays in a process.
- The Report task enables a report to be generated via any report application based on a previously defined template.
- The Web report/form task is used to hold the path for Java Server Pages/forms to run across the Web. Utilises Ajax to ensure once only entry of information with intelligent grids.
- The Calculation task can contain calculations involving almost anything including dates, numbers and strings. As well as Procession specific calculations, SQL commands can be placed here to manipulate the database directly.
- The Sub-Process task allows the process to move to another, or 'sub', process.
- The Event task bundles the same task together in multiple runs and waits for another process to action it. .
- The VB Script task allows the use of Visual Basic code. This task can have many differ-

ent functions according the requirement of the developer but only used in client server environments.

- The Finish task tells Procession that the process has ended. As far as the user is concerned, the "run" of the process will disappear from the trays. At this point, that particular run will be placed into the "Process History" tray of the manager.
- The Import/Export task handles the movement of "bulk" loaded data into and out of the database. It can be both completed by the user and/or the system according the specifications.
- The Server Side Message Queue task handles communication between Procession (and therefore the database) and many other external systems, such as legacy systems. It's more popular use is in the sending of e-mails from inside Procession. This is a very versatile and important task.

These tasks were created as "objects" and expressed as data inside a relational database. To make all work in the required order linking capability was incorporated. It was recognised such linking could be used to make decisions within the flow of work. It was designed to allow full audit trail of work to be tracked and all aspects could be reused as required. This visual interpretation gives a good idea of how the core design architecture was built inside the database (see Figure 1).

This core capability was built but the challenge remained how to make it easy for business to use. A user friendly interface was built to set up the database (see Figure 2).

It was decided that a Graphical Process Designer (GPD) should be built over the core code and icons were created that mirrored established ways of mapping out work activity. It was built to allow the GPD to "declare" through to the preformatted database the custom requirements. This technique removed code generation and compiling

Figure 1. An overview of the architecture

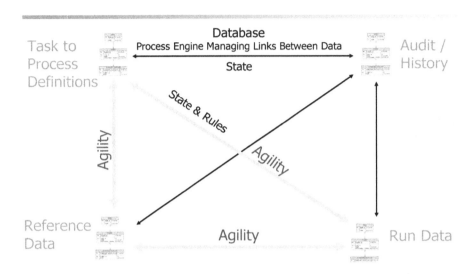

Figure 2. The front end to set up the database fields

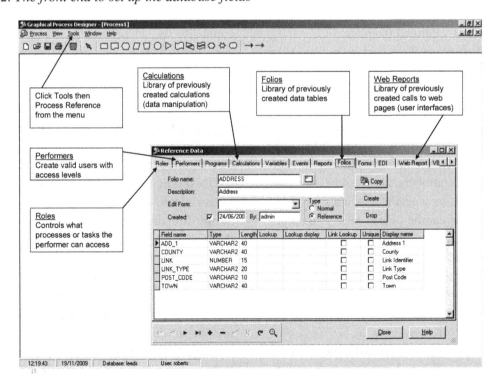

allowing rapid change in build and in the future. A version control capability was added to include flexibility as to where in a process change is adopted (see Figure 3).

Each task opens up a form to allow the parameters to be inserted.

TASK LINKS

Although these links are very powerful, they are quite easy to learn and understand. They contain much of the logic that drives a process application. They are the workflow of the application allowing rules to be built and great flexibility supporting asynchronous work.

A link can be thought of, in its simplest form, as a bridge between two or more tasks. There always has to be at least one *Source* task and at least one *Target* task and a link spans the two.

There are only two types of links to remember, the True link and the False link (see Figure 4).

A link, whether True or not, is made up of a *Source* part and a *Target* part. The *Source* part includes the *Source Node* and the *Target* part includes the *Target Node* and the Red/Blue line.

The two types of nodes or link points are where other links are joined or diverged. For example if you have two source tasks, there would need to be two links (one from each task) coming together at the *source node*. The *target node* relates to the *target task*. If there were to be two target tasks, there would be two links emanating from the *target node,* as shown in Figure 5.

The True/False links are the main type of link, but within these are various other types, as shown below. Each type is represented by a coloured line.

However, if all you want to do is link tasks together in a simple A-B scenario with very simple branches, e.g. a 'Yes/No' decision, then you would use the default 'Normal' link (see Figure 6).

Now we can introduce slightly more complex parts to these links to give processes more func-

Figure 3. The build graphical tool with traditional icons for types of task

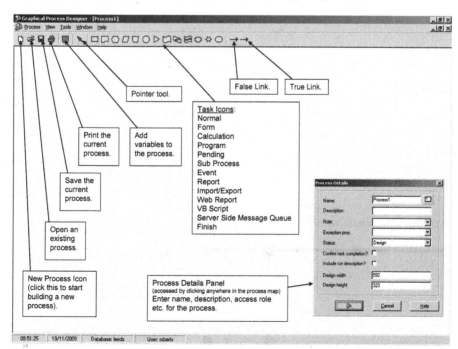

Figure 4. The links creating the workflow collaboration

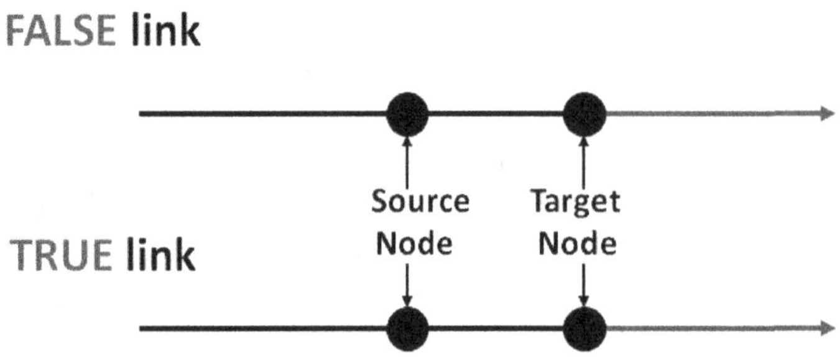

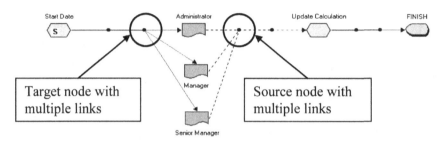

Figure 5. How one to many and many to one is built

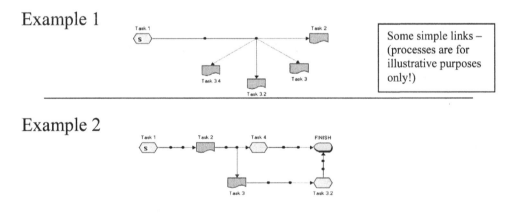

Figure 6. Simple or complex links

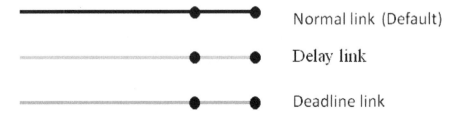

Figure 7. Delays and deadlines capability

tionality. These attributes are deadlines and delays. They are very useful in processes and are used a great deal. Further discussion is needed and will be covered in more detail later on. It is sufficient for now to explain the different types (see Figure 7).

As you can see, the links have different colours, representing the different link types. The default type is always the "normal" type and would be used most of the time.

DELAYS

If a delay value is inserted into the delay field on the *target node* of a link, the black line changes to a green colour. This denotes in the GPD that a delay has been applied.

Delays are handled in "days" or divisions thereof. Thus putting a delay in a process so that the process "halts" for six months would require that '180' be written into the delay field on the dialogue box. In the example below, the process would delay for one day (see Figure 8).

To access the dialogue box, move the mouse over the *target node* and right click.

DEADLINES

If a deadline is used in a process, the black line (*source* side) turns purple, denoting in the GPD that a deadline is used. In order for the deadline to work, both the delay and the condition fields must be filled. The condition tests the delay and this brings about the "deadline". For example, a date can be set and placed into a folio field. This, in turn, is used in a calculation, which is placed into the delay field on the "Link Details" dialogue box. A condition is then applied in the "Condition" field, whereby the delay is tested. It the answer is false, nothing happens. However, if the answer is true then the next *target task* is triggered. This gives the desired result of waiting for a date deadline to pass before anything else happens.

This is an automatic trigger. However, a process can be moved onto the next *target task* before a deadline has passed, by manual means as required (see Figure 9).

Figure 8. Applying values to delays

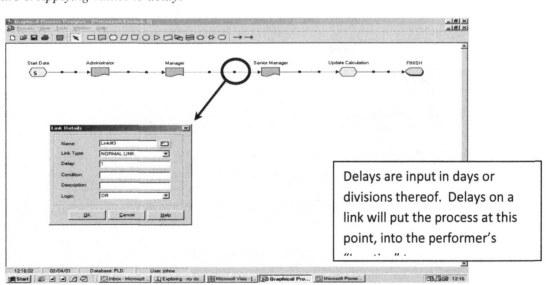

Figure 9. Applying conditions for deadlines

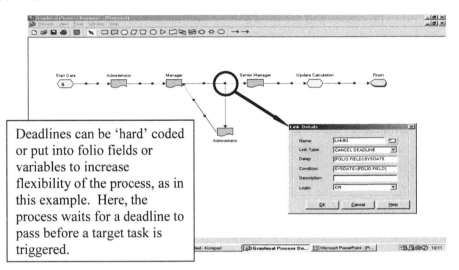

Start Date
Administrator
Manager
Senior Manager
Update Calculation
Finish
Administrator
Deadlines can be 'hard' coded or put into folio fields or variables to increase flexibility of the process, as in this example. Here, the process waits for a deadline to pass before a target task is triggered.
Link Details
Name:
Link#3
Link Type:
CANCEL DEADLINE
Delay:
{FOLIO.FIELD}-SYSDATE
Condition:
SYSDATE>{FOLIO.FIELD}
Description:
Logic:
OR
OK
Cancel
Help

BUILD

See Figure 10 for actual build using swim lane.

DEPLOYMENT

- By activating and saving a process
  - The Process Engine breaks down the designed process into its constituent elements
  - Through a declarative technique these are saved into Oracle tables.
- At run time
  - The engine interrogates these tables to decide

Figure 10. The actual build taking place using swim lanes

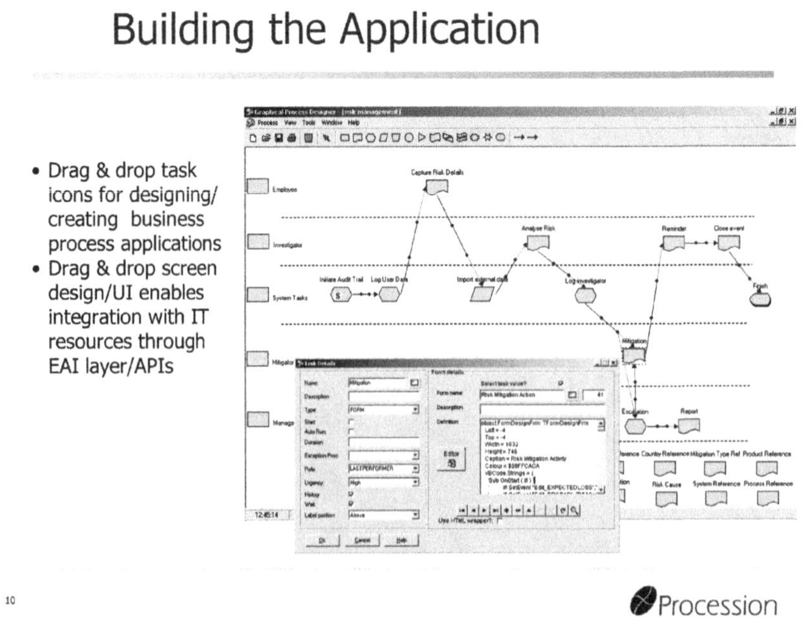

Building the Application
• Drag & drop task icons for designing/ creating business process applications
• Drag & drop screen design/UI enables integration with IT resources through EAI layer/APIs
Capture Risk Details
Analyse Risk
Reminder
Close event
Log investigator
Mitigation
Exception
Report
10
Procession

Figure 11. The "clicks" to deploy application

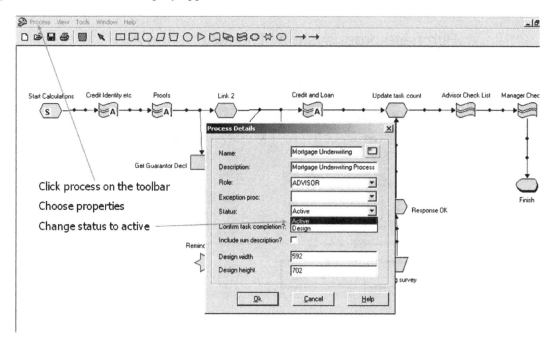

○ Who does what, when and how in the application.

• No code is compiled to enable this to happen (see Figure 11).

As can be seen it is possible to customise the icons to suit. This GPD is the deployed application and as such is the new Object Model "code" (see Figure 12).

THE USER INTERFACE

The most important task type is the user interface/form. Whilst the core Object Model Engineering remains solid and robust there was a need to recognise forms need to be simple yet handle the user requirements at work This was addressed by having a library of template forms stored in the database which are dynamically populated with data required by that user recognising the specific instance of the task being undertaken and data required to be used or created.

The arrival of the Web brought new challenges and to improve functionality and performance a comprehensive a library of capabilities was built to continue the ease of build of Web forms by business analyst skills. This includes;

• Information across an organisation can be used effectively by combining data from separate sources into a single form—also known as 'mash-ups'—supporting a Master Data Management (MDM) approach,

• The use of AJAX and the built-in data cache ensures users do not need to wait for information to load, and only parts of the form that need refreshing are reloaded,

• The data cache makes data manipulation, such as sorting and filtering.

This is continuing to evolve to keep pace with new delivery technologies such as cloud and mobile.

Figure 12. The visual view of a completed application using custom icons

Purchase order system "The MAP is the APP"

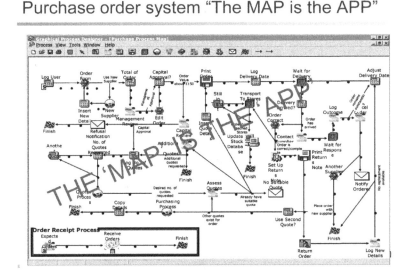

OTHER ISSUES

One of the big challenges has been expressing what had been built in context of the industry which has seen business software evolved in a rather disjointed manner. The core thinking with a data centric architecture has by default created a unified capability including the following:

- Process engine *to ensure all works to plan*
- Rules engine *reflecting real world of work and compliance*
- Calculation engine *automating system work*
- State/instance engine *real time feed back*
- Workflow *everything connected in right order and supporting asynchronous work*
- Integrated forms *dynamically and custom created for who, when & where as required*
- Audit trail, events, escalations *supporting control with empowerment*
- Roles and performers *people and machines ready to work*

- Management hierarchy *who sees what re-allocation of work as required*
- Version control *easy install of required changes with no disruption*

The result is producing new functionality that includes the ability to incorporate intelligence into applications recognising user decisions to dynamically change future actions. It is just the start of a new journey for Enterprise Software.

THE REAL CHALLENGE!

It is legitimate to describe this Object Model Development / Engineering as the start of commoditisation of business software with one tool environment handling all key requirements that are currently in multiple tool sets that sit with IT and beyond comprehension of business. As such the gate keepers are still "IT" who focus on technical requirements. This new way is driven by business; indeed all early adopters have been business lead. It has the potential to change Enterprise Software that ...*will take the applications that support knowledge work well outside of the box, connecting us to way we really work* as described by Tom Koulopoulos (2012)

The objectives set out in the early days have been achieved and maybe more as it has been a learning experience for the developers and early adopters as new capability is recognised. However the simplicity to allow business people to build applications and the speed of build is a big challenge set against the complexity of business software as it has evolved over the decades. Therein is the real challenge as many vested interests will resist.

But cutting out the interpretation gap between users and traditional "IT" will have significant benefits for all.

REFERENCES

Bloom, N. (2013). *The future of HRM software: Agile, models-driven, definitional.* Retrieved from http://infullbloom.us/?p=3222

Koulopoulos, T. (2012). *ACMLive.* Retrieved from http://www.acmlive.tv/agenda.html

Chapter 14
Consistency Checking of Specification in UML

P. G. Sapna
Coimbatore Institute of Technology, India

Hrushikesha Mohanty
University of Hyderabad, India

Arunkumar Balakrishnan
Coimbatore Institute of Technology, India

ABSTRACT

The increasing use of software is giving rise to the development of highly complex software systems. Further, software systems are required to be of high quality as a defect can have catastrophic effect on business as well as human life. Testing is defined as the process of executing a program with the intention of finding errors. Software testing is an expensive process of the software development life cycle consuming nearly 50% of development cost. Software testing aims not only to guarantee consistency in software specification but also to validate its implementation meeting user requirements. On the whole, it is observed that in general, errors in software systems set in at the early stages of the software development cycle (i.e. while gathering user requirements and deciding on specification of intended software). Even though formal specification in B and Z assures a provable system, its use has become less popular due to mathematical rigor. The Unified Modeling Language (UML), a semi-formal language with graphical notations consisting of various diagrams has caught software developers' imaginations and, it has become popular in industry. UML, with its several diagrams, helps to develop a model of intended software, and the model behaviour is simulated and tested to the satisfaction of both developer as well as users. As a UML model includes specifications of different aspects of a software system through several diagrams, it is essential to maintain consistency among diagrams so that quality of the model is maintained, and through inconsistency checking and removal, the model moves toward completeness. The works reported in literature on this topic are reviewed here.

DOI: 10.4018/978-1-4666-4494-6.ch014

INTRODUCTION

Specification is the genesis of a software system and maintaining its correctness is of prime concern for ensuring quality of system under development. Software system testing includes both checking specification as well as its implementation. Formulating software specification (requirements gathering) precedes design. Both need a method for concrete as well as unambiguous specifications so that the testing team can trace implementation to requirements. Though formal specification with Z or B leads to provable systems, they are not commonly used due to mathematical rigour.

Responding to the want of a concrete as well as acceptable technique for professionals in industry, the Object Management Group defines the Unified Modelling Language (UML) as a general-purpose visual modeling language that is used to specify, visualize, construct and document artifacts of a software system. UML captures information about the static structure as well as dynamic behaviour of a system. The static structure defines objects as well as the relationship between objects that are part of the system implementation usually represented using use case, class and component diagram. Dynamic behaviour of the system is specified by the activity, sequence and state diagrams.

The semi-formal nature of UML has both advantages and disadvantages: the advantage primarily lies in its ease of use as well as understandability by various stakeholders of the system. Also, different diagrams can be used to model varying aspects of the system. The same leads to difficulties in the form of maintaining completeness and consistency within and between UML diagrams. Specification based testing using UML needs consistent and complete UML diagrams. Again for testing, scenarios representing the working of intended system are extracted and studied for the purpose.

The focus of this chapter is to look at how the Unified Modeling Language aids in exploring the issue of consistency checking which forms the basis for testing. The Unified Modeling Language is discussed, with the use as well as advantages and disadvantages. Next, the issue of checking consistency in UML models is explored followed by a discussion on Model-driven Testing. Comparison of work in the area is presented followed by scope for research.

THE UNIFIED MODELING LANGUAGE

What is UML?

The Unified Modeling Language (UML) is a general-purpose visual modeling language used to specify, visualize, construct, and document artifacts of a software system. Developed and propagated by the OMG group, UML can be used across all phases of the software development process (requirement, analysis and design, testing, and documentation). One or more diagrams can be used to represent the system. UML models can be classified as static models and dynamic models. Static models represent the structure of the system, whereas dynamic models are used to represent the behaviour of the system. Thus, a combination of the models may be used to suit the type and domain of the software to be developed. A UML diagram is not refined to provide all relevant aspects of an application. The semi-formal nature of UML leads to ambiguities in representation and interpretation of stated requirements. To overcome this, the Object Constraint Language (OCL) is used to write constraints on model elements. OCL expressions are used to specify invariants on classes, define pre- and post conditions on operations and methods, describe guards, constraints on methods as well as specify derivation rules for attributes for

Figure 1. Representation of class "employee"

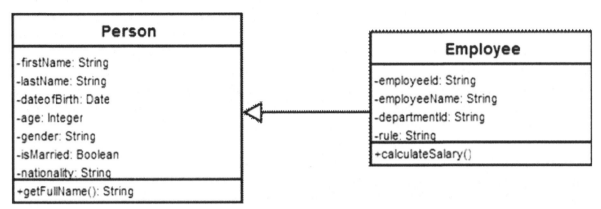

an expression over a UML model. Hence, OCL is used along with UML to make up for the lack of formalism.

UML and the Object Constraint Language

The Object Constraint Language (OCL), is a formal language used to describe expressions on UML models. UML diagrams are not expressive enough to provide all aspects of a specification. Hence, there is need to describe additional constraints about objects in a model. OCL is used as a specification language along with UML to augment the expressiveness of UML. Constraints are expressed as an OCL rule of the form:

```
context <model element>
invariant <constraint name>:
OCL expression
```

OCL can be used as a query language, to specify invariants, to describe pre and post conditions on operations and to describe guards. Consider the example given in Figure 1. The class 'Person' with attributes and method is shown. Suppose, it is required that an employee must be above 18 to be employed by the company, then this constraint on the age of the person cannot be represented through the class represented in the class diagram.

Therefore, there is need for a language to describe additional constraints about the objects of the class 'Person'. The OCL invariant to specify the constraint is given below:

```
context Person
invariant checkAge:
self.age > 18
```

OCL also provides constraint and object query expressions on any Meta Object Facility (MOF) model or meta-model which otherwise cannot be expressed by the diagrammatic notation. Inter-model diagram navigation involves following links from one object to locate another object or a collection of objects. OCL can be used for the purpose of navigation between model elements. A constraint on a model can be defined as a restriction on one or more values of a model. Thus, OCL constraints are used to define constraints on a model.

Use of UML

Semiformal modeling languages are a powerful means of describing requirements. UML has been accepted as the standard for object-oriented analysis and design and finds use in modelling requirements and design of a system through the various diagrams. UML consists of 13 diagrams, both static and dynamic, with different diagrams conveying different information. A survey of UML usage finds class diagrams, sequence diagrams, use case diagrams, use case narratives, activity diagrams, state chart diagrams and collaboration diagrams(in order) to be the most used of UML diagrams.

Different diagrams have been used to describe system functionality as they convey different information. Also, usage of diagrams depends on the ease of understanding and level of detail that a diagram can represent. Use case narratives, activity diagrams and use case diagrams find highest usage in verifying and validating requirements with clients while class, sequence and use case narratives are used more in specifying requirements for programmers. For communication among technical team members, sequence and class diagrams find highest usage whereas class diagrams are used more to document for future maintenance and enhancements. Also, the style and rigor in modelling using UML depends on factors like analyst's knowledge, client requests, time constraints and system domain.

Some UML diagrams find minimal usage in software projects. The reasons include that the diagrams:

- Are not well understood by analysts
- Are not useful for most projects
- Provide insufficient value to justify cost
- Capture redundant information
- Are not useful with clients
- Are not useful with programmers

User/client involvement in various phases of system development has been considered a crucial factor in the success of projects. UML diagrams find varied usage in development, review, and approval activities where customer involvement is high. Use case narratives, use case diagrams, and activity diagrams find highest usage where client involvement is essential. Development, review, and

approval of use case narratives and use case diagrams are activities clients were actively involved with. Also, activity diagrams are the easiest for clients to understand besides use case narratives and use case diagrams. Class diagrams are the preferred UML model used by project teams to communicate requirements whereas collaboration diagrams are least used. Also, a use case-driven approach is supported due to the involvement of users in establishing requirements of a system. However, user/client involvement also extends beyond use cases. With respect to testing, the use of UML models for testing is still limited due to uncertainty regarding productivity impact of using UML for testing. This maybe due to the fact that use of UML models for creating testing plans is not a common practice.

Thus, UML diagrams find varied levels of use in development of software. Also, only a subset of UML diagrams finds widespread use in practice. Another interesting finding is the degree to which UML diagrams are used in development, review and approval activities that involve clients. This level of involvement indicates that UML is not solely used by software professionals and strengthens the view that it facilitates communication between users and analysts which is crucial for the success of a software product. According to Nugroho et al, majority of respondents find UML helpful in making design, analysis and implementation activities productive. However, respondents of the survey were uncertain about the productivity impact of using UML on testing with only 38% considering UML helpful in testing and maintenance activities. However, they conclude that the use of UML models has to be considered for testing as they will provide a simpler, structured and more formal approach to the construction of functional testing and non-formal specification. Also, the development of test cases from specification has the additional benefit of producing test cases before implementation.

Advantages and Disadvantages of Testing using UML

Testing using UML as a modeling language has both advantages and disadvantages.

Advantages of testing using UML include:

1. **Ease of Understanding Requirements:** Requirements represented using UML are easily understandable by the stakeholders of the system when compared to other modelling approaches like FSM, Petrinets and formal methods like Z and B due to its semi-formal nature.

2. **Scalability:** Software is subject to change as requirements change, or are added throughout the software life cycle. UML models are easy to change and scale as the software grows.

3. **Specification Document as Reference:** Specification is used as the reference point across all phases, namely, analysis and design, testing and documentation.

4. **Customer Viewpoint:** Testing must be done from the customer's viewpoint, keeping in mind the need of the customer. UML captures user's perspective of system through the use-case diagram.

5. **Independent Test Generation:** Testing activity is independent of influence as the development and test team work independently though using the same specification document.

6. **Tools:** A large number of tools exist for creating and editing UML models.

Disadvantages of using UML as a specification language are:

1. **Nature of UML:** UML is a semi-formal modeling language. Hence, it leads to ambiguities in representation and misinterpretation. Also, the semantics of communication within UML is only partially defined. However, disadvantage due to semi-formal nature of UML is overcome by writing constraints and guard conditions using OCL.

2. **Choice of Diagrams:** UML provides a number of diagrams and the choice of diagrams is an important factor.

3. **Creating Models:** The language provides features that can be used to create models of deceptive complexity.

CONSISTENCY CHECKING

The objective of software development being delivery of quality software to the customer, the need to capture requirements without errors is of primary importance. The area of software specification has two classes of techniques: formal techniques (e.g. Z which emphasizes formality at the cost of ease of use and understanding) and rigour based techniques (e.g. UML, which emphasizes ease of use and understanding, at the cost of formality). As a rigour based technique lacks mathematical foundation, it is free to be interpreted differently by different people, leading to inconsistency. UML models presented by different diagrams view a system from different perspectives or from different abstraction levels. Thus, the various UML models of the same system are not independent specifications but strongly overlapping i.e. they depend on each other in many ways. Therefore, a change in one model affects another. The usage of several diagrams within the UML to capture specification leads to inconsistency that may arise due to mismatch between different diagrams. This has given rise to the need for checking consistency of requirements captured using different diagrams. Inconsistency management has been defined in Finkelstein et al. as 'the process by which inconsistencies between software models are handled so as to support the goals of the stakeholders concerned'. A number of techniques and tools have been developed for capturing requirements to suit the needs of different domains and applications.

A model is consistent when it conforms to the semantics of all the domains involved in the development process. e.g. application domain, modelling domain, language domain. Inconsistency arises when some model expression violates a principle like same name for attribute and method. Erroneous models have a huge impact on the development process in terms of added cost, time and effort. Thus, consistency management is a process comprising activities like specification and management of consistency handling policy, consistency checking, detection of overlaps, diagnosis and consistency handling and tracking.

UML AND CONSISTENCY CHECKING

The Unified Modeling Language (UML) is used as a standard for modeling object oriented software. As mentioned before, UML helps in modeling different aspects of a system through the use of various diagrams, like use case diagrams, activity diagrams, sequence diagrams, class diagrams and component diagrams. Each aspect of a system is represented using a particular type of UML diagram and a set of such diagrams is termed as a model.

UML diagrams represent two different views of a system model, namely, static (structural) view and dynamic (behavioural) view. Static view emphasizes the structure of the system while the dynamic view tries to project the run time interactions of the system. Static diagrams include class diagram, object diagram, component diagram, deployment diagram, package diagram, composite structure diagram and profile diagram. Dynamic diagrams include use case diagram, activity diagram, sequence diagram, state chart diagram, collaboration diagram, interaction overview diagram and timing diagram. The advantage of using UML is the fact that different diagrams can be used to model varying aspects of a system. i.e. the diagram that best represents an aspect of the

system to be captured can be used. This same aspect leads to difficulties too, in the form of maintaining completeness and consistency within and between UML diagrams. This problem is further magnified as software systems are built today in a manner that requires collaborative and collective effort of people physically distributed across many locations. The phases of requirements gathering, analysis and design involve more than one UML model. Any error in these initial phases of software development will be carried over through the rest of the software development phases and will require major effort to be corrected. The following are the factors that have been found to lead to the problem of inconsistency when modelling using UML:

- Omission
- Lack of Standardization
- Multiplicity of Stakeholders
- Addition of new features during system evolution
- Interdependency of diagrams
- Overlapping of model elements

The Unified Modeling Language (UML) defines well formedness rules as constraints on the modeling primitives to prevent errors in modelling. However, they do not suffice in detecting inconsistencies, especially those across diagrams. Therefore, additional constraints are required to ensure consistency and completeness to help detect and prevent errors during specification of a system.

There are different dimensions to consistency checking: intra-model Vs inter-model, structural/static Vs behavioural/dynamic, horizontal Vs vertical. Intra-model consistency is concerned with the issue of consistency between diagrams of the same type e.g. consistency between use case diagrams. Inter-model consistency looks at the way each model is designed with reference to the other models so as to be meaningful. Structural/static consistency concerns consistency among the structural elements used in representing requirements. Behavioural/-dynamic consistency concerns checking consistency during run time. Horizontal consistency checking involves checking consistency between diagrams at the same level of abstraction whereas vertical consistency involves checking consistency between diagrams at different levels.

Example

An example for inconsistency that may arise between UML diagrams is shown in Figure 2. Figure 2a shows a sequence diagram for the scenario, Customer places an order. Also, the classes involved in the scenario are shown using the class diagram in Figure 2b. A customer places an order. The order is checked with the items in the inventory and a notification is given. The customer then confirms the order after which the inventory is updated.

A comparison of elements between Figure 2a and Figure 2b show that the sequence diagram

does not match the operations stated in the class diagram. A consistency rule defined in Sapna et al (2007) states that a class in a class diagram must have an operation that corresponds to a message in a sequence diagram. In the example, the message, *confirm_order( )* is not specified in the class diagram. This is a case of inconsistency between the sequence diagram and class diagram. There is a need to add the method *confirm_order( )* in the class diagram. UML defines constraints on models as well formedness rules. These well formedness rules are applicable at the intra-model level and not at the inter-model level. Current work in literature handles inconsistency between UML diagrams based on the interaction specified between models in the UML superstructure specification. As navigation between metamodels of all diagrams is not present, it is not possible to check for consistency of specification in a wholesome manner (e.g. no relation between use case and sequence diagram).

Thus, UML being a semi-formal modelling language, it is not possible to define all aspects

Figure 2. Consistency checking between UML diagrams

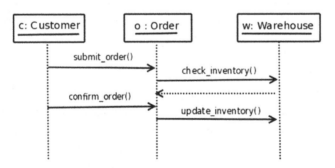

a. Sequence Diagram for 'Customer places an Order'

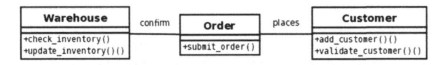

b. Fragment of Class Diagram related to 'Customer places an Order'

of a specification. To aid specifying constraints on modelling elements, the Object Constraint Language (OCL) is used. UML diagrams can therefore be enriched with assertions and constraints written in OCL, a specification language.

MODEL-DRIVEN TESTING

Work in literature differentiates model based testing from specification based testing. Specification-Based Testing (SBT) refers to the process of testing a program to check if the implementation conforms to the specification. i.e. testing is done irrespective of the implementation. Also, test cases are built before or in parallel to the actual development of the software. The test suite is developed based on the specification to satisfy a test criterion, like coverage or fault detection. The objective of specification based testing is to check if the developed software meets specification as defined by the customer.

Specification based testing differs from Model-Based Testing (MBT). In the former, specification document is used as the basis for all development activities including testing. i.e. test suite is developed from specification. Model based Testing is similar to specification based testing except that a separate model is built specifically for the purpose of testing, using specification as basis. This model therefore represents some aspects of the specification and is used as basis for testing only. The advantage of the former lies in the fact that specification is used to derive the test suite ensuring that software matches customer requirements. It is the reference point for test case generation. In case of the latter, the model is derived from the specification and is built separately. This may cause defects to be introduced in the model due to wrong understanding and misinterpretation of specification. Also, additional effort is involved in building the model. In both of the above cases, consistency of models plays an important role.

Here, both specification using models, and testing based on models shall be considered under that gambit of model-driven testing.

UML and Model-Driven Testing

UML can be used at various levels of testing, namely, unit, integration, system and regression testing. Table 1 shows different testing levels as well as the UML diagrams that can be used for testing at different levels with the coverage criteria. Specification using UML helps in testing by providing information required to generate test cases, select test cases, measure test adequacy as well as check correctness of output. Besides, test cases can be created in parallel with the development process and can be used when the software is developed. Also, inconsistencies and ambiguities in specification can be identified, which can be resolved early saving time and cost.

Approaches to Consistency Checking

There are two approaches to consistency checking, namely, the Direct Approach and the Transformational Approach. Direct approach involves comparing design models directly. Constraints defined on UML using OCL are subjected to syntactic and view dependent analysis. Also, in-

Table 1. Testing phases and corresponding UML diagrams

Type of Testing	Coverage Criteria	UML Diagrams
Unit	Statement, Class	Class & State Diagrams
Integration (Feature Testing)	Between Classes & between modules	Class diagrams & Interaction Diagrams
System		Use case & Activity diagrams
Regression	Usage Scenarios	Use case & Activity diagrams

formation about the metamodel (entities, attributes and links) are used. Constraints can be defined at different levels namely, general and specific on different layers like the system model layer, model layer and diagram layer. Another way of directly checking consistency is by using pre and post conditions and state invariants.

The transformational approach involves transforming design models from one to another or into some intermediate model, for comparison. Examples of intermediate models used in the transformational approach include graphical models, formal methods, petrinets and knowledge representation based techniques. Graphical models are the most commonly used approach. Formal methods like Z, Object Z and LOTOS

which are used to define integrity constraints of graphical models like UML also help in checking consistency.

Transformational approach to consistency checking is done in two ways as shown in Figure 3. In the first case, source diagrams are transformed into the diagram type of the target. This is done so that transformations of the source diagrams are conceptually close to target diagrams they need to be compared with.

Then, consistency checking is done to compare the transformation of the source diagram with the target diagram. For example, if a sequence diagram is to be compared with a class diagram, then the sequence diagram would be transformed to an 'interpreted' class diagram which is then compared

Figure 3. Direct and transformational approach to consistency checking

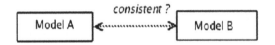

a) Is model A consistent with model B ?

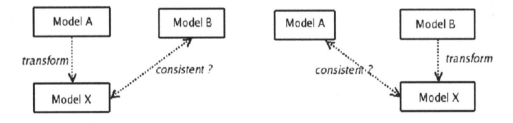

b) Model transformation followed by consistency checking

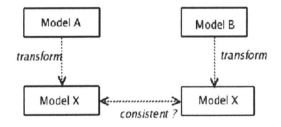

c) Model transformation followed by consistency checking

with the target class diagram or vice versa. The class, sequence, collaboration, object and state chart diagrams are taken into consideration. The second approach is to convert both the source diagrams into an intermediate model (e.g. Petrinets) and then checking for consistency.

The advantage provided by the transformational approach is in having the same intermediate model for all the diagrams (e.g. Petrinets). Also, once the rules for transforming a model are decided, it becomes easier by using automation techniques. Models can be forward and reverse engineered. The disadvantage of using the transformational approach is the need to transformation the model in addition to consistency checking, resulting in larger overheads.

Classification of Consistency Techniques

Various techniques have been proposed to handle the issue of consistency. Broadly, the techniques can be classified based on various factors such as level, type, approach and dimension as discussed below.

1. **Level:** Consistency between UML diagrams can be checked at the intra-model or inter-model level. Intra-model level is concerned with checking consistency within model elements. For example, checking consistency of elements within the class diagrams is intra-model. Inter model consistency is concerned with the way each model is designed with reference to the other models so as to be meaningful.

2. **Type:** There are two perspectives to consistency: syntactic/structural consistency, and behavioural consistency. Syntactic consistency requires that a model conforms to its abstract syntax (specified by its metamodel). Semantic consistency requires that models' behaviour be semantically compatible. It is essential that models be syntactically and structurally consistent before they can be checked for behavioural consistency. Also, there are two ways of checking consistency: active and passive. Active consistency checking is done as the diagrams capture requirements thereby indicating inconsistency in real time. They monitor and control models edition for preventing inconsistencies (Fombelle, Blanc, Rioux, & Gervais, 2006). However, the disadvantage is that strict enforcement of consistency limits the modeller's possibilities for exploring conflicting or trade-off solutions (Fombelle, Blanc, Rioux, & Gervais, 2006). Thus, it may be an impediment to the process of capturing requirements. Passive consistency checking involves checking all or a subset of models capturing requirement. The advantage of passive consistency checking is that rules guiding consistency are applied at all models and results obtained. E.g. IBM's Rose model checker uses activity consistency checking.

3. **Approach:** Consistency checking is done on both static and dynamic diagrams using one of the following approaches: a) Direct approach b) Trans-formational approach. The first approach uses the constructs of UML and OCL to check for inter model consistency. The transformational approach involves transforming one model to another or to a common notation before checking for consistency.

4. **Dimension:** UML diagrams must be consistent vertically and horizontally. A model consists of different submodels because a system is modeled from different viewpoints allowing different aspects to be captured in varied models. However, different viewpoint specifications must be consistent. This type of consistency problem is called horizontal consistency. Also, models are used to record specification at different levels of abstraction. A model can be transformed into another model by replacing with one or more sub-

models. Then, the replaced submodel must be a refinement of a previous submodel. This type of consistency problem is called vertical consistency.

SCOPE FOR FUTURE RESEARCH

Research in the area of checking consistency of specification using UML finds importance given the need for automating the process of testing. To understand issues, an idea of existing work based on varied parameters is shown in Table 2.

Comparison of different techniques is based on the following parameters: technique used, version of UML, kind of UML diagrams, formal method, use of OCL, approach followed, level at which consistency is checked, type of consistency checking, automated support and CASE tools available. Based on the study, the following conclusions can be made:

- Most approaches use a simple, basic or restricted version of UML consisting of main primitives i.e. basic features. Also, most work use version 1.x of UML. UML 2.0 brought in significant changes compared to UML 1.x. Activity diagrams were delineated from state diagrams and additional primitives were added. Another factor influencing use of UML models is the understanding and knowledge in UML of users including architects, designers and developers. Currently, the beta version of UML 2.5 has been released and there is need to explore the relationships between metamodels defined.

- Most work in consistency checking techniques use a formal method. A model checker like SPIN is used to simulate behaviour and check for inconsistencies. The advantage of using a formal approach is that the nature of mathematics being precise it is possible to represent requirements

unambiguously. Formal techniques though find limited use in industry due to lack of expertise among practitioners. Besides, the cost of training personnel, difficulty in use and the time involved in applying a formal technique, limitation in terms of feedback and the difficulty in understanding by non-experts stand as barriers to their use.

- Consistency checking using class, state diagrams and sequence diagrams is the focus of most work due to its widespread use. Activity diagrams and use case diagrams are considered by relatively fewer works. Interestingly, collaboration diagram which is similar to sequence diagram is not widely used. Also, there is very few work that involves use case diagrams showing the lack of connect in UML between use case and other diagrams.

- Transformation approach to check for inconsistency using either formal methods like Object Z (Rasch & Wehrheim, 2002; Derrick, Akehurst, & Boiten, 2002; Kim & Carrington, 2004), model checkers like PVS (Krishnan, 2001), and SPIN (Inverardi, Muccini, & Pelliccione, 2001), Petrinets (Saldhana & Shatz, 2000; Bernardi, Donatelli, & Merseguer, 2002; Shinkawa, 2006; Yao & Shatz, 2006), graphical approaches (Kosters & Winter, 2001; Hausmann, Heckel, & Sauer, 2002), writing rules (Bellur & Vallieswaran, 2006; Lavanya, Balakishore, Mohanty, & Shyamasundar, 2005), or by representing in a knowledge base (Simmonds, Bastarrica, Hitschfeld-Kahler, & Rivas, 2008). Direct approach is less used due to the varying nature of UML diagrams capturing different perspectives (Lano, Clark, & Androutsopoulos, 2002; Litvak, Tysvberowicz, & Yehudai, 2003; Bellur & Vallieswaran, 2006).

- Consistency checking is done within a diagram where specification is captured or be-

Table 2. Comparison of work on consistency checking

Ref #	UML Ver	Technique	UC	SD	AD	CD	STD	CLD	Other	FM	OCL	Trf	Dir	Intra	Inter	Str	Beh	AS	CASE
			UML Diagrams									**Appr**		**Level**		**Type**			
Periyasamy(1999)	B	Z, CLIPS	✗	✓	✓	✓	✓	✗	✗	✓	✗	✓	✗	✗	✓	✓	✗	✗	✗
Krishnan(2000)	S	PVS	✓	✓	✓	✓	✗	✗	✗	✓	✗	✓	✗	✗	✓	✗	✓	✗	✗
Ritter(2000)	1.1	ORDBMS	✗	✗	✗	✓	✗	✗	✗	✗	✓	✓	✗	✓	✗	✓	✗	✓	✗
Glinz(2000)			✓(T)	✗	✗	✓	✗	✗	✓(UCN)	✗	✗	?	?	✓	✗	✓	✗	✗	✗
Saldhana et al(2000)		Object Petrinets	✗	✗	✗	✓	✓	✗	✗	✓	✗	✓	✗	✗	✓	✓	✗	✗	✗
Mens et al(2003)	S	Desc. Logic	✗	✓	✗	✓	✓	✗	✗	✓	✗	✓	✗	✗	✓	✓	✗	✗	✗
Li et al(2001)	2.0	CSP	✗	✓	✗	✓	✓	✗	✗	✓	✗	✓	✗	✗	✓	✗	✓	✗	✗
FeiXie et al(2004)	xUML	S/R COSPAN	✗	✗	✗	✗	✗	✗	✗	✓	✗	?	✗	?	?	?	?	✗	✗
Six et al(2001)		Activity Graphs	✓	✗	✗	✓	✗	✗	✗	✗	✗	✓	✗	✗	✓	✓	✗	✗	✗
Zisman et al(2001)		Knowledge Based	✓	✓	✓	✓	✓	✓	✓	✓	✗	✓	✗	✗	✓	✓	✗	✗	✓
Paolo Inverardi et al(2001)	S	SPIN	✗	✓	✗	✗	✓	✗	✗	✓	✗	✓	✗	✗	✓	✗	✓	✗	✓
Egyed et al(2001)	1.3(B)		✓	✓	✓	✓	✓	✗	✗	✗	✗	✓	✗	✗	✓	✓	✓	✗	✗
Bougmila et al(2002)	1.3	OCL	✗	✗	✓	✓	✓	✗	✗	✗	✓	✓	✗	✗	✓	✓	✗	✗	✗
John Derrick et al(2002)	B	Object Z, LOTOS	✗	✓	✗	✓	✓	✗	✗	✓	✗	✓	✗	✗	✓	✗	✓	✗	✗
Simona Bernardi et al(2002)		Petrinet	✗	✓	✗	✗	✓	✗	✗	✓	✗	✓	✗	✗	✓	✓	✗	✗	✗
Holger Rasch et al(2002)	1.5	Object Z, CSP	✗	✗	✗	✓	✓	✗	✗	✓	✗	✓	✗	✗	✓	✓	✓	✗	✗
Lano et al(2002)			✗	✗	✗	✓	✓	✗	✗	✗	✗	✗	✓	✗	✓	✗	✗	✗	
Jan Hendrik Hausmann (2002)	1.4(B)	Graphs	✗	✓	✗	✓	✗	✗	✓(OD)	✗	✗	✓	✗	✓	✗	✓	✗	✗	✗
Boris Litvak (2003)	BVUML		✗	✓	✗	✗	✓	✗	✗	✗	✗	✗	✓	✓	✗	✗	✗	✓	✗
Il-Kyu Ha (2003)	1.4(B)	CPN	✗	✓	✓	✓	✓	✗	✗	✓	✗	✓	✗	✗	✓	✓	✓	✗	✗

continued on following page

Table 2. Continued

Ref #	UML Ver	Technique	UML Diagrams							FM	OCL	Appr		Level		Type		AS	CASE
			UC	SD	AD	CD	STD	CLD	Other			Trf	Dir	Intra	Inter	Str	Beh		
Dan Chiorean (2004)	1.3	OCL	x	x	x	✓	x	x	x	x	✓	✓	x	✓	■	✓	x	x	✓
Kim et al (2004)	S	Object Z	x	x	x	x	✓	x	x	✓	✓	✓	x	x	✓		✓	x	x
Zhaoxia Hu(2004)		CPN	x	x	x	x	✓	✓	x	✓	x	✓	x	x	✓	✓	x	x	x
Francisco (2005)	2.0	Maude Spec	x	x	x	✓	x	x	✓	✓	x	✓	x	x	✓	✓	x	x	x
Marcel Kyasa (2004)	1.5(R)	PVS	x	x	x	✓	✓	x	✓	✓	✓	✓	x	x	✓	✓	x	✓	x
Lavanya (2005)	1.5	Rule based	x	✓	x	✓	■	x	x	x	✓	✓	x	x	✓	✓	x	✓	x
Shuzhen Yao (2006)	S	Petrinet-ECPN	x	✓	x	x	✓	x	x	✓	✓	✓	x	x	✓	x	✓		x
Zhao (2006)	S	Split Automata	x	✓	x	x	✓	x	x	✓	x	✓	x	x	✓	x	✓	✓	✓
Umesh Bellur (2006)	1.5(R)	Rule based	✓	✓	x	x	■	x	x	x	x	x	✓	x	✓	✓	x	✓	x
Shinkawa (2006)	2.0	Petrinet CPN	✓	✓	✓	x	✓	x	x	✓	x	✓	x	x	✓	✓	x	x	x
Egyed (2007)	1.3(B)	UML Analyzer	x	✓	x	✓	✓	x	x	x	x	x	✓	x	✓	✓	✓	✓	✓
Carlos Zapata (2007)		OCL	✓	x	x	✓	x	x	x	x	✓	✓	x	x	✓	✓	x	✓	✓
Sapna et al(2007)	2.0	DBMS	✓	✓	✓	✓	x	x	x	x	✓	x	x	✓	✓	✓	x	✓	x
Jocelyn Simmonds (2008)	2.0(B)	Knowledge Based	x	✓	✓	✓	✓	x	x	x	x	✓	x	x	✓	✓	✓	x	✓

✓ - means use of technique/method and x indicates absence of technique/method; ? – not known to author; A – specification annotated with details; C – UML used as base model and converted to another; B – Basic UML model; S – Simple UML model; R – restricted model of UML, CSP – Communication Sequential Process, CPN – Colored Petri Nets, ECPN – Extended Colored Petri Nets; A blank indicates that data could not be gleaned.

tween diagrams (two or more). Checking consistency of specification within a diagram is easier. However, UML diagrams capture various aspects of the specification and hence, it is not possible to completely transform aspects captured in one diagram to another. Also, both structural and behavioural aspects of the specification are checked though behavioural consistency checking finds minimal techniques.

- Automated support and CASE integration are limited. Automated prototypes are available which can be used to check for consistency.

Thus, despite work done in the area, there are problems that need to be solved:

- UML superstructure specification document defines rules for intra-model consistency while constraints for elements across models have not been specified.
- Even though using constructs of UML (meta-model) for checking consistency reduces the cost and time involved in the process, the relationship between the models is such that it does not allow for direct checking between all models. Similarly, though the OCL can be used to specify elements and relations, it is difficult to use the same in specifying the consistency rules between diagrams.
- The transformational approach converts the UML models into a secondary model and use an intermediate means like trees and graphs for verification. The advantage provided by the transformational approach is in having the same intermediate model for all the diagrams (e.g. Petrinets). Also, once the rules for transforming a model are decided, it becomes easier by using automation techniques. Also, models can be forward and reverse engineered. The disadvantage of using the transformational approach is having to implement the trans-

formation in addition to consistency checking resulting in larger overheads.

- Common CASE (Computer Aided Software Engineering) tools do not support consistency checking completely.

CONCLUSION

Model-driven testing involves using a model for capturing requirements that in turn can be used for testing. Given that model-driven testing aids in automating the testing process, models play a crucial role in the software development life cycle. Inconsistent models lead to issues in specification, design and testing. Hence, checking consistency of models becomes a precursor for developing software as it ensures correctness of specification and helps in removing defects early.

In this chapter, the focus has been on issues related to consistency checking of specification captured using UML. The nature of UML, its use in testing and the need to check consistency of the model has been elaborated. Consistency checking has been discussed looking at levels and techniques applicable for the same. Various work in literature has been summarized to give an idea of future work that can be done in the area.

More work is needed in the area given that consistency checking is limited in tools that are available. Also, the nature and complexity of software systems today require that testing be done diligently. For this, consistency of specification captured using UML becomes a crucial issue.

REFERENCES

Bellur, U., & Vallieswaran. (2006). On OO design consistency in iterative development. In Proceedings of the 3rd International Conference on Information Technology: New Generations, (pp. 46-51). IEEE Computer Society. doi:doi:10.1109/ITNG.2006.102 doi:10.1109/ITNG.2006.102.

Bernardi, S., Donatelli, S., & Merseguer, J. (2002). From UML sequence diagrams and statecharts to analysable petrinet models. In *Proceedings of the 3rd International Workshop on Software and Performance*, (pp. 35-45). New York, NY: ACM.

Chiorean, D., Pasca, M., Carcu, A., Botiza, C., & Moldovan, S. (2004). Ensuring UML models consistency using the OCL environment. *Electronic Notes in Theoretical Computer Science, 102*, 99–110. doi:10.1016/j.entcs.2003.09.005.

Dalal, S. R., Jain, A., Karunanithi, N., Leaton, J. M., Lott, C. M., Patton, G. C., & Horowitz, B. M. (1999). Model-based testing in practice. In *Proceedings of International Conference on Software Engineering (ICSE 99)*, (pp. 285-294). IEEE Computer Society.

de Fombelle, G., Blanc, X., Rioux, L., & Gervais, M.-P. (2006). Finding a path to model consistency. In *Proceedings of the Second European Conference on Model Driven Architecture - Foundations and Applications (ECMDA-FA 2006)*, (pp. 101-112). Berlin: Springer-Verlag.

Demuth, B., & Hussmann, H. (1999). Using UML/OCL constraints for relational database design. In *Proceedings of UML'99: The Unified Modeling Language - Beyond the Standard, Second International Conference*, (pp. 598 – 613). Springer-Verlag.

Demuth, B., Hussmann, H., & Loecher, S. (2001). OCL as a specification language for business rules in database applications. In *Proceedings of the 4th International Conference on The Unified Modeling Language: Modeling Languages, Concepts, and Tools*, (pp. 104-117). Springer-Verlag.

Derrick, J., Akehurst, D., & Boiten, E. (2002). A framework for UML consistency. In L. Kuzniarz, G. Reggio, J. L. Sourrouille, & Z. Huzar (Eds.), 2002 Workshop on Consistency Problems in UML-Based Software Development, (pp. 30-45). Springer.

Egyed, A. (2001). Scalable consistency checking between diagrams - The VIEWINTEGRA approach. In *Proceedings of the 16ᵗʰ IEEE International Conference on Automated Software Engineering (ASE)*, (pp. 387-390). IEEE Computer Society.

Egyed, A. (2006). Instant consistency checking for the UML. In *Proceedings of the 28ᵗʰ International Conference on Software Engineering (ICSE2006)*, (pp. 381-390). ACM.

Egyed, A. (2007). Fixing Inconsistencies in UML design models. In *Proceedings of the 29ᵗʰ International Conference on Software Engineering (ICSE 2007)*, (pp. 292-301). IEEE Computer Society.

Glinz, M. (2000). Problems and deficiencies of UML as a requirements specification language. In *Proceedings of the 10th International Workshop on Software Specification*. IEEE Computer Society.

Ha, I.-K., & Kang, B.-W. (2003). Meta-validation of UML structural diagrams and behavioral diagrams with consistency rules. In *Proceedings of the IEEE Pacific Rim Conference on Communications, Computers and signal Processing (PACRIM 2003)*, (Vol. 2, pp. 679-683). IEEE Computer Society.

Hausmann, J. H., Heckel, R., & Sauer, S. (2002). Extended model relations with graphical consistency conditions. In *Proceedings of the Workshop on Consistency Problems in UML-Based Software Development (UML 2002)*, (pp. 61-74). UML.

Hnatkowska, B., Huzar, Z., Kuzniarz, L., & Tuzinkiewicz, L. (2002). A systematic approach to consistency within UML based software development process. In *Proceedings of Workshop on Consistency Problems in UML-Based Software Development (UML 2002)*, (pp. 16-29). UML.

Hu, X., & Shatz, S. M. (2004). Mapping UML diagrams to a petri net notation for system simulation. In *Proceedings of the Sixteenth International Conference on Software Engineering & Knowledge Engineering (SEKE'2004)*, (pp. 213-219). SEKE.

Inverardi, P., Muccini, H., & Pelliccione, P. (2001). Automated check of architectural models consistency using SPIN. In *Proceedings of the 16th International Conference on Automated Software Engineering (ASE'01)*, (p. 346). IEEE Computer Society.

Khovich, L. O., & Koznov, D. V. (n.d.). OCL-based automated validation method for UML specifications. In *Programming and Computer Software*, 29(6), 323-327.

Kim, S. K., & Carrington, D. (2004). A formal object-oriented approach to define consistency constraints for UML models. In *Proceedings of the Australian Software Engineering Conference (AWSEC 2004)* (pp. 87-94). IEEE Computer Society.

Krishnan, P. (2001). Consistency checks for UML. In *Proceedings of the 7th Asia Pacific Software Engineering Coference (APSEC '00)*, (pp. 162-169). IEEE Computer Society.

Kyasa, M., Fechera, H., de Boera, F. S., Jacoba, J., Hoomana, J., van der Zwaaga, M., et al. (2004). Formalizing UML models and OCL constraints in PVS. In *Proceedings of the Second Workshop on Semantic Foundations of Engineering Design Languages (SFEDL 2004)*, (Vol. 115, pp. 39-47). SFEDL.

Lano, K., Clark, D., & Androutsopoulos, K. (2002). Formalising inter-model consistency of the UML. In *Proceedings of the Workshop on Consistency Problems in UML-Based Software Development (UML 2002)*. UML.

Lavanya, K. C., Balakishore, K. V., Mohanty, H., & Shyamasundar, R. K. (2005). How good is a UML diagram? A tool to check it. In *Proceedings of TENCON*, (pp. 386-391). IEEE Computer Society.

Li, X., Liu, Z., & He, J. (2001). Formal and use-case driven requirement analysis in UML. In *Proceedings of the 25th Annual International Computer Software and Applications Conference*, (pp. 215-224). IEEE.

Litvak, B., Tyszberowicz, S., & Yehudai, A. (2003). Behavioral consistency validation of UML diagrams. In *Proceedings of the First International Conference on Software Engineering and Formal Methods (SEFM 2003)*, (pp. 118-125). SEFM.

Martnez, F. J. L., & Alvarez, A. T. (2005). A precise approach for the analysis of the UML models consistency. In *Proceedings of the 24th International Conference on Perspectives in Conceptual Modeling*, (pp. 74-84). Berlin: Springer-Verlag.

Mens, T., Van Der Straeten, R., & Simmonds, J. (2003). Maintaining consistency between UML models with description logic tools. In *Proceedings of the Sixth International Conference on the Unified Modelling Language - The Language and its Applications, Workshop on Consistency Problems in UML-Based Software Development II*. UML.

Periyasamy, K., Alagar, V. S., & Muthiayen, D. (1999). International conference on verification and validation techniques of object-oriented software systems. *Technology of Object-Oriented Languages*, 413.

Rasch, H., & Wehrheim, H. (2002). Consistency between UML classes and associated state machines. In *Proceedings of the Workshop on Consistency Problems in UML-Based Software Development(UML 2002)*. UML.

Ritter, N., & Steiert, H. P. (2000). Enforcing modeling guidelines in an ORDBMS-based UML-repository. In *Proceedings of the International Resource Management Association International Conference on Challenges of Information Technology Management in the 21ˢᵗ Century* (pp. 269-273). Hershey, PA: IGI Global.

Saldhana, J. A., & Shatz, S. M. (2000). UML diagrams to object petri net models: An approach for modeling and analysis. In *Proceedings of the International Conference on Software Engineering and Knowledge Engineering (SEKE)*. SEKE.

Shan, H. Z. L. (2006). Well-formedness, consistency and completeness of graphic models. [Oxford, UK: UKSIM.]. *Proceedings of, UKSIM06*, 47–53.

Shinkawa, Y. (2006). Inter-model consistency in UML based on CPN formalism. In *Proceedings of the 13ᵗʰ Asia-Pacific Software Engineering Conference(APSEC)*, (pp. 411-418). IEEE Computer Society.

Simmonds, J., Bastarrica, M. C., Hitschfeld-Kahler, N., & Rivas, S. (2008). A tool based on DL for UML model consistency checking. *International Journal of Software Engineering and Knowledge Engineering*, *18*(6), 713–735. doi:10.1142/S0218194008003829.

Six, H. W., Kosters, G., & Winter, M. (2001). Coupling use cases and class models as a means for validation and verification of requirements specification. *Requirements Engineering*, *6*(1).

Tsiolakis, A., & Ehrig, H. (2000). Consistency analysis of UML class and sequence diagrams using attributed graph grammars. In *Proceedings of the Workshop on Graph Transformation Systems(GraTra)*, (pp. 77-86). GraTra.

Xie, F., Levin, V., Kurshan, R. P., & Browne, J. C. (2004). Translating software designs for model checking. In *Proceedings of Fundamental Approaches to Software Engineering (FASE 2004)* (LNCS), (vol. 2984, pp. 324-338). Berlin: Springer.

Yao, S., & Shatz, S. M. (2006). Consistency checking of UML dynamic models based on petri net techniques. In *Proceedings of the 15ᵗʰ International Conference on Computing (CIC '06)*, (pp. 289-297). IEEE Computer Society.

Zapata, C. M., Gonzlez, G., & Gelbukh, A. (2007). A rule-based system for assessing consistency between UML models. In *Proceedings of the 6ᵗʰ Mexican International Conference on Artificial Intelligence (MICAI '07)* (LNAI), (pp. 215-224). Berlin: Springer.

Zhao, X., Long, Q., & Qiu, Z. (2006). Model checking dynamic UML consistency. In *Proceedings of the 8ᵗʰ International Conference on Formal Engineering Methods, ICFEM 2006*, (LNCS), (pp. 440-459). Berlin: Springer.

Zisman, A., & Kozlenkov, A. (2001). Knowledge base approach to consistency management of UML specifications. In *Proceedings of the 16ᵗʰ International Conference on Automated Software Engineering (ASE'01)*. IEEE Computer Society.

Zisman, A., & Kozlenkov, A. (2001). Knowledge base approach to consistency management of UML specifications. In *Proceedings of the 16ᵗʰ International Conference on Automated Software Engineering (ASE'01)*, (pp. 359-363). IEEE Computer Society.

Chapter 15
Viewpoint–Based Modeling:
A Stakeholder–Centered Approach for Model–Driven Engineering

Klaus Fischer
German Research Center for Artificial Intelligence, Germany

Dima Panfilenko
German Research Center for Artificial Intelligence, Germany

Julian Krumeich
German Research Center for Artificial Intelligence, Germany

Marc Born
ikv++ technologies, Germany

Philippe Desfray
SOFTEAM, France

ABSTRACT

Viewpoint-based modeling is an important recent development in software engineering. It is likely to boost the wider use of modeling techniques because it allows the tailoring of existing tools with respect to the different stakeholders in software design. This chapter reports the results of the project ViBaM, in which viewpoint concepts are investigated. In doing so, the authors give an overview of the most important contributions from literature regarding viewpoint concepts, from which they derive the position that they take in the ViBaM project. After presenting ViBaM's position, the authors derive features that they consider important for tools that support viewpoints. Afterwards, they present use cases, in which the viewpoint concepts are illustrated and discuss which of the viewpoint features are relevant in these use cases.

DOI: 10.4018/978-1-4666-4494-6.ch015

INTRODUCTION

In recent years, the complexity of software development has been continuously growing (Lehtola, Gause, Dumdum, & Barnes, 2011), since developers have to face more and more technical and business requirements in shorter time periods while dealing with an increasing number of internal and external stakeholders and participants (Bosch & Bosch-Sijtsema, 2010; Mohagheghi, Fernandez, Martell, Fritzsche, & Gilani, 2009). As a result, a new software development concept called Model-driven Software Engineering (MDSE) has recently emerged in research, but is also more and more used in business practice. MDSE aims at a stronger and more efficient incorporation of domain knowledge into the software development process to overcome the just mentioned difficulties (Schatz, 2011). In doing so, MDSE considers models as a basis for illustrating the system to be implemented. The basic idea is to create and refine the models of the desired system by using elaborated tools. As one of the main objectives of MDSE (Schatz, 2011), special attention is given to a clear distinction between the specification of the modeled system and its technical implementation in order to foster the consideration of domain knowledge.

The Model Driven Architecture (MDA) (Object Management Group, 2003) of the Object Management Group (OMG) has established as the most common approach to MDSE. To follow the idea of MDSE, MDA introduces a strong separation of concerns regarding modeling a system at different abstraction levels. This starts from computational (CIM) and platform-independent (PIM) models and uses transformations to produce the actual code for the selected programming language and platform (PSM).

In present days modeling techniques are, not only due to MDSE, widely used—in particular standard modeling languages such as the Business Process Modeling Notation (BPMN) (Object Management Group, 2011a) or the Unified Modeling Language (UML) (Object Management Group,

2011b) but also proprietary and domain specific modeling languages. However, according to our estimation, usage of these techniques has already reached a significant threshold in software development but a wider usage and diffusion needs more added values and perceived ROI by end users. Modeling can gain a stronger acceptance by supporting cooperative and collaborative system design in a more efficient and simplified manner. In such settings, different stakeholders work together to design a system. However, each of them has his or her own perspective to the system design and would like to view and manipulate the models or model fragments according to his or her own needs. In addition, the usage of models to support method frameworks (most of them using viewpoints) will be much improved through this new capacity.

To foster the modeling usage with regard to these demands, the concept of Viewpoint-based Modeling is an emerging approach in the course of software development. However, the concept does not have its origins in software development, but its general idea can already be found in the overall context of constructing large-scale systems several years ago (Finkelstein, Easterbrook, Kramer, & Nuseibeh, 1992; Wood-Harper, Antill, & Avison, 1985). Since 1990, this concept is largely used in method frameworks such as RM-ODP (International Organization for Standardization, 1996), RUP (Kruchten, 2003), or TOGAF (The Open Group, 2011). Accordingly, its major goal is to reduce complexity by adapting the overarching model to stakeholder specific fragments in a successful manner. In doing so, stakeholders are being more put into the focus of the modeling process, which results in a higher perceived value by them. By utilizing stakeholder-specific viewpoints on a model, its overall understanding and productivity increases (Finkelstein et al., 1992). As a direct consequence, the viewpoint concept leads also to better conceptual models, which was proven by several studies. For example, according to the study conducted by Easterbrook et al. (2005) applying the viewpoint concept significantly helps to cope with the overall size of a given problem domain.

To foster Viewpoint-based Modeling, the Eurostars-funded research project ViBaM was launched to provide a scientific foundation for Viewpoint-based Modeling as well as viewpoint-support in the tools of the ViBaM project partners. In ViBaM, viewpoint definitions given in literature were collected and evaluated against the position of the ViBaM partners regarding their modeling tool development. Hence, we report on results of the ViBaM project, in which the viewpoint concept has been investigated and based on these findings integrated into the commercial modeling tools medini analyze and Modelio of the project partners ikv++ technologies and SOFTEAM. The rest of this book chapter is organized according to the following structure:

- **Viewpoint Concept:** Provides a basic definition of the viewpoint concept as it is used in the underlying book chapter.
- **Related Work:** Gives an overview of approaches to use viewpoints in (model driven) system design as they were already published in literature. Furthermore, we link and evaluate our understanding of the viewpoint concepts with existing approaches.
- **Viewpoint Features:** Introduces and discusses some features that are necessary in modeling tools in order to successfully support the viewpoint concept.
- **Applications of the Viewpoint Concept:** Presents use cases in which the use presented viewpoint concepts have been evaluated.
- **Conclusion and Outlook:** Summarizes the chapter and give an outlook to future work on viewpoints.

VIEWPOINT CONCEPT: BASIC DEFINITIONS

According to our observations, most of the definitions we found have something in common: they define the concept of "viewpoint" as a guideline for constructing views. This can for example be observed in the *IEEE 1471-2000* standard definition (Software Engineering Standards Committee of the IEEE Computer Society, 2000) for viewpoints, which was adopted by ISO/IEC as an *ISO/IEC 42010:2007* standard in 2007 (International Organization for Standardization, 2007). Accordingly, a viewpoint is "a specification of the conventions for constructing and using a view." This definition is most widely accepted in general system engineering, but also in software development. A related viewpoint definition is given by (Ainsworth, Cruickshank, Wallis, & Groves, 1994).

Another common feature of most of those definitions is that a viewpoint explicitly specifies one or more stakeholders, whose point of view it represents. Furthermore, some definitions explicitly note that a viewpoint should be as much self-contained as possible. In our position, a viewpoint can be seen as a pattern that defines a set of views. What we are particularly interested in and what we will investigate in this section is whether viewpoints are defined in terms of a metamodel, as well as whether this metamodel is self-contained or related to other metamodels. Having relations or guidelines, which specify a viewpoint life cycle and viewpoint interactions as well as (de-)centralization of viewpoint underlying metamodels, are further aspects in this section. Furthermore, another feature we are interested in is whether a viewpoint directly reflects the needs of a particular stakeholder.

Figure 1. Viewpoint metamodel

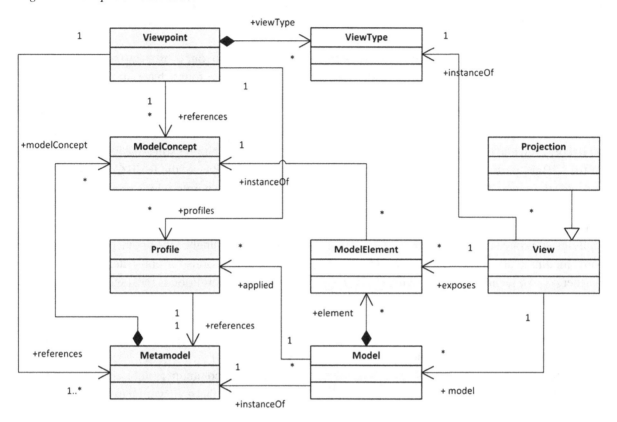

Towards Defining the Viewpoint Concept

Due to the large amount of the related work and definitions found in the literature (we found more than a dozen of definitions and additionally methods for dealing with viewpoints providing their own definitions), we first clarify the terms and definitions which we take as a basis. These terms are stated in the lists below and visualized in Figure 1.

An intuitive reading of the metamodel depicted in Figure 1 is that viewpoints are either directly defined on metamodels or on top of profiles. We stress that this viewpoint metamodel is not meant to define concepts for which we want to create models and model instances. It describes the terms which we consider important when it comes to viewpoint-based modelling and how these concepts relate to each other. This viewpoint

metamodel is meant to give a guideline to tool developers regarding how viewpoint concepts should be used and implemented in tools which support a viewpoint-based modelling process. The selected applications show that the resulting tools are useful in specific application domains.

We give a brief characterization of the basic terms where the first five below were derived from the definitions given in OMG's Model Driven Architecture (MDA) (Object Management Group, 2003). The definitions characterize the respective terms regarding their use in methodologies which support viewpoint-based modeling and regarding how tool support for such methodologies should support these terms:

- **Metamodel:** A metamodel defines a frame and a set of rules for creating models by introducing concepts and their relationships as well as constraints that should be applied to them. A metamodel serves as a basis for models instantiating it. Related modeling concepts usually belong to a certain metamodel.

- **Model Concept:** A model concept (we also use modeling concept) is a part of a metamodel and the basis for model elements in model instances.

- **Model Element:** A model element is a concrete instance of a modeling concept, and thus it represents either a domain object or a relationship between two or more objects. These elements are a part of a model instance and are being exposed in certain views belonging to certain viewpoint instances.

- **Profile:** A profile is an extension of a metamodel, which uses the metamodel as a reference for redefining existing modeling concepts and thus targeting a metamodel towards a given application domain. It refers to a certain metamodel and can be applied to various models for domain alignment.

- **Model:** A model is an instance of a metamodel. It contains a concrete set of model elements, which adhere to the rules defined in the corresponding metamodel. Models can apply certain profiles and thus represent model elements accordingly.

The just given basic ideas for the defined terms provide a foundation for creating models for specific application domains. To be able to support different stakeholders we need to extend the list given above by additional terms. In Figure 1 these additional terms are already included and we define them as follows:

- **Viewpoint** (cf. Champeau, Mekerke, and Rochefort, 2003; The Open Group, 2011; Lankhorst, 2013): The purpose of a viewpoint is to support a stakeholder in contributing to system design from a specific

perspective. A viewpoint defines which modeling concepts and relations can be used to define, view, or manipulate model instances within this viewpoint and thus separates modeling concerns and guarantees consistency regarding the information in the model instance. It is therefore related to a (set of) metamodel(s), a (set of) profile(s) or a part of them. The viewpoint in this sense can restrict the original metamodel(s) but it can also correspond to a metamodel 1:1.

- **View Type** (cf. Goldschmidt, Becker, and Burger, 2012): A view type serves a basis for view instantiation and offers a specific slice of system perspective to the stakeholders (i.e. human users). A collection of view types is defined for each viewpoint.

- **View** (cf. Champeau et al., 2003; Lankhorst, 2013): A view is an instance of a view type and defines the presentation of model elements to a stakeholder to whom the viewpoint belongs and the way(s) how model elements can be modified (this is usually achieved by diagram types together with a tool box for manipulations of model elements). It enables the user to interact with particular aspects of one or more models that adhere to the viewpoint's metamodel. Consistency between views is dealt with at the level of the model instances, i.e. when changes in a view are stored, the model instance is checked and in case of inconsistencies these are alerted to the user in the given view.

- **Projection** (cf. Kruchten, 2003; Praxeme Institute, 2011): Function that maps a model instance μ to another model instance μ' where μ' is a restriction of μ in the sense that it contains only elements that are also contained in μ and both μ and μ' are instances of the same metamodel. A projection is a specialization of a view.

During development of the viewpoint metamodel we noticed that our viewpoint metamodel design could be seen in the light of the Model-View-Controller (MVC) design pattern (Gamma, Helm, Johnson, & Vlissides, 1995):

- **MVC-Model:** The model part of the MVC corresponds primarily with the metamodel and model instances. Our viewpoint concept consists of the *viewpoints*, *modeling concepts*, *profiles* and *metamodels*. The cardinality between *viewpoint* and *metamodel* allows for centralized and decentralized metamodel constellations. *Viewpoints* refer to a defined set of *model concepts* highlighting the flexibility in the role assignments of different stakeholders. Another way of connecting *viewpoints* to *metamodels* is profiling, where one *profile* always belongs to a certain *metamodel* and a *viewpoint* may refer to more than one *profile*. This allows for dedicated exploitation and organizational adjustments due to the usage versatility of the *viewpoints* with the aid of the profiling mechanism.
- **MVC-Controller:** Corresponds to the mediator between models containing information and the views presenting this information to the stakeholders. The controller is responsible for the two directions of the information propagation: either from the views to the model propagating the changes made by the respective stakeholder; or from the model to the views to show the changes made, thus notifying other stakeholders. The controller is also responsible for creating view instances from view types.
- **MVC-View:** The view part of ViBaM's viewpoint concept consists of *views* and their possible *projections*. The *views* themselves refer to multiple *models* and *model elements*, thus allowing for spanning multiple *modeling concepts* from different *metamodels* over their model instance. For representation purposes, the projection concept allows for versatile *model concepts* representation, projecting and filtering of the *model concepts*, which are displayed in the views.

Based on the general terms and the analysis of existing definitions, we derived a harmonized viewpoint definition that serves as a basis for the work in the ViBaM project.

Definition: A viewpoint is defined in relation to one or more metamodels. For each viewpoint a non-empty set of view types is defined. In a viewpoint instance any number of views for each of the view types can be dynamically created.

More intuitively one can state that viewpoints offer a set of views which allow the stakeholder for whom the viewpoint was defined to access and manipulate the model instances.

RELATED WORK

Towards Viewpoints in General System Design

Without doubt, the concept of viewpoints is not novel. According to Lankhorst (2013), already in 1985, the *MultiView* approach proposed by Wood-Harper et al. (1985) forms the concept's origin. *MultiView* aims at supporting the development process of (computerized) information systems by splitting its complex process into five different perspectives or viewpoints, respectively: Human Activity System, Information Modeling, Socio-Technical System, Human-computer Interface, and Technical System. A decade later, the *MultiView* framework has been revised to *MultiView2* (Avison, Wood-Harper, Vidgen, & Wood, 1998). Another early-published and frequently-cited work was conducted by Finkelstein et al. (1992).

Even though it was published in the *International Journal of Software Engineering and Knowledge Engineering*, the focus of their work is not limited to software development but can be applied to multiple kinds of artifacts which need to be constructed through a non-trivial engineering process. Hence, this corroborates the multi-domain spanning interest of the viewpoint concept, which will be further outlined in the remainder of this section.

With a stronger focus on software systems, Kruchten (1995) presents a model for describing the architecture of software-intensive systems, which it is often referred to as the *"4+1" View Model of Software Architecture*. This approach aims at the design and the implementation of the high-level structure of software systems. The architecture model is composed of multiple, concurrent perspectives called views. A view addresses a specific set of concerns looking on the system from the perspective of a particular stakeholder (group) (Kruchten, 1995). Also in the context of software development, viewpoints are utilized in the domain of requirements engineering. For example, Kotonya and Sommerville (1996) present a *Viewpoint-Oriented Requirements Definition* (VORD) approach.

Another domain in which the viewpoint technique is frequently applied is the Enterprise Architecture (EA) domain. In the course of this, *The Zachman Framework* (initial called as *Framework for Information Systems Architecture*), which is one of the earliest-published and most known frameworks, consists of a two-dimensional classification matrix (Zachman, 1997). The first dimension differentiates six different viewpoints: Planner, Owner, Designer, Builder, Subcontractor, and Functioning System view. Orthogonal to this dimension, the Zachman framework differentiates between six different aspects: Data, Function, Network, People, Time and Motivation description. Another EA framework that is frequently used in practice is *The Open Group Architecture Framework* (TOGAF) (The Open Group, 2011).

TOGAF differentiates—without providing a concrete definition—between four viewpoints: Business, Data, Application, and Technology Architecture.

While the previously mentioned concepts of viewpoints rather follow the idea of separating an artifact of interest in a vertical dimension, i.e. a union of all viewpoints would provide an overall vision on this artifact based on a specific level of detail, there is also a popular concept following the orthogonal direction (Goldschmidt et al., 2012). The *Model Driven Architecture* (MDA) (Object Management Group, 2003) follows the idea of introducing a strong separation of concerns regarding modeling a system at different abstraction levels in a horizontal manner. This starts from computational (CIM) and platform-independent (PIM) models and uses transformations to produce the actual code for the selected programming language and platform (PSM). In contrast to vertical-centric viewpoints, a viewpoint in MDA encompasses the whole underlying system, but on its specific level of abstraction regarding the distance to a concrete system implementation. Of course, even in MDA, a separation of concerns—preferable by using viewpoints—within one abstraction level seems necessary to reduce complexity.

Even though this section outlined the fact that viewpoints as a concept are already frequently used in multiple domains, the concept still either lacks a definitional or scientific foundation (e.g. The Open Group (2011)) or is strongly focused on a specific domain. Furthermore, knowledge on requirements and implications for modeling tools aiming at realizing the viewpoint concept for practical usage is missing. One first approach towards this direction was recently published by Goldschmidt et al. (2012). Even though their work is closely related to the one we present in this paper, it mainly focuses on a feature-based classification of view-based domain-specific modeling concepts, while in contrast we follow a more general approach; nevertheless, this recent paper shows that there is still research needed in particular regarding modeling tools that provide means for using the viewpoint technique.

Consideration of Viewpoint Concepts in Literature

For better understanding of the interrelations between the terms described in Section 2 and the viewpoint definitions found in the literature we present Table 1 (cf. IEEE 1471-2000 [Software Engineering Standards Committee of the IEEE Computer Society, 2000], RM-ODP [International Organization for Standardization, 1996], ISO/IEC 42010 [International Organization for Standardization, 2011b], TOGAF [The Open Group, 2011], PRAXEME [Praxeme Institute, 2011], RUP [Kruchten, 2003; Steen, Akehurst, Doest, and Lankhorst, 2004], 4+1 [Kruchten, 1995; Rozanski and Woods, 2005], Zachman Framework [Zachman, 2009; Nuseibeh, 1994; Ainsworth et al., 1994; Dijkman, Quartel, and van Sinderen, 2008; Finkelstein et al., 1992]). Regarding the meaning of the different symbols in the table we want to add that: the green tick mark means that the respective term from the rows on the left is explicitly mentioned in the viewpoint definition the columns stands for; the red cross means that the term is not mentioned; yellow slash line means that the respective term is not literally mentioned but rather implied by the given definition.

The first two terms specified in Table 1, namely *metamodel* and *model concept*, are referred by the listed viewpoint definitions almost in the same way. Most of the definitions mention these terms at least in an implied way.

The term *model* is so common that the majority of the viewpoint definitions refer to it. Only two of them are not speaking of it explicitly—ISO/IEC 42010 standard and PRAXEME method. RUP is only implicitly concerned about the models in the viewpoints, but rather concentrates on the development processes.

Almost none of the viewpoint definitions are explicitly concerned about the *model element* term. The level of detail in these definitions is obviously not fine-grained enough to mention the model element level, although the most of them imply its existence.

Regarding the term *profile*, only IEEE 1471-2000, ISO/IEC 42010 and RUP have implicit definitions of this term.

Table 1. Viewpoint terms and their relation to viewpoint definitions

Key: ✔ = Yes ✗ = No / = Implied	IEEE 1471-2000	RM-ODP	ISO/IEC 42010	TOGAF	PRAXEME	RUP	Steen et al.	4+1 Rozanski-Woods	Zachman Framework	Nuseibeh	Ainsworth et al.	Dijkman et al.	Finkelstein et al.
Metamodel	✔	✔	✔	✔	/	/	/	/	✔	/	/	/	✔
Model Concept	✔	✔	✔	✔	/	/	/	/	/	/	/	/	✔
Model	✔	✔	✔	✔	/	/	✔	✔	✔	✔	✔	✔	✔
Model Element	/	/	/	✔	/	/	/	/	/	/	/	/	/
Profile	/	✗	/	✗	✗	/	✗	✗	✗	✗	✗	✗	✗
Viewpoint	✔	✔	✔	✔	✔	/	✔	✔	/	✔	✔	✔	✔
View Type	/	/	/	✗	✗	✗	/	/	✗	✗	✗	✗	✗
View	✔	✔	✔	✔	/	/	✔	✔	/	✔	✔	✔	✔
Projection	/	✗	/	✗	✔	✔	/	✗	✗	✗	✗	✗	✗

The most prominent term is of course *viewpoint*, because each and every viewpoint definitions should use it. Only RUP and Zachman Framework use a different wording, though speaking of it throughout the model description.

The *view type* is thought as a mediator between viewpoints and views, providing the means for view creation. View types might be implicitly included in most of the viewpoint definitions at hand referring to the relation between viewpoints and views, however not mentioning this mediator explicitly (International Organization for Standardization (1996), International Organization for Standardization (2011b), Kruchten (1995), Lankhorst (2013), and Steen et al. (2004) mention it implicitly, though).

The second most important term for the comparison is the *view*. Its essence to facilitate for every stakeholder the system observation from a certain perspective and constituting a viewpoint in general is a statement in multitude of definitions. Thus, is no surprise that the term view is mentioned in almost every viewpoint definition explicitly (although PRAXEME, RUP, and Zachman Framework do not focus on defining views explicitly).

A specialization of a view is a *projection*, being as such a non-bijective mapping of its ancestor class and adjusting at the same time in the first place the visual aspects of a view. It is mentioned in PRAXEME and RUP methods, but apart from implicit referral in Lankhorst (2013), International Organization for Standardization (2011b) and Steen et al. (2004) in no other viewpoint definition.

During related work analysis, our understanding on viewpoints has been influenced by several features of viewpoint definitions. In the following we illustrate and evaluate those features of viewpoint definitions which we consider important in this regard. In doing so, Table 2 shows whether these features are considered within existing definitions.

A viewpoint is responsible for a partial specification of a system: A viewpoint creates and manipulates certain functional descriptions and information types which are implemented and

Table 2. Viewpoint definition features

Key: ✔ = Yes ✗ = No / = Implied	IEEE 1471-2000	RM-ODP	ISO/IEC 42010	TOGAF	PRAXEME	RUP	Steen et al.	4+1/Rozanski-Woods	Zachman Framework	Nuseibeh	Ainsworth et al.	Dijkman et al.	Finkelstein et al.
A viewpoint is a partial specification of a system	✔	✔	✔	✔	/	/	/	✗	/	✔	✔	/	✔
A viewpoint is composed of one or more views	✔	✗	✔	✗	✗	✗	✗	✗	✔	✗	✗	✗	/
A viewpoint is a specification for creating views	✔	✗	✔	✔	/	/	✔	✔	✗	/	✗	✔	✗
A viewpoint is defined by means of a Metamodel	✔	✔	✔	✔	/	✗	/	/	✔	/	/	/	✔
Metamodels are centralized	✗	✗	✗	✗	/	✗	✗	✗	✗	✗	✗	✗	✗
Metamodels are decentralized	/	/	/	/	✗	✗	/	/	/	/	/	/	✔
There is assignment of stakeholders	✔	✔	✔	✔	✔	✔	✔	✔	✔	✗	✗	✔	✔
There is a method which adopts the definition	✔	✔	✔	✔	✔	✔	✗	✔	✔	✗	✗	✗	✗

used in the developed system at run-time. As we can see from Table 2, almost all of the research shares this opinion on the viewpoint definition.

A viewpoint is composed of one or more views: Only a few authors (Finkelstein et al., 1992; International Organization for Standardization, 2011b; Software Engineering Standards Committee of the IEEE Computer Society, 2000; Zachman, 2009) support this idea regarding *viewpoints* in their definitions, whereas our opinion is that this leads us to a more comprehensive and consistent system description.

A viewpoint is a specification for creating views: A viewpoint offers view types which are patterns or templates from which individual views can be derived. Thus a collection of views is a concrete instance of a viewpoint. Most of the authors see the viewpoint like this.

A viewpoint is defined by means of a metamodel: A viewpoint is derived from a metamodel. As in the previous point, almost all authors agree that metamodels are a basis for viewpoints defining their target domain usage.

Metamodels are centralized: This means viewpoints are based on one and only one metamodel. Each viewpoint governs which kind of model element can be represented, the consistency rules and completeness rules that need to be applied, and the different views that can be provided. Only PRAXEME method partly defends this point of view.

Metamodels are decentralized: This means viewpoints can be based on loosely coupled metamodels and encapsulate partial knowledge about the system and domain, specified in a particular, suitable representation scheme. Apart from PRAXEME and RUP other methods are supporting this interpretation of a viewpoint concept.

There is assignment of stakeholders: Each viewpoint targets a specific group of stakeholders. Each stakeholder is then responsible for designing his model part using the tools provided for the viewpoint. This is agreed upon by all listed references except for Ainsworth et al. (1994) and Nuseibeh (1994).

There is a method which adopts the definition: The question is whether a method in research or industry exists which adopts the given viewpoint definition. If a viewpoint definition is not used in any methods, this makes the definition a pure academic matter.

VIEWPOINT FEATURES

In the following, we give a list of features which we consider important for tooling aspects regarding the support of specific methods and present a table which summarizes the methods supporting them. The methods under consideration are: IEEE 1471-2000 (Software Engineering Standards Committee of the IEEE Computer Society, 2000), 4+1 (Kruchten, 1995), RM-ODP (International Organization for Standardization, 1996), ISO/IEC 42010 (International Organization for Standardization, 2011b), SysML (Object Management Group, 2013), Zachman Framework (2009), MODAF (Ministry of Defence, 2012), TOGAF (The Open Group, 2011), Boiten, Bowman, Derrick, Linington, and Steen (2000), PRAXEME (Praxeme Institute, 2011), and RUP (Kruchten, 2003). Later in this chapter, we will revisit the list of these features in the light of the use cases.

The first set, shown in Table 3, is called the viewpoint general features set. The listed features are being observed in the analyzed viewpoint methods and relate to the common viewpoint-based system design and its communication with external systems.

Support for Predefined Viewpoints: A predefined viewpoint is defined prior to the application of a method. Usually, there is a fixed set of predefined viewpoints which are not dedicated to a specific application domain, i.e. they are defined by the method itself (e.g. RM-ODP). As a consequence, this feature can imply a limitation on versatility of the method's application. One example in this regard is TOGAF which is a dedicate enterprise architecture. Apart from SysML, all of the methods are defining viewpoints in advance.

Table 3. Viewpoint general features

Key: ✔ = Yes ✗ = No / = Implied	IEEE 1471-2000	4+1	RM-ODP	ISO/IEC 42010	SysML	Zachman Framework	MODAF	TOGAF	Boiten et al.	PRAXEME	RUP
Support for predefined viewpoints	/	✔	✔	/	✗	✔	✔	✔	✔	✔	✔
Support for addressing specific stakeholders	/	✔	✔	/	✗	✔	✔	✔	✔	✔	✔
Support for adaptable presentation formalisms	✗	✗	✔	✗	✔	✔	✔	✔	✔	✗	✔
Support for transformation rules between different viewpoints	✔	✗	✗	✔	✗	✗	✗	✗	✗	✔	✗
Support for ad hoc viewpoint creation	✗	✗	✗	✗	✔	✗	✗	✔	✗	✗	✗
Support for dynamic viewpoint creation	✗	✗	✗	✗	✔	✗	✗	✗	✗	✗	✗
Support for dedicated exploitations	✗	✗	/	✗	/	✔	✔	✔	/	✗	✔
Support for adaptation to the organization context	✔	✗	✗	✔	✔	✔	✔	✔	✗	✔	✔
Support for relationship between viewpoints and development lifecycle	✗	/	✗	✗	✗	/	/	/	✗	✔	✔

Support for Addressing Specific Stakeholders: This feature defines whether the viewpoint definition in a method targets specific stakeholders and hence proposes specific concepts for these stakeholders. This implies a number of predefined user groups, which in turn means targeting specific domains in advance resulting in limitation of versatility and more specific stakeholder targeting. All considered methods except for SysML are defining stakeholders a priori and therefore support this feature. However, in IEEE 1471-2000 the feature is only implied.

Support for Adaptable Presentation Formalisms: Adaptable presentation formalisms provide the ability to adjust the presentation of model elements to the needs of certain users or to conform to certain viewpoints. This can be realized by a profile. Apart from IEEE, 4+1, ISO/IEC and PRAXEME, all methods are designed for inclusion of a flexible definition of visual representation of views.

Support for Transformation Rules between Different Viewpoints: This feature expresses the existence of constructive rules that allow deriving model elements for a particular viewpoint out of model elements from another viewpoint. These rules can be understood as a model transformation between different viewpoints. Although almost all of the viewpoint definitions support decentralized viewpoints, not every method provides transformation rules—except for IEEE, ISO/IEC and PRAXEME.

Support for Ad Hoc Viewpoint Creation: To address a specific application domain, some methods recommend defining viewpoints during a project's preparation phase to deal with stakeholder groups unspecified in the method. Only SysML and TOGAF see the need for introducing this feature.

Support for Dynamic Viewpoint Creation: Creating a viewpoint completely dynamically in a sense of creating a viewpoint on the fly after a project has already begun to serve at project setup time unforeseen needs of specific users. SysML is the only method that provides means for introducing dynamic viewpoints.

Support for Dedicated Exploitation: Adaptation of the usage of a specific viewpoint, which might be necessary while a project is already active (e.g. a viewpoint with deactivated dependencies to other viewpoints for protecting sensible information). Most of the methods support this at least partially.

Support for Adaptation to the Organization Context: This feature defines whether the existing viewpoint of a method can be adapted to a specific organizational context in order to suit the intended usage of the viewpoint. The three methods—4+1, RM-ODP and Boiten et al.—do not see the need for adapting the viewpoints to specific organizations.

Support for Relationship between Viewpoints and Development Lifecycle: This features outlines whether the method provides guidance or recommendations to relate viewpoints with the development lifecycle of the system under development. Only two of the listed methods, namely PRAXEME and RUP, are looking into realizing this feature to the full extent.

The second set of features, shown in Table 4, is called the viewpoint dedicated features set. The listed features are being observed in the analyzed viewpoint methods and describe more precisely which properties for the viewpoints the intended tool support should implement.

Contains an Own Viewpoint Definition: This feature signifies whether the method provides its own viewpoint definition. SysML, MODAF, and Boiten et al. do not provide own definitions.

Contains Impact Analysis Features: The question here is whether a traceability methodology is available and, if yes, whether it allows impact analysis of model changes. All the methods except for 4+1 and Zachman Framework do explicitly ask for this feature.

Contains Projection Features: This feature is derived from the projection definition in Section 2. Hence, the idea is to let different views edit the same model, whereas certain constructs are represented in different ways in each of the views. None of the methods apart from PRAXEME and RUP include this feature.

Contains Filtering Features: In their filtering capacities, viewpoints filter out model concepts from one or more metamodels, view types filter model concepts from the chosen model concepts in the viewpoint definition, and additional rules can be used in views to filter out model elements that should not be exposed at run time. Only the two standards—IEEE and ISO/IEC—as well as the PRAXEME method ask for this feature.

Table 4. Viewpoint dedicated features

Key: ✔ = Yes ✗ = No / = Implied	IEEE 1471-2000	4+1	RM-ODP	ISO/IEC 42010	SysML	Zachman Framework	MODAF	TOGAF	Boiten et al.	PRAXEME	RUP
Contains an own viewpoint definition	✔	✔	✔	✔	✗	/	✗	✔	✗	/	✔
Contains impact analysis features	✔	✗	/	✔	✔	✗	✔	✔	/	✔	✔
Contains projection features	✗	✗	✗	✗	✗	✗	✗	✗	✗	✔	✔
Contains filtering features	✔	✗	✗	✔	✗	✗	✗	✗	✗	/	✗
Contains consistency rules between viewpoints	/	/	✔	✔	✔	/	/	/	/	/	/
Contains consistency rules between views	✔	/	✔	✔	✔	/	/	/	/	/	/
Contains consistency rules within a view	✔	/	✔	✔	✔	/	/	/	/	/	/

Contains Consistency Rules between Different Viewpoints: If a system is modeled using different viewpoints, the model elements which are created in the underlying model instance are usually not completely independent from each other. There might be certain rules that need to be obeyed to ensure the overall consistency of the underlying model. All of the methods support the consistency rules feature.

Contains Consistency Rules between Views: In the same manner as it was for consistency between different viewpoints, certain rules may need to be obeyed in order to ensure the consistency in a certain viewpoint between its views. All of the methods ask for consistency between the views.

Contains Consistency Rules within a View: As in the two features before, there might be a threat regarding consistency inside a certain view due to editing from the different view instances. As a consequence, certain instance level consistency rules may have to be obeyed. All methods require this feature.

APPLICATIONS OF THE VIEWPOINT CONCEPT

The viewpoint concept as introduced in the previous sections is a generic concept that is intended to be applied in various contexts supporting different usage scenarios. It will help stakeholders in executing their development tasks. Generally, by facilitating the viewpoint concept it is possible to better focus on the important information for a certain activity while unnecessary information is hidden from the user. Depending on the context in which the viewpoint concept is applied, significant productivity and quality increases can be expected.

Since the viewpoint concept deals mainly with the appropriate presentation of relevant model information to potential users, its realization is strongly related to suitable tool support. Without such tool support, benefits of the viewpoint concept will be hard to achieve. On the other

hand, modeling tools that support viewpoints will have a strategic advantage over those which work without viewpoints as they will offer an increased productivity to their users.

In the following, we will show with concrete application cases how the viewpoint concept is used. In order to demonstrate the generic applicability of the viewpoint concept, we will outline two examples from the domain of embedded systems and one example from the domain of enterprise systems modeling. The modeling languages in these domains as well as the used tools differ but nevertheless, the generic concepts of viewpoint support are applicable and lead in both cases to increased productivity and quality of the work products.

Functional Safety of Embedded Systems in Automotive: Introduction to the Domain

Ensuring functional safety of cars is an important aspect of the vehicle development. Functional safety is the part of overall system safety that is related to the functionality of the system. The task is to deliver products (cars) that in the ideal case will not cause any harm to humans resulting from a malfunctioning behavior of the system. As this is technically not feasible, the development engineers have at least to reduce the remaining risk of potential malfunctions to a tolerable level. Examples for safety issues in the context of functional safety are: accidents caused by sudden explosion of an airbag, loss of control over the car caused by a blocking of an electric steering support system, and unintended acceleration of a car caused by a malfunction of the electronic gas pedal.

Issues not in the scope of functional safety are e.g. the use of toxic material or a fire caused by improper usage of materials.

The consideration of functional safety is by definition strongly related to the design of the system itself – the functionality of the system. To cope with functional safety in the development process, additional (safety analysis) activities are included. In order to have a guideline for and a control of safety analysis activities that are necessary to appropriately reduce the remaining risk, international standards have been defined. ISO 26262 (International Organization for Standardization, 2011a) is a recently published standard which regulates safety analysis activities for road vehicles. It defines the acceptable level for the remaining risk and it provides requirements upon the development processes that are applied.

Since over the recent years more and more safety related functions in vehicles are realized by electronics and software based systems, all stakeholders in the development process need to be aware of safety related aspects and therefore shall apply appropriate functional safety analysis methods during requirements elicitation, development, verification, and validation. Preliminary and detailed safety analysis methods have to be applied at the conceptual, system, component configuration and software levels in order to identify harmful behavior and take appropriate risk reduction actions. Hazard analysis and risk assessment, Fault Tree Analysis (FTA), Failure Mode and Effects Analysis (FMEA), and hardware diagnostic coverage metrics are prominent examples for such safety analysis methods.

Modern development processes put architecture and design models in the center of system engineering activities; standardized modeling languages such as SysML/Unified Modeling Languages (UML) are applied (even for Hardware), typically in concert with requirement management environments and team support platforms. ISO 26262 requires safety analysis procedures to be tightly integrated with the engineering activities. Moreover it is essential to provide capabilities to execute safety analysis procedures iteratively and to trace the safety aspects to the Hardware (HW)/Software (SW) artifacts.

Within the scope of ISO 26262, we will demonstrate the applicability and usefulness of the viewpoint concept in two different scenarios, which are elaborated in the following. The first

scenario deals with the strong requirements upon the assessment of safety related work products imposed by the standard. Normally, these work products have to undergo an in-depth review process to ensure their conformity with the standard regulations. Such reviews are time-consuming tasks and involve several independent persons (at least the reviewer and the developer). Furthermore, there might be iterations and all results, issues etc. must be documented. We are going to show that the application of viewpoints will help both reviewers and developers to increase their productivity.

A second aspect is the typical distributed work organization in the automotive industry. Today we find a deep supply chain involving besides the OEM many suppliers. Whereas formally the OEM is responsible for the safety of the products that it brings to the market, the suppliers get besides their functional requirements also safety related requirements that they need to implement in their parts. Furthermore, suppliers need to deliver also safety related information (e.g. evidences and argumentations) back to their customers so that at the end the OEM has all information necessary to evaluate the overall functional safety of the car. We will show with the second use case that the application of viewpoints will help to improve the information exchange between members of the supply chain in both directions with respect to the completeness and to the confidentiality of such information.

As stated before, the application of the viewpoint concept is strongly related to the provision of suitable tool support. We are going to demonstrate the use cases of the viewpoint concept in the domain of functional safety using the tool medini analyze of ikv++ technologies as an example. Please note, that other modeling and safety analysis tool could also benefit from the generic viewpoint concept and that this tool is only used for demonstration purposes.

Medini analyze is an integrated tool which implements efficiently core activities of the functional safety analysis and integrates them with the existing processes. Target users are safety managers and experts as well as development engineers and quality managers involved in the development of electronic and software based components in the automotive industry. Medini analyze is a model based tool that means all information is available via the different types of models that are supported. Because of this fact, the viewpoint concept is applicable to medini analyze and realized in this tool.

Use Case 1: Functional Safety of Embedded Systems in Automotive: Review of Safety-Related Work Products

This Use Case relates to the area of safety analysis of electronic vehicle functions according to ISO 26262. The structured safety activities prescribed by the standard lead to several safety-related work products, that have to be provided to complete the so called safe case for the system under development. After certain development activities have been executed, the standard also requires reviews of the produced work products and artifacts. This review typically focuses not on all safety related information which is available for the system, instead it requires a clear focus. It is essential for an effective review to have a support for defining this focus as a view and make this view available to the participants in the review.

Scenario Description

In general, we can differentiate between two phases of this use case. In the first phase, the review process is initiated and the review is conducted.

Figure 2. Review of safety-related work products

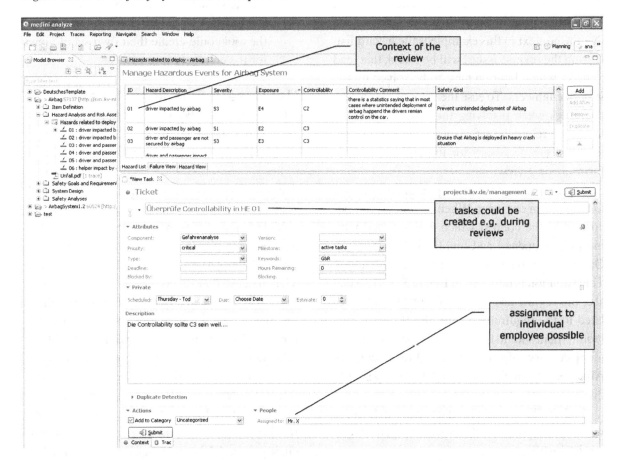

In the review, a typical set of models and model elements (depending on the type of the review) are under inspection. During the review, issues (or tasks) are created that have a specific context (in terms of the involved models and model elements and are subject for later work.

In the second phase the created tasks are resolved. For each task, it is necessary to work on the context that has been set in the first phase.

The use case is conducted by a group of safety experts, typically engineers and a safety manager. They select a set of models or model elements for a review. A context for a review could be for example the Hazard and Risk Analysis (HARA) or the set of functional safety requirements. Ideally, the tool would allow for the definition of pre-defined context definitions (views) which can be selected for the review purpose. Then, the tool would only present the model elements which are relevant for the review. The review activity itself would be just navigation through the usual editors like it is shown in Figure 2. One can see that the context in this example is the hazard list. During the review, issues will be found, that lead to tasks which later on have to be solved by the system engineers. To ease the work of the system engineers when resolving the task, it would be helpful to relate the set of model elements to the task that are necessary for its fulfillment. This forms again a view.

Each review comment is presented as a task to the system engineer. When clicking on a specific task, the context which is associated with the task is opened and highlighted in the tool. Other elements which are not relevant are hided. This situation is depicted in Figure 3. As one can see, the review has resulted in a task that is related to the requirement with the ID R02. This requirement should be opened with the editor and the necessary information should be visualized. All other information that is not necessary should disappear.

In the following, we list the viewpoint features which are required for this use case together with a brief justification:

- **Predefined Viewpoints:** It should be possible to define predefined viewpoints for a set of typical reviews that are prescribed by ISO 26262.
- **Address Specific Stakeholder:** The stakeholders are the Reviewers and the System engineers that have to resolve the issues.
- **Select Data from the Entire Domain Description:** Depending on the kind of review, multiple models may be involved. An example is a review of the "Functional Safety Concept" which involves HARA, System Architecture, Functional Safety Requirements and System Definitions.
- **Allow Ad Hoc Viewpoint Creation:** When a task is created, the context forms a view, which has to be created ad hoc (during the review process). The context depends on the model and cannot be foreseen.
- **Allow Dynamic Viewpoint Creation:** Not all viewpoints for reviews can be set beforehand. It would be good, if the users of the tool could create their own set of views for the conduction of reviews.

Use Case 2: Functional Safety of Embedded Systems in Automotive: Exchange of Safety Artifacts in the Supply Chain

This Use Case relates to the organization of supply chains in the area of safety analysis of electronic vehicle functions according to ISO 26262. Safety activities are usually executed in the whole lifecycle of the system (item) under development. That means, that safety related work products have to be exchanged between the organizations involved in the development and safety analysis process. For automotive systems, it is typical that a deep supply chain is involved in the development consisting of multiple organizations with different roles and responsibilities. In this Use Case, we consider an exchange of safety related information between different organizations in the supply chain; say an OEM and a tier 1 supplier. Involved

Figure 3. Review results of the safety-related products

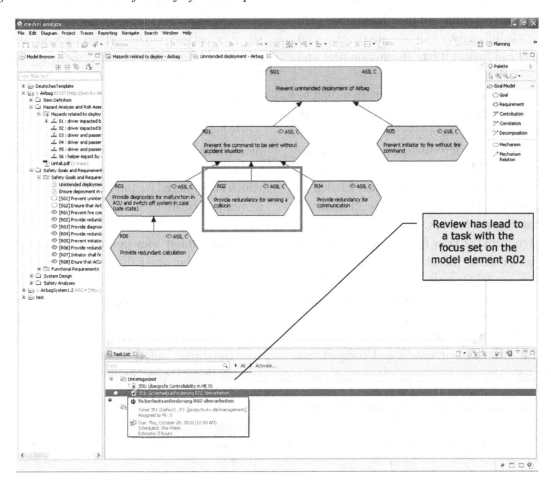

stakeholders are the safety teams at the OEM and this supplier. The used tool is medini analyze and the intention is to define a view for the supplier on the overall safety case managed by the OEM that allows to restrict the model elements visible to the supplier and also to allow a flexible exchange of these model elements between OEM and supplier

Scenario Description

In this scenario, we assume that the OEM after the finalization of the functional safety concept assigns the responsibility for an architecture element to a supplier. In order to do this, the OEM needs to provide to the supplier the following safety related information:

- A part of the intended architecture that contains the component that the supplier has to realize. The boundaries have to be selected because sometime it is necessary to also include some surrounding of the component.

- The safety requirements that are allocated to the specific architecture element(s).

- The safety goals which are connected to the included safety requirements.

- Any supplementary information like attached document etc.

This forms the view for the supplier. After the selection process (see Figure 4), the view is created (i.e. all necessary model elements are collected) and it is exported, so that it can be submitted to the supplier. It is supposed that the supplier also uses the medini analyze tool, obtains the exported view and can import it into its environment and start its work.

Subsequently, the supplier develops a component based on the information that was handed-in by the OEM. After the completion of the development work, certain information has to be passed back to the OEM. This information needs to be related to the information which was handed in by the OEM, but it is not necessarily all information that the supplier has. Hence a viewpoint is needed to form the view for the OEM. After the selection process, the view is created (i.e. all necessary model elements are collected) and it is exported, so that it can be submitted to the OEM. It is supposed that the OEM also uses the medini analyze tool, obtains the exported view and can import it into its environment and complete its work.

Finally, the OEM gets back an information package with safety related information form the supplier and needs to import this back into its own project. During the import, consistency has to be ensured. That means, the information passed back needs to be consistently integrated into the information which is present on the OEM side. Appropriate actions (possibly with user involvement) have to be triggered if this is not possible.

In the following, we list the viewpoint features, which are required for this use case together with a brief justification:

- **Predefined Viewpoints:** There are typical borders at which the change of responsibility is executed. An example is the border at the functional safety concept level where the OEM hands over the responsibility for a certain component to a supplier after the functional safety concept is established. In that case, a "supplier viewpoint on the

Figure 4. Exchange of safety artifacts in the supply chain

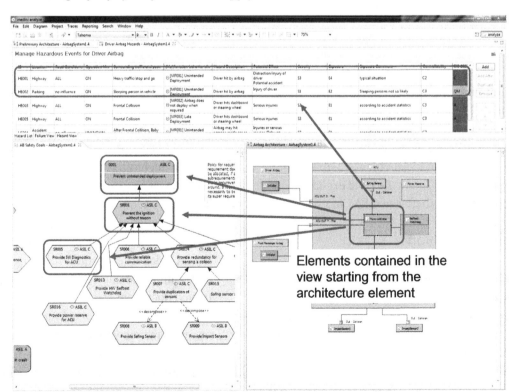

functional safety concept" could be pre-defined and parameterized by e.g. the system elements for which the supplier gets responsibility.

- **Select Data from the Entire Domain Description:** The exchange of information covers multiple involved models and model elements.
- **Support Consistency Rules Between Different Viewpoints:** There is a consistency problem when the information from the supplier needs to be integrated into the OEM models or when the OEM sends an update of the supplier view to the supplier.
- **Allow Dynamic Viewpoint Creation:** It might be the case that an arbitrary export viewpoint is to be defined by the OEM if a predefined viewpoint is not available.

Enterprise Systems Modeling: Introduction to the Domain

One of the most widespread frameworks for Enterprise (Architecture) Modeling is The Open Group Architecture Framework (TOGAF), which was developed by the vendor- and technology-independent consortium The Open Group in the 1990s (Winter & Fischer, 2006). While it was initially developed as a framework for the design of technical architectures, it has evolved to an EAM-framework (Lankhorst, 2013). In contrast to many other EAM-frameworks, TOGAF was not developed as a theoretical model, but it is based on best practices of more than 300 companies. In 2011 version 9.1 of TOGAF has been released, which represents the current state (The Open Group, 2011). TOGAF defines views and viewpoints as follows: "A viewpoint defines the perspective from which a view is taken. More specifically, a viewpoint defines: how to construct and use a view (by means of an appropriate schema or template); the information that should appear in the view; the modeling techniques for expressing and analyzing the information; and a rationale for

these choices (e.g., by describing the purpose and intended audience of the view)."

Architecture views are representations of the overall architecture in terms meaningful to stakeholders. They enable the architecture to be communicated to and understood by the stakeholders, so they can verify that the system will address their concerns. Therefore, specific viewpoints can be created under TOGAF. By default, we can consider that the four architecture domains—Business, Data, Application and Technology—are predefined viewpoints for TOGAF.

Hence, the TOGAF architecture domains are themselves viewpoints that address stakeholders' concerns as follows:

- The Business Architecture domain addresses the needs of users, planners, and business management.
- The Data Architecture domain addresses the needs of database designers, database administrators, and system engineers.
- The Application Architecture domain addresses the needs of system and software engineers.
- The Technology Architecture domain addresses the needs of acquirers, operators, administrators, and managers.

During the Architecture Vision phase, the viewpoints relevant to the project have to be identified, considering which architecture views and viewpoints need to be developed to satisfy the various stakeholder requirements. One major usage of viewpoints is to enable the architect to demonstrate how the stakeholder concerns are being addressed in the envisaged architecture. Under TOGAF, the stakeholders need to be identified, from which the viewpoints to develop will be determined.

For a given viewpoint, appropriate kinds of diagrams, tools and methods have to be defined. When applying TOGAF within a particular enterprise or project, it may be necessary to take each of these stakeholder types (e.g., user, database administrator, and acquirer) and create an explicit set of stakeholder concerns. These concerns can then be used to refine and enhance the basic viewpoints provided by TOGAF.

Definition of the Four Predefined Default Viewpoints Specified by TOGAF

In the modeling tool Modelio, the four standard Viewpoints of TOGAF—Business Architecture, Information System Architecture, Data Architecture, and Technical Architecture—are predefined.

The Data Architecture is actually distributed between the Business and the Application Architecture. Therefore, the data architecture has been separated into two levels: Business Entities, which represents the data at the conceptual level under the Business Architecture, and Data Architecture under the Application Architecture. At this level, data is logically defined, and is intended to correspond to persistent data for repositories such as RDBs.

In doing so, five viewpoints result from the split of the Data Architecture: Business Entities, Business Architecture, Application Architecture, Data Architecture, and Technical Architecture. To support these Viewpoints adequately, Modelio provides 30 different types of diagrams that belong to these different Viewpoints; thus, enabling a simpler GUI for each kind of stakeholder.

Utilizing Viewpoints in Modelio

To enable a successful utilization of the viewpoints within an Enterprise Systems Modeling project, Modelio provides different kinds of services. These services are for instance consistency check services and model transformation services; the latter one for example helps to move from one viewpoint to another. In addition, there will be services to create messages from Business Entities, or services creating data models from Business

Entities. These services will need to be refined, in order to define the appropriate wizards that help the modeling task. Following the TOGAF method, there is also a support of various services that generate recommended artifacts, such as matrixes and catalogs.

Regarding consistency check services, it can be distinguished between two different types: consistency rules have structural aspects that are checks which guarantee that each model element is physically well established, and semantics aspects that are checks which guarantee that the model is semantically sound. For example, a structural consistency check will verify that each link or dependency has two allocated ends; a semantic check can verify that there is no circular generalization graph.

Definition of Ad Hoc Viewpoints that are not Pre-Specified by TOGAF

In order to create in Modelio a specific viewpoint dedicated to specific stakeholder concerns, one can select from the predefined five viewpoints the services, consistency checks and kind of diagrams that are relevant for the viewpoint under development. Afterwards, the models within a viewpoint can be defined and the necessary model transformation or pattern applications realized. At this point it should be mentioned that it is intended that new viewpoints can be built by reusing elements of existing viewpoints. This will be an easier and faster activity for method/process engineers.

While evaluating the usage of Modelio, we frequently noticed that only parts of an enterprise architecture are being modeled. One usual practice is to focus on the application architecture, using a SOA-based architecture approach. Another usual practice is to focus on business process modeling, which is only a part of the Business Architecture viewpoint. SOA Architecture and Business Architecture are therefore two specifically defined viewpoints.

Use Case 3: Enterprise Systems Modeling: Using Viewpoint in TOGAF

Once the viewpoints for a TOGAF supported project are defined, one can start with modeling the underlying enterprise. In this regard, the current use case will outline how Modelio realizes the modeling of an example company. In doing so, the use case will further validate whether the ViBaM viewpoint definition is appropriate and which of the viewpoint features are addressed by the tool and the underlying use case.

Scenario Description

This use case will outline how Modelio utilizes the Viewpoint concept in the domain of Enterprise Architecture Modeling. For the scenario we chose a travel agency as the underlying enterprise. Travel agencies represent a well-known business, consisting in selling travels, managing bookings, managing partners (airlines, hotels, etc.). In doing

so, this example focuses on a small travel agency that has to face the necessity of providing Internet access to increase its revenues, which represents one of their business goals.

At the very beginning of the Enterprise Architecture initiative, a requirement document will be created. Based on this document the models and artifacts can be established.

After a certain Viewpoint is selected during the startup of Modelio, the modeling environment will be adapted to the Viewpoint specifics. In doing so, the Modelio explorer, which comprises different diagram types and models, will be adapted to the Viewpoint features. In addition, the context menu of the depicted models will also change, so that only those features are provided that are necessary for the Viewpoint. On top of this, also the diagram itself will be adapted to the chosen Viewpoint and hence provide means for a dedicated exploitation of the Enterprise Architecture models.

Consequently, Modelio allows a very user, respectively Viewpoint-centric, modeling of the

Figure 5. Enterprise modeling based on TOGAF

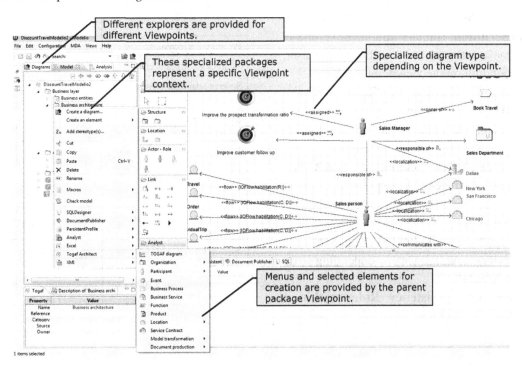

views and models in accordance to the underlying requirement document. As mentioned in the previous subsection, Modelio further allows the adaption of the pre-defined Viewpoints to the needs of the respective modeling project or enterprise.

In Figure 5 the Modelio modeling tool suite is illustrated. An organization role diagram is shown that illustrates the actors within an organization (such as here "Sales Manager" and "Sales Person"), their hierarchical links, their goals, their responsibilities, localization, etc. In doing so, Figure 5 relates to the Business Architecture viewpoint.

In the following, we list the viewpoint features which are required for this use case together with a brief justification:

- **Predefined Viewpoints:** To support the TOGAF methodology appropriately within Modelio, the different Viewpoints are given in a predefined fashion.
- **Address Specific Stakeholder:** Each viewpoint is related to specific stakeholders (e.g. user, database administrator, and acquirer).
- **Select data from the Entire Domain Description:** In the initialization phase, some existing data is visible to the user, depending on the viewpoint's rules.
- **Support Consistency Rules between Different Viewpoints:** Consistency will in particular be checked between the preceding viewpoints and the current one.
- **Supports Transformation Rules between Different Viewpoints:** Moving from a viewpoint to another necessitates this feature.
- **Support for Dedicated Exploitations:** Services will vary depending on viewpoints.
- **Contains Consistency Rules within a View:** For example in Figure 5, there are

specific kinds of dependencies (assigned, responsible of, etc.) which have specific allowed kinds of elements as origin and destination. Completeness can be checked by identifying orphan elements.

Viewpoint Features Required to Realize Use Cases

Table 5 lists several viewpoint features—which have already been introduced in Section 4 (cf. Table 3)—and indicates whether they are required to realize the use cases illustrated above. The use cases nicely extend each other in the sense that together they require a significant number of the introduced features and that they do not purely duplicate their requirements.

Table 5. Viewpoint features coverage in use cases

Viewpoint features	Use Case 1	Use Case 2	Use Case 3
Predefined Viewpoints	Yes	Yes	Yes
Address specific stakeholder	Yes	No	Yes
Support adaptable presentation formalisms	No	No	Yes
Support consistency rules between viewpoints	No	Yes	Yes
Supports transformation rules between viewpoints	No	Yes	Yes
Allow ad hoc viewpoint creation	Yes	No	No
Allow dynamic viewpoint creation	Yes	Yes	No
Support for dedicated exploitation	No	No	Yes
Contains consistency rules within a view	No	No	Yes

CONCLUSION AND OUTLOOK

In this chapter, we reported on results of the ViBaM project in which viewpoint concepts are investigated. We presented an overview of the current state-of-the-art on viewpoint definitions, concepts, and methods. Derived from the definitions we found in literature, we presented definitions for the list of concepts on the basis of which we defined the position that we take in the ViBaM project regarding viewpoint concepts.

After we discussed ViBaM's position on viewpoints, we presented a list of features that we consider important for tool support of the viewpoint definition presented in this paper.

In the last section we presented three use cases in which the viewpoint concepts are illustrated. We explained how the viewpoint concept is applied in these use cases and listed the viewpoint features that are relevant for each use case.

The next step in our work is to complete the integration of the defined concepts in the commercial modeling tools medini analyze and Modelio. As it was shown in the use cases, a significant number of the presented features are already available and ready for use. With respect to the underlying technologies the implementation of dynamic viewpoint creation is rather challenging. Already the definition of views for example on the basis of EMF/GMF is cumbersome if one is not satisfied with the default behavior that is offered for this technology stack. We will investigated what changes would be needed to make the use of the basic technologies more flexible. Additionally, we will investigate whether the concepts of task-task oriented interfaces will allow us to introduce views more dynamically. The idea is to use a recommender system together with an issue tracking system and to support the user by making suggestions for a proper context (i.e. a view) to work on a given task.

ACKNOWLEDGMENT

This work has been funded by the German Federal Ministry of Education and Research (FKZ 01QE1106B/C) and by the French OSEO through ICT Project ViBaM (Viewpoint-Based Modeling) which is running in the context of the European Eurostars Program (E!5529). The authors wish to acknowledge EUREKA and the Commission for their support. We also thank the reviewers for their valuable comments.

REFERENCES

Ainsworth, M., Cruickshank, A. H., Wallis, P. J. L., & Groves, L. J. (1994). Viewpoint specification and Z. *Information and Software Technology*, *36*(1), 43–51. doi:10.1016/0950-5849(94)90007-8.

Avison, D., Wood-Harper, A., Vidgen, R., & Wood, J. (1998). A further exploration into information systems development: The evolution of Multiview2. *Information Technology & People*, *11*(2), 124–139. doi:10.1108/09593849810218319.

Boiten, E., Bowman, H., Derrick, J., Linington, P., & Steen, M. (2000). Viewpoint consistency in ODP. *Computer Networks*, *34*(3), 503–537. doi:10.1016/S1389-1286(00)00114-6.

Bosch, J., & Bosch-Sijtsema, P. (2010). From integration to composition: On the impact of software product lines, global development and ecosystems. *Journal of Systems and Software*, *83*(1), 67–76. doi:10.1016/j.jss.2009.06.051.

Champeau, J., Mekerke, F., & Rochefort, E. (2003). Towards a clear definition of patterns, aspects and views in MDA. In J. Ralyté & C. Rolland (Eds.), *Proceedings of the First International Workshop on Engineering Methods to Support Information Systems Evolution (EMSISE)* (pp. 29–36). Geneva, Switzerland: EMSISE.

Dijkman, R. M., Quartel, D. A. C., & van Sinderen, M. J. (2008). Consistency in multi-viewpoint design of enterprise information systems. *Information and Software Technology, 50*(7–8), 737–752. doi:10.1016/j.infsof.2007.07.007.

Easterbrook, S., Yu, E., Ar, J., Fan, Y., Horkoff, J., Leica, M., & Qadir, R. A. (2005). Do viewpoints lead to better conceptual models? An exploratory case study. In *Proceedings of the 13th IEEE International Conference on Requirements Engineering (RE '05)* (pp. 199–208). Los Alamitos, CA: IEEE Computer Society.

Finkelstein, A., Easterbrook, S., Kramer, J., & Nuseibeh, B. (1992). *Requirements engineering through viewpoints*. London: Imperial College London, Department of Computing. Retrieved February 2, 2013, from http://discovery.ucl. ac.uk/854/1/3.3_dra.pdf

Gamma, E., Helm, R., Johnson, R., & Vlissides, J. (1995). *Design patterns: Elements of reusable object-oriented software*. Boston, MA: Addison-Wesley.

Goldschmidt, T., Becker, S., & Burger, E. (2012). Towards a tool-oriented taxonomy of view-based modelling. In E. J. Sinz & A. Schürr (Eds.), GI-LNI: Vol. 201: Modellierung 2012 (pp. 59–74). Bonn, Germany: Gesellschaft für Informatik.

International Organization for Standardization. (1996). *ISO/IEC 10746-3:1996: Information technology – Open distributed processing – Reference model: Architecture (RM-ODP)*. Retrieved February 2, 2013, from http://www.iso.org/iso/ home/store/catalogue_ics/catalogue_detail_ics. htm?csnumber=20697

International Organization for Standardization. (2007). *ISO/IEC 42010:2007: Systems and software engineering -- Recommended practice for architectural description of software-intensive systems*. Retrieved February 2, 2013, from http://www.iso.org/iso/catalogue_detail. htm?csnumber=45991

International Organization for Standardization. (2011a). *ISO 26262-1:2011: Road vehicles -- Functional safety -- Part 1: Vocabulary*. Retrieved February 2, 2013, from http://www.iso.org/iso/ catalogue_detail?csnumber=43464

International Organization for Standardization. (2011b). *ISO/IEC/IEEE 42010:2011: Systems and software engineering -- Architecture description*. Retrieved February 2, 2013, from http://www.iso. org/iso/catalogue_detail.htm?csnumber=50508

Kotonya, G., & Sommerville, I. (1996). Requirements engineering with viewpoints. *Software Engineering Journal, 11*(1), 5–18. doi:10.1049/ sej.1996.0002.

Kruchten, P. (1995). Architectural blueprints — The 4+1 view model of software architecture. *IEEE Software, 12*(6), 42–50. doi:10.1109/52.469759.

Kruchten, P. (2003). *Rational unified process: An introduction* (3rd ed.). Boston, MA: Addison-Wesley.

Lankhorst, M. (2013). *Enterprise architecture at work: Modelling, communication and analysis* (3rd ed.). Berlin, Germany: Springer. doi:10.1007/978-3-642-29651-2.

Lehtola, L., Gause, D., Dumdum, U., & Barnes, R. J. (2011). The challenge of release planning. In *Proceedings of the 5th International Workshop on Software Product Management (IWSPM 2011)* (pp. 36–45). New York: IEEE.

Ministry of Defence. (2012). *MOD architecture framework (MODAF)*. Retrieved February 2, 2013, from https://www.gov.uk/mod-architecture-framework

Mohagheghi, P., Fernandez, M., Martell, J., Fritzsche, M., & Gilani, W. (2009). MDE adoption in industry: Challenges and success criteria. In Chaudron, M. (Ed.), *Models in Software Engineering* (pp. 54–59). Berlin, Germany: Springer. doi:10.1007/978-3-642-01648-6_6.

Nuseibeh, B. A. (1994). *A multi-perspective framework for method integration*. (Doctoral dissertation). Imperial College of Science, Technology and Medicine, University of London, Department of Computing, London, UK.

Object Management Group. (2003). *MDA guide version 1.0.1*. Retrieved February 2, 2013, from http://www.omg.org/cgi-bin/doc?omg/03-06-01. pdf

Object Management Group. (2011a). *Business process model and notation (BPMN) version 2.0*. Retrieved February 2, 2013, from http://www. omg.org/spec/BPMN/2.0/PDF

Object Management Group. (2011b). *Unified modeling language (UML) version 2.4.1*. Retrieved February 2, 2013, from http://www.omg.org/spec/ UML/2.4.1/

Object Management Group. (2013). *OMG systems modeling language (OMG SysML) version 1.3*. Retrieved February 2, 2013, from http://www. omg.org/spec/SysML/1.3/PDF Praxeme Institute. (2011). *PRAXEME – Opus, the product*. Retrieved February 2, 2013, from http://www.praxeme.org/ index.php?n=Opus.Opus?userlang=en

Rozanski, N., & Woods, E. (2005). *Software systems architecture: Working with stakeholders using viewpoints and perspectives* (2nd ed.). Boston, MA: Addison-Wesley.

Schatz, B. (2011). 10 years model-driven – What did we achieve? In M. Popović & V. Vranić (Eds.), *Proceedings of the 2nd Eastern European Regional Conference (EERC) on the Engineering of Computer Based Systems (ECBS)*. Bratislava, Slovakia: IEEE Computer Society.

Software Engineering Standards Committee of the IEEE Computer Society. (2000). *1471-2000 – IEEE recommended practice for architectural description for software-intensive systems (IEEE Standard)*. Los Alamitos, CA: IEEE Computer Society.

Steen, M., Akehurst, D., Doest, H., & Lankhorst, M. (2004). Supporting viewpoint-oriented enterprise architecture. In *Proceedings of the 8th IEEE International Enterprise Distributed Object Computing Conference (EDOC'04)* (pp. 201–211). Monterey, CA: IEEE Computer Society.

The Open Group. (2011). *Welcome to TOGAF version 9.1, an open group standard*. Retrieved February 2, 2013, from http://pubs.opengroup.org/ architecture/togaf9-doc/arch/index.html

Winter, R., & Fischer, R. (2006). Essential layers, artifacts, and dependencies of enterprise architecture. In *Proceedings of the 10th IEEE International Enterprise Distributed Object Computing Conference Workshop (EDOCW '06), Workshop on Trends in Enterprise Architecture Research (TEAR 2006)*. Hong Kong: IEEE Computer Society.

Wood-Harper, A. T., Antill, L., & Avison, D. E. (1985). *Information systems definition: The multiview approach*. Oxford, UK: Blackwell Scientific Publications.

Zachman, J. A. (1997). *Concepts of the framework for enterprise architecture*. Retrieved February 2, 2013, from http://www.ies.aust.com/papers/ zachman3.htm

Zachman, J. A. (2009). *John Zachman's concise definition of the Zachman framework*. Retrieved February 2, 2013, from http://www.zachman.com/ about-the-zachman-framework

Chapter 16
Analyzing Mobile Application Software Power Consumption via Model–Driven Engineering

Chris Thompson
Vanderbilt University, USA

Jules White
Vanderbilt University, USA

Douglas C. Schmidt
Vanderbilt University, USA

ABSTRACT

Smartphones are mobile devices that travel with their owners and provide increasingly powerful services. The software implementing these services must conserve battery power since smartphones may operate for days without being recharged. It is hard, however, to design smartphone software that minimizes power consumption. For example, multiple layers of abstractions and middleware sit between an application and the hardware, which make it hard to predict the power consumption of a potential application design accurately. Application developers must therefore wait until after implementation (when changes are more expensive) to determine the power consumption characteristics of a design. This chapter provides three contributions to the study of applying model-driven engineering to analyze power consumption early in the lifecycle of smartphone applications. First, it presents a model-driven methodology for accurately emulating the power consumption of smartphone application architectures. Second, it describes the System Power Optimization Tool (SPOT), which is a model-driven tool that automates power consumption emulation code generation and simplifies analysis. Third, it empirically demonstrates how SPOT can estimate power consumption to within ~3-4% of actual power consumption for representative smartphone applications.

DOI: 10.4018/978-1-4666-4494-6.ch016

INTRODUCTION

Emerging Trends and Challenges

Smart devices, such as smartphones and tablet computers have been steadily increasing in capabilities and performance. Modern devices commonly contain multi-core CPUs running at clock speeds well over 1GHz. For example, the Galaxy Nexus from Google contains a 1.2 GHz, dual-core processor with 1GB of RAM and other Android smartphones now have quad-core processors.

The increased performance of smart devices comes at a cost, however. Due to their mobility, smartphones and tablets run exclusively on battery power, so every increase in performance must be balanced with an increase in battery capacity to maximize the time between charges. The design of applications running on the device can also impact battery life, *e.g.*, frequently sampling the GPS or making network requests can consume a significant amount of power.

To balance functionality with power consumption effectively, developers must understand various trade-offs. A key trade-off is between performance and battery life, as well as the implications of the application and infrastructure software architecture on power consumption. Functional requirements, such as minimum application response time, can conflict with power consumption optimization needs.

For example, a traffic accident detection application (White, Clarke, Dougherty, Thompson, & Schmidt, 2010) must be able to detect sudden accelerations indicative of a car accident. To detect acceleration events that indicate accidents, the application must sample device sensors and perform numerous calculations at a high rate. Conflicts occur between the *functional requirements* (*e.g.*, the minimum sensor sampling rate needed to accurately detect accidents) and *non-functional*

requirements (*e.g.*, sustaining operations on the mobile device without frequent battery recharging).

Due to complex middleware, Operating System (OS), and networking layers, it is hard to predict the effects of application software architecture decisions on power consumption *a priori*, *i.e.*, without actually *implementing* a design, which makes it hard to analyze the power consumption of design until late in the development cycle, when changes are more expensive (Kang, Park, Seo, Choi, & Hong, 2008). For example, a developer may elect to use HTTPS instead of HTTP to satisfy a security requirement by making communication between the application and server more confidential. It is currently hard, however, to predict how much additional power is consumed by the added encryption and decryption of data without actually implementing the system.

It is also hard to quantify the trade-off between power consumption and security, as well as many other design decisions. Moreover, certain design decisions, such as data transmission policies (*e.g.*, should an application transmit immediately or wait for a specific medium like a Wi-Fi or 3G cellular connection) are hard to analyze early in the design cycle, due to their variability. For example, if an application only sends data intermittently, it may be beneficial to transmit small amounts of data over cellular connections due to the decreased power consumption of 3G cellular connection compared to Wi-Fi (Agarwal, Chandra, Wolman, Bahl, Chin, & Gupta, 2007). If a large amount of data must be sent, however, the total time required to transmit it over 3G may negate the benefit of using the less power consumptive connection. The cellular connection will take longer to transmit the data, therefore, which may in turn consume more total power than the Wi-Fi radio that can transmit the data faster.

Solution Approach → Power Consumption Emulation of Mobile Software Architectures with Model-Driven Testing and Auto-Generated Code

By using *Model-Driven Engineering* (MDE) (Schmidt, 2006), we allow developers to specify a *Domain-Specific Modeling Language* (DSML) (Ledeczi, Bakay, Maroti, Volgyesi, Nordstrom, Sprinkle, & Karsai, 2001) to capture key software architecture elements related to power consumption. Developers can then use automated code generators to produce emulation code from this model. Developers can also execute the generated emulation code on target hardware, collect power consumption information, and analyze the application's power consumption prior to actual deployment.

This emulation code allows developers to analyze proposed software architectures prior to investing significant time and effort in a complete implementation of the design. The auto-generation of emulation code also enables developers to compare a range of different designs quantitatively during initial phases of a development process (Hill, Schmidt, Slaby, & Porter, 2008), which allows developers to select designs that satisfy both functional and non-functional requirements while minimizing overall power consumption. This analysis can also occur early in the software lifecycle (*e.g.*, at design time), thereby avoiding costly changes to optimize power consumption later downstream.

This paper describes the *System Power Optimization Tool* (SPOT), which uses MDE techniques to analyze the power consumption of mobile software architectures. SPOT allows developers to create high-level models of candidate software architectures using the *System Power Optimization Modeling Language* (SPOML) that capture key software components related to power consumption. SPOT generates emulation code from the model that can be executed on target devices.

As the SPOT-generated emulation code runs it is instrumented to collect power consumption data. This data can later be downloaded and analyzed offline. Developers can apply this emulation and analysis cycle to understand power consumption implications of their designs *without* expensive and time consuming manual programming using third-generation languages, such as C#, C/C++, and Java.

SPOT's generated emulation code mimics key power-consuming aspects of a proposed software architecture. Key power consumptive components of mobile application software architectures include GPS, acceleration, orientation, sensor data consumers, and network bandwidth consumers (Pering, Agarwal, Gupta, & Want, 2006). Focusing SPOT on these components allows developers to model the most significant power expenditures of their applications. Moreover, as applications are constructed, the generated emulation code can be replaced incrementally with actual application components, allowing developers to refine the accuracy of the analysis continuously throughout the software lifecycle.

This chapter extends our prior work on model-driven application power consumption analysis (Thompson, White, Dougherty, & Schmidt, 2009) by meeting the following objectives: (1) surveying related work on optimizing power consumption, domain-specific power optimizations, software-based power consumption estimation, and mathematical power estimation models (2) quantitatively analyzing the accuracy of a SPOT model based on only key power-consuming components and showing how modeling only power consuming components allows developers to avoid the accidental complexities of non-power consuming architectural aspects, (3) discussing the implications of the simplified model on application software architecture and design, (4) empirically evaluting the effects of design decisions, such as sensor sample rate, on overall application power consumption, and (5) describing how data produced by SPOT can be used to refine and optimize application power consumption.

BACKGROUND

This section provides a survey of prior work that compares SPOT with related research in the following categories:

- **System Execution Modeling Tools**: Which model and evaluate performance related information early in an application's development process,
- **Domain-Specific Power Optimizations**: Which are approaches for reducing the power consumption of specific application functions, such as geo-localization,
- **Power Instrumentation and Mathematical Estimation of Power Consumption:** Which attempt to directly measure or predict power consumption of an application,
- **Hardware-Based Power Consumption Optimization:** Which involves specialized firmware or low power physical hardware,
- **Network Protocol Optimizations:** Are made in the network stack to reduce the overall amount of data transmitted as well as contention on the physical link,
- **Post-Implementation Power Consumption Analysis:** Requires developers to completely implement an application before analyzing its power consumption.
- **Model-Driven Engineering:** Which uses domain-specific modeling languages and code generation to reduce development time and cost.

As discussed below, these methods of power consumption optimization are effective at reducing overall power consumption and are complimentary to SPOT. SPOT does not optimize the power consumption of an application, but rather allows developers to analyze the power consumption of key power consuming elements of an architecture at design-time.

System Execution Modeling Tools

The *Component Utilization Test Suite* (CUTS) (Hill, Schmidt, Slaby, & Porter, 2008) is a system execution modeling tool (Smith & Williams, 2001) that allows developers to model key architectural elements of distributed systems that determine performance. CUTS allows developers to model a distributed software architecture and then generate emulation code to predict performance. Although CUTS and SPOT share common domain-specific modeling patterns, CUTS's modeling abstractions focus on performance-related aspects of distributed systems, whereas SPOT's modeling abstractions focus on important power consumers in mobile applications, such as GPS usage. Moreover, it is not possible to capture power consumption information from CUTS emulation code or to generate mobile emulation applications for Android or iPhone.

Domain-Specific Power Optimizations

Prior work has examined domain-specific power optimizations that are geared towards a specific type of software feature. For example, Kjasrgaard (2012) have evaluated specialized approaches for optimizing the power consumption of location-based services using both client and server modifications. They propose lowering the accuracy of position information based on application-specific information, such as the current zoom level of a map. Other optimizations they propose include caching position and then continuing to report that position as long as the device is expected to be within a specified threshold of that cached position.

Many optimizations proposed by Kjasrgaard et al. are obtained through power profiling of applications. SPOT is complementary to this work and provides tools for profiling software applications early in their lifecycle. SPOT's goal is to aid developers in discovering the types of power optimization strategies described by Kjasrgaard.

Another area of mobile power consumption that studied by Thiagarajan et al. (Thiagarajan, Aggarwal, Nicoara, Boneh, & Singh, 2012) is the energy expended downloading and rendering Web pages using a mobile browser. They have conducted extensive instrumentation of mobile devices and Webkit-based browsers to profile exactly how much energy is consumed during each phase of accessing a mobile site. For example, they measure the variabilities of energy consumption across popular sites, such as Apple and Amazon, as well as exploring how images, javascript, and overall rendering consume power on the device. Their research highlights the importance of known optimizations, such as minimization of javascript files and scoping of CSS to a particular page, that are not always employed as they should be by developers.

The work by Thiagarajan et al. is synergistic with SPOT and looks at power consumption within an individual existing application on a device. SPOT also facilitates the collection of power consumption data from applications, but earlier in the design cycle when the implementation is may not be complete. Both prior approaches aid developers in improving design decisions to reduce mobile power consumption but are focused on different domains (*e.g.*, mobile Website design vs. early analysis of mobile applications).

Context-aware power optimizations have been proposed for sensor networks to ensure that sensors are only sampled in contexts where it makes sense. Herrmann et al. (Herrmann, Zappai, & Rosing, 2012) propose using optimizations, such as detecting indoor and outdoor transitions to determine when an outdoor sensor should be used. They demonstrate through modeling that these types of context-based optimizations can have significant power savings. SPOT helps evaluate these types of context-aware power optimizations by allowing developers to model architectures that exhibit different optimizations and experiment with them early in the design cycle.

Due to the high power consumption of many mobile sensors, another important area of research involves optimizing sensor sampling rates and times to minimize power consumption. For example, Wang et al. (Wang, Krishnamachari, & Annavaram, 2012) use Markov models to determine the best sampling approaches for sensor networks based on smartphones. These approaches are orthogonal to SPOT and focus on a different aspect of power optimizations. SPOT focuses on generating applications that behave like the planned architectures of specific mobile applications to facilitate early experimentation and reasoning about power consumption.

Power Instrumentation

A challenge on mobile devices is accurately collecting detailed and real-time power consumption information without external hardware. Yoon et al. (Yoon, Kim, Jung, Kang, & Cha, 2012) have proposed a more accurate and real-time framework for monitoring power consumption on Android devices. Their approach, called AppScope, intercepts key system calls to the underlying operating system, such as calls to the Android Binder, and then use information about the types and frequency of calls being made to infer power usage.

AppScope could be used to further improve the power consumption predictions made by SPOT. In particular, SPOT focuses on synthesizing mobile application code to run experiments that mimic how a particular mobile application architecture will behave. These synthetic applications can then be used to profile the app and collect power consumption information. Yoon et al.'s framework could be used in place of the Android APIs leveraged by SPOT to improve the fidelity of the power consumption data obtained from running the synthetic applications.

Mathematical Estimation of Power Consumption

Mathematical models of power consumption have been developed to help estimate the cost of using smartphone sensors, such as GPS, and processing capabilities. Kim et al. (Kim, Kong, & Chung, 2012) have developed a mathematical model for estimating the power consumption of smartphones with multicore CPUs. These types of mathematical analyses share a similar aim to SPOT and attempt to provide early prediction of power consumption before the implementation of an application may be completely ready. SPOT, however, assumes that real-world power consumption may vary substantially between devices and may depend on application-specific design details.

Hardware-Based Power Optimizations

Conventional techniques for reducing mobile device power consumption have focused on hardware- or firmware-level modifications to achieve the target consumption (Pering, Agarwal, Gupta, & Wang, 2006). These techniques are highly effective, but are limited to environments in which the hardware can be modified by the end user or developer. Moreover, these modifications tend to result in a tightly coupled architecture which makes them hard to use outside of the context in which they were developed. In other words, a solution might reduce power consumption for one application or in one environment, but may not have the same effect if any of those assumptions change. Moreover, such modifications are typically beyond the scope of software engineering projects and require substantial understanding of low-level systems.

In some instances, hardware-level modifications can actually hurt power consumption by increasing overhead. These techniques are also useful for reducing overall power consumption but

do not help in power consumption analysis that is necessary when developing power-conscious applications. SPOT is complimentary to these approaches in that developers can use SPOT to identify the most effective method to minimize power consumption without requiring extensive hardware knowledge or restricting the optimizations a single hardware platform.

Network Protocol and Interface Optimization

Due to the limited battery power available to mobile and embedded systems, much prior work has focused on optimizing system power consumption. In particular, network interfaces consume a large portion of overall device power (Krashinsky & Balakrishnan, 2005), so reducing the power consumption of networking components has been a research priority. Ultimately, the amount of power consumed is directly proportional to the amount of data transmitted (Feeney & Nilsson, 2001); in some instances require 100 times the power consumed by one CPU instruction to transmit one byte of data (Liu, Sadler, Zhang, & Martonosi, 2004). The power consumption of the network interface can thus be reduced by reducing the amount of data transmitted. Moreover, utilizing MAC protocols that reduce contention can significantly reduce power consumption (Chen, Sivalingam, Agrawal, & Kishore, 1998). While MAC-layer modification is effective, it is typically beyond the scope of software-engineering projects, which is common with mobile application development.

Using a custom MAC protocol or specifically selecting a pre-existing one based on the needs of the network may be possible when designing a custom sensor network, it is not possible when developing an application for a pre-existing platform, such as Android. Similarly, other work (Mohapatra, Cornea, Dutt, Nicolau, & Venkatasubramanian, 2003) achieved significant power reductions by using middleware to optimize the data transmitted. These approaches require both the device and any access points to which it connects to implement the same MAC protocol, which is infeasible for mobile application developers. Other work accomplished similar goals through transport layer modifications (Kravets & Krishnan, 2000), but the same limitations apply as with modifications to the MAC layer.

SPOT seeks to accomplish similar goals by modifying the data transmitted by the application layer, rather than attempting to modify the underlying network stack. SPOT helps developers analyze the data they transmit to maximize throughput of relevant data (*e.g.* actual data versus markup or framing overhead) thereby reducing power consumption. In addition, SPOT can function in a complimentary role allowing developers to analyze the power consumption of network protocol optimizations to identify the most effective configuration.

Post-Implementation Power Consumption Analysis

Previous work (Creus & Kuulusa, 2007) not only identified software as a key player in mobile device power consumption (*e.g.*, Symbian devices), but also sought to allow developers to analyze the power consumption of applications during execution. Moreover, other work (Landsiedel, Wehrle, & Gotz, 2005) utilized a similar approach to analyze the power consumption of sensor nodes in a wireless sensor network.

While post-implementation power consumption analysis can provide highly accurate results, it suffers from the pitfalls of post-implementation testing, including increased cost of refactoring if problems are discovered. To prevent costly post-implementation surprises, SPOT allows developers to analyze designs before any code is written. It also allows them to perform continuous integration testing through the use of custom code components to further refine the accuracy of the model as development progresses.

Model-Driven Engineering

Modeling and code generation have been combined in a number of model-driven engineering (Schmidt, 2006) approaches to reduce the number of development errors and required manual code creation. Prior work has investigated environments, such as GME (Ledeczi, Bakay, Maroti, Volgyesi, Nordstrom, Sprinkle, & Karsai, 2001) and GEMS (White, Schmidt, & Mulligan, 2007), for rapidly creating domain-specific modeling language tools and code generation infrastructure. Where as GME and GEMS are generic modeling platforms for creating domain-specific languages and tooling, SPOT provides a specific domain-specific language targeted to the domain of early analysis of mobile architecture power consumption. SPOT builds upon the prior work and provides a specific instantiation of the concepts and new and novel techniques targeted at mobile power consumption.

MAIN FOCUS OF THE CHAPTER

Issues, Controversies, and Problems

This section of the chapter describes the challenges associated with developing power-aware mobile software, using a case study to highlight the key issues, controversies, and problems in this domain. High-level mobile application development SDKs, such as Google Android or Apple iPhone, simplify mobile application development, but do not simplify power consumption prediction during application design. In fact, the abstractions present in these SDKs make it hard for developers to understand the power implications of software architecture decisions until their designs have been implemented [20]. Interaction with sensors, such as accelerometers or GPS receivers and network interaction, can also result in a significant amount of power consumption.

Motivating Example: The WreckWatch Case Study

Below we describe *WreckWatch* (White, Clarke, Dougherty, Thompson, & Schmidt, 2010), which is an open-source[1] mobile application we built on the Android smartphone platform to detect automobile accidents. We use WreckWatch as a case study throughout this paper to demonstrate key complexities of predicting the power consumption of mobile software architectures.

As shown in Figure 1, WreckWatch operates by (1) monitoring smartphone sensors (such as GPS receivers and accelerometers) for sudden acceleration/deceleration events that are indicative of an accident. Data about the event are then (2) uploaded to a remote server over HTTP where first-responders and/or other motorists can access the information via a Web browser or Android client. WreckWatch allows bystanders (3) to upload images of the accident to the same Web server, thereby increasing the information first-responders possess before arriving at the scene.

Information, such as images of the accident and force of impact, provides first-responders with immediate knowledge of an accident. It also seeks to provide them with richer situational awareness to better prepare for conditions at the accident scene. Moreover, accident location information helps alleviate potential congestion due to an accident by providing motorists a means to immediately identify potential areas of congestion and avoid the area altogether (*e.g.* this accident location information can be combined with a navigation package, such as the Google Maps API, to alter routing information dynamically so motorists can avoid the accident location). When an accident occurs, both first-responders and emergency contacts specified by accident victims can be notified via integrated VoIP infrastructure, as well as by SMS messages and/or email.

To detect traffic accidents accurately, WreckWatch runs continuously as a background service and continuously consumes a great deal of accelerometer and GPS data. The application must therefore be conservative in its power consumption. If not designed properly, WreckWatch can significantly decrease smartphone battery life.

Challenge 1: Accurately Predicting Power Consumption of Framework API Calls

Each line of code executed results in a specific amount of power consumed by the hardware. In the simplest case, this power consumption results from a small series of CPU operations, such as reading from memory or adding numbers. In some cases, however, a single line of code can result in a chain of hardware interactions, such as activa-

Figure 1. WreckWatch operation

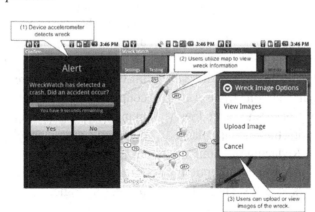

tion of the GPS receiver and increasing the rate at which the screen is redrawn. Moreover, although the higher levels of abstraction provided by modern smartphone SDKs make it easier for developers to implement mobile application software, they also complicate predictions of the effects on the hardware.

For example, WreckWatch heavily utilizes the Google Maps API and the "MyLocation" map overlay, which provides end users with a marker indicating their current GPS location. The use of the "MyLocation" is typically accomplished with fewer than 10 lines of code, but results in substantial power consumption. This is because the overlay is redrawn at a high rate to achieve a "flashing" effect, and because the overlay enables and heavily utilizes the GPS receiver on the device, which further increases power expenditure. It is hard to predict how using arbitrary API calls, such as this overlay, will affect application power consumption without implementing a particular design and testing it on a particular target device.

This abstraction in code makes power-consumption analysis on arbitrary segments of code hard. Predicting power usage from high-level design abstractions, such as a UML diagram, is even harder. Later in this chapter we describe the MDE and emulation approach used to address this challenge.

Challenge 2: Accurately Predicting Power Consumption of Sensor Usage Architectures

In applications utilizing sensor data, the most accurate sensor data is obtained by sampling as often as possible. Sampling at high rates, however, incurs high power consumption (Krause, et al., 2005) by not allowing sensors to enter low power modes and by increasing the amount of data processed by applications. Reducing the sample rate can decrease application power consumption considerably, but also reduces accuracy. The trade-offs between power consumption and accuracy at a given sampling rate are hard to determine without empirical tests on a target device due to the high-degree of layering in modern smartphone APIs and system architectures.

For example, WreckWatch was originally designed to collect GPS data every 500 milliseconds and consume accelerometer data at Android's predefined *NORMAL* rate setting. During the development of WreckWatch, it was clear that reducing WreckWatch's GPS sampling rate would reduce overall power consumption, but it was unclear to what degree. Moreover, it was hard to predict what sample rate would provide sufficient accuracy and still allow the phone to operate for days between charges. Later in this chapter we describe how automatic code generation is used to create emulated applications that accurately analyze the power consumption of a candidate sensor utilization architecture without incurring the substantial time and effort to manually implement the architecture.

Challenge 3: Accurately Assessing the Effects of Different Communication Protocols on Power Consumption Prior to Implementation

Each application and network communication protocol has a specific overhead associated with it and can result in significant power consumption (Heinzelman, Chandrakasan, & Balakrishnan, 2000). Certain protocols require more development overhead to implement, but have low runtime overhead (*e.g.* bandwidth consumption, message processing time, etc.). For example, communicating with UDP packets requires implementing flow control, error detection, etc., whereas TCP communication requires setup and teardown time for communication channels, as well as the processing and data associated with these operations, in the form of additional protocol message overhead and processing logic.

More advanced and robust protocols, such as HTTP, are easier to use as communication protocols and are more attractive to developers despite the increased overhead (Anand, Manikopoulos, Jones, & Borcea, 2007). For example, communicating over HTTP simplifies establishing a communication channel between client and server. There are also many libraries available to extract data from an HTTP message. HTTP supports a great deal of functionality that may not be necessary for a particular application, however, and using this functionality incurs extra overhead that consumes more power.

For a small number of messages the overhead of HTTP will likely have little impact on application power consumption. With large numbers of operations, however, the overhead of HTTP can have a significant impact on power consumption. It is hard to determine early (*e.g.*, at design time) in an applications lifecycle, however, how this overhead will affect power consumption and whether the number of messages transmitted will be substantial enough to impact power consumption significantly. This challenge is exacerbated if certain network operations consume more power than others, *e.g.*, receiving data often consumes more power than transmitting data (Wang, Hempstead, & Yang, 2006).

For example, to provide the most accurate situational awareness to first responders—and provide the most accurate congestion information to motorists—the WreckWatch application must periodically request wreck information from the central Web server. These updates must be done periodically and were originally intended to run over HTTP. Using HTTP results in a significantly less developer effort but results in a considerable amount of communication overhead from the underlying TCP and HTTP protocols, which ultimately transmits substantial amounts of data that have no relevance to the application. It is hard to determine at design time if/how this additional data transmission will significantly impact power consumption. Below we show how MDE code generation was used to implement and analyze potential communication protocols rapidly.

Solutions and Recommendations

This section describes the structure and functionality of the *System Power Optimization Tool* (SPOT), which is an MDE tool that allows developers to model potential mobile application software architectures to predict their power consumption, generate code to emulate that architecture, and then systematically analyze its power consumption properties. SPOT addresses the challenges

Figure 2. SPOT modeling and analysis process

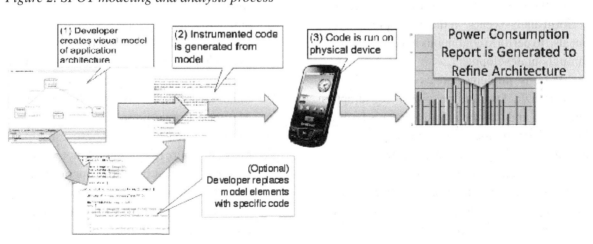

351

described earlier in this chapter by allowing developers to understand the implications of their software architecture decisions at design time.

SPOT's development process enables developers to generate visual, high-level models rapidly, as shown in Figure 2.

These models can then be used to analyze the power consumption of mobile application software architectures (step 1 of Figure 2). SPOT thus helps overcome key challenges of predicting power consumption by generating device logic that can be used to gather power consumption information on physical hardware during early phases of an application's software lifecycle, which helps minimize expensive redesign/refactoring costs in later phases.

SPOT uses models shown in Figure 2 to generate instrumented emulation code for the given platform or device (step 2 of Figure 2). When executed on actual hardware (step 3 of Figure 2), this generated code collects power consumption and system state information. This power consumption data can then be downloaded and analyzed offline to provide developers with application power utilization at key points throughout the lifecycle and execution path (step 4 of Figure 2).

SPOT also supports the use of custom code modules. These models allow developers to replace automatically generated code with actual application logic while still providing the same analytical capabilities available when using generated code. SPOT therefore not only allows developers to perform analysis early in the development cycle, but also to perform continuous integration testing throughout development.

SPOT is implemented as a plugin for the Eclipse IDE. Its runtime power consumption emulation and capture infrastructure is built using predefined, user-configurable classes that emulate specific power consuming components, such as GPS data consumers. This infrastructure captures power consumption information during executing by using the device's power API. For example, power data collection on the Android

platform is performed by interfacing with the OS application power API, *i.e.*, the Android power consumption API as implemented by the "Fuel-Gauge" application.

The remainder of this section describes how SPOT's DSML, emulation code generation, and performance measurement infrastructure help application developers address the challenges presented earlier.

Mobile Application Architecture Modeling and Power Consumption Estimation

SPOT describes key power-consuming aspects of a mobile application via a DSML with specific language elements. This DSML allows developers to specify their software architecture visually with respect to power consuming components, as shown in Figure 2. Prior work (Thompson, White, Dougherty, & Schmidt, 2009; White, Clarke, Dougherty, Thompson, & Schmidt, 2010; Turner, White, Thompson, Zienkiewicz, Champbell, & Schmidt, 2009) showed how the following components are often significant power consumers in mobile applications:

- **CPU Consumers:** Are used to represent CPU-intensive code segments such as calculations on sensor data. Developers can specify the amount of CPU time that should be consumed by specifying the number of loop iterations of a square root calculation that should be run. For example, WreckWatch developers can model the mathematical calculation time to determine the current G-forces on the phone.

- **Memory Consumers:** Generate dynamically allocated memory. These consumers allow developers to analyze not only the power consumed by actual operations, but also their impact (such as the frequency and duration of garbage collector sweeps) on garbage collection. Developers can specify the amount of memory to consume as bytes. For example, WreckWatch developers can model the effects of caching accident images of varying sizes.

- **Accelerometer Consumers:** Which interact with system accelerometers and consume accelerometer data. These consumers can be configured to utilize the full range of system-supported sample rates. For example, WreckWatch developers can model the sensor interaction needed to accurately detect car accidents.
- **GPS Consumers:** Interact with the device's GPS receiver. These consumers can be configured with custom sample rates as well as a minimum distance between points, *i.e.*, the sensor will only return a data point if the distance between the current point and the last point is greater than a specified value. GPS consumers allow developers to analyze the impact of using a location service configuration on power consumption. For example, WreckWatch developers use this capability to model how polling for a vehicle's location at different rates impacts power consumption.
- **Network Consumers:** Emulate application network interaction by periodically transmitting and receiving data. Network consumers allow users to supply SPOT with sample data that is then transmitted at the interval specified. For example, WreckWatch developers can provide a URI along with server and port information to configure SPOT to make a specific request. These consumers can also be configured to execute at varying times to emulate periodic updates.
- **Screen Drawing Agents:** Utilize graphics libraries, such as OpenGL, to emulate a graphics-intensive application, such as a game or streaming video to a first responder's WreckWatch client. Users can configure these consumers by specifying

the types and size of objects to draw on the screen, along with any transformations that should be performed on the object. For example, WreckWatch developers can use the drawing agents to show how the use of images and video for situational awareness impacts battery life.

- **Custom Code Modules:** Allow developers to specify their own code to run against the profiling package. This capability allows developers to extend SPOT's functionality to meet their needs, as well as incrementally replace the emulation code with actual application logic as it becomes available. Replacing the emulation logic allows developers to perform testing as development progresses and increase the accuracy of the evaluation and analysis. For example, WreckWatch developers can use these consumers to include a service for uploading multimedia content about an accident to a central Web server.

The metamodel for SPOT's DSML, called the *System Power Optimization Modeling Language* (SPOML), allows application developers to build software architectural specifications that determine power consumption from key power consuming components. SPOML was created using the metamodeling features of the Generic Eclipse Modeling System (GEMS) (White, Hill, Tambe, Gray, Gokhale, & Schmidt, 2009), which is a tool for rapidly generating visual modeling tools atop the Eclipse Modeling Framework (EMF). GEMS is built atop Ecore, which provides metamodeling and modeling facilities for Eclipse (Budinsky, Steinberg, Merks, Ellersick, & Grose, 2003).

Figure 3 shows SPOML's metamodel.

The primary application serves as the root element of the model. Power consumption modules can exist within either *activities* (which are basic building block components for Android applications and represent a "screen" or "view" that provides a single, focused thing a user can

Figure 3. SPOT's metamodel

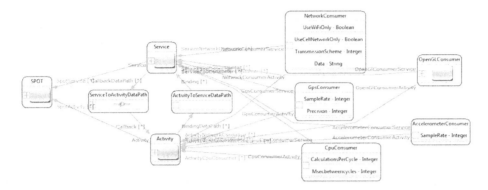

do) or *services* (which are background processes that run without user intervention and do not terminate when an application is closed).

Each activity or service can contain one or more power consumer modeling elements described above. Developers can therefore emulate potential decisions that they make when designing a mobile device application, which allows them to emulate a wide range of application software architectures. For example, Figure 4 shows a SPOML model of the WreckWatch application's sensor usage design.

Acceleration and GPS consumers run in independent services, while the network consumer runs in the application's activity. This model results in an application that monitors GPS and accelerometer values at all times regardless of what the user is doing. It only utilizes the network connection, however, when the user has the WreckWatch application open and running.

Generating Software Architecture Emulation Code

Predicting the power consumption of an arbitrary design decision is hard. SPOT addresses this challenge by generating application emulation code automatically to execute on the underlying device hardware. SPOT's automatic generation of emulation code allows application developers to reduce the time required to write enough code to analyze system power consumption accurately. This emulation code is instrumented so the architecture's power expenditures can be examined after a test run.

In addition to instrumenting the code, SPOT has the potential to apply the same model for multiple target platforms, such as Android and iPhone, as long as configurable code is available for the power-consuming elements. This emulation and analysis cycle allows developers

Figure 4. WreckWatch model

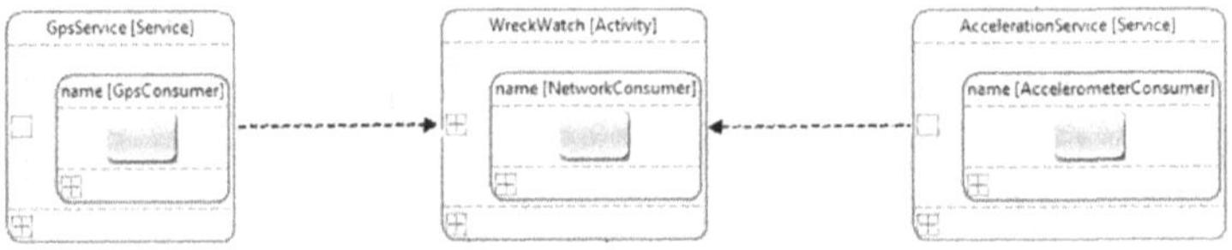

to observe the power consumption of their design very early in the development cycle, as well as evaluate their software designs across multiple hardware configurations to assess how changes in hardware affect application power consumption. For example, even though the Motorola Droid and Google Nexus One both run the Android platform, each possesses key hardware differences, such as the type and size of the display, that impact power consumption.

The generated emulation code allows developers to address the remaining challenges of selecting an optimal communication protocol and optimizing sensor polling rates, as described earlier. Generated emulation code allows developers to evaluate the power consumption of a potential design empirically, rather than simply guessing its power consumption or waiting until an application implementation is complete before running tests. Moreover, developers can quantitatively compare the power consumption effects of choosing different networking protocols and can evaluate the power consumption of different sensor sampling rates.

To accomplish this mobile software architectural emulation, SPOT uses a set of predefined code blocks that can be configured at runtime to perform power consuming operations. SPOT uses an XML configuration file to perform any necessary configuration and create an application that emulates the power consumption of the desired software architecture. To generate this XML configuration file, SPOT interprets the model shown in Figure 4. Users of SPOT define the model and tweak configuration parameters, and SPOT can then compile the model and parameters into an intermediate XML format which is utilized to configure prebuilt implementations of each power consuming element shown in Figure 4.

Figure 5 shows a sample of the XML configuration file generated for the WreckWatch model shown in Figure 4.

The XML shown in Figure 5 represents a configuration with two background services running an accelerometer consumer and a GPS consumer. The GPS consumer samples every 500 milliseconds, with a minimum radius between sample points set to 0. The accelerometer consumer is set to sample at the *NORMAL* rate, which

Figure 5. Sample WreckWatch emulation XML

```
1   <?xml version="1.0" encoding="UTF-8"?>
2   <spot>
3   <application package="org.vuphone.wwatch.android">
4     <activity>
5       <network delay="30" repeat="true"
6          host="dre.vanderbilt.edu" port="8080"
7          protocol="http">
8          <![CDATA[uri=/wreckwatch/notifications?type
9          =info&latbr=35458867&lonbr=-95189644&
10         lattl=36838778&lontl=-96683784&maxtime=0]]>
11      </network>
12    </activity>
13    <service process="main">
14      <gps sampleRate="500" precision="0" />
15    </service>
16    <service process="main">
17      <accelerometer sampleRate="NORMAL" />
18    </service>
19    <service class="org.vuphone.wwatch.android.media.
20                  MediaUploadService" />
21  </application>
22  </spot>
```

is a constant defined by Android. There is also a network consumer transmitting sample data every 30 seconds and repeating infinitely. The network consumer is configured to connect to a specific host on a specific port and utilize the HTTP protocol.

The predefined power consumers have configurable options (such as the sample rate of the GPS and data to transmit over the network) provided by the XML configuration file. These power consumption elements are generic and applicable to a wide range of platforms, such as Android, iPhone, or BlackBerry. The predefined power consumers are implemented on the Android platform as follows:

- **CPU Consumers:** Are implemented as a set of nested loops that perform basic mathematical operations, such as square root calculations, on randomly generated data. This module utilizes the CPU while minimizing utilization of other system resources, such as memory that could skew power consumption information. For example, this consumer uses primitive types to avoid allocating dynamic memory. Users can adjust the length of the loops via a configurable parameter. Various end device processors result in same-length loops performing differently on divferent devices, in a manner similar to CPU-intensive algorithmic performance on end-user devices.
- **Memory Consumers:** Are implemented by dynamically allocating custom objects that wrap byte arrays. To analyze the frequency of garbage collection, a Java WeakReference object is used to inform the garbage collector that they can be reclaimed, despite having active references within running code. The object's finalize() method (which is called immediately before the object is reclaimed by the Android Dalvik virtual machine) is

overridden to record the time of garbage collection, thereby allowing developers to analyze the frequency of garbage collection runs. The WeakReference object will thus be reclaimed during every garbage collection run. Due to the limitations of the Android instrumentation API, garbage collection and memory usage must be inferred through analysis of the frequency and duration of garbage collection runs, rather than through direct power consumption measurement. Although this limitation prevents developers from including memory statistics in the data along with CPU, sensor, and network utilization, they can still examine how their design uses memory. Users can also configure the amount of memory and frequency of allocation, as well as supply custom objects (such as WreckWatch's image caches) to use rather than the byte arrays used by default.

- **GPS Consumers:** Are implemented by code that registers to receive location updates from the GPS receiver at specific intervals. To emulate an application's interaction with the GPS receiver properly, SPOT provides developers with all configuration options, such as polling frequency and minimum distance between points, presented by the Android GPS API. This configuration information—along with generic sensor setup and tear-down code—is then inserted into the appropriate location in the emulation application.
- **Accelerometer Consumers:** Are implemented using the configuration specified in the XML file, along with generic setup code to establish a connection to the appropriate hardware. SPOT provides developers with access to device accelerometers and all configuration options the underlying framework presents, such as sample rates.

- **Network Consumers:** Are implemented as emulation code containing a timer that executes a given networking operation, such as an HTTP operation at a user defined interval. The purpose of the network consumer is to emulate applications that use network resources to store information that the phone must periodically retrieve. Developers configure SPOT by providing the data to transmit along with the frequency of transmission. They can also specify whether the data should always be transmitted or whether the application should wait for the availability of a specific transmission medium, such as Wi-Fi or 3G.
- **Screen Drawing Agents:** Allow users to specify 3D and 2D graphics to draw on the screen. Developers will specify object contents along with any potential motion or actions.
- **Custom code Modules:** Allow developers to supply their own code blocks to extend the functionality of SPOT and enhance emulation accuracy as the development cycle progresses by substituting the *faux* emulation code with actual application logic. SPOT allows developers to supply class files to load into the emulation application dynamically, as well as method "hooks" to allow the emulation code to interact with the custom code properly.

Power Consumption Instrumentation

SPOT uses an instrumentation system to capture power consumption statistics as the emulation code is executed, as shown in Figure 6.

Components in the instrumentation system are either Collectors, Recorders, or Event Handlers. Collectors interface directly with the specific power API on the system and pass collected data to Recorders, which persist the data collected by Collectors by writing it to a file or other storage

Figure 6. SPOT instrumentation system

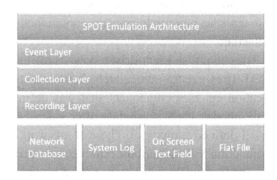

medium. Event Handlers respond to the events fired by entering or leaving emulation code blocks.

These components are dynamically loaded via Java reflection to ensure extensibility and flexibility. For instance, developers can implement a custom Collector to monitor which components are in memory at any given time. Alternatively, developers could define Recorders to log power consumption information to another data storage medium, such as a local or network database rather than a flat file.

To analyze an architecture effectively, SPOT records battery state information over time to allow developers to pinpoint specific locations in their application's execution path that demand the most power. To collect power consumption information accurately, SPOT uses an event-driven architecture that precisely identifies the occurrence of each major application-state change, such as registering or unregistering a GPS listener and SPOT takes a "snapshot" of the power consumption when the application performs these operations. This event-driven architecture allows developers to understand the power consumption of the application before, during, and after key power-intensive components.

In addition to event-triggered power snapshots, SPOT also periodically collects power consumption information. This information allows developers to trace overall power consumption or power consumption within a given block. The power information Collector that collects snapshots and periodic samples can be configured to run in a separate process to prevent contamination of the data.

To accomplish event-driven power profiling, SPOT fires events immediately before an application enters a component that was defined in the model and immediately after it leaves a model-defined component. These events work in conjunction with the periodic power consumption updates to provide developers with a complete description of how their software architecture elements consume power. SPOT's event-driven model of collecting power consumption data also allows developers to identify precisely what the application was doing when key power consumption spikes occur, further helping them optimize their designs.

SPOT's emulation infrastructure currently runs on the Android mobile platform and uses the power consumption API utilized by the "FuelGauge" application in the core Android installation. The power consumption API provides application developers with access to the amount of time the application utilizes the CPU, sensors, wake-locks, and other system resources, in addition to the overall power consumption.

Android's power consumption API provides power-consumption information on a per-package basis. By implementing SPOT in a different package, developers can analyze power consumption without influencing it. Collecting power consumption information in this manner increases accuracy. Moreover, SPOT can be implemented simply as a collector to analyze existing applications without modifying their source code.

Pros and Cons of SPOT's MDE-Based Approach

SPOT's focus on MDE provides a number of capabilities and restrictions on the breadth and accuracy of what can be modeled. A key ben-

efit of SPOT's MDE-based approach is that the models are easy to construct and generation of emulating mobile application code does not require any manual development effort. This approach enables early and low-cost analysis of the impact of architectural decisions on power consumption. Moreover, SPOT's domain-specific modeling language (SPOML) shields developers from lower-level details that are not relevant to optimizing power consumption.

The key downside of SPOT's MDE-based approach, however, is that SPOML cannot express every possible application behavior and thus may not provide perfect fidelity to a mobile application that is eventually implemented. For example, an application that provides alerts based on complex calculations involving the accelerometer and GPS may not be expressible in SPOML; which would make it hard to reason about how the algorithm itself affects the power consumption profile of the app. For these types of scenarios, SPOT allows the integration of hand-coded application components into an application to provide greater fidelity. The downside of this flexibility, of course, is that components must be manually implemented if they are not expressible in SPOML.

Empirical Results

This section analyzes the results of experiments that empirically evaluate SPOT's MDE-based capabilities presented earlier. These experiments measure SPOT's ability to collect power consumption information on a given model, as well as accurately model the power consumption of a proposed application software architecture. These results show how SPOT can assess and analyze power consumption information gathered through the Android's power consumption API and evaluate SPOT's accuracy in predicting power consumption information about a software architecture at design time.

Hardware/Software Testbed

All tests were performed on a Google Nexus One with a 1Ghz CPU, 512MB of RAM, 512MB of ROM and a 4 GB SD card running the default installation of Android 2.1 (Eclair). The SPOT application was the only third-party application running at the time of experimentation. The same power consumption information gathering logic was used to collect information on emulation code, as well as the sample applications. The information was analyzed in Excel based on power consumption data written to the device's SD card in the form of a CSV file.

To assess the consumption characteristics of different designs, the current SPOT version generates an Android application package. It then periodically samples the battery usage statistics from the OS writing these values to a CSV file on the SD card. SPOT also fires events when the application's state changes, *e.g.*, when the GPS is started or a sensor is disconnected. These events allow SPOT users to examine the power consumption of active hardware, in addition to the overall consumption of the application.

SPOT uses an XML-based configuration file that is generated from the SPOML model shown in Figure 4. This XML file is loaded on to the device's SD card and parsed at startup. Due to the way that the power consumption API collects information, the data gathered reflects only power consumed by the SPOT application and does not include any power consumed by system processes, such as the GUI display or garbage collector.

Experiment 1: Empirical Evaluation of SPOT's Emulation Code Accuracy

This experiment quantitatively compares the power consumption of two Android applications and the power consumption of the emulation code derived from their respective SPOT models. Ensuring SPOT's accuracy is important since it allows developers to compare the power consump-

tion of key power consuming components in their mobile architecture.

The applications used in this experiment are the WreckWatch case study application and OpenGPSTracker (open-gpstracker.googlecode.com), which is an open-source Android application that uses GPS to track the coordinates of the phone and display it on a Google Map on the device. The GPS points, and other information about the route, are stored on the device to allow the user to replay the route later. OpenGPSTracker also determines device speed as GPS points are collected.

Hypothesis

SPOT is intended to provide developers with an estimate of how a proposed application software architecture will consume power. We therefore hypothesized that SPOT could provide power consumption information to within 25% of the actual power consumption of WreckWatch and OpenGPSTracker. Based on prior work (Thompson, White, Dougherty, & Schmidt, 2009; White, Clarke, Dougherty, Thompson, & Schmidt, 2010; Turner, White, Thompson, Zienkiewicz, Campbell, & Schmidt, 2009), we also hypothesized that the components we chose represented the key factors in mobile application power consumption and would be adequate to provide this level of accuracy.

WreckWatch Results

Figure 7 shows the graph of the actual power consumption of the WreckWatch application compared with the power consumption of the WreckWatch emulation code generated by SPOT.

The emulation code's power consumption follows the same trend as that of the application and provided a final power consumption value that was within 3% of the actual power consumed by WreckWatch. The SPOT model consisted solely of GPS and accelerometer consumers and did not attempt to model any additional aspects of the WreckWatch application. The model was

Figure 7. Comparison of WreckWatch application logic and emulation code

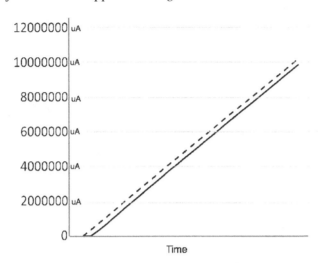

accurate due to the substantial amount of power required by the GPS receiver. This result confirms that the GPS, sensor, and network modules that SPOT provides are key determinants of mobile power consumption. Although WreckWatch performs a number of CPU-intensive calculations on sensor data to determine if an accident occurred, these calculations are minor power consumers compared to sensor users.

OpenGPSTracker Results

Figure 8 shows the graph of the actual power consumption of the OpenGPSTracker applica-

tion compared with the power consumption of the emulation code generated by SPOT. As with the WreckWatch emulation code, the OpenGP-STracker emulation code consumes power at a rate similar to the actual application. Over the same time period, the emulation code for the OpenGPSTracker application was accurate at predicting power consumption to within 4%. The SPOT model for the OpenGPSTracker application only used a GPS consumer and did not attempt to model any Google Maps functionality (or requisite network functionality) or any processing required to determine speed or store the location informa-

Figure 8. Comparison of OpenGPSTracker application logic and emulation code

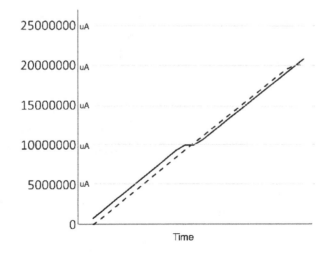

Figure 9. GPS power consumption

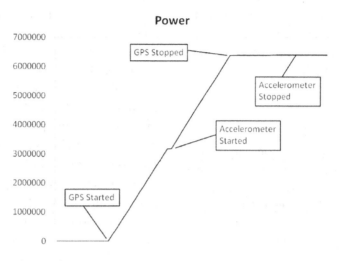

tion. In this instance, the GPS power consumption was sufficient to model the power consumption of the entire application.

With both applications, SPOT modeled and predicted the power consumption of the actual application to within 4% of the actual power consumption. This result confirms our hypothesis that SPOT can provide an accurate prediction of power consumption by modeling key components of a mobile application.

Experiment 2: Evaluating the Accuracy of Simplified Architectural Models

This experiment shows that modeling the power consumption of the largest power consumer can be sufficient to accurately estimate power consumption, even though there are significant differences in the power consumed by the various sensors, CPU operations, and networking components of an application. SPOT provides a limited set of key power consumption elements for modeling a mobile software architecture. Although there is a finite set of elements for modeling power consumption in SPOT, a small number of these elements can often accurately determine the

power consumption of a variety of applications, *e.g.*, modeling the power consumption of the GPS device is nearly as accurate as modeling the power consumption of both the accelerometer, CPU, and GPS of an application because the GPS receiver consumes so much power.

For experiment 2, we modeled the GPS sensor utilization of an app that used two sensors: the GPS and the accelerometer. The GPS was configured to sample every 500 milliseconds while the accelerometer was configured to use the FASTEST Android sample rate constant. We compared the power consumption of the emulation code, which only used GPS, to the actual app that used both the GPS receiver and the accelerometer. Modeling the largest power consumers allows users to model their designs even earlier by allowing them to ignore implementation details during the modeling process.

Hypothesis

We hypothesized that certain hardware components, such as GPS, consume so much power that emulating them with SPOT will consume nearly the same amount of power as the actual application. For example, an application that uses a GPS sample frequency of one sample per 500 milliseconds will have roughly the same power consumption as another application using the sampling the GPS every 500 milliseconds and receiving accelerometer updates.

Results

Figure 9 shows the results of experiment 2.

As shown in the figure, the callouts represent the times that correspond with the entrance and exit from different components of the software architecture. This graph shows that the accelerometer had little effect on the overall power consumption in relation to the GPS. Despite running for 2.5 minutes after the deactivation of the GPS, the power consumption increases only a small amount in relation to the total power consumed by the application.

Figure 10 shows the relative power consumption of the several Google Nexus One components, such as the screen, accelerometer, Bluetooth radio, Wi-Fi module, GPS receiver and the CPU per time unit.

As shown in the figure, certain hardware components consume a significantly larger amount of power than others. Moreover, the usage char-

acteristics of these hardware components increase the difference in power consumption. For example, despite consuming almost as much power as the GPS, usually the Wi-Fi or Bluetooth is only active for short bursts of time whereas the GPS is usually left in the active state for much longer periods of time. Due to these usage characteristics, the slightly greater power consumption per time unit of the GPS allows the GPS to quickly become the largest consumer of power on the device.

The results shown in Figures 9 and 10 support our hypothesis that mobile application power consumption depend largely on a handful of components, so SPOT can accurately predict mobile application power consumption by modeling these key power-intensive components.

Experiment 3: Qualitative Comparison of Sensor Sample Rates

This experiment evaluates the effects of sensor sample rates on an application's overall power consumption. The rate at which sensor data is consumed can have a significant impact on application power consumption. For example, Android's accelerometer provides four enumerations

Figure 10. Relative power of hardware components

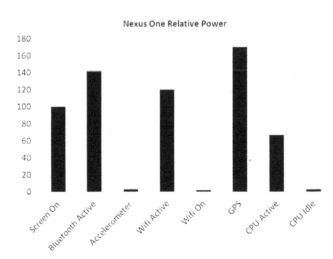

for sample rate: NORMAL, UI, GAME, and FASTEST. Although these sample rates provide varying levels of QoS, the trade-off between a given level of QoS and power consumption is not readily apparent at design time. The enumeration names give developers a clue to potential uses, as well as rank these sample rates according to rate and consequently power consumption. Alternatively, the GPS receiver allows developers to specify a value as the delay, in milliseconds, between samples.

SPOT allows developers to evaluate the power consumption of potential sensor sample rates. For experiment 3, we compared the power consumption of the GPS receiver while sampling at once every 2 seconds, once every 30 seconds, and once every 500 milliseconds.

Hypothesis

We hypothesized that SPOT could capture the power consumption effects of sensor sampling rate changes. In particular, we believed sampling rate changes of a few hundred milliseconds would produce power consumption changes that SPOT could detect.

Results

Figure 11 show SPOT's output for three different sample rates for the GPS sensor.

The dashed line represents the power consumption of the application when the sensor was sampled every 500 milliseconds, the solid line represents a sample taken every 2 seconds, and the dotted line represents the power consumption of the application sampling the sensor twice per minute. The samples in this graph were collected over 5 minutes and support the following observations:

- Power consumption during the first several seconds is uniform regardless of sample rate. Each graph is approximately equivalent for the first several seconds of data gathered during the GPS sampling, which implies that if developers need access to GPS for a short period of time, the greatest benefit would come from using a higher sample rate.

- The greatest improvement in power consumption due to a lower sample rate will occur over time. Although the graphs dem-

Figure 11. GPS sample rate comparison

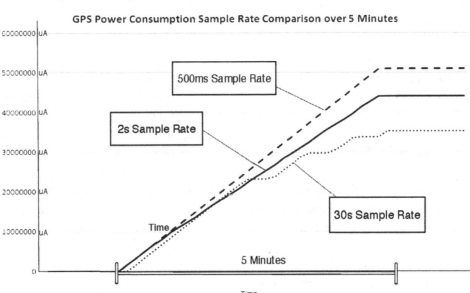

onstrate a noticeable difference in power consumption over the 5-minute sample period, the improvement in battery life from a change in sample rate will only be realized when sampling occurs over an extended period of time.

Ultimately, the amount of power consumed by the GPS receiver is directly proportional to the sampling rate of the application. Reducing the sampling rate of the GPS receiver is an effective means to reduce the overall power consumption of an application if the receiver is active for longer than ~2 minutes.

Summary and Analysis of Results

The results of the two experiments presented above show how SPOT can accurately analyze and predict an application's power consumption based on a model of the application software architecture. This capability allows developers to understand the implications of their design decisions early in the software lifecycle, *i.e.*, without implementing the complete application. The emulation code SPOT generates is accurate, *e.g.*, for our WreckWatch application it could predict power consumption within 3% of the actual application's power consumption. SPOT's accuracy stems in part from the significant power consumption of hardware components, such as the GPS receiver, that consume significantly more power than other hardware components on the device, such as the CPU or even accelerometers.

FUTURE RESEARCH DIRECTIONS

Based on our experiences developing evaluating SPOT, the following are our assessment of future and emerging research trends and directions:

- Sensor sample rates play an important role in long-term power consumption. The power consumed by the device sensors is largely uniform over the first several minutes of activation regardless of sample rate. It is only when these sensors are used for an extended period that the benefit of lower sample rates is realized. Developers must therefore consider the amount of time to activate the sensor in addition to the overall sample rate.

- Certain hardware components draw significantly more power than others. In the case of utilizing the GPS receiver, the power consumed by the location (GPS) service is so substantial that it becomes difficult to identify the effects of enabling or disabling other hardware components. Due to this "masking" effect, developers may overlook a significant consumer of power in an application. In general, small changes in GPS sample rate can have significant impacts on overall application power consumption.

- Power consumption of an application can be accurately modeled by a few key components. The power consumed by certain components, such as the GPS receiver and accelerometers is so significantly greater than the power consumed by running the CPU, displaying images on the screen, etc. that SPOT can effectively model the power consumption of an application simply by modeling those components. The most effective power consumption optimization can be realized, therefore, with changes to only a small number of components.

- Certain system-related operations such as garbage collection are not reflected in data gathered by SPOT. Through the current method of data collection SPOT is only able to gather power consumption information about operations that it performs such as CPU, memory or sensor operations that it specifically requests. Our future work will therefore develop mechanisms for capturing the impact of these services on power consumption.

- Power consumption of networking hardware is dependent on data transmitted which is often dependent on user interaction. The power consumption of hardware, such as the Bluetooth or Wi-Fi interfaces, is dependent on the data transmitted and received on the device. This data is often generated at run-time by user interaction with the program. While it is effective for the developer to generate some sample data and provide it to SPOT, it would be more effective if developers could simply describe the data, *e.g.*, via a schema file. Our future work is extending SPOT so it can process a data schema file and generate data that is random, yet still valid to the application.
- Although GPS is a major consumer of power, not all applications rely on GPS. Although GPS is a major consumer of power on today's smartphone devices, it is still important to understand the power consumption of applications that do not use the GPS, even if their power consumption is less drastic. Our future work is therefore analyzing SPOT's accuracy with mobile applications (such as 3D games with acceleration-based controls, streaming video players, and audio recording/processing applications) that do not use GPS, such as 3D games, feed readers, and multimedia applications.

CONCLUSION

The *System Power Optimization Tool* (SPOT) is an MDE tool that allows developers to evaluate the power consumption of potential mobile application architectures early in the software lifecycle, *e.g.*, at design time. Our experiments indicate that SPOT provides a high degree of accuracy, *e.g.*, it predicted the power consumption of the WreckWatch and OpenGPSTracker applications to within ~3-4%. SPOT is available in open-source form at syspower.googlecode.com.

REFERENCES

Agarwal, Y., Chandra, R., Wolman, A., Bahl, P., Chin, K., & Gupta, R. (2007). Wireless wakeups revisited: energy management for voip over wi-fi smartphones. *ACM MobiSys, 7*.

Anand, A., Manikopoulos, C., Jones, Q., & Borcea, C. (2007). A quantitative analysis of power consumption for location-aware applications on smart phones. In *Proceedings of the 2007 IEEE International Symposium on Industrial Electronics*, (pp. 1986–1991). IEEE.

Budinsky, F., Steinberg, D., Merks, E., Ellersick, R., & Grose, T. J. (2003). *Eclipse modeling framework*. Reading, MA: Addison-Wesley.

Chen, J., Sivalingam, K., Agrawal, P., & Kishore, S. (1998). A comparison of MAC protocols for wireless local networks based on battery power consumption. []. IEEE.]. *Proceedings - IEEE INFOCOM, 1*, 150–157.

Creus, G., & Kuulusa, M. (2007). Optimizing mobile software with built-in power profiling. In Fitzek, F., & Reichert, F. (Eds.), *Mobile Phone Programming and its Application to Wireless Networking*. Berlin: Springer. doi:10.1007/978-1-4020-5969-8_25.

Feeney, L., & Nilsson, M. (2001). Investigating the energy consumption of a wireless network interface in an ad hoc networking environment. []. IEEE.]. *Proceedings - IEEE INFOCOM, 3*, 1548–1557.

Heinzelman, W., Chandrakasan, A., & Balakrishnan, H. (2000). Energy-efficient communication protocol for wireless microsensor networks. In *Proceedings of the 33rd Hawaii International Conference on System Sciences*, (vol. 8, p. 8020). IEEE.

Herrmann, R., Zappi, P., & Rosing, T. (2012). Context aware power management of mobile systems for sensing applications. In *Proceedings of the 2nd International Workshop on Mobile Sensing*. IEEE.

Hill, J., Schmidt, D. C., Slaby, J., & Porter, A. (2008). CiCUTS: Combining system execution modeling tools with continuous integration environments. In *Proceeedings of 15th Annual IEEE International Conference and Workshops on the Engineering of Computer Based Systems (ECBS)*. Belfast, Ireland: ECBS.

Kang, J., Park, C., Seo, S., Choi, M., & Hong, J. (2008). User-centric prediction for battery lifetime of mobile devices. In *Proceedings of the 11th Asia-Pacific Symposium on Network Operations and Management: Challenges for Next Generation Network Operations and Service Management*, (pp. 531–534). Springer.

Kim, M., Kong, J., & Chung, S. (2012). An online power estimation technique for multi-core smartphones with advanced display components. In *Proceedings of the 2012 IEEE International Conference on Consumer Electronics (ICCE)*, (pp. 666–667). IEEE.

Kjasrgaard, M. (2012). Location-based services on mobile phones: Minimizing power consumption. *IEEE Pervasive Computing / IEEE Computer Society [and] IEEE Communications Society*, *11*(1), 67–73. doi:10.1109/MPRV.2010.47.

Krashinsky, R., & Balakrishnan, H. (2005). Minimizing energy for wireless web access with bounded slowdown. *Wireless Networks*, *11*(1), 135–148. doi:10.1007/s11276-004-4751-z.

Krause, A., Ihmig, M., Rankin, E., Leong, D., Gupta, S., & Siewiorek, S. … Sengupta, S. (2005). Trading off prediction accuracy and power consumption for context-aware wearable computing. In *Proceedings of the Ninth IEEE International Symposium on Wearable Computers*, (pp. 20–26). IEEE Computer Society.

Kravets, R., & Krishnan, P. (2000). Application-driven power management for mobile communication. *Wireless Networks*, *6*(4), 263–277. doi:10.1023/A:1019149900672.

Landsiedel, O., Wehrle, K., & Gotz, S. (2005). Accurate prediction of power consumption in sensor networks. In *Proceedings of The Second IEEE Workshop on Embedded Networked Sensors (EmNetS-II)*. IEEE.

Ledeczi, A., Bakay, A., Maroti, M., Volgyesi, P., Nordstrom, G., Sprinkle, J., & Karsai, G. (2001). Composing domain-specific design environments. *Computer*, 44–51. doi:10.1109/2.963443.

Liu, T., Sadler, C., Zhang, P., & Martonosi, M. (2004). Implementing software on resource-constrained mobile sensors: Experiences with impala and zebranet. In *Proceedings of the 2nd International Conference on Mobile Systems, Applications, and Services*, (pp. 256–269). ACM.

Mohapatra, S., Cornea, R., Dutt, N., Nicolau, A., & Venkatasubramanian, N. (2003). Integrated power management for video streaming to mobile handheld devices. In *Proceedings of the Eleventh ACM International Conference on Multimedia*, (pp. 582–591). ACM.

Parikh, D., Skadron, K., Zhang, Y., Barcella, M., & Stan, M. (2002). Power issues related to branch prediction. In *Proceedings of the Eighth International Symposium on High-Performance Computer Architecture*, (pp. 233–244). IEEE.

Pering, T., Agarwal, Y., Gupta, R., & Want, R. (2006). Coolspots: Reducing the power consumption of wireless mobile devices with multiple radio interfaces. In *Proceedings of the 4th International Conference on Mobile systems, Applications and Services*, (p. 232). ACM.

Schmidt, D. C. (2006). Model-driven engineering. *IEEE Computer*, *39*(2), 25–31. doi:10.1109/MC.2006.58.

Smith, C., & Williams, L. (2001). *Performance solutions: A practical guide to creating responsive, scalable software*. Boston: Addison-Wesley Professional.

Thiagarajan, N., Aggarwal, G., Nicoara, A., Boneh, D., & Singh, J. (2012). Who killed my battery? Analyzing mobile browser energy consumption. In *Proceedings of the 21st International Conference on World Wide Web*, (pp. 41–50). ACM.

Thompson, C., White, J., Dougherty, B., & Schmidt, D. (2009). Optimizing mobile application performance with model-driven engineering. In *Proceedings of the 7th IFIP Workshop on Software Technologies for Future Embedded and Ubiquitous Systems*. IFIP.

Turner, H., White, J., Thompson, C., Zienkiewicz, K., Campbell, S., & Schmidt, D. (2009). Building mobile sensor networks using smartphones and web services: Ramifications and development challenges. In Cruz-Cunha, M. M., & Moreira, F. (Eds.), *Handbook of Research on Mobility and Computing: Evolving Technologies and Ubiquitous Impacts*. Hershey, PA: IGI Global.

Wang, Q., Hempstead, M., & Yang, W. (2006). A realistic power consumption model for wireless sensor network devices. In *Proceedings of the Third Annual IEEE Communications Society Conference on Sensor, Mesh and Ad Hoc Communications and Networks (SECON)*. IEEE.

Wang, Y., Krishnamachari, B., & Annavaram, M. (2012). Semi-Markov state estimation and policy optimization for energy efficient mobile sensing. In *Proceedings of Sensor, Mesh and Ad Hoc Communications and Networks (SECON)*, (pp. 533–541). IEEE. doi:10.1109/SECON.2012.6275823.

White, J., Clarke, S., Dougherty, B., Thompson, C., & Schmidt, D. (2010). R&D challenges and solutions for mobile cyber-physical applications and supporting internet services. *Springer Journal of Internet Services and Applications*, *1*(1), 45–56. doi:10.1007/s13174-010-0004-9.

White, J., Hill, J., Tambe, S., Gray, J., Gokhale, A., & Schmidt, D. C. (2009, July/August). Improving domain-specific language reuse through software product-line configuration techniques. *IEEE Software*. doi:10.1109/MS.2009.95.

White, J., Schmidt, D., & Mulligan, S. (2007). The generic eclipse modeling system. In *Proceedings of the Model-Driven Development Tool Implementors Forum at TOOLS 2007*. TOOLS.

Yoon, C., Kim, D., Jung, W., Kang, C., & Cha, H. (2012). Appscope: Application energy metering framework for android smartphone using kernel activity monitoring. In *Proceedings of USENIX Annual Technical Conference*. USENIX.

ENDNOTES

[1] WreckWatch is available from vuphone. googlecode.com.

Compilation of References

Accellera. (2002). *SystemVerilog 3.0: Accellera's extensions to verilog.* San Francisco, CA: Accellera.

Agarwal, Y., Chandra, R., Wolman, A., Bahl, P., Chin, K., & Gupta, R. (2007). Wireless wakeups revisited: energy management for voip over wi-fi smartphones. *ACM MobiSys, 7.*

Agarwal, N. K. (2011). Verifying survey items for construct validity: A two-stage sorting procedure for questionnaire design in information behavior research. *Proceedings of the American Society for Information Science and Technology, 48*(1), 1–8. doi:10.1002/meet.2011.14504801166.

Agile Modeling (AM) Home Page. (n.d.). *Effective practices for modeling and documentation.* Retrieved January 13, 2013, from http://www.agilemodeling.com/

Ainsworth, M., Cruickshank, A. H., Wallis, P. J. L., & Groves, L. J. (1994). Viewpoint specification and Z. *Information and Software Technology, 36*(1), 43–51. doi:10.1016/0950-5849(94)90007-8.

Alanen, M., & Porres, I. (2003). Difference and union of models. InStevens, , Whittle, , & Booch, (Eds.), *UML 2003 - The Unified Modeling Language. Modeling Languages and Applications*LNCS), (Vol. *2863*, pp. 2-17). Berlin: Springer-Verlag. doi:10.1007/978-3-540-45221-8_2.

Alexander, P. (2006). *System level design with rosetta.* San Francisco, CA: Morgan Kaufmann Publishers Inc..

Alexander, P., Barton, D., & Kong, C. (2000). *Rosetta usage guide.* Lawrence, KS: The University of Kansas.

Alhir, S. (2003). Understanding the model driven architecture. *Methods & Tools: An international Software Engineering Digital Newsletter, 11*(3), 17-24.

Allen, J. F. (1983). Maintaining knowledge about temporal intervals. *Communications of the ACM, 26*(11), 832–843. doi:10.1145/182.358434.

Allen, R., & Garlan, D. (1997). A formal basis for architectural connection. *ACM Transactions on Software Engineering and Methodology, 6*(3), 213–249. doi:10.1145/258077.258078.

Alpaydin, E. (2010). *Introduction to machine learning* (2nd ed.). Cambridge, MA: The MIT Press.

Ambler, S. (2005). *AmbySoft agile unified process page.* Retrieved January 13, 2013, from http://www.ambysoft.com/unifiedprocess/agileUP.html

Ambler, S. (2008). *Agile adoption rate survey results.* Retrieved January 13, 2013, from http://www.ambysoft.com/surveys/agileFebruary2008.html

Ambler, S. (2009). *The agile scaling model (ASM), adapting agile methods for complex environments.* Somers, NY: IBM Corporation Software Group. Retrieved from ftp://ftp.software.ibm.com/common/ssi/sa/wh/n/raw14204usen/RAW14204USEN.PDF

Ambler, S. (2010). *Context count: Position paper for SOMAT.* Retrieved January 13, 2013, from www.semat.org/pub/.../WorkshopPositions/SEMAT_position_Ambler.doc

Ambler, S. (2002). *Agile modeling: Effective practices for extreme programming and the unified process.* New York: John Wiley & Sons.

Anand, A., Manikopoulos, C., Jones, Q., & Borcea, C. (2007). A quantitative analysis of power consumption for location-aware applications on smart phones. In *Proceedings of the 2007 IEEE International Symposium on Industrial Electronics,* (pp. 1986–1991). IEEE.

Anli, A. (2006). *Méthodologie de développement des systèmes d'information personnalisés. (PhD Thésis).* Valenciennes, France: University of Valenciennes et du Hainaut-Cambrésis.

Appelo, J. (2009). *A big agile practices survey.* Retrieved January 13, 2013, from http://www.noop.nl/2009/04/the-big-agile-practices-survey.html

Arabshianand, K., & Schulzrinne, H. (2006). Distributed context-aware agent architecture for global service discovery. In *Proceedings of SWUMA 2006.* Trentino, Italy: SWUMA.

Aßmann, U., & Zschaler, S. (2006). Ontologies, metamodels, and the model-driven paradigm. In Calero, C., Ruiz, F., & Piattini, M. (Eds.), *Ontologies for Software Engineering and Software Technology* (pp. 249–273). Berlin: Springer-Verlag. doi:10.1007/3-540-34518-3_9.

Astesiano, E., Bidoit, M., Kirchner, H., Krieg-Brückner, B., Mosses, P. D., Sannella, D., & Tarlecki, A. (2002). CASL: The common algebraic specification language. *Theoretical Computer Science, 286*(2), 153–196. doi:10.1016/S0304-3975(01)00368-1.

Atkinson, C., & Kuhne, T. (2003). Model-driven development: A metamodeling foundation. *IEEE Software, 20*(5), 36–41. doi:10.1109/MS.2003.1231149.

Atkinson, T. K. (2003, September). Model-driven development: A metamodeling foundation. *IEEE Software.* doi:10.1109/MS.2003.1231149.

ATL. (2012). *Atlas transformation language (ATL) documentation.* Retrieved September 24, 2012, from http://www.eclipse.org/atl/documentation/

Avison, D., Wood-Harper, A., Vidgen, R., & Wood, J. (1998). A further exploration into information systems development: The evolution of Multiview2. *Information Technology & People, 11*(2), 124–139. doi:10.1108/09593849810218319.

Ayse, K. (2002). *Component-oriented modeling of land registry and cadastre information system using COSEML.* (Master Thesis). Orta Doğu Teknik Üniversitesi, Ankara, Turkey.

Bacha, F., Oliveira, K., & Abed, M. (2011). Providing personalized information in transport systems: A model driven architecture approach. In *Proceedings of the First IEEE International Conference on Mobility, Security and Logistics in Transport* (pp. 452-459). Hammamet, Tunisia: IEEE.

Bacha, F., Oliveira, K., & Abed, M. (2011). Using context modeling and domain ontology in the design of personalized user interface. *International Journal on Computer Science and Information Systems, 6,* 69–94.

Baldauf, M., Dustdar, S., & Rosenberg, F. (2007). A survey on context-aware systems. *International Journal of Ad Hoc and Ubiquitous Computing, 2*(4), 263–277. doi:10.1504/IJAHUC.2007.014070.

Balmelli, L., Brown, D., Cantor, M., & Mott, M. (2006). Model-driven systems development. *IBM Systems Journal, 45*(3), 569–585. doi:10.1147/sj.453.0569.

Baniassad, E., & Clarke, S. (2004). An approach for aspect-oriented analysis and design. In *Proceedings of the 26th International Conference on Software Engineering.* IEEE.

Barra, E., Genova, G., & Llorens, J. (2004). An approach to aspect modelling with UML 2.0. In *Proceedings of the 5th Wsh. Aspect-Oriented Modeling (AOM'04).* Lisboa, Portugal: AOM.

Basin, D. A., Doser, J., & Lodderstedt, T. (2006). Model driven security: From UML models to access control infrastructures. *ACM Transactions on Software Engineering and Methodology, 15*(1), 39–91. doi:10.1145/1125808.1125810.

Batchel, O., et al. (2012). *Translational medicine ontology.* Retrieved from http://translationalmedicineontology.googlecode.com/svn/trunk/ontology/tmo.owl

Baudoin, C., Covnot, B., Kumar, A., LaCrosse, K., & Shields, R. (2010). Business architecture: The missing link between business strategy and enterprise architecture. In B. Michelson (Ed.), *EA2010 Working Group, SOA Consortium.* Retrieved October 15, 2012, from http://www.soa-consortium.org/EA2010.htm

Beck, K., Beedle, M., & van Bennekum, et al. (2001). *Principles behind the agile manifesto.* Retrieved October 15, 2012, from http://agilemanifesto.org/principles.html

Becker, S., Gruschko, B., Goldschmidt, T., & Koziolek, H. (2007). A process model and classification scheme for semi-automatic meta-model evolution. In GiTO-Verlag (Ed.), *Proceeding of the 1st Workshop MDD, SOA and IT-Management (MSI'07)*, (pp. 35-46). MSI.

Becker, C., & Dürr, F. (2005). On location models for ubiquitous computing. *Personal and Ubiquitous Computing, 9*(1), 20–23. doi:10.1007/s00779-004-0270-2.

Beck, K., & Andres, C. (2004). *Extreme programming explained: Embrace change* (2nd ed.). Reading, MA: Addison-Wesley Professional.

Bellur, U., & Vallieswaran. (2006). On OO design consistency in iterative development. In Proceedings of the 3rd International Conference on Information Technology: New Generations, (pp. 46-51). IEEE Computer Society. doi:doi:10.1109/ITNG.2006.102 doi:10.1109/ITNG.2006.102.

Bernardi, S., Donatelli, S., & Merseguer, J. (2002). From UML sequence diagrams and statecharts to analysable petrinet models. In *Proceedings of the 3rd International Workshop on Software and Performance*, (pp. 35-45). New York, NY: ACM.

Bernard, Y. (2012). Requirements management within a full model-based engineering approach. *Systems Engineering, 15*(2), 119–139. doi:10.1002/sys.20198.

Bernhardt, T., & Vasseur, A. (2007). Esper: Event stream processing and correlation. *O'Reilly OnJava*. Retrieved from http://onjava.com/pub/a/onjava/2007/03/07/esper-event-stream-processing-and-correlation.html

Berti, S., Mori, G., Paternò, F., & Santoro, C. (2004). TERESA: A transformation-based environment for designing multi-device interactive applications. In *Proceedings of CHI 2004* (pp. 793-794). Wien, Austria: ACM Press.

Bertot, Y., & Castéran, P. (2004). *Interactive theorem proving and program development: Coq'Art: The calculus of inductive constructions*. New York: Springer. doi:10.1007/978-3-662-07964-5.

Bézivin, J. (2006). Model driven engineering: An emerging technical space. In Lämmel, Saraiva, & Visser (Eds.), Generative and Transformational Techniques in Software Engineering (LNCS), (vol. 4143, pp. 36-64). Berlin: Springer. doi:doi:10.1007/11877028_2 doi:10.1007/11877028_2.

Bézivin, J. (2004). In search of a basic principle for model driven engineering. *UPGRADE, 5*(2), 21–24.

Bézivin, J. (2005). On the unification power of models. *Software & Systems Modeling, 4*(2), 171–188. doi:10.1007/s10270-005-0079-0.

Bhasin, S. (2012). Quality assurance in agile – A study towards achieving excellence. In *Proceedings of AGILE INDIA* (pp. 64-67). AGILE.

Bidoit, M., & Mosses, P. (2004). *CASL user manual- Introduction to using the common algebraic specification language (LNCS)* (Vol. 2900). Berlin: Springer-Verlag.

Bieszczad, A., & Bieszczad, K. (2006). Contextual learning in the neurosolver. In *Proceedings of ICANN*, (pp. 474-484). Berlin: Springer.

Blair, G., Bencomo, N., & France, R. B. (2009). Models@ run.time. *Computer, 42*(10), 22–27. doi:10.1109/MC.2009.326.

Blanes, D., & Insfran, E. (2012). A comparative study on model-driven requirements engineering for software product lines. *Revista de Sistemas e Computação, 2*(1), 1–11.

Bloom, N. (2013). *The future of HRM software: Agile, models-driven, definitional*. Retrieved from http://infulbloom.us/?p=3222

Boehm, B. (1981). *Software engineering economics*. Englewood Cliffs, NJ: Prentice Hall.

Boiten, E., Bowman, H., Derrick, J., Linington, P., & Steen, M. (2000). Viewpoint consistency in ODP. *Computer Networks, 34*(3), 503–537. doi:10.1016/S1389-1286(00)00114-6.

Bosch, J., & Bosch-Sijtsema, P. (2010). From integration to composition: On the impact of software product lines, global development and ecosystems. *Journal of Systems and Software, 83*(1), 67–76. doi:10.1016/j.jss.2009.06.051.

Bosch, J., & Molin, P. (1997). *Software architecture design: Evaluation and transformation*. Karlskrona.

Bouchelliga, W., Mahfoudi, A., & Abed, M. (2012). A model driven engineering approach toward user interfaces adaptation. *International Journal of Adaptive, Resilient and Autonomic Systems, 3*, 65–86. doi:10.4018/jaras.2012010104.

Bouchelligua, W., Mahfoudhi, A., Mezhoudi, N., Daassi, O., & Abed, M. (2010). User interfaces modelling of workflow information systems. In Barjis, J. (Ed.), *Enterprise & Organizational Modeling and Simulation (LNBIP)* (*Vol. 63*). Berlin: Springer. doi:10.1007/978-3-642-15723-3_10.

Boudreau, M., Gefen, D., & Straub, D. W. (2001). Validation in information systems research: A state-of-the-art assessment. *Management Information Systems Quarterly, 25*(1), 1–16. doi:10.2307/3250956.

Brooks, F. (1986). No silver bullet—Essence and accident in software engineering. In *Proceedings of the IFIP Tenth World Computing Conference*, (pp. 1069–1076). IFIP.

Brooks, F. P. Jr. (1995). *The mythical man-month: Essays on software engineering: Anniversary Ed.* Boston, MA: Addison-Wesley.

Brossard, A., Abed, M., & Kolski, C. (2011). Taking context into account in conceptual models using a model driven engineering approach. *Information and Software Technology, 53*, 1349–1369. doi:10.1016/j.infsof.2011.06.011.

Brun, C., & Pierantonio, A. (2008). Model differences in the eclipse modeling framework. *UPGRADE: The European Journal for the Informatics Professional, 9*(2), 29–32.

Budinsky, F., Steinberg, D., Merks, E., Ellersick, R., & Grose, T. J. (2003). *Eclipse modeling framework.* Reading, MA: Addison-Wesley.

Cabot, J., & Yu, E. (2008). *Improving requirements specifications in model-driven development processes.* Paper presented at the 1st International Workshop on Challenges in Model-Driven Software Engineering (MoDELS'08). Toulouse, France.

Calvary, G., Coutaz, J., Thevenin, D., Limbourg, Q., Bouillon, L., & Vanderdonckt, J. (2003). A unifying reference framework for multi-target user interfaces. *Interacting with Computers, 15*(3), 289–308. doi:10.1016/S0953-5438(03)00010-9.

Carlson, J. (2004). *An intuitive and resource-efficient event detection algebra.* (Unpublished doctoral dissertation). Mälardalen University, Eskilstuna, Sweden.

CASE MDA. (2012). *Committed companies and their products.* Retrieved September 24, 2012 from www.omg.org/mda/committed-products.htm

Champeau, J., Mekerke, F., & Rochefort, E. (2003). Towards a clear definition of patterns, aspects and views in MDA. In J. Ralyté & C. Rolland (Eds.), *Proceedings of the First International Workshop on Engineering Methods to Support Information Systems Evolution (EMSISE)* (pp. 29–36). Geneva, Switzerland: EMSISE.

Cheng, B., & Atlee, J. (2007). Research directions in requirements engineering. In *Proceedings of FOSE 07 Future of Software Engineering* (pp. 285-303). Los Alamitos, CA: IEEE Computer Society Press.

Chen, H., Perich, F., Finin, T., & Jochi, A. (2004). *SOUPA: Standard ontology for ubiquitous and pervasive applications.* IEEE Computer Society, 258-267. doi:10.1109/MOBIQ.2004.1331732.

Chen, J., Sivalingam, K., Agrawal, P., & Kishore, S. (1998). A comparison of MAC protocols for wireless local networks based on battery power consumption.[]. IEEE.]. *Proceedings - IEEE INFOCOM, 1*, 150–157.

ChinaLabs. (2002). *Business system infrastructure platform (Technical report).* Beijing, China: ChinaLabs.

Chiorean, D., Pasca, M., Carcu, A., Botiza, C., & Moldovan, S. (2004). Ensuring UML models consistency using the OCL environment. *Electronic Notes in Theoretical Computer Science, 102*, 99–110. doi:10.1016/j.entcs.2003.09.005.

Cicchetti, A., Di Ruscio, D., & Di Salle, A. (2007). Software customization in model driven development of web applications. In *Proceedings of the 2007 ACM Symposium on Applied Computing* (pp. 1025-1030). ACM Press.

Cicchetti, A., Di Ruscio, D., Eramo, R., & Pierantonio, A. (2008). Automating co-evolution in model-driven engineering. In IEEE Computer Society (Ed.), *Proceedings of the 2008 12th International IEEE Enterprise Distributed Object Computing Conference (EDOC 2008)* (pp. 222-231). Washington, DC. IEEE.

Cicchetti, A., Di Ruscio, D., & Pierantonio, A. (2010). Model patches in model-driven engineering.[LNCS]. *Proceedings of Models in Software Engineering, 6002*, 190–204. doi:10.1007/978-3-642-12261-3_19.

Cirstea, C., Kurz, A., Pattinson, D., Schröder, L., & Venema, Y. (n.d.). Modal logics are coalgebraic. *The Computer Journal, 54*(1), 31–41.

Clarke, S., Harrison, W., Ossher, H., & Tarr, P. (1999). Subject-oriented design: Towards improved alignment of requirements, design and code. In *Proceedings of the Object-Oriented Programming, Systems, Languages and Applications*. Denver, CO: IEEE.

Clerckx, T., Luyten, K., & Coninx, K. (2005). Dynamo-aid: A design process and a runtime architecture for dynamic model-based user interface development. In *LNCS* (pp. 77–95). Berlin: Springer-Verlag. doi:10.1007/11431879_5.

Cockburn, A. (2004). *Crystal clear. A human-powered methodology for small teams, including the seven properties of effective software projects*. Retrieved from http://st-www.cs.illinois.edu/users/johnson/427/2004/crystalclearV5d.pdf

Costabile, M., Mussio, P., Piccinno, A., & Provenza, L. (2008). End users as unwitting software developers. In *Proceedings of the International Conference on Software Engineering, ISSU*, (pp. 6-10). ISSU.

Creus, G., & Kuulusa, M. (2007). Optimizing mobile software with built-in power profiling. In Fitzek, F., & Reichert, F. (Eds.), *Mobile Phone Programming and its Application to Wireless Networking*. Berlin: Springer. doi:10.1007/978-1-4020-5969-8_25.

Cuenca, L., Boza, A., & Ortiz, A. (2010). Enterprise engineering approach for business and IS/IT strategic alignment. In *Proceedings of the MOSIM 2010 Conference*. Lavoisier.

Cypher, A., Halbert, D. C., Kurlander, D., Lieberman, H., Maulsby, D., Myers, B. A., & Turransky, A. (Eds.). (1993). *Watch what I do: Programming by demonstration*. Cambridge, MA: MIT Press.

Czarnecki, K., & Helsen, S. (2006). Feature-based survey of model transformation approaches. *IBM Systems Journal, 45*, 621–645. doi:10.1147/sj.453.0621.

Czarnecki, U. (2000). *Generative programming*. Reading, MA: Addison-Wesley.

Dalal, S. R., Jain, A., Karunanithi, N., Leaton, J. M., Lott, C. M., Patton, G. C., & Horowitz, B. M. (1999). Model-based testing in practice. In *Proceedings of International Conference on Software Engineering (ICSE 99)*, (pp. 285-294). IEEE Computer Society.

Davey, B., & Cope, C. (2008). Requirements elicitation - What's missing? *Issues in Informing Science and Information Technology, 5*(1), 543–551.

Dávid, I. (2012). A model-driven approach for processing complex events. In *Proceedings of EDCC 2012 - Fast Abstracts & Student Forum*. Retrieved from http://arxiv.org/abs/1204.4428

Davis, J., & Chang, E. (2011). Automatic application update with user customization integration and collision detection for model driven applications. In *Proceedings of the World Congress on Engineering and Computer Science 2011* (pp. 1081-1086). Newswood Limited.

Davis, J., & Chang, E. (2011). Lifecycle and generational application of automated updates to MDA based enterprise information systems. In *Proceedings of the Second Symposium on Information and Communication Technology (SoICT '11)* (pp. 207-216). ACM Press.

Davis, J., & Chang, E. (2011). Temporal meta-data management for model driven applications - Provides full temporal execution capabilities throughout the meta-data EIS application lifecycle. In *Proceedings of the 13th International Conference on Enterprise Information Systems (ICEIS 2011)* (pp. 376-379). SciTePress.

Davis, J., & Chang, E. (2011). variant logic meta-data management for model driven applications applications - Allows unlimited end user configuration and customisation of all meta-data EIS application features. In *Proceedings of the 13th International Conference on Enterprise Information Systems (ICEIS 2011)* (pp. 395-400). SciTePress.

Davis, J., Tierney, A., & Chang, E. (2004). Meta data framework for enterprise information systems specification - Aiming to reduce or remove the development phase for EIS systems. In *Proceedings of the 6th International Conference on Enterprise Information Systems (ICEIS 2004)* (pp. 451-456). SciTePress.

Davis, J., Tierney, A., & Chang, E. (2005). Merging application models in a MDA based runtime environment for enterprise information systems. In *Proceedings of INDIN 2005: 3rd International Conference on Industrial Informatics, Frontier Technologies for the Future of Industry and Business* (pp. 605-610). IEEE Computer Society.

De Alwis, B., & Sillito, J. (2009). Why are software projects moving from centralized to decentralized version control systems? In *Proceedings of the 2009 ICSE Workshop on Cooperative and Human Aspects on Software Engineering* (pp. 36-39). IEEE Computer Society.

de Fombelle, G., Blanc, X., Rioux, L., & Gervais, M.-P. (2006). Finding a path to model consistency. In *Proceedings of the Second European Conference on Model Driven Architecture - Foundations and Applications (ECMDA-FA 2006)*, (pp. 101-112). Berlin: Springer-Verlag.

de Lemos, R., Giese, H., Müller, H., Shaw, M., Andersson, J., Baresi, L., & Wuttke, J. (2011). Software engineering for self-adaptive systems: A second research roadmap. In de Lemos, , Giese, , Müller, , & Shaw, (Eds.), *Software Engineering for Self-Adaptive Systems*. Dagstuhl, Germany: Schloss Dagstuhl - Leibniz-Zentrum fuer Informatik.

Decker, G., Grosskopf, A., & Barros, A. (2007). A graphical notation for modeling complex events in business processes. *Business Information Systems*, *4439*, 29–40. doi:10.1007/978-3-540-72035-5_3.

Demuth, B., & Hussmann, H. (1999). Using UML/OCL constraints for relational database design. In *Proceedings of UML'99: The Unified Modeling Language - Beyond the Standard, Second International Conference*, (pp. 598– 613). Springer-Verlag.

Demuth, B., Hussmann, H., & Loecher, S. (2001). OCL as a specification language for business rules in database applications. In *Proceedings of the 4th International Conference on The Unified Modeling Language: Modeling Languages, Concepts, and Tools*, (pp. 104-117). Springer-Verlag.

Derrick, J., Akehurst, D., & Boiten, E. (2002). A framework for UML consistency. In L. Kuzniarz, G. Reggio, J. L. Sourrouille, & Z. Huzar (Eds.), 2002 Workshop on Consistency Problems in UML-Based Software Development, (pp. 30-45). Springer.

Desel, J. (2002). Formalization and validation - An iterative process in model synthesis. In *Proceedings of the Workshop on Foundations for Modeling and Simulation Verification and Validation in the 21st Century* (pp. 1-18). The Society for Modeling and Simulation International.

Dey, A. (2001). Understanding and using context. *Journal of Personal and Ubiquitous Computing*, *5*, 4–7. doi:10.1007/s007790170019.

Dijkman, R. M., Quartel, D. A. C., & van Sinderen, M. J. (2008). Consistency in multi-viewpoint design of enterprise information systems. *Information and Software Technology*, *50*(7–8), 737–752. doi:10.1016/j.infsof.2007.07.007.

Diskin, Z., & Wolter, U. (2008). A diagrammatic logic for object-oriented visual modeling. *Electronic Notes in Theoretical Computer Science*, *203*(6), 19–41. doi:10.1016/j.entcs.2008.10.041.

Dittrich, Y., Vaucouleur, S., & Giff, S. (2009). ERP customization as software engineering: knowledge sharing and cooperation. *IEEE Software*, *26*(1), 41–47. doi:10.1109/MS.2009.173.

Djuric, D., Devedzic, V., & Gasevic, D. (2007). Adopting software engineering trends in AI. *IEEE Intelligent Systems*, *22*, 59–66. doi:10.1109/MIS.2007.2.

Dobson, I., & Hietala, J. (Eds.). (2011). *Risk management: The open group guide*. Norfolk, UK: Van Haren Publishing.

Dolques, X., Huchard, M., & Nebut, C. (2009). From transformation traces to transformation rules: Assisting model driven engineering approach with formal concept analysis. In *Proceedings of ICCS*, (pp. 93-106). Moscow, Russia: ICCS.

Douglas, S., & Moran, T. P. (1983). Learning text-editing semantics by analogy. In *Proceedings of ACM Human Factors in Computing Systems (CHI)*. Boston, MA: ACM.

DSDM Atern Handbook v.1.0. (2008). Retrieved January 13, 2013, from http://www.dsdm.org/atern-handbook/flash.html

Dubray, J.-J. (2011). Why did MDE miss the boat? *InfoQ*. Retrieved October 15, 2012, from http://www.infoq.com/news/2011/10/mde-missed-the-boat

Dyche, J. (2002). *The CRM handbook: A business guide to customer relationship management*. Reading, MA: Addison-Wesley Educational Publishers.

Easterbrook, S., Yu, E., Ar, J., Fan, Y., Horkoff, J., Leica, M., & Qadir, R. A. (2005). Do viewpoints lead to better conceptual models? An exploratory case study. In *Proceedings of the 13th IEEE International Conference on Requirements Engineering (RE '05)* (pp. 199–208). Los Alamitos, CA: IEEE Computer Society.

Eclipse. (2012). *The eclipse modeling framework*. Retrieved September 24, 2012 from http://www.eclipse.org/emf/

Eclipse-M2M. (2010). *The model to model (M2M) transformation framework*. Retrieved June 22, 2010, from http://www.eclipse.org/m2m/

Efftinge, S., & Völter, M. (2006). oAW xText: A framework for textual DSLs. In *Proceedings of the Workshop on Modeling Symposium at Eclipse Summit*. IEEE.

Egyed, A. (2001). Scalable consistency checking between diagrams - The VIEWINTEGRA approach. In *Proceedings of the 16th IEEE International Conference on Automated Software Engineering (ASE)*, (pp. 387-390). IEEE Computer Society.

Egyed, A. (2006). Instant consistency checking for the UML. In *Proceedings of the 28th International Conference on Software Engineering (ICSE2006)*, (pp. 381-390). ACM.

Egyed, A. (2007). Fixing Inconsistencies in UML design models. In *Proceedings of the 29th International Conference on Software Engineering (ICSE2007)*, (pp. 292-301). IEEE Computer Society.

Elmasri, R., & Navathe, S. B. (2003). *Fundamentals of database systems* (4th ed.). Reading, MA: Addison Wesley.

Erwig, M. (2003). Toward the automatic derivation of XML transformations. In *Proceedings of XSDM*, (pp. 342-354). Berlin: Springer.

Esposito, F., Semeraro, G., Fanizzi, N., & Ferilli, S. (2000). Multistrategy theory revision: Induction and abduction in inthelex. *Machine Learning*, *38*(1-2), 133–156. doi:10.1023/A:1007638124237.

Essaidi, M., & Osmani, A. (2009). Data warehouse development using MDA and 2TUP. In *Proceedings of SEDE*, (pp. 138-143). ISCA.

Essaidi, M., & Osmani, A. (2010). Towards model driven data warehouse automation using machine learning. In *Proceedings of IJCCI*, (pp. 380-383). Valencia, Spain: SciTePress.

Essaidi, M., Osmani, A., & Rouveirol, C. (2011). Transformation learning in the context of model-driven data warehouse: An experimental design based on inductive logic programming. In *Proceedings of ICTAI*, (pp. 693-700). IEEE.

Essaidi, M., & Osmani, A. (2010). Model driven data warehouse using MDA and 2TUP. *Journal of Computational Methods in Sciences and Engineering*, *10*, 119–134.

Essaidi, M., & Osmani, A. (2012). Business intelligence-as-a-service: Studying the functional and the technical architectures. In *Business Intelligence Applications and the Web: Models, Systems and Technologies* (pp. 199–221). Hershey, PA: IGI Global.

Evans, E. (2003). *Domain-driven design: Tackling complexity in the heart of software*. Reading, MA: Addison-Wesley.

Fabra, J., Pena, J., Ruiz-Cortez, A., & Ezpeleta, J. (2008). Enabling the evolution of service-oriented solutions using an UML2 profile and a reference Petri nets execution platform. In *Proceedings of the 3rd International Conference on Internet and Web Applications and Services* (pp. 198-204).). IEEE Computer Society.

Falleri, J. R., Huchard, M., Lafourcade, M., & Nebut, C. (2008). Metamodel matching for automatic model transformation generation. In Busch, Ober, Bruel, Uhl, & Völter (Eds.), *Proceedings of the 11th International Conference on Model Driven Engineering Languages and Systems (MoDELS '08)* (LNCS), (Vol. 5301, pp. 326–340). Berlin: Springer-Verlag.

Falvo, M. C., Lamedica, R., & Ruvio, A. (2012). An environmental sustainable transport system: A trolley-buses line for Cosenza city. In *Proceedings of the International Symposium on Power Electronics Power Electronics, Electrical Drives, Automation and Motion*, (pp. 1479–1485). IEEE.

Favre, L. (2009). A formal foundation for metamodeling. In *Proceedings of the 14th ADA-Europe International Conference on Reliable Software Technologies* (LNCS), (Vol. 5570, pp. 177-191). Berlin: Springer-Verlag.

Favre, J.-M. (2004). *Foundations of model (driven) (reverse) engineering - Episode I: Story of the fidus papyrus and the solarus*. Language Engineering for Model-Driven Software Development.

Fawcett, T. (2004). *Roc graphs: Notes and practical considerations for researchers (Technical report)*. HP Laboratories.

Feeney, L., & Nilsson, M. (2001). Investigating the energy consumption of a wireless network interface in an ad hoc networking environment.[). IEEE.]. *Proceedings - IEEE INFOCOM, 3*, 1548–1557.

Fernández Sáez, A. (2009). *Un análisis crítico de la aproximación model-driven architecture. (Máster en Investigación)*. Madrid, Spain: Informática Facultad de Informática, Universidad Complutense de Madrid.

Fernández-Medina, E., Jürjens, J., Trujillo, J., & Jajodia, S. (2009). Model-driven development for secure information systems. *Information and Software Technology, 51*(5), 809–814. doi:10.1016/j.infsof.2008.05.010.

Finkelstein, A., Easterbrook, S., Kramer, J., & Nuseibeh, B. (1992). *Requirements engineering through viewpoints*. London: Imperial College London, Department of Computing. Retrieved February 2, 2013, from http://discovery.ucl.ac.uk/854/1/3.3_dra.pdf

FIPA. (2001). *Device ontology specification*. Retrieved from http://www.fipa.org/specs/fipa00091/PC00091A.html

Formula. (2012). *Formula - Modeling foundation*. Retrieved from http://research.microsoft.com/en-us/projects/formula/

Forza, C. (2002). Survey research in operations management: A process-based perspective. *International Journal of Operations & Production Management, 22*(2), 152. doi:10.1108/01443570210414310.

Fouad, A., Phalp, K., Kanyaru, J., & Jeary, S. (2011). Embedding requirements within model-driven architecture. *Software Quality Journal, 19*(2), 411–430. doi:10.1007/s11219-010-9122-7.

Fowler, M. (1999). *Refactoring: Improving the design of existing programs*. Reading, MA: Addison-Wesley.

Fowler, M. (2010). *Domain-specific languages*. Boston, MA: Addison-Wesley Professional.

Fox, M. S., & Gruninger, M. (1998). Enterprise modeling. *AI Magazine, 19*(3), 109–121.

France, R., & Bieman, J. M. (2001). Multi-view software evolution: A UML-based framework for evolving object-oriented software. In *Proceedings of the IEEE International Conference on Software Maintenance* (ICSM 2001), (pp. 386-395). IEEE.

France, R., & Rumpe, B. (2007). Model-driven development of complex software: A research roadmap. In *Proceedings of the 2007 Future of Software Engineering*, (pp. 37-54). Washington, DC: IEEE Computer Society.

Frankel, D. S. (2003). *Model driven architecture: Applying MDA to enterprise computing*. New York: Wiley Publishing, Inc..

Frisby, N., Peck, M., Snyder, M., & Alexander, P. (2011). Model composition in rosetta. In *Proceedings of the IEEE Conference and Workshops on the Engineering of Computer Based Systems* (pp. 140–148). IEEE Computer Society.

Fundation, E. (2001). *Eclipse*. Retrieved from http://www.eclipse.org.

Gajos, K., Weld, D., & Wobbrock, J. (2010). Automatically generating personalized user interfaces with Supple. *Artificial Intelligence, 174*(12), 910–950. doi:10.1016/j.artint.2010.05.005.

Gama, J. (1998). Combining classifiers by constructive induction. In *Proceedings of ECML*, (pp. 178-189). Springer.

Gama, J. A., & Brazdil, P. (2000). Cascade generalization. *Machine Learning, 41*, 315–343. doi:10.1023/A:1007652114878.

Gamma, E., Helm, R., Johnson, R., & Vlissides, J. (1995). *Design patterns: Elements of reusable object-oriented software*. Boston, MA: Addison-Wesley.

Garg, R. M., & Dahiya, D. (2011). *Integrating aspects and reusable components: An archetype driven methodology*. Paper presented in the 4th International Conference on Contemporary Computing (IC3-2011). Noida, India.

Garg, R. M., Dahiya, D., Tyagi, A., Hundoo, P., & Behl, R. (2011). *Aspect oriented and component based model driven architecture*. Paper presented in the International Conference on Digital Information and Communication Technology and it's Applications (DICTAP-2011). Paris, France.

Garg, R. M. (2011). *Aspect oriented component based archetype driven ontogenesis: Combining aspect-oriented and model-driven development for better software development*. LAP Lambert Academic Publishing.

Garía-Barrios, V., Mödritscher, F., & Gütl, C. (2005). Personalisation versus adaptation? A user-centred model approach and its application. In K. Tochtermann & H. Maurer (Eds.), *Proceedings of the International Conference on Knowledge Management* (pp. 120-127). Graz, Austria: IEEE.

Georgitzikis, V., Akribopoulos, O., & Chatzigiannakis, I. (2012). Controlling physical objects via the internet using the arduino platform over 802.15.4. networks. *Revista IEEE America Lantina, 10*(3), 1686–1689. doi:10.1109/TLA.2012.6222571.

Gerber, A., Lawley, M., Raymond, K., Steel, J., & Wood, A. (2002). Transformation: The missing link of MDA. In *Proceedings of ICGT*, (pp. 90-105). Springer.

Glinz, M. (2000). Problems and deficiencies of UML as a requirements specification language. In *Proceedings of the 10th International Workshop on Software Specification*. IEEE Computer Society.

Goldschmidt, T., Becker, S., & Burger, E. (2012). Towards a tool-oriented taxonomy of view-based modelling. In E. J. Sinz & A. Schürr (Eds.), GI-LNI: Vol. 201: Modellierung 2012 (pp. 59–74). Bonn, Germany: Gesellschaft für Informatik.

Gönczy, L., Csertán, G., Urbanics, G., Khelil, A., Ghani, H., & Suri, H. (2011). Monitoring and evaluation of semantic rooms. In *Collaborative Financial Infrastructure Protection: Tools, Abstractions, and Middleware* (pp. 99–116). Berlin: Springer.

Grötker, T., Liao, S., Martin, G., & Swan, S. (2002). *System design with SystemC*. New York: Springer.

Gruber, T. R. (1993). Toward principles for the design of ontologies used for knowledge sharing. *International Journal of Human-Computer Studies, 43*, 907–928. doi:10.1006/ijhc.1995.1081.

Gruschko, B., Kolovos, D., & Paige, R. (2007). Towards synchronizing models with evolving metamodels. In *Proceeding of Workshop on Model-Driven Software Evolution (MODSE 2007)*. Amsterdam, The Netherlands: MODSE.

Gu, H., & Wang, D. (2009). A content-aware fridge based on RFID in smart home for home-healthcare. In *Proceedings of the Advanced Communication Technology Conference,* (vol. 2, pp. 987–990). IEEE.

Gustafson, S. M., & Hsu, W. H. (2001). Layered learning in genetic programming for a cooperative robot soccer problem. In *Proceedings of EuroGP* (pp. 291–301). London, UK: Springer-Verlag. doi:10.1007/3-540-45355-5_23.

Ha, I.-K., & Kang, B.-W. (2003). Meta-validation of UML structural diagrams and behavioral diagrams with consistency rules. In *Proceedings of the IEEE Pacific Rim Conference on Communications, Computers and signal Processing (PACRIM 2003)*, (Vol. 2, pp. 679-683). IEEE Computer Society.

Hachani, S., Chessa, S., & Front, A. (2009). Une approche générique pour l'adaptation dynamique des IHM au contexte. In *Proceedings of the 21st International Conference on Association Francophone d'Interaction Homme-Machine* (pp. 89-96). Grenoble, France: IEEE.

Hafner, M., Breu, R., Agreiter, B., & Nowak, A. (2006). Sectet: An extensible framework for the realization of secure inter-organizational workflows. *Internet Research, 16*(5), 491–506. doi:10.1108/10662240610710978.

Hagen, C., & Brouwers, G. (1994). Reducing software life-cycle costs by developing configurable software. In *Proceedings of the Aerospace and Electronics Conference,* (Vol. 2, pp. 1182-1187). IEEE Press.

Hagen, P., Manning, H., & Souza, R. (1999). *Smart personalization*. Washington, DC: Forrester Research.

Hao, C., Lei, X., & Yan, Z. (2012). The application and Implementation research of smart city in China. In *Proceedings of System Science and Engineering* (pp. 288–292). IEEE. doi:10.1109/ICSSE.2012.6257192.

Happel, H.-J., & Seedorf, S. (2006). Applications of ontologies in software engineering. In *Proceedings of the 2nd International Workshop on Semantic Web Enabled Software Engineering*. IEEE.

Hausmann, J. H., Heckel, R., & Sauer, S. (2002). Extended model relations with graphical consistency conditions. In *Proceedings of the Workshop on Consistency Problems in UML-Based Software Development (UML 2002)*, (pp. 61-74). UML.

Hausmann, D., Mossakowski, T., & Schröder, L. (2006). A coalgebraic approach to the semantics of the ambient calculus. *Theoretical Computer Science*, *366*(1), 121–143. doi:10.1016/j.tcs.2006.07.006.

Hegedus, A., Horvath, A., Rath, I., Ujhelyi, Z., & Varro, D. (2011). Implementing efficient model validation in EMF tools. In *Proceedings of the 26th IEEE/ACM International Conference on Automated Software Engineering* (pp. 580-583). IEEE.

Heinzelman, W., Chandrakasan, A., & Balakrishnan, H. (2000). Energy-efficient communication protocol for wireless microsensor networks. In *Proceedings of the 33rd Hawaii International Conference on System Sciences*, (vol. 8, p. 8020). IEEE.

Helms, J., Schaefer, R., Luyten, K., Vermeulen, J., Abrams, M., Coyette, A., & Vanderdonckt, J. (2009). Human-centered engineering with the user interface markup language. In Seffah, , Vanderdonckt, , & Desmarais, (Eds.), *Human-Centered Software Engineering*, (pp. 141-173). London: Springer. doi:10.1007/978-1-84800-907-3_7.

Herrmann, R., Zappi, P., & Rosing, T. (2012). Context aware power management of mobile systems for sensing applications. In *Proceedings of the 2nd International Workshop on Mobile Sensing*. IEEE.

Herrmannsdoerfer, M. (2011). COPE – A workbench for the coupled evolution of metamodels and models. In Malloy, Staab, & van den Brand (Eds.), *Software Language Engineering (SLE 2010)* (LNCS), (Vol. 6563, pp. 286-295). Berlin: Springer-Verlag.

Herrmannsdoerfer, M., Benz, S., & Juergens, E. (2009). COPE - Automating coupled evolution of metamodels and models. In *Proceedings of the ECOOP 2009 – Object-Oriented Programming* (LNCS), (Vol. 5653, pp. 52-76). Berlin: Springer-Verlag.

Herrmannsdoerfer, M., Vermolen, S., & Wachsmuth, G. (2010). An extensive catalog of operators for the coupled evolution of metamodels and models. In *Proceedings of the 3rd International Conference on Software Language Engineering (SLE' 10)*, (pp. 163-182). Berlin: Springer-Verlag.

Herrmannsdoerfer, M., Benz, S., & Juergens, E. (2008). Automatability of coupled evolution of metamodels and models in practice. In Czarnecki, , Ober, , Bruel, , Uhl, , & Völter, (Eds.), *Model Driven Engineering Languages and Systems (MoDELS '08)* LNCS), (vol. *5301*, pp. 645-659). Berlin: Springer-Verlag. doi:10.1007/978-3-540-87875-9_45.

Hill, J., Schmidt, D. C., Slaby, J., & Porter, A. (2008). CiCUTS: Combining system execution modeling tools with continuous integration environments. In *Proceedings of 15th Annual IEEE International Conference and Workshops on the Engineering of Computer Based Systems (ECBS)*. Belfast, Ireland: ECBS.

Hnatkowska, B., Huzar, Z., Kuzniarz, L., & Tuzinkiewicz, L. (2002). A systematic approach to consistency within UML based software development process. In *Proceedings of Workshop on Consistency Problems in UML-Based Software Development (UML 2002)*, (pp. 16-29). UML.

Hobbs, J., & Pan, F. (2006). *Time ontology in OWL*. Retrieved from http://www.w3.org/TR/owl-time/

Holt, J. (2012). *Model-based requirements engineering*. London: The Institution of Engineering and Technology.

Hongying, G., & Cheng, Y. (2011). A customizable agile software quality assurance model.[NISS]. *Proceedings of Information Science and Service Science*, *2*, 382–387.

Hribernik, K. A., Ghrairi, Z., Hans, C., & Thoben, K. (2011). Co-creating the internet of things - First experiences in the participatory design of intelligent products with arduino. In *Proceedings of Concurrent Enterprising*. IEEE.

Hu, X., & Shatz, S. M. (2004). Mapping UML diagrams to a petri net notation for system simulation. In *Proceedings of the Sixteenth International Conference on Software Engineering & Knowledge Engineering (SEKE'2004)*, (pp. 213-219). SEKE.

Hui, B., Liaskos, S., & Mylopoulos, J. (2003). Requirements analysis for customisable software: A goals-skills-preferences framework. In *Proceedings of the 11th IEEE International Requirements Engineering Conference*, (pp. 117-126). IEEE Press.

Hull, E., Jackson, K., & Dick, J. (2011). *Requirements engineering*. Dordrecht, The Netherlands: Springer. doi:10.1007/978-1-84996-405-0.

IEEE. (1994). *VHDL language reference manual*. New York: IEEE.

IEEE. (1995). *Standard verilog hardware description language reference manual*. New York: IEEE.

Inmon, W., Strauss, D., & Neushloss, G. (2008). *DW 2.0: The architecture for the next generation of data warehousing*. San Francisco, CA: Morgan Kaufmann Publishers Inc..

International Organization for Standardization. (1996). *ISO/IEC 10746-3:1996: Information technology – Open distributed processing – Reference model: Architecture (RM-ODP)*. Retrieved February 2, 2013, from http://www.iso.org/iso/home/store/catalogue_ics/catalogue_detail_ics.htm?csnumber=20697

International Organization for Standardization. (2007). *ISO/IEC 42010:2007: Systems and software engineering -- Recommended practice for architectural description of software-intensive systems*. Retrieved February 2, 2013, from http://www.iso.org/iso/catalogue_detail.htm?csnumber=45991

International Organization for Standardization. (2011). *ISO 26262-1:2011: Road vehicles -- Functional safety -- Part 1: Vocabulary*. Retrieved February 2, 2013, from http://www.iso.org/iso/catalogue_detail?csnumber=43464

International Organization for Standardization. (2011). *ISO/IEC/IEEE 42010:2011: Systems and software engineering -- Architecture description*. Retrieved February 2, 2013, from http://www.iso.org/iso/catalogue_detail.htm?csnumber=50508

Inverardi, P., Muccini, H., & Pelliccione, P. (2001). Automated check of architectural models consistency using SPIN. In *Proceedings of the 16th International Conference on Automated Software Engineering (ASE'01)*, (p. 346). IEEE Computer Society.

IoBridge. (2013). *Thingspeak*. Retrieved from http://www.thingspeak.com

ISO/IEC 12207:2008. (2008). Software and systems engineering – Software lifecycle processes. Geneva, Switzerland: ISO.

ISO/IEC/IEEE 24765:2010. (2010). Systems and software–Vocabulary. Geneva, Switzerland: ISO.

Jackson, D., & Gibbons, A. P. (2007). Layered learning in boolean GP problems. In *Proceedings of EuroGP* (pp. 148–159). Berlin: Springer-Verlag.

Jackson, J. (1957). *Networks of waiting lines*. Academic Press.

Jacobs, B., & Poll, E. (2003). Coalgebras and monads in the semantics of Java. *Theoretical Computer Science*, *291*(3), 329–349. doi:10.1016/S0304-3975(02)00366-3.

Jacobs, B., & Rutten, J. (1997). A tutorial on (co) algebras and (co) induction. *Bulletin-European Association for Theoretical Computer Science*, *62*, 222–259.

Jacobson, I., Booch, G., & Rumbaugh, J. (1999). *The unified software development process*. Reading, MA: Addison-Wesley Professional.

Jacobson, I. (2003). Case for aspects - Part I. *Software Development Magazine*, 32-37.

Jacobson, I. (2003). Case for aspects - Part II. *Software Development Magazine*, 42-48.

Jansen, H. (2010). The logic of qualitative survey research and its position in the field of social research methods. *Forum Qualitative Sozial Forschung, 11*(2).

Janus, A., Schmietendorf, A., Dumke, R., & Jager, J. (2012). The 3C approach for agile quality assurance. In *Proceedings of Emerging Trends in Software Metrics (WETSoM)* (pp. 9-13). WETSoM. doi:10.1109/WET-SoM.2012.6226998.

Jones, J. (2006). An introduction to factor analysis of information risk. *Norwich Journal of Information Assurance, 2*(1), 67–76.

Jouault, F., & Bézivin, J. (2006). KM3: A DSL for metamodel specification. In *Proceedings of FMOODS*, (pp. 171-185). Springer.

Jouault, F., & Kurtev, I. (2005). Transforming models with ATL. In *Proceedings of MoDELS Satellite Events*, (pp. 128-138). Springer.

Jouault, F., Allilaire, F., Bézivin, J., & Kurtev, I. (2008). Atl: A model transformation tool. *Science of Computer Programming, 72*(1-2), 31–39. doi:10.1016/j.scico.2007.08.002.

Jouault, F., & Kurtev, I. (2005). Transforming models with ATL. InBruel, (Ed.), *Model Driven Engineering Languages and Systems (MoDELS 2005)* LNCS), (Vol. *3844*, pp. 128-138). Berlin: Springer-Verlag.

Kaindl, H. (1999). Difficulties in the transition from OO analysis to design. *IEEE Software, 16*(5), 94–102. doi:10.1109/52.795107.

Kaldeich, C., & Sá, J. O. (2004). Data warehouse methodology: A process driven approach. In *Proceedings of CAiSE*, (pp. 536-549). Springer.

Kalibatiene, D., Vasilecas, O., & Guizzardi, G. (2010). Transforming ontology axioms to information processing rules – An MDA based approach. In *Proceedings of the 3rd International Workshop on Ontology, Conceptualization and Epistemology for Information Systems, Software Engineering and Service Science*. Amsterdam, The Netherlands: IEEE.

Kallel, S., Charfi, A., Mezini, M., Jmaiel, M., & Sewe, A. (2009). A holistic approach for access control policies: From formal specification to aspect-based enforcement. *International Journal of Information and Computer Security, 3*(3-4), 337–354. doi:10.1504/IJICS.2009.031044.

Kang, J., Park, C., Seo, S., Choi, M., & Hong, J. (2008). User-centric prediction for battery lifetime of mobile devices. In *Proceedings of the 11th Asia-Pacific Symposium on Network Operations and Management: Challenges for Next Generation Network Operations and Service Management*, (pp. 531–534). Springer.

Kapsammer, E., Kargl, H., Kramler, G., Reiter, T., Retschitzegger, W., & Wimmer, M. (2006). Lifting metamodels to ontologies - A step to the semantic integration of modeling languages. In *Proceedings of MoDELS/UML*, (pp. 528-542). Springer.

Kardoš, M., & Drozdová, M. (2010). Analytical method of CIM to PIM transformation in model driven architecture (MDA). *Journal of Information and Organizational Sciences, 34*(1), 89–99.

Karow, M., Gehlert, A., Becker, J., & Esswein, W. (2006). On the transition from computation independent to platform independent. In *Proceedings of AMCIS 2006*. AMCIS.

Kaur, P., & Singh, H. (2009). Version management and composition of software components in different phases of the software development life cycle. *ACM Sigsoft Software Engineering Notes, 34*(4), 1–9. doi:10.1145/1543405.1543416.

Kent, S. (2002). Model driven engineering. In Butler, Petre, & Sere (Eds.), Integrated Formal Methods (LNCS), (vol. 2335, pp. 286-298). Berlin: Springer. doi:doi:10.1007/3-540-47884-1_16 doi:10.1007/3-540-47884-1_16.

Kent, S. (2003). Model driven language engineering. *Electronic Notes in Theoretical Computer Science, 72*(4). doi:10.1016/S1571-0661(04)80621-2.

Kerievsky, J. (2004). *Refactoring to patterns*. Reading, MA: Addison-Wesley.

Kersulec, G., Cherfi, S. S.-S., Comyn-Wattiau, I., & Akoka, J. (2009). Un environnement pour l'évaluation et l'amélioration de la qualité des modèles de systèmes d'information. In *Proceedings of INFORSID*, (pp. 329-344). INFORSID.

Kessentini, M., Sahraoui, H., & Boukadoum, M. (2008). Model transformation as an optimization problem. In *Proceedings of MoDELS*, (pp. 159-173). Berlin: Springer-Verlag.

Kessentini, M., Wimmer, M., Sahraoui, H., & Boukadoum, M. (2010). Generating transformation rules from examples for behavioral models. In *Proceedings of BM-FA*, (pp. 2:1-2:7). ACM.

Kessentini, M., Sahraoui, H., & Boukadoum, M. (2009). *Méta-modélisation de la transformation de modèles par l'exemple: Approche par méta-heuristiques*. LMO.

Khovich, L. O., & Koznov, D. V. (n.d.). OCL-based automated validation method for UML specifications. In *Programming and Computer Software*, *29*(6), 323-327.

Kim, E., & Choi, J. (2006). An ontology-based context model in a smart home. *Computational Science and Its Applications*, 11-20.

Kim, M., Kong, J., & Chung, S. (2012). An online power estimation technique for multi-core smartphones with advanced display components. In *Proceedings of the 2012 IEEE International Conference on Consumer Electronics (ICCE)*, (pp. 666–667). IEEE.

Kim, S. K., & Carrington, D. (2004). A formal object-oriented approach to define consistency constraints for UML models. In *Proceedings of the Australian Software Engineering Conference (AWSEC 2004)* (pp. 87-94). IEEE Computer Society.

Kimball, R., & Ross, M. (2002). *The data warehouse toolkit: The complete guide to dimensional modeling*. New York: John Wiley & Sons, Inc..

Kimball, R., & Ross, M. (2010). *The Kimball group reader: Relentlessly practical tools for data warehousing and business intelligence*. New York: John Wiley & Sons, Inc..

Kimmell, G., Komp, E., Minden, G., Evans, J., & Alexander, P. (2008). Synthesizing software defined radio components from rosetta. In *Proceedings of the Forum on Specification, Verification and Design Languages, 2008, FDL 2008* (pp. 148-153). New York: IEEE.

Kitchenham, B. A., & Pfleeger, S. L. (2002). Principles of survey research: Part 3: Constructing a survey instrument. *ACM SIGSOFT Software Engineering Notes*, *27*(2), 20–24. doi:10.1145/511152.511155.

Kjasrgaard, M. (2012). Location-based services on mobile phones: Minimizing power consumption. *IEEE Pervasive Computing / IEEE Computer Society [and] IEEE Communications Society*, *11*(1), 67–73. doi:10.1109/MPRV.2010.47.

Kolovos, D. S., Rose, L. M., Paige, R. F., & Polack, F. A. C. (2009). Raising the level of abstraction in the development of GMF-based graphical model editors. In *Proceedings of the 2009 ICSE Workshop on Modeling in Software Engineering*, (pp. 13–19). IEEE.

Kolovos, D., Di Ruscio, D., Paige, R., & Pierantonio, A. (2009). Different models for model matching: An analysis of approaches to support model differencing. In *Proceedings of the 2009 ICSE Workshop on Comparison and Versioning of Software Models (CVSM'09)*, (pp. 1-6). ICSE.

Kolovos, D., Paige, R., & Polack, F. A. (2008). The epsilon transformation language. In Vallecillo, Gray, & Pierantonio (Eds.), *Theory and Practice of Model Transformations, First International Conference, ICMT 2008* (LNCS), (Vol. 5063, pp. 46-60). Berlin: Springer.

Kong, C., Alexander, P., & Menon, C. (2003). Defining a formal coalgebraic semantics for the rosetta specification language. *Journal of Universal Computer Science*, *9*(11), 1322–1349.

Korpipää, P., Mäntyjärvi, J., Kela, J., Keränen, H., & Malm, E. (2003). Managing context information in mobile devices. *IEEE Pervasive Computing / IEEE Computer Society [and] IEEE Communications Society*, *2*(3), 42–51. doi:10.1109/MPRV.2003.1228526.

Kosanke, K., & Martin, R. (Eds.). (2008). SC5 glossary. ISO/TC 184/SC5 N994 Version 2 (2008-11-02).

Kostadinov, D. (2008). *Personnalisation de l'information: une approche de gestion de profils et de reformulation de requêtes*. (PhD Thesis). University of Versailles Saint-Quentin –en-Yvelines, Versailles, France.

Kotonya, G., & Sommerville, I. (1996). Requirements engineering with viewpoints. *Software Engineering Journal, 11*(1), 5–18. doi:10.1049/sej.1996.0002.

Koulopoulos, T. (2012). *ACMLive*. Retrieved from http://www.acmlive.tv/agenda.html

Krashinsky, R., & Balakrishnan, H. (2005). Minimizing energy for wireless web access with bounded slowdown. *Wireless Networks, 11*(1), 135–148. doi:10.1007/s11276-004-4751-z.

Krause, A., Ihmig, M., Rankin, E., Leong, D., Gupta, S., & Siewiorek, S. … Sengupta, S. (2005). Trading off prediction accuracy and power consumption for context-aware wearable computing. In *Proceedings of the Ninth IEEE International Symposium on Wearable Computers*, (pp. 20–26). IEEE Computer Society.

Kravets, R., & Krishnan, P. (2000). Application-driven power management for mobile communication. *Wireless Networks, 6*(4), 263–277. doi:10.1023/A:1019149900672.

Krishnan, P. (2001). Consistency checks for UML. In *Proceedings of the 7ᵗʰ Asia Pacific Software Engineering Coference (APSEC '00)*, (pp. 162-169). IEEE Computer Society.

Kruchten, P. (1995). Architectural blueprints — The 4+1 view model of software architecture. *IEEE Software, 12*(6), 42–50. doi:10.1109/52.469759.

Kruchten, P. (1999). *Rational unified process—An introduction*. Reading, MA: Addison-Wesley.

Kruchten, P. (2003). *Rational unified process: An introduction* (3rd ed.). Boston, MA: Addison-Wesley.

Kühne, T. (2006). Matters of (meta-) modeling. *Software & Systems Modeling, 5*(4), 369–385. doi:10.1007/s10270-006-0017-9.

Kulkarni, V., Reddy, S., & Rajbhoj, A. (2010). Scaling up model driven engineering - Experience and lessons learnt. In *Proceedings of MoDELS*, (pp. 331-345). Springer.

Kulkarni, V., & Reddy, S. (2003). Separation of concerns in model-driven development. *IEEE Software, 20*, 64–69. doi:10.1109/MS.2003.1231154.

Kurtev, I. (2007). Metamodels: Definitions of structures or ontological commitments? In *Proceedings of the Workshop on Towers of Models*, (pp. 53–63). York, UK: University of York.

Kurz, A. (2001). *Coalgebras and modal logic*. Retrieved from http://www.cs.le.ac.uk/people/akurz/CWI/public_html/cml.ps.gz

Kyasa, M., Fechera, H., de Boera, F. S., Jacoba, J., Hoomana, J., van der Zwaaga, M., et al. (2004). Formalizing UML models and OCL constraints in PVS. In *Proceedings of the Second Workshop on Semantic Foundations of Engineering Design Languages (SFEDL 2004)*, (Vol. 115, pp. 39-47). SFEDL.

Laddad, R. (2004). *AspectJ in action* (2nd ed.). Manning Publication.

Landsiedel, O., Wehrle, K., & Gotz, S. (2005). Accurate prediction of power consumption in sensor networks. In *Proceedings of The Second IEEE Workshop on Embedded Networked Sensors (EmNetS-II)*. IEEE.

Lankhorst, M. (2013). *Enterprise architecture at work: Modelling, communication and analysis* (3rd ed.). Berlin, Germany: Springer. doi:10.1007/978-3-642-29651-2.

Lano, K., Clark, D., & Androutsopoulos, K. (2002). Formalising inter-model consistency of the UML. In *Proceedings of the Workshop on Consistency Problems in UML-Based Software Development (UML 2002)*. UML.

Larman, C. (2004). *Applying UML and patterns: An introduction to object-oriented analysis and design and iterative development* (3rd ed.). Reading, MA: Addison Wesley Professional.

Larman, C., & Basili, V. R. (2003). Iterative and incremental developments: A brief history. *Computer, 36*(6), 47–56. doi:10.1109/MC.2003.1204375.

Lavanya, K. C., Balakishore, K. V., Mohanty, H., & Shyamasundar, R. K. (2005). How good is a UML diagram? A tool to check it. In *Proceedings of TENCON*, (pp. 386-391). IEEE Computer Society.

Lavrac, N., & Dzeroski, S. (1994). *Inductive logic programming: Techniques and applications*. New York: Ellis Horwood.

Lazowska, J. Z. (1984). *Quantitative system performance*. Englewood Cliffs, NJ: Prentice Hall.

Ledeczi, A., Bakay, A., Maroti, M., Volgyesi, P., Nordstrom, G., Sprinkle, J., & Karsai, G. (2001). Composing domain-specific design environments. *Computer*, 44–51. doi:10.1109/2.963443.

Lehtola, L., Gause, D., Dumdum, U., & Barnes, R. J. (2011). The challenge of release planning. In *Proceedings of the 5th International Workshop on Software Product Management (IWSPM 2011)* (pp. 36–45). New York: IEEE.

Li, X., Liu, Z., & He, J. (2001). Formal and use-case driven requirement analysis in UML. In *Proceedings of the 25th Annual International Computer Software and Applications Conference*, (pp. 215-224). IEEE.

Liles, D. H., & Presley, A. R. (1996). Enterprise modeling within an enterprise engineering framework. In *Proceedings of the 96 Winter Simulation Conference*. IEEE.

Liles, D. H., Johnson, M. E., & Meade, L. (1996). The enterprise engineering discipline. In *Proceedings of the 5th Industrial Engineering Research Conference*, (pp. 479–484). IEEE.

Limbourg, Q., Vanderdonckt, J., Michotte, B., Bouillon, L., & Lopez, V. (2005). UsiXML: A language supporting multi-path development of user interfaces. In *Proceedings of 9th IFIP Working Conference on Engineering for Human-Computer Interaction jointly with 11th International Workshop on Design, Specification, and Verification of Interactive Systems EHCI-DSVIS'2004* (LNCS), (vol. 3425, pp. 200-220). Berlin: Springer-Verlag.

Lin, X., Li, S., Xu, J., Shi, W., & Gao, Q. (2005). An efficient context modeling and reasoning system in pervasive environment: using absolute and relative context filtering technology.[LNCS]. *Proceedings of Advances in Web-Age Information Management*, *3739*, 357–367. doi:10.1007/11563952_32.

List, B., Schiefer, J., & Tjoa, A. M. (2000). Process-oriented requirement analysis supporting the data warehouse design process - A use case driven approach. In *Proceedings of DEXA*, (pp. 593-603). Springer.

Litvak, B., Tyszberowicz, S., & Yehudai, A. (2003). Behavioral consistency validation of UML diagrams. In *Proceedings of the First International Conference on Software Engineering and Formal Methods (SEFM 2003)*, (pp. 118-125). SEFM.

Liu, T., Sadler, C., Zhang, P., & Martonosi, M. (2004). Implementing software on resource-constrained mobile sensors: Experiences with impala and zebranet. In *Proceedings of the 2nd International Conference on Mobile Systems, Applications, and Services*, (pp. 256–269). ACM.

Liu, A. F. (2005). *Design-level performance prediction of component-based applications*. Washington, DC: IEEE.

LogMeIn. (2013). *COSM*. Retrieved January 15, 2013, from https://cosm.com/

Lohoefener, J. (2011). *A methodology for automated verification of rosetta specification transformations*. (PhD thesis). University of Kansas, Lawrence, KS.

Luján-Mora, S., Trujillo, J., & Song, I.-Y. (2006). A UML profile for multidimensional modeling in data warehouses. *Data & Knowledge Engineering*, *59*(3), 725–769. doi:10.1016/j.datak.2005.11.004.

Martin, J. (1995). *The great transition: Using the seven disciplines of enterprise engineering to align people, technology, and strategy*. New York: AMACOM.

Martnez, F. J. L., & Alvarez, A. T. (2005). A precise approach for the analysis of the UML models consistency. In *Proceedings of the 24th International Conference on Perspectives in Conceptual Modeling*, (pp. 74-84). Berlin: Springer-Verlag.

Mazón, J.-N., & Trujillo, J. (2008). An MDA approach for the development of data warehouses. *Decision Support Systems*, *45*, 41–58. doi:10.1016/j.dss.2006.12.003.

MDA. (2012). *The model-driven architecture*. Retrieved September 24, 2012, from http://www.omg.org/mda/

Mehmood, K., Cherfi, S. S.-S., & Comyn-Wattiau, I. (2009). Data quality through conceptual model quality - Reconciling researchers and practitioners through a customizable quality model. In *Proceedings of ICIQ*, (pp. 61-74). HPI/MIT.

Mehta, N. R. (2000). Towards a taxonomy of software connectors. In *Proceedings of the 22nd International Conference on Software Engineering*, (pp. 178-187). IEEE.

Melnik, S., Garcia-Molina, H., & Rahm, E. (2002). Similarity flooding: A versatile graph matching algorithm and its application to schema matching. In *Proceedings of ICDE*, (pp. 117-128). IEEE Computer Society.

Mens, T., Van Der Straeten, R., & Simmonds, J. (2003). Maintaining consistency between UML models with description logic tools. In *Proceedings of the Sixth International Conference on the Unified Modelling Language - The Language and its Applications, Workshop on Consistency Problems in UML-Based Software Development II*. UML.

Meyer, B. (1992). Applying 'design by contract'. *IEEE Computer*, *25*(10), 40–51. doi:10.1109/2.161279.

Michaelson, D. (2006). *Setting best practices in public relations research*. Retrieved January 13, 2013, from http://www.instituteforpr.org/iprwp/wp-content/uploads/Michaelson_09IPRRC.pdf

Microsoft. (2011). *Microsoft AdventureWorks 2008R2*. Retrieved September 8, 2011, from http://msftdbprodsamples.codeplex.com/

Miller, G. A. (1956). The magical number seven, plus or minus two: Some limits on our capacity for processing information. *Psychological Review*, *63*, 81–97. doi:10.1037/h0043158 PMID:13310704.

Miller, J., & Mukerji, J. (2003). *MDA guide version 1.0.1 (Technical report). Object Management Group*. OMG.

Ministry of Defence. (2012). *MOD architecture framework (MODAF)*. Retrieved February 2, 2013, from https://www.gov.uk/mod-architecture-framework

Mitchell, T. M. (1982). Generalization as search. *Artificial Intelligence*, *18*, 203–226. doi:10.1016/0004-3702(82)90040-6.

Mitchell, T. M. (1997). *Machine learning*. New York: McGraw-Hill.

Mnkandla, E., & Dwolatzky, B. (2006). Defining agile software quality assurance. In *Proceedings of Software Engineering Advances* (p. 36). IEEE.

ModelWare. (2007). *ModelWare information society technologies (IST) sixth framework programme glossary*. Retrieved 2008-01-06, from http://www.modelware-ist.org/index.php?option=com_rd_

Modularity. (2009). *Composite application guidance for WPF and silverlight*. Retrieved from http://msdn.microsoft.com/en-us/library/dd490825.aspx

Moebius, N., Stenzel, K., & Reif, W. (2009). Generating formal specifications for security-critical applications - A model-driven approach. In *Proceedings of IWSESS*. IWSESS.

MOF. (2011). *Meta object facility (MOF) core specification version 2.4.1* (OMG Document Number: formal/2011-08-07). Retrieved September 24, 2012, from http://www.omg.org/spec/MOF/2.4.1

Mohagheghi, P., Fernandez, M., Martell, J., Fritzsche, M., & Gilani, W. (2009). MDE adoption in industry: Challenges and success criteria. In Chaudron, M. (Ed.), *Models in Software Engineering* (pp. 54–59). Berlin, Germany: Springer. doi:10.1007/978-3-642-01648-6_6.

Mohapatra, S., Cornea, R., Dutt, N., Nicolau, A., & Venkatasubramanian, N. (2003). Integrated power management for video streaming to mobile handheld devices. In *Proceedings of the Eleventh ACM International Conference on Multimedia*, (pp. 582–591). ACM.

Morrison, M., & Morgan, M. S. (1999). *Models as mediating instruments*. Cambridge, UK: Cambridge University Press.

Mossakowski, T., Roggenbach, M., & Schröder, L. (2003). CoCasl at work—Modelling process algebra. *Electronic Notes in Theoretical Computer Science*, *82*(1), 206–220. doi:10.1016/S1571-0661(04)80640-6.

Mostafa, A., Ismall, M., El-Bolok, H., & Saad, E. (2007). Toward a formalisation of UML2.0 metamodel using Z specifications. In *Proceedings of the 8th International Conference on Software Engineering, Artificial Intelligence, Networking, and Parallel*[). IEEE Computer Society.]. *Distributed Computing, 1,* 694–701.

Muggleton, S. (1993). Optimal layered learning: A PAC approach to incremental sampling. In *Proceedings of ALT,* (pp. 37-44). London, UK: Springer-Verlag.

Muggleton, S. (1991). Inductive logic programming. *New Generation Computing, 8,* 295–318. doi:10.1007/BF03037089.

Muggleton, S., & Raedt, L. D. (1994). Inductive logic programming: Theory and methods. *The Journal of Logic Programming, 19/20,* 629–679. doi:10.1016/0743-1066(94)90035-3.

Muggleton, S., & Road, K. (1994). Predicate invention and utilisation. *Journal of Experimental & Theoretical Artificial Intelligence, 6,* 6–1. doi:10.1080/09528139408953784.

Mulla, N., & Girase, S. (2012). Comparison of various elicitation techniques and requirement prioritisation techniques. *International Journal of Engineering, 3*(3), 51–60.

Nebot, V., & Llavori, R. B. (2010). Building data warehouses with semantic data. In *Proceedings of EDBT/ICDT Workshops.* ACM.

Nebulon. (2005). Retrieved from http://www.featuredrivendevelopment.com/

Nguyen, S. H., Bazan, J. G., Skowron, A., & Nguyen, H. S. (2004). Layered learning for concept synthesis. *Transactions on Rough Sets, 3100,* 187–208.

Nienhuys-Cheng, S.-H., & de Wolf, R. (1997). *Foundations of inductive logic programming.* Berlin: Springer. doi:10.1007/3-540-62927-0.

Noble, J., Schmidmier, A., Pearce, D. J., & Black, A. P. (2007). Patterns of aspect-oriented design. In *Proceedings of European Conference on Pattern Languages of Programs.* Irsee, Germany: IEEE.

Norell, U. (2009). Dependently typed programming in Agda. *Advanced Functional Programming,* 230-266.

Nuseibeh, B. A. (1994). *A multi-perspective framework for method integration.* (Doctoral dissertation). Imperial College of Science, Technology and Medicine, University of London, Department of Computing, London, UK.

Nuseibeh, B., Kramer, J., & Finkelstein, A. (2003). ViewPoints: Meaningful relationships are difficult! In *Proceedings of the 25th International Conference on Software Engineering* (pp. 676-681). New York: IEEE.

Object Management Group. (2003). *MDA guide version 1.0.1.* Retrieved February 2, 2013, from http://www.omg.org/cgi-bin/doc?omg/03-06-01.pdf

Object Management Group. (2011). *Business process model and notation (BPMN) version 2.0.* Retrieved February 2, 2013, from http://www.omg.org/spec/BPMN/2.0/PDF

Object Management Group. (2011). *Unified modeling language (UML) version 2.4.1.* Retrieved February 2, 2013, from http://www.omg.org/spec/UML/2.4.1/

Object Management Group. (2013). *OMG systems modeling language (OMG SysML) version 1.3.* Retrieved February 2, 2013, from http://www.omg.org/spec/SysML/1.3/PDF Praxeme Institute. (2011). *PRAXEME – Opus, the product.* Retrieved February 2, 2013, from http://www.praxeme.org/index.php?n=Opus.Opus?userlang=en

Object Management Group/ODM. (2009). *The ontology definition metamodel (ODM) specification.* Retrieved April 19, 2009, from http://www.omg.org/spec/ODM/

Object Management Group/QVT. (2010). *The query/view/transformation (QVT) specification.* Retrieved June 14, 2010, from http://www.omg.org/spec/QVT/

Object Management Group/Security. (2008). *The OMG security specifications catalog.* Retrieved December 13, 2008, from http://www.omg.org/technology/documents/formal/omg_security.htm

OCL. (2012). *OCL: Object constraint language, version 2.3.1* (OMG Document Number: formal/2012-01-01). Retrieved September 24, 2012, from http://www.omg.org/spec/OCL/2.3.1/

OMG. (2003). *MDA guide.* Retrieved from http://www.omg.org/cgi-bin/doc?omg/03-06-01

OMG. (2006). *Business process modeling notation specification.* Retrieved from http://www.omg.org

OMG. (2006). *Modeldriven architecture- Specifications.* Retrieved June 2, 2012, from http://www.omg.org/mda/mda_files/Model-Driven_Architecture.pdf

OMG. (2007). *Unified modeling language: Infrastructure version 2.1.1.* Retrieved January 13, 2013, from http://www.omg.org/cgi-bin/doc?formal/07-02-06

OMG. (2012). *OMG model driven architecture.* Retrieved May 13, 2012, from http://www.omg.org/mda/

Ono, K., Koyanagi, T., Abe, M., & Hori, M. (2002). Xslt stylesheet generation by example with wysiwyg editing. In *Proceedings of SAINT,* (pp. 150-161). Washington, DC: IEEE Computer Society.

Oracle. (2009). *Oracle.* Retrieved January 20, 2013, from http://www.oracle.com

Ortiz, G., & Bordbar, B. (2009). Aspect-oriented quality of service for web services: A model-driven approach. In *Proceedings of the IEEE International Conference on Web Services* (pp. 559-566). IEEE Computer Society.

Palmer, S. R., & Felsing, J. M. (2002). *A practical guide to feature-driven development.* Upper Saddle River, NJ: Prentice Hall PTR.

Parikh, D., Skadron, K., Zhang, Y., Barcella, M., & Stan, M. (2002). Power issues related to branch prediction. In *Proceedings of the Eighth International Symposium on High-Performance Computer Architecture,* (pp. 233–244). IEEE.

Paternò, F. (1999). *Model-based design and evaluation of interactive applications.* Berlin: Springer.

Peck, W. (2011). *Hardware/software co-design via specification refinement.* (PhD thesis). University of Kansas, Lawrence, KS.

Pedersen, T. B. (2007). Warehousing the world: A few remaining challenges. In *Proceedings of DOLAP,* (pp. 101-102). New York, NY: ACM.

Pérez, J. M., Berlanga, R., & Aramburu, M. J. (2009). A relevance model for a data warehouse contextualized with documents. *Information Processing & Management, 45*(3), 356–367. doi:10.1016/j.ipm.2008.11.001.

Pering, T., Agarwal, Y., Gupta, R., & Want, R. (2006). Coolspots: Reducing the power consumption of wireless mobile devices with multiple radio interfaces. In *Proceedings of the 4th International Conference on Mobile systems, Applications and Services,* (p. 232). ACM.

Periyasamy, K., Alagar, V. S., & Muthiayen, D. (1999). International conference on verification and validation techniques of object-oriented software systems. *Technology of Object-Oriented Languages,* 413.

Pfleeger, S. L., & Kitchenham, B. A. (2001). Principles of survey research: Part 1: Turning lemons into lemonade. *ACM SIGSOFT Software Engineering Notes, 26*(6), 16–18. doi:10.1145/505532.505535.

Piras, A., Carboni, D., Pintus, A., & Features, D. M. T. (2012). A platform to collect, manage and share heterogeneous sensor data. In *Proceedings of the 9th International Conference on Networked Sensing Systems.* IEEE.

Poe, V., Brobst, S., & Klauer, P. (1997). *Building a data warehouse for decision support.* Upper Saddle River, NJ: Prentice-Hall, Inc..

Prat, N., Akoka, J., & Comyn-Wattiau, I. (2006). A UML-based data warehouse design method. *Decision Support Systems, 42*(3), 1449–1473. doi:10.1016/j.dss.2005.12.001.

Prat, N., Akoka, J., & Comyn-Wattiau, I. (2012). An MDA approach to knowledge engineering. *Expert Systems Application, 39*(12), 10420–10437. doi:10.1016/j.eswa.2012.02.010.

Pressman, R. (2004). *Software engineering: A practitioner's approach* (6th ed.). Columbus, OH: McGraw-Hill Science/Engineering/Math.

Preuveneers, D., Bergh, J., Wagelaar, D., Georges, A., Rigole, P., Clerckx, T., & Berbers, Y. (2004). Towards an extensible context ontology for ambient intelligence. *Ambient Intelligence,* 148-159.

Project Management Institute. (2004). *PMBOK: A guide to the project management body of knowledge* (3rd ed). Retrieved from http://ebookee.org/A-Guide-to-the-Project-Management-Body-of-Knowledge-Fourth-Ed._282107.html

Qiao, B., Yang, H., Xu, B., & Chu, W. (2003). Bridging legacy systems to model driven architecture. In IEEE Computer Society Staff (Ed.), *Proceedings 27th Annual International Computer Software and Applications Conference* (pp. 304-309). IEEE Digital Library.

Rajkovic, P., Jankovic, D., Stankovic, T., & Tosic, V. (2010). Software tools for rapid development and customization of medical information systems. In *Proceedings of 12th IEEE International Conference on e-Health Networking Applications and Services* (pp. 119-126). IEEE Computer Society.

Rasch, H., & Wehrheim, H. (2002). Consistency between UML classes and associated state machines. In *Proceedings of the Workshop on Consistency Problems in UML-Based Software Development(UML 2002)*. UML.

Ráth, I., Ökrös, A., & Varró, D. (2010). Synchronization of abstract and concrete syntax in domain-specific modeling languages. *Software & Systems Modeling, 9*(4), 453–471. doi:10.1007/s10270-009-0122-7.

Reddy, R., France, R., Ghosh, S., Fleurey, F., & Baudry, B. (2005). Model composition - A signature-based approach. In *Proceedings of Aspect Oriented Modeling (AOM)*. Montego Bay, Jamaica: MoDELS.

Reina, A. M., Toress, J., & Toro, M. (2004). Towards developing generic solutions with aspects. In *Proceedings of the Workshop in Aspect Oriented Modeling*. IEEE.

Ren, Y., Xing, T., Quan, Q., & Zhao, Y. (2010). Software configuration management of version control study based on baseline. In *Proceedings of 3rd International Conference on Information Management, Innovation Management and Industrial Engineering,* (Vol. 4, pp. 118-121). IEEE Press.

Repenning, A., & Perrone, C. (2000). Programming by example: programming by analogous examples. *Communications of the ACM, 43*(3), 90–97. doi:10.1145/330534.330546.

Rilling, J., Meng, W., Charland, P., & Witte, R. (2008). Story-driven approach to software evolution. *IET Software, 2*(4), 304–320. doi:10.1049/iet-sen:20070095.

Rios, R., & Matwin, S. (1998). Predicate invention from a few examples. In *Proceedings of AI,* (pp. 455-466). London, UK: Springer-Verlag.

Ritter, N., & Steiert, H. P. (2000). Enforcing modeling guidelines in an ORDBMS-based UML-repository. In *Proceedings of the International Resource Management Association International Conference on Challenges of Information Technology Management in the 21ˢᵗ Century* (pp. 269-273). Hershey, PA: IGI Global.

Rose, L. M., Herrmannsdoerfer, M., Williams, J., Kolovos, D., Garces, K., Paige, R., & Polack, F. (2010). A comparison of model migration tools. In Petriu, Rouquette, & Haugen (Eds.), *Proceedings of the 13th International Conference on Model Driven Engineering Languages and Systems (MoDELS)* (LNCS), (Vol. 6394, pp. 61-75). Berlin: Springer.

Rose, L., Paige, R., Kolovos, D., & Polack, F. (2009). An analysis of approaches to model migration. In *Proceeding of the 1st International Workshop on Model Co-Evolution and Consistency Management* (pp. 6-15). IEEE.

Rose, L. M., Kolovos, D. S., Paige, R. F., & Polack, F. A. C. (2010). Model migration with epsilon flock. In *Theory and Practice of Model Transformation (LNCS)* (Vol. 6142, pp. 184–198). Berlin: Springer. doi:10.1007/978-3-642-13688-7_13.

Roser, S., & Bauer, B. (2006). An approach to automatically generated model transformations using ontology engineering space. In *Proceedings of SWESE*. SWESE.

Rothensee, M. (2007). A high-fidelity simulation of the smart fridge enabling product-based services. In *Proceedings of the 3rd IET International Conference on Intelligent Environments (IE 07),* (pp. 529–532). IEE.

Rousseau, B., Browne, P., Malone, P., & Ofughlu, M. (2004). User Profiling for content personalisation in information retrieval. In *Proceedings of the ACM Symposium on Applied Computing*. Nicosia, Chypre: ACM Press.

Rozanski, N., & Woods, E. (2005). *Software systems architecture: Working with stakeholders using viewpoints and perspectives* (2nd ed.). Boston, MA: Addison-Wesley.

Rukzio, E., Noda, C., De Luca, A., Hamard, J., & Coskun, F. (2008). Automatic form filling on mobile devices. *Pervasive and Mobile Computing, 4*(2), 161–181. doi:10.1016/j.pmcj.2007.09.001.

Russom, P. (2009). *Next generation data warehouse platforms*. Retrieved October 26, 2009, from http://www.oracle.com/database/docs/tdwi-nextgen-platforms.pdf

Rutle, A., Rossini, A., Lamo, Y., & Wolter, U. (2009). A diagrammatic formalisation of mof-based modelling languages. In *Proceedings of TOOLS*, (pp. 37-56). Springer.

Rutle, A., Wolter, U., & Lamo, Y. (2008). A diagrammatic approach to model transformations. In *Proceedings of EATIS*. EATIS.

Saldhana, J. A., & Shatz, S. M. (2000). UML diagrams to object petri net models: An approach for modeling and analysis. In *Proceedings of the International Conference on Software Engineering and Knowledge Engineering (SEKE)*. SEKE.

Schatz, B. (2011). 10 years model-driven – What did we achieve? In M. Popović & V. Vranić (Eds.), *Proceedings of the 2nd Eastern European Regional Conference (EERC) on the Engineering of Computer Based Systems (ECBS)*. Bratislava, Slovakia: IEEE Computer Society.

Schilit, B., Adams, N., & Want, R. (1994). Context-aware computing applications. In *Proceedings of the International Workshop on Mobile Computing Systems and Applications,* (pp. 85-90). IEEE Computer Society.

Schmidt, A., Beigl, M., & Gellersen, H. (1999). There is more to context than location. *Computers & Graphics Journal, 23*(6), 893–902. doi:10.1016/S0097-8493(99)00120-X.

Schmidt, D. (2006). Introduction: Model-driven engineering. *IEEE Computer Science, 39*(2), 25–31. doi:10.1109/MC.2006.58.

Schmidt, D. C. (2006). Model-driven engineering. *IEEE Computer, 39*(2), 25–31. doi:10.1109/MC.2006.58.

Schwaber, K. (2003). *Agile project management with scrum.* Seattle, WA: Microsoft Press.

Scott, K. (2001). *The unified process explained.* Reading, MA: Addison-Wesley Professional.

Seehusen, F., & Stølen, K. (2007). Maintaining information flow security under refinement and transformation. In *Proceedings of FAST*. FAST.

Seehusen, F., & Stølen, K. (2008). A transformational approach to facilitate monitoring of high-level policies. In *Proceedings of POLICY*, (pp. 70-73). IEEE Computer Society.

Selic, B. (2003). The pragmatics of model-driven development. *IEEE Software, 20*(5), 19–25. doi:10.1109/MS.2003.1231146.

Shan, H. Z. L. (2006). Well-formedness, consistency and completeness of graphic models.[Oxford, UK: UKSIM.]. *Proceedings of, UKSIM06,* 47–53.

Sharifi, H. R., Mohsenzadeh, M., & Hashemi, S. M. (2012). CIM to PIM transformation: An analytical survey. *International Journal of Computer Technology and Applications, 3*(2), 791–796.

Shinkawa, Y. (2006). Inter-model consistency in UML based on CPN formalism. In *Proceedings of the 13th Asia-Pacific Software Engineering Conference(APSEC)*, (pp. 411-418). IEEE Computer Society.

Simitsis, A. (2005). Mapping conceptual to logical models for ETL processes. In *Proceedings of DOLAP*, (pp. 67-76). ACM.

Simmonds, D. (2008). Aspect-oriented approaches to model driven engineering. In *Proceedings of the International Conference on Software Research and Practice*. Las Vegas, NV: IEEE.

Simmonds, D., Reddy, R., Song, E., & Grant, E. (2009). A comparison of aspect-oriented approaches to model driven engineering. In *Proceedings of Conference on Software Engineering Research and Practice*, (pp. 327-333). IEEE.

Simmonds, J., Bastarrica, M. C., Hitschfeld-Kahler, N., & Rivas, S. (2008). A tool based on DL for UML model consistency checking. *International Journal of Software Engineering and Knowledge Engineering, 18*(6), 713–735. doi:10.1142/S0218194008003829.

Six, H. W., Kosters, G., & Winter, M. (2001). Coupling use cases and class models as a means for validation and verification of requirements specification. *Requirements Engineering, 6*(1).

Smith, B. C. (1985). The limits of correctness. *SIGCAS Computer Society, 14*(15), 18–26. doi:10.1145/379486.379512.

Smith, C., & Williams, L. (2001). *Performance solutions: A practical guide to creating responsive, scalable software.* Boston: Addison-Wesley Professional.

Sochova, Z. (2009). *Agile adoption survey.* Retrieved January 13, 2013, from http://soch.cz/AgileSurvey.pdf

Software Engineering Institute. (2006). *CMMI® for development, version 1.2*. 2006. Retrieved from http://www.sei.cmu.edu/publications/documents/06.reports/06tr008.html

Software Engineering Standards Committee of the IEEE Computer Society. (2000). *1471-2000 – IEEE recommended practice for architectural description for software-intensive systems (IEEE Standard)*. Los Alamitos, CA: IEEE Computer Society.

Sottet, J. S., Ganneau, V., Calvary, G., Coutaz, J., Demeure, A., Favre, J. M., & Demumieux, R. (2007). Model-driven adaptation for plastic user interfaces. *Human-Computer Interaction*, *4662*, 39–410.

Spinellis, D. (2001). Notable design patterns for domain-specific languages. *Journal of Systems and Software*, *56*(1), 91–99. doi:10.1016/S0164-1212(00)00089-3.

SRI FormalWare. (2011). *PVS specification and verification system*. Retrieved from http://pvs.csl.sri.com/

Srinivasan, A. (2006). *A learning engine for proposing hypotheses (aleph)*. Retrieved from http://web.comlab.ox.ac.uk/oucl/research/areas/machlearn/Aleph

Stahl, I. (1994). On the utility of predicate invention in inductive logic programming. In *Proceedings of ECML*, (pp. 272-286). Springer.

Stahl, I. (1995). The appropriateness of predicate invention as bias shift operation in ILP. *Machine Learning*, *20*, 95–117. doi:10.1007/BF00993476.

Stallings, W. (2008). *Operating systems internals & design*. Upper Saddle River, NJ: Prentice Hall.

State of Agile Survey. (2011). *Agile methods and practices*. Retrieved January 13, 2013, from http://www.versionone.com/state_of_agile_development_survey/10/page3.asp

Steen, M., Akehurst, D., Doest, H., & Lankhorst, M. (2004). Supporting viewpoint-oriented enterprise architecture. In *Proceedings of the 8th IEEE International Enterprise Distributed Object Computing Conference (EDOC'04)* (pp. 201–211). Monterey, CA: IEEE Computer Society.

Stein, D., Hanenberg, S., & Unland, R. (2002). A UML-based aspect- oriented design notation for aspect. In *Proceedings of Aspect-Oriented Software Development (AOSD 2002)*. AOSD.

Steinholtz, B., & Walden, K. (1987). Automatic identification of software system differences. *IEEE Transactions on Software Engineering*, *13*(4), 493–497. doi:10.1109/TSE.1987.233186.

Stevens, P. (2010). Bidirectional model transformations in QVT: Semantic issues and open questions. *Software & Systems Modeling*, *9*(1), 7–20. doi:10.1007/s10270-008-0109-9.

Stone, P., & Veloso, M. M. (2000). Layered learning. In *Proceedings of ECML*, (pp. 369-381). Springer.

Streb, J., & Alexander, P. (2006). Using a lattice of coalgebras for heterogeneous model composition. In *Proceedings of the MoDELS Workshop on Multi-Paradigm Modeling* (pp. 27-38). MoDELS.

Streb, J., Kimmell, G., Frisby, N., & Alexander, P. (2006). Domain specific model composition using a lattice of coalgebras. In *Proceedings of the OOPSLA* (Vol. 6). OOPSLA.

Strommer, M., Murzek, M., & Wimmer, M. (2007). Applying model transformation by-example on business process modeling languages. In *Proceedings of ER*, (pp. 116-125). Berlin: Springer-Verlag.

Sun, Y., White, J., & Gray, J. (2009). Model transformation by demonstration. In *Proceedings of MoDELS*, (pp. 712-726). Springer.

Szatmári, Z., Izsó, B., Polgár, B., & Majzik, I. (2010). Ontology-based assessment of software models and development processes for safety-critical systems. In *Monographs of System Dependability* (Vol. *2*). Wroclaw.

Taconet, C., & Kazi Aoul, Z. (2008). Context-awareness and model driven engineering: Illustration by an e-commerce application scenario. In *Proceedings of ICDIM*. ICDIM.

Talevski, A., Chang, E., & Dillon, T. S. (2003). Meta model driven framework for the integration and extension of application components. In *Proceedings of the 9th IEEE International Workshop on Object-Oriented Real-Time Dependable Systems* (pp. 255-261). IEEE Computer Society.

Tamm, T., Seddon, P. B., Shanks, G., & Reynolds, P. (2011). How does enterprise architecture add value to organisations? *Communications of the Association for Information Systems, 28*(1).

Tang, Y., Feng, K., Cooper, K., & Cangussu, J. (2009). Requirement engineering techniques selection and modeling an expert system based approach. In M. A. Wani, et al. (Eds.), *International Conference on Machine Learning and Applications, 2009, ICMLA'09* (pp. 705-709). Los Alamitos, CA: IEEE.

Tesoriero, R., & Vanderdonckt, J. (2010). Extending UsiXML to support user-aware interfaces. In *Proceedings of 3rd IFIP Conf. on Human-Centred Software Engineering HCSE 2010* (LNCS), (vol. 6409, pp. 95-110). Berlin: Springer-Verlag.

The Open Group. (2011). *TOGAF version 9.* Retrieved October 15, 2012, from http://pubs.opengroup.org/architecture/togaf9-doc/arch/

The Open Group. (2011). *Welcome to TOGAF version 9.1, an open group standard.* Retrieved February 2, 2013, from http://pubs.opengroup.org/architecture/togaf9-doc/arch/index.html

Thiagarajan, N., Aggarwal, G., Nicoara, A., Boneh, D., & Singh, J. (2012). Who killed my battery? Analyzing mobile browser energy consumption. In *Proceedings of the 21st International Conference on World Wide Web,* (pp. 41–50). ACM.

Thompson, C., White, J., Dougherty, B., & Schmidt, D. (2009). Optimizing mobile application performance with model-driven engineering. In *Proceedings of the 7th IFIP Workshop on Software Technologies for Future Embedded and Ubiquitous Systems.* IFIP.

Ting, K. M., & Witten, I. H. (1997). Stacked generalization: When does it work? In *Proceedings of IJCAI,* (pp. 866-871). San Francisco, CA: Morgan Kaufmann.

Ting, K. M., & Witten, I. H. (1999). Issues in stacked generalization. *Journal of Artificial Intelligence Research, 10,* 271–289.

Toulmé, A. (2006). *Presentation of EMF compare utility.* Paper presented at 10th Eclipse Modeling Symposium. New York, NY.

Tsiolakis, A., & Ehrig, H. (2000). Consistency analysis of UML class and sequence diagrams using attributed graph grammars. In *Proceedings of the Workshop on Graph Transformation Systems(GraTra),* (pp. 77-86). GraTra.

Turban, E., Sharda, R., & Delen, D. (2010). *Decision support and business intelligence systems.* Englewood Cliffs, NJ: Prentice Hall.

Turner, H., White, J., Thompson, C., Zienkiewicz, K., Campbell, S., & Schmidt, D. (2009). Building mobile sensor networks using smartphones and web services: Ramifications and development challenges. In Cruz-Cunha, M. M., & Moreira, F. (Eds.), *Handbook of Research on Mobility and Computing: Evolving Technologies and Ubiquitous Impacts.* Hershey, PA: IGI Global.

Turney, P. D. (1993). Exploiting context when learning to classify. In *Proceedings of ECML,* (pp. 402-407). London, UK: Springer-Verlag.

UML. (2011). *Unified modeling language: Infrastructure, version 2.4.1* (OMG Specification formal/2011-08-05). Retrieved September 24, 2012, from http://www.omg.org/spec/UML/2.4.1/

UML. (2011). *Unified modeling language: Superstructure, version 2.4.1* (OMG Specification: formal/2011-08-06). Retrieved September 24, 2012, from http://www.omg.org/spec/UML/2.4.1/

UMO (User Model Ontology). (2003). Retrieved from http://www.u2m.org/2003/02/UserModelOntology.daml

University of Vienna. (2013). *European smart cities.* Retrieved November 26, 2012, from http://www.smart-cities.eu

Use. (2011). *A UML-based specification environment.* Retrieved September 24, 2012, from http://sourceforge.net/apps/mediawiki/useocl/

UsiXML. (2007). *User interface extensible markup language) (version 1.8)*. Louvain, Belgium: Université Catholique de Louvain.

van Amstel, M., van den Brand, M., & Engelen, L. (2010). An excercise in iterative domain-specific language design. In *Proceedings of the Joint ERCIM Workshop on Software Evolution (EVOL) and International Workshop on Principles of Software Evolution (IWPSE)* (pp. 48-57). ERCIM.

Vanderperren, Y., & Dehaene, W. (2005). SysML and systems engineering applied to UML-based SoC design. In *Proceedings of DAC UML-SoC Workshop*. DAC UML-SoC.

Varró, D. (2006). Model transformation by example. In *Proceedings of MoDELS*, (pp. 410-424). Genova, Italy: Springer.

Varró, D., & Balogh, Z. (2007). Automating model transformation by example using inductive logic programming. In *Proceedings of SAC*, (pp. 978-984). New York, NY: ACM.

Vermolen, S., & Visser, E. (2008). Heterogeneous coupled evolution of software languages. In *Proceeding of the 11th International Conference on Model Driven Engineering Languages and Systems (MODELS'08)* (LNCS), (Vol. 5301, pp. 630-644). Berlin: Springer.

Vermolen, S., Wachsmuth, G., & Visser, E. (2011). Reconstructing complex metamodel evolution. In Sloane & Aßmann (Eds.), *Proceeding of the 4th International Conference on Software Language Engineering (SLE 2011)* (LNCS), (Vol. 6940, pp. 201-221). Berlin: Springer.

Villarroel, R., Fernández-Medina, E., & Piattini, M. (2005). Secure information systems development - A survey and comparison. *Computers & Security*, 24(4), 308–321. doi:10.1016/j.cose.2004.09.011.

Vogel, T., Seibel, A., & Giese, H. (2011). The role of models and megamodels at runtime. In Dingel & Solberg (Eds.), Models in Software Engineering (LNCS), (vol. 6627, pp. 224-238). Berlin: Springer. doi:doi:10.1007/978-3-642-21210-9_22 doi:10.1007/978-3-642-21210-9_22.

W3C. (2009). *Delivery context ontology (DCO)*. Retrieved from http://www.w3.org/TR/2009/WD-dcontology-20090616/

Wachsmuth, G. (2007). Metamodel adaptation and model co-adaptation. In *Proceeding of ECOOP 2007-Object Oriented Programming* (LNCS), (Vol. 4609, pp. 600-624). Berlin: Springer.

Wagelaar, D., & Bergmans, L. (2002). Using a concept-based approach to aspect-oriented software design. In *Proceedings of Aspect-Oriented Design workshop*. AOSD.

Walden, K., & Nerson, J. (1994). *Seamless object-oriented software architecture: Analysis and design of reliable systems*. New York: Prentice Hall.

Walls, C., & Breidenbach, R. (2004). *Spring in action* (2nd ed.). Dreamtech Press.

Wampler, D. (2003). *The role of aspect-oriented programming in OMG's model-driven architecture*. Aspect Programming Inc..

Wang, Q., Hempstead, M., & Yang, W. (2006). A realistic power consumption model for wireless sensor network devices. In *Proceedings of the Third Annual IEEE Communications Society Conference on Sensor, Mesh and Ad Hoc Communications and Networks (SECON)*. IEEE.

Wang, X., Gu, T., Zhang, D., & Pung, H. (2004). Ontology based context modeling and reasoning using OWL. In *Proceedings of the Second IEEE Annual Conference on Pervasive Computing and Communications Workshops*, (pp. 18-22). IEEE.

Wang, Y., Krishnamachari, B., & Annavaram, M. (2012). Semi-Markov state estimation and policy optimization for energy efficient mobile sensing. In *Proceedings of Sensor, Mesh and Ad Hoc Communications and Networks (SECON)*, (pp. 533–541). IEEE. doi:10.1109/SECON.2012.6275823.

Weichel, B. (2001). *Introduction for ASAM-MCD-2MC audience*. Retrieved from http://www.msr-wg.de/medoc/download/msrsw/v222/msrsw-tr-intro/msrsw-tr-intro.pdf

Weißenberg, N. (2004). Using ontologies in personalized mobile applications. In *Proceedings of the 12th Annual ACM International Workshop on Geographic Information Systems* (pp. 2-11). ACM Press.

Westerman, P. (2001). *Data warehousing: Using the Wal-Mart model*. San Francisco, CA: Morgan Kaufmann Publishers Inc..

White, J., Schmidt, D., & Mulligan, S. (2007). The generic eclipse modeling system. In *Proceedings of the Model-Driven Development Tool Implementors Forum at TOOLS 2007*. TOOLS.

White, J., Clarke, S., Dougherty, B., Thompson, C., & Schmidt, D. (2010). R&D challenges and solutions for mobile cyber-physical applications and supporting internet services. *Springer Journal of Internet Services and Applications, 1*(1), 45–56. doi:10.1007/s13174-010-0004-9.

White, J., Hill, J., Tambe, S., Gray, J., Gokhale, A., & Schmidt, D. C. (2009, July/August). Improving domain-specific language reuse through software product-line configuration techniques. *IEEE Software*. doi:10.1109/MS.2009.95.

Wimmer, M., Strommer, M., Kargl, H., & Kramler, G. (2007). Towards model transformation generation by-example. In *Proceedings of HICSS*, (p. 285b). Washington, DC: IEEE Computer Society.

Winter, R., & Fischer, R. (2006). Essential layers, artifacts, and dependencies of enterprise architecture. In *Proceedings of the 10th IEEE International Enterprise Distributed Object Computing Conference Workshop (EDOCW '06), Workshop on Trends in Enterprise Architecture Research (TEAR 2006)*. Hong Kong: IEEE Computer Society.

Wolpert, D. H. (1992). Stacked generalization. *Neural Networks, 5*, 241–259. doi:10.1016/S0893-6080(05)80023-1.

Woodcock, J., & Davies, J. (1996). *Using Z: Specification, refinement, and proof* (Vol. 1). Englewood Cliffs, NJ: Prentice Hall.

Wood-Harper, A. T., Antill, L., & Avison, D. E. (1985). *Information systems definition: The multiview approach*. Oxford, UK: Blackwell Scientific Publications.

Wrembel, R., & Koncilia, C. (2007). *Data warehouses and OLAP: Concepts, architectures and solutions*. Hershey, PA: IGI Global.

Wu, L., Barash, G., & Bartolini, C. (2007). A service-oriented architecture for business intelligence. *Service Oriented Computing and Applications*, 279–285.

Xi, H., & Pfenning, F. (1999). Dependent types in practical programming. In *Proceedings of the 26th ACM SIGPLAN-SIGACT Symposium on Principles of Programming Languages* (pp. 214-227). ACM.

Xie, F., Levin, V., Kurshan, R. P., & Browne, J. C. (2004). Translating software designs for model checking. In *Proceedings of Fundamental Approaches to Software Engineering (FASE 2004)* (LNCS), (vol. 2984, pp. 324-338). Berlin: Springer.

Xie, Z. (2006). Several speed-up variants of cascade generalization. In *Proceedings of FSKD*, (pp. 536-540). Xi'an, China: Springer.

Xing, Z., & Stroulia, E. (2005). UMLDiff: An algorithm for object-oriented design differencing. In *Proceedings of the 20th IEEE/ACM International Conference on Automated Software Engineering (ASE'05)*, (pp. 54-65). ACM.

Yamanoue, T., Oda, K., & Shimozono, K. (2012). A M2M system using Arduino, Android, and Wiki software. In *Proceedings of the 2012 IIAI International Conference on Advanced Applied Informatics*, (pp. 123–128). IEEE. Retrieved January 13, 2013, from http://ieeexplore.ieee.org/lpdocs/epic03/wrapper.htm?arnumber=6337170

Yan, J., & Zhang, B. (2009). Support multi-version applications in SaaS via progressive schema evolution. In *Proceedings of the IEEE 25th International Conference on Data Engineering* (pp. 1717-1724). IEEE Computer Society.

Yan, L. L., Miller, R. J., Haas, L. M., & Fagin, R. (2001). Data-driven understanding and refinement of schema mappings. In *Proceedings of SIGMOD*, (pp. 485-496). New York, NY: ACM.

Yao, S., & Shatz, S. M. (2006). Consistency checking of UML dynamic models based on petri net techniques. In *Proceedings of the 15th International Conference on Computing (CIC '06)*, (pp. 289-297). IEEE Computer Society.

Yoon, C., Kim, D., Jung, W., Kang, C., & Cha, H. (2012). Appscope: Application energy metering framework for android smartphone using kernel activity monitoring. In *Proceedings of USENIX Annual Technical Conference.* USENIX.

Yu, T.-Y. (1999). Toward 21st century new generation of enterprise information systems. *Enterprise Engineering Forum.* Retrieved October 15, 2012, from http://www.ee-forum.org/eis21c.html

Yu, T.-Y. (2002). Model-driven software architecture and hierarchical principle to complex systems. *Enterprise Engineering Forum.* Retrieved October 15, 2012, from http://www.ee-forum.org/pub/1998-2009/hm.html

Yu, T.-Y. (2002). Emerging enterprise engineering. In *Proceedings of the Enterprise Engineering Conference 2001.* Retrieved October 15, 2012, from http://www.ee-forum.org/eee/

Yu, T.-Y. (2005). New generation of enterprise information system: From essential requirements analysis and research to model-driven system. *Enterprise Engineering Forum.* Retrieved October 15, 2012, from http://www.qiji.cn/eprint/abs/3641.html

Yu, T.-Y. (2007). Lecture on enterprise engineering, model, and information system. *Enterprise Engineering Forum.* Retrieved October 15, 2012, from http://www.ee-forum.org/pub/1998-2009/downloads/ty_jnu070917a.pps

Zachman, J. A. (1997). *Concepts of the framework for enterprise architecture.* Retrieved February 2, 2013, from http://www.ies.aust.com/papers/zachman3.htm

Zachman, J. A. (2009). *John Zachman's concise definition of the Zachman framework.* Retrieved February 2, 2013, from http://www.zachman.com/about-the-zachman-framework

Zapata, C. M., Gonzlez, G., & Gelbukh, A. (2007). A rule-based system for assessing consistency between UML models. In *Proceedings of the 6th Mexican International Conference on Artificial Intelligence (MICAI '07)* (LNAI), (pp. 215-224). Berlin: Springer.

Zaremski, A., & Wing, J. (1997). Specification matching of software components. *ACM Transactions on Software Engineering and Methodology, 6*(4), 333–369. doi:10.1145/261640.261641.

Zepeda, L., Celma, M., & Zatarain, R. (2008). A mixed approach for data warehouse conceptual design with MDA. In *Proceedings of ICCSA,* (pp. 1204-1217). Perugia, Italy: Springer-Verlag.

Zhang, D., & Tsai, J. J. P. (2007). *Advances in machine learning applications in software engineering.* Hershey, PA: IGI Global.

Zhao, X., Long, Q., & Qiu, Z. (2006). Model checking dynamic UML consistency. In *Proceedings of the 8th International Conference on Formal Engineering Methods, ICFEM 2006,* (LNCS), (pp. 440-459). Berlin: Springer.

Zhu, X., & Wang, S. (2009). Software customization based on model-driven architecture over SaaS platforms. In *Proceedings of International Conference on Management and Service Science* (pp. 1-4). CORD.

Zhu, Y. L. (2007). Revelor: Model driven capacity planning tool suite. In *Proceedings of the 29th International Conference on Software Engineering,* (pp. 797–800). IEEE.

Zisman, A., & Kozlenkov, A. (2001). Knowledge base approach to consistency management of UML specifications. In *Proceedings of the 16th International Conference on Automated Software Engineering (ASE'01),* (pp. 359-363). IEEE Computer Society.

Zloof, M. M. (1975). Query-by-example: The invocation and definition of tables and forms. In *Proceedings of VLDB,* (pp. 1-24). New York, NY: ACM.

Zschaler, S. (2008). Formal specification of non-functional properties of component-based software systems -- A semantic framework and some applications thereof. In *Software and Systems Modeling* (pp. 161–201). Academic Press.

About the Contributors

Vicente García Díaz is an associate professor in the Computer Science Department of the University of Oviedo. He has a PhD from the University of Oviedo in computer engineering. His research interests include model-driven engineering, domain specific languages, technology for learning and entertainment, project risk management, software development processes and practices. He graduated in Prevention of Occupational Risks and is a Certified Associate in Project Management through the Project Management Institute.

Juan Manuel Cueva Lovelle graduated from Oviedo Mining Engineers Technical School in 1983 (Oviedo University, Spain). He has a PhD from Madrid Polytechnic University, Spain (1990). From 1985, he has been a professor at the languages and computers systems area in Oviedo University (Spain), and is an ACM and IEEE voting member. His research interests include object-oriented technology, language processors, human-computer interface, Web engineering, modeling software with BPM, DSL, and MDA.

Begoña Cristina Pelayo García-Bustelo is a lecturer in the Computer Science Department of the University of Oviedo. She has a PhD from the University of Oviedo in computer engineering. Her research interests include object-oriented technology, Web engineering, eGovernment, modeling software with BPM, DSL, and MDA.

Oscar Sanjuán Martínez is a lecturer in the Computer Science Department of the Carlos III University of Madrid. He has a PhD from the Pontifical University of Salamanca in computer engineering. His research interests include object-oriented technology, Web engineering, software agents, modeling software with BPM, DSL, and MDA.

* * *

Mourad Abed obtained his Ph.D in 1990 and HDR in 2001. He is specialized in human-computer interaction, software engineering for interactive systems, Context-Aware MDA-Ontology Approach, and intelligent interaction. He is professor in Computer Engineering at the University of Valenciennes and member of the "Human- Computer Interaction and Automated Reasoning" research group in the LAMIH. He is chair or co-chair of international conferences or special sessions, author or co-author of numerous book chapters, journal articles, papers in international congresses. He is coordinator of several European projects. He is involved in several research networks, projects and associations.

Perry Alexander is a Professor in the EECS Department and Director of the Information and Telecommunication Technology Center at The University of Kansas. His is the chief architect of the Rosetta specification languages and leads its ongoing standards efforts. His research interests include formal methods, system-level design, verification, and synthesis and high-assurance systems.

Firas Bacha is a PhD Student at the University of Valenciennes and Hainaut Cambrésis, in France, and member of the "Human-Computer Interaction and Automated Reasoning" research group in the LAMIH laboratory. His current research interests are in the area of context-aware Human Computer Interaction design and Model Driven software Engineering. He graduated from the National School of Computer Sciences (Tunisia) as a Computer Engineer, and he is currently working as an R&D engineer within one of the leading companies in France in the field of Model Driven Architecture (MDA).

Arunkumar Balakrishnan is a Senior Consultant at Mettler Toledo Turing Softwares where he is responsible for Software Quality Assurance along with Software Project Management. He is also a Professor (on sabbatical leave) in the Department of Computer Applications at Coimbatore Institute of Technology. He has published work in International Journals and Conferences, in the fields of Artificial Intelligence (in Machine learning and Intelligent Tutoring Systems) and in Software Testing.

Marc Born is CTO of ikv++ technologies ag and in his position responsible for the technical development and the product strategy of the company since 2005. His main interest is in development and functional safety analysis of embedded systems in the automotive domain. Marc has deep knowledge in functional safety standards as well as in the related safety analysis methods. Before joining ikv Marc was with the research institute Fraunhofer Fokus in Berlin, where he founded and led the business unit Model Driven System Integration and Development of Competence Center Platin. In 2002, Marc received his PhD from the Humboldt Universität zu Berlin for his work on a method for component-based development of distributes telecommunication systems. Marc studied Computer Science at the Humboldt Universität zu Berlin and received his master degree in 1996.

David Chassels is Scottish trained CA, and currently Chief Executive Officer, Procession plc. He was the Founder Chairman of Procession in 1994 and became CEO in December 2001. He has written a number of papers about the importance of technology aligning people and their processes. A founder member of a group called "the Unreasonable Learners" recognising need for changes in society to encourage empowerment of people as opposed to old command and control management style. Previously David was a Partner in International Accounting firm BDO and spent 20 years with ICFC/3i where he was actively involved with many early stage companies and advising on fund raising and M&A transactions.

Elizabeth Chang has over 20 years of experience in academia and industry. She is an expert on ontologies, XML, and semi-structured databases, data mining for business intelligence, collaborative systems, human system Interfaces, and service-oriented computing. She has supervised 35 PhD students to completion. She has over 500 publications including 6 books, 25 book chapters, 90 journal articles, and 35 Keynote speeches, with a h-index of 20 (Google Scholar). She initiated a new tier 1 research institute at Curtin University, the Digital Ecosystems and Business Intelligence Institute, the largest research institute in ICT in Australia. Institute achievements include: Dean's award, Chancellor Awards,

VC and DVC awards, each year over the last 6 years. As one of the chief investigators, Professor Chang has obtained over $11.5 million in competitive research funds. Professor Chang has a PhD and research Master in Computer Science and Software Engineering from La Trobe University, Australia.

István Dávid holds a Masters degree in Computer Engineering from the Budapest University of Technology and Economics, Hungary (BUTE). His professional focus includes language design, models@run.time, complex event processing, and the Eclipse platform. He has been employed as a software engineer at one of the world-leading automotive electronic design companies, which allowed him to gain significant practical experience in the field of model-driven engineering, especially AUTOSAR and related standards. He also contributed to several R&D projects (e.g. SENSORIA EU FP6) and courses as a teaching assistant at the BUTE.

Jon Davis has over 25 years of industry and academic experience. For many years, he has been developing software solutions for the heavy engineering, manufacturing, and mining industries, plus the public and university sector, his main focus being on software modelling and reusability. He is also a project management specialist lecturing to university, public, and private sector organisations. In recent years, he has paid special attention to research in the scope of generic models for Enterprise Information Systems definition, in which he has analyzed the advantages of using model-driven automated application generation and led the development of a new model and framework publishing numerous papers in the field. Jon Davis graduated in Science then Mathematics at Melbourne and Newcastle University, followed by an Honours degree in Computer Science at Newcastle University and is currently in the final stages of completing a PhD at Curtin University.

Philippe Desfray has been involved for 22 years in model-driven development research, standardization, and tooling. Co-founder of the SOFTEAM company (650 people in 2012), he has been precursor to OO Modeling (inventor of the Class Relation OO model, 1990, supported already at that time by a case tool), to MDA (1994, Addison Welsey book on the matter, dedicated tool), one of the main contributor in the UML standardization team at the OMG since the origin, and contributing to many modeling language standardization efforts(BPMN, SOA, ...). Philippe serves as VP for R&D at SOFTEAM, is one key expert in the modeling and methodology fields, is driving international R&D projects and is leading the Modelio (previously Objecteering) UML/BPMN/MDA modeling tool development team (www.modeliosoft.com; wwww.modelio.org). Philippe has also worked on enterprise architecture methodologies and modeling activities. He recently wrote a book titled *TOGAF in Practice: Enterprise Architecture Modeling*.

Iwona Dubielewicz received MSc degree and PhD degree in Computer Science in 1972 and 1977, respectively, both from Wroclaw University of Technology, Poland. Her PhD was associated with the use of formal languages in software engineering. Since 1977, she has been working as an Assistant Professor at the Institute of Informatics, Wroclaw University of Technology. Her main scientific interests include, but are not limited to software development methodologies, modeling languages, and quality of the software systems and processes. She is a member of Polish Committee for Standardization. Since 1994, she has been involved in the development of several international standards for Polish Committee. Iwona Dubielewicz has over 40 publications in the international journals and conference proceedings from different areas of software engineering.

Moez Essaidi is a Ph.D student, member of the department of computer and information sciences (LIPN - UMR CNRS 7030) at University of Paris 13. His thesis is supervised by Pr. Aomar Osmani and Pr. Céline Rouveirol. The thesis was financed in part by the ANRT (French National Association for Research and Technology), Grant CIFRE-1341/2008, and Intelligence Power Company. In 2008, He received a M.Sc. degree in computer science from University of Paris 1 - Panthéon Sorbonne. His main research interests are: data warehouse design, model-driven development, machine learning for engineering and cloud-based architectures.

Liliana Favre is a full professor of Computer Science at Universidad Nacional del Centro de la Provincia de Buenos Aires in Argentina. She is also a researcher of CIC (Comisión de Investigaciones Científicas de la Provincia de Buenos Aires). Her current research interests are focused on model driven development, model driven architecture, and formal approaches, mainly on the integration of algebraic techniques with MDA-based processes. She has been involved in several national research projects about formal methods and software engineering methodologies. Currently, she is research leader of the Software Technology Group at Universidad Nacional del Centro de la Provincia de Buenos Aires. She has published several book chapters, journal articles, and conference papers. She has acted as editor of the book *UML and the Unified Process*. She is the author of the book *Model Driven Architecture for Reverse Engineering Technologies: Strategic Directions and System Evolution*.

Klaus Fischer is DFKI Research Fellow and head of the Multiagent Systems (MAS) research Group at DFKI. He holds a diploma and doctoral degree from Technische Universität (TU) in München. In 1992, he joined the MAS research group at DFKI in Saarbrücken and assumed the responsibility of group leader in November 1993. At DFKI he has a long record of successfully managed projects for industrial partners (e.gl. Siemens, Daimler, Saarstahl) as well as for public authorities (e.g. EU: ATHENA, SHAPE, COIN, ViBaM). His main research interest is in agent technologies for virtual worlds and model driven design of agent-based system in virtual worlds as well as in commercial applications. Klaus has published more than 100 scientific papers at international conferences and workshops as well as in international journals. He serves in programme committee of major international conferences (AAMAS, IAT, ICAART) and in the editorial board of international journals (JAAMAS, APIN).

Steven Gibson has been engaged in software analysis and development for over twenty years. As owner of Superant Computing, he applies his interest in using formal approaches to computer engineering, including functional programming methods and mathematical models of computational systems. He has developed Small Linux, an open source package of small applications for operation on older x86 hardware and has published and presented works ranging from managing risk in computing to modeling artificial intelligence approaches. In addition, he has delivered papers and articles about Bayesian and Process Algebra approaches to modeling cognition and machine learning. During his career, Steven has assisted in the development of database and framework solutions for business applications. He is currently engaged in graduate work in Communications Studies at California State University in Northridge, California, focusing in part on the social science applications of computational methods and research into information gathering and communication.

László Gönczy completed his PhD studies at BUTE in 2006. He holds an MSc in CS, and an MSc in Engineer-economics from Corvinus University of Budapest. His professional interest includes model-driven development of SOA and dependability analysis based on high level system models. He led SOA-driven analysis of business processes in SENSORIA EU FP6. In CoMiFin FP7 he participated in the design of a global monitoring architecture for critical financial infrastructures. He has many papers on modelling and V&V of system level dependability and security attributes, SOA-based system integration and testing.

Rubén González Crespo is a research manager in accessibility and usability in Web environments for people with severe disabilities. BS in Computer Management from the Polithecnical University of Madrid and Computer Engineering from the Pontifical University of Salamanca. Master in Project Management from the same university. PhD from the Pontifical University of Salamanca in 2008 and PhD Extraordinary Award given by the UPSA where is professor at the School of Computing (Madrid Campus) since 2005. Currently is the Director of the Master in Project Management, is professor of doctoral programs organized by the University and Master's programs in GIS, Software Engineering and Security.

Cristian González García, Technical Engineering in Computer Systems from School of Computer Engineering of Oviedo in 2011 (University of Oviedo, Spain). Currently studying a Master in Web Engineering from the University of Oviedo and doing his Master 's Thesis. His research interests are in the field of Internet of Things, Web Engineering, Modeling Software with DSL and MDA.

Bogumila Hnatkowska received MSc degree and PhD degree in Computer Science in 1992 and 1997, respectively, both from the Wroclaw University of Technology, Poland. Her PhD dissertation was associated with the use of formal methods in software engineering. Since 1998, she has been working as a professor assistant at the Institute of Informatics, Wroclaw University of Technology. Her main scientific interests include, but are not limited to software development processes, modeling languages, model-driven development, model transformations, and quality of the software products. She is a member of programme committees of several international conferences. Bogumila Hnatkowska has over 60 publications in international journals and conference proceedings from different areas of software engineering.

Zbigniew Huzar received the M.Sc., Ph.D. and habilitation degrees in Computer Science from Wrocław University of Technology, Poland, in 1969, 1974, and 1990, respectively. During 1978-1984 he was deputy director of Computer Center, during 1984-2003 he served as a head of Informatics Center, during 2004-2008 as director of the Institute of Applied Informatics, and since 2008 as director of the Institute of Informatics, Wrocław University of Technology, Poland. The scope of his scientific interests concerns software engineering, in particular, covers methods of formal specification and design of real-time systems, and model-based software development. He is author and co-author of 10 books. He is a member of the Polish Information Processing Society and editor-in-chief of the e-Informatica Software Engineering Journal.

Julian Krumeich is researcher at the Institute for Information Systems (IWi) at the German Research Center for Artificial Intelligence (DFKI). He studied Information Systems at Saarland University and holds a Master's of Science Degree in this subject. His main research areas include Model-Driven Development of Enterprise Systems (with a particular interest in enterprise modeling and business process modeling), Business Process Management (with a focus on process flexibility and process recommender systems), as well as Business Model Research, in which he researched in several national and international research projects. Julian published a number of scientific papers and articles in internationally renowned conferences (e.g. International Conference on Information Systems and Americas Conference on Information Systems) and journals (e.g. *Knowledge Management & E-Learning*).

Káthia Marçal de Oliveira is associate professor at the University of Valenciennes and Hainaut-Cambrésis, in France. She received her PhD in 1999 in software engineering from the Federal University of Rio de Janeiro. Her interests include software quality assurance, ontologies, and human-computer interaction design and evaluation.

Liliana Martinez is an assistant professor in computer science area at the Facultad de Ciencias Exactas, Universidad Nacional del Centro de la Provincia de Buenos Aires (UNCPBA), Tandil, Argentina. She is a member of the Software Technology Group, which develops its activities at the INTIA Research Institute at the UNCPBA. She has a Master degree in Software Engineering from Universidad Nacional de La Plata, Argentina. Her research interests are focused on system modernization, reverse engineering in particular. She has published book chapters, journal articles, and conference papers. She has been member of the program committee of international conferences related to software engineering.

Gustavo Millán Garcia is Computer Engineering from the Pontifical University of Salamanca, where is professor at the School of Computing (Madrid Campus) since 2003. His research interests include Software Architectures, software patterns and software engineering.

Rachit Mohan Garg is currently working as Systems Engineer in the Education and Research Department at Infosys Ltd., Mysore, India. He has B.Tech and M.Tech degrees in Computer Science from Vishveshwrya Institute of Engineering and Technology (VIET), Ghaziabad, and Jaypee University of Information Technology (JUIT), Waknaghat, respectively. Rachit has over 2 years of experience in IT Industry and Academics in India. He has conducted many practical sessions in colleges. In the IT Industry, he has been the part of training activities in Infosys, Mysore, India. Rachit's interdisciplinary research interests span over Software Engineering and IT Managament. He is member of many organizations, some of which are International Association of Computer Science & Information Technology (IACSIT), Academy & Industry Research Collaboration Center (AIRCC), Indian Science Congress (ISC), International Association of Engineers (IAENG), etc.

Hrushikesha Mohanty is a Professor at the School of Computer and Information Sciences, University of Hyderabad, Hyderabad. His research interests include Mobile Computing, Software Engineering, and Computational Social Science. He has published papers in International Journals and Conferences. He has also penned three volumes of Odia poems.

Aomar Osmani studied artificial intelligence and pattern recognition at Paris 6 university and got his PhD in 1999 at University Paris 13. He has a permanent research position (associate professor) at LIPN-UMR CNRS 7030 since 2000. Aomar Osmani is particularly interested in (1) qualitative temporal and spatial reasoning including reasoning about finding maximal tractable classes in several interval algebras models and updating data in geographical information systems, (2) representing and reasoning about temporal sequences and simulating dynamic in systems, (3) diagnostic of industrial systems particularly using model based approaches including finding deeps explanation of alarms in telecommunication networks (Transpac, ATM, SDH, WDM, etc.) and fault diagnostic in automobiles systems, (4) modeling of complexes systems including work an metamodeling approaches (UML, CWM, MDA, etc.), and (5) complexity analysis of machine learning algorithms. His research has led to over 50 publications. He has been involved in several research projects.

Dima Panfilenko is researcher at the Centre for e-Learning Technology (CeLTech) at the German Research Center for Artificial Intelligence (DFKI). His main research areas are model driven architectures and development, recommender systems, business process modeling, and domain specific languages. He has successfully worked for more than five years in European and domestic research and software development projects. Currently, he is the group leader for services modeling in the diverse national projects. He received a bachelor's degree in Economics from Kiev National University in Ukraine and master's degree in Information Systems from University of Saarland in Germany.

Jordán Pascual Espada, research scientist at Computer Science Department of the University of Oviedo. Ph.D. from the University of Oviedo in Computer Engineering, B.Sc. in Computer Science Engineering, and a M.Sc. in Web. He has published several articles in international journals and conferences, he has worked in several national research projects. His research interests include the Internet of Things, exploration of new applications and associated human computer interaction issues in ubiquitous computing and emerging technologies, particularly mobile and Web applications.

Megan Peck received her Masters of Science Degree in 2012 from The University of Kansas. Her research interests include specification languages, coalgebraic semantics, and component composition. She is currently continuing her PhD studies at The University of Kansas.

Claudia Pereira is an assistant professor in computer science area at the Facultad de Ciencias Exactas, Universidad Nacional del Centro de la Provincia de Buenos Aires (UNCPBA), Tandil, Argentina. She is a member of the Technology Software Group at the INTIA Research Institute at the UNCPBA. She has a Master degree in Software Engineering from Universidad Nacional de La Plata, Argentina. Her main research interests are focused on system modernization, refactoring in particular. She has published book chapters, journal articles, and conference papers. She has been member of the program committee of international conferences related to software engineering.

Céline Rouveirol got her PhD in 1991 at Université Paris 11 (Orsay), on Logical models of supervised Machine Learning. She has then been Assistant Professor at Université Paris 11 until 2006. She defended her "Habilitation à Diriger les Recherches" in 2001 and she is now a Professor of Computer Science in LIPN, Université Paris 13. Her research interests include Inductive Logic Programming, Bioinformatics and Systems Biology, Link Prediction in Large Complex Networks.

P. G. Sapna is an Associate Professor at the Department of Computer Technology and Applications at Coimbatore Institute of Technology. Sapna received her Ph.D. in Computer Science from the University of Hyderabad. Her research interests include software testing, requirements engineering, and software engineering education.

Douglas C. Schmidt is a Professor of Computer Science, Associate Chair of the Computer Science and Engineering program, and a Senior Researcher at the Institute at Software Integrated Systems, all at Vanderbilt University. Dr. Schmidt has published 10 books and more than 500 technical papers covering a wide range of software-related topics, including patterns, optimization techniques, and empirical analyses of object-oriented frameworks and domain-specific modeling environments that facilitate the development of DRE middleware and mission-critical applications running over data networks and embedded system interconnects. In addition to his academic research, Dr. Schmidt has led the development of ACE, TAO, CIAO, and CoSMIC for the past two decades. These technologies are open-source DRE middleware frameworks and model-driven tools used successfully by thousands of companies and agencies worldwide in many domains, including national defense and security, datacom/telecom, financial services, medical engineering, and massively multiplayer online gaming.

Chris Thompson graduated of Vanderbilt University in Nashville, Tennessee, with a Bachelor of Science degree in Computer Engineering. While at Vanderbilt University, Chris worked on projects ranging from mobile technologies to distributed real-time and embedded systems. Since graduating, he co-founded the automotive safety company splitsecnd (www.splitsecnd.com) based on his research on WreckWatch at Vanderbilt University. He has also contributed to helping improve the Nashville high-technology ecosystem.

余彤鹰 **(Yu Tong-Ying)** is an independent consultant from China in the fields of enterprise application, business process improvement and transition, management; has a wealth of experience from industrial and the experiences on each main phase in the lifecycle of enterprise applications. He is committed to the innovative development and research of the application for enterprises and information systems for a long term. The research interests include model-driven applications and model working mechanisms, architectures, enterprise modeling, enterprise engineering, and general modeling theory. Since 1998, he has published many influential articles through EE-Forum.org, and so on; in 2001, launched the first enterprise engineering seminar in China.

Lech Tuzinkiewicz received Msc degree and PhD degree in Computer Science in 1976 and 1982, respectively, both from Wroclaw University of Technology, Poland. His PhD dissertation was associated with the automation of the design process of industrial electrical networks and electrical equipments–formalization issues. Since 1983, he has been working as an Assistant Professor at the Institute of Informatics, Wroclaw University of Technology. His main scientific interests include, but are not limited to databases, warehouses, data modeling. Software development processes, modeling languages, model driven development, model transformations, and quality of software products. Lech Tuzinkiewicz has over 70 publications in international journals and conference proceedings from different areas of software engineering.

Jules White is an Assistant Professor of Computer Science in the Dept. of Electrical Engineering and Computer Science at Vanderbilt University. His research focuses on securing, optimizing, and leveraging data from mobile cyber-physical systems. His mobile cyber-physical systems research spans four key focus areas: (1) mobile security and data collection, (2) high-precision mobile augmented reality, (3) mobile device and supporting cloud infrastructure power and configuration optimization, and (4) applications of mobile cyber-physical systems in multi-disciplinary domains, including energy-optimized cloud computing, smart grid systems, healthcare/manufacturing security, next-generation construction technologies, and citizen science. Dr. White has published over 80 papers and has won four "Best Paper" awards. His research has also been transitioned to industry, where it won an Innovation Award at CES 2013, was a finalist for the Technical Achievement Award at SXSW Interactive, and was a top three finalist for mobile in the Accelerator Awards at SXSW 2013.

Index

A

abstract syntax 109, 113, 128, 178, 244, 259, 309
agile software development 60, 155, 158, 175
Application Programming Interfaces (APIs) 6
Aspect Oriented Modeling (AOM) 36, 199
Aspect Oriented Software Development (AOSD) 39
AtlanMod Matching Language (AML) 185
ATLAS Transformation Language (ATL) 97

B

bisimilarity 235
business-IT gap 54-55, 68
business-IT strategic alignment 60
Business Process Management (BPM) 60
Business Process Model (BPM) 64, 90-91, 96-98, 341
Business Process Model Notation (BPMN) 91, 318
Business Process Reengineering (BPR) 60

C

cascade learning 242, 245
coalgebraic structure 209-211, 218, 225, 229, 235
coinduction 235
Common Warehouse Metamodel (CWM) 4, 242, 256-257, 262
Complex Event Description Language (CEDL) 108-110, 115, 119
Complex Event Processing (CEP) 109, 133
Component Utilization Test Suite (CUTS) 345
composition operations
 conjunctive 202, 205-207, 215
 disjunctive composition 202, 204, 207, 213
 structural 202, 204, 207, 228, 236
Computer-Aided Software Engineering (CASE) 58, 313

C

Computer Integrated Manufacturing (CIM) 60
Configuration Management Databases (CMDB) 111, 120, 125
context learning 242, 245
cross-cutting 40, 42-44, 47, 50-51

D

Database Management System (DBMS) 64
Dependent-Concept Learning (DCL) 242, 246
Domain-Driven Development (DDD) 53
Domain Specific Language (DSL) 75
 context domain 107
 problem domain 107
Domain-Specific Modeling Language (DSML) 344
Domain Specific Transformation Language (DSTL) 181

E

Eclipse Modeling Framework (EMF) 185, 200, 353
Eclipse Modeling Project 269
entanglement 53, 60-61, 68-69
Enterprise Architecture (EA) 53, 60-61, 68, 323
Enterprise Information Systems (EIS) 1
 Meta-Data EIS (MDEIS) 2
Enterprise Integration (EI) 60
Enterprise Model Driven Application (EMDA) 53-54, 69
Enterprise Modeling (EM) 53, 60
Enterprise Resource Planning (ERP) 2
Entity-Relationship (ER) 243
Epsilon Transformation Language (ETL) 181
event POJO 116
Extended Backus-Naur Form (EBNF) 115